3rd International Conference on

MOLTEN SLAGS AND FLUXES

27 - 29 June 1988
University of Strathclyde, Glasgow

3rd International Conference on MOLTEN SLAGS AND FLUXES

27 - 29 June 1988
University of Strathclyde, Glasgow

Organised by the Ironmaking and Steelmaking Committee
of the Institute of Metals
and held at the University of Strathclyde
on 27 June 1988, in collaboration with
Socièté Française de Metallurgie
The Indian Institute of Metals
The Institute of Mining and Metallurgy
The Iron and Steel Institute of Japan
The Metallurgical Society of AIME
The Metallurgical Society of CIM
Verein Deutscher Eisenhuttenleute

THE INSTITUTE OF METALS

1989

Book Number 455

Published in 1989 by The Institute of Metals
1 Carlton House Terrace, London SW1Y 5DB

and
The Institute of Metals
North American Publications Center
Old Post Road, Brookfield VT 05036
U S A

British Library Cataloguing in Publication Data

International Conference on Molten Slags and
Fluxes (3rd: 1988: University of
Strathclyde).
3rd International conference on molten slags
and fluxes
1. Molten slag. Molten flux
I. Title. II. Institute of Metals 1985
(Ironmaking and Steelmaking Committee
III. Societe Francaise de Metallurgie
669' .84
I S B N 0-901462-54-3

Compiled by the Institute's CRC unit from original typescripts
and illustrations provided by the authors

Printed in Great Britain by M & A Thomson Litho Ltd, East Kilbride, Scotland

Contents

POSTERS

Foreword

These conferences were initiated by Dr C R Masson in 1980 and the first was held in Halifax as part of the annual Canadian Metallurgist's Conference. The proposal was to hold them at four year intervals in a different country each time and the second conference was held in the USA in 1984. At that conference it was decided to hold the third conference in Glasgow, Scotland in 1988 and the present volume is the proceedings of that conference. The reason for choosing Glasgow was to mark the long interest in research in liquid slags of the Metallurgy Department at Strathclyde University and the proceedings were held in that University.

The present volume represents the advances made in the understanding of the properties of liquid slags since the previous conference. The aim of these conferences is to bring both theory and practical work together and I suggest this has been done in the present volume.

The organizing committee are grateful to all those who participated in the conference as plenary lecturers, authors, chairmen, and as discussion participants.

H B Bell
University of Strathclyde,
Glasgow

Henry B Bell - An Appreciation

It was entirely appropriate that the 3rd International Conference on Molten Slags and Fluxes should have been held at the University of Strathclyde in Glasgow in June 1988 to mark their contribution to the science of molten slags over 50 years. It is particularly gratifying that the Institute of Metals have decided to dedicate the Proceedings volume to Professor Henry B. Bell.

Harry Bell started his illustrious academic career at the Royal Technical College, Glasgow in 1950. He joined a department which had already established a reputation in the physical chemistry of slags through the work of Tom Carter and John Taylor. Throughout his career, Harry Bell has added to the international respect of the Glasgow school with his wide variety of research in chemical metallurgy. He has worked and published extensively in this field, particularly on the thermodynamic properties of metal and slag solutions, the thermodynamics of smelting and refining reactions in both ferrous and non-ferrous metallurgy and on the physical chemistry of welding fluxes. Professor Bell has received many international honours for his work including the Kroll medal of the Metals Society.

Well known as he is for his research, Harry Bell will probably be remembered best by hundreds of students for his teaching in Glasgow. He has inspired many students who have gone out throughout the world and made significant contributions in both academic and industrial circles. The success enjoyed by many of his former students can be attributed in part to Harry Bell's interest, encouragement and guidance. His text book written with Colin Bodsworth is a standard text for all interested in ferrous extraction metallurgy.

I am sure all the other conference delegates join with me in thanking Harry Bell for his scientific contributions, his wise advice and friendship over the years. We are proud to have contributed to a volume bearing his dedication.

Paul Grieveson

Reaction of liquid steel with slag during furnace tapping

E T TURKDOGAN

Consultant: Pyrometallurgy and Thermochemistry, Pittsburgh, Pennsylvania, USA.

ABSTRACT

Reactions that occur between liquid steel and slag in the ladle during furnace tapping may have deleterious or beneficial effects on the subsequent ladle refining of the steel. Three types of reactions of practical importance are discussed in this paper: (i) the reaction of steel with the furnace slag carried into the ladle, (ii) slag-aided partial deoxidation of steel in the tap ladle, and (iii) steel desulphurization with an exothermic slag mixture during furnace tapping. An analysis of plant data is presented to demonstrate how the furnace slag carry over affects the recovery of aluminium and silicon added to the tap ladle and rephosphorization of steel in the tap ladle. Plant data from production heats are given to demonstrate how the calcium aluminate ladle slag enhances the extent of partial deoxidation of steel when only the ferro-manganese is added to the tap ladle as a deoxidant. The analysis of plant data for Al-killed steels indicates that the concentration of residual dissolved oxygen in liquid steel is close to that in equilibrium with the calcium aluminate inclusions retained in the steel. An account is given of a recent development in the ladle desulphurization of Al-killed steel during furnace tapping, using an exothermic slag mixture, EXOSLAG, that is self ignited in the preheated ladle prior to furnace tapping. Lime-saturated molten calcium aluminate slag generated by the alumino-thermic reaction in the preheated ladle is shown to be capable of removing 40 to 50 percent of the sulphur from the steel during furnace tapping. Plant data are given to show that with this practice, the tap temperature of the steel can be lowered by about 24°C.

INTRODUCTION

Advances made during the past 10 to 15 years in the technology of pneumatic steelmaking and ladle refining processes focuses an increasing awareness to the deleterious effects of the slag carry over during furnace tapping. The devices for BOF and Q-BOP such as refractory sphere, cube, 'tetron' and 'stinger' are being used to varying degrees of success in controlling the slag carry over into the tap ladle. For the electric-arc furnaces, EAF, the slag control is achieved by eccentric bottom tapping, or using a slide-gate system.

In some practices, the burnt lime and pre-fused calcium aluminate are added to the tap ladle to improve deoxidation and to desulphurize steel during furnace tapping.

In this lecture, I shall discuss three types of reactions of liquid steel with slag in the ladle during furnace tapping:

(i) reactions with furnace slag

(ii) slag-aided partial deoxidation of steel

(iii) steel desulphurization with an exothermic slag mixture.

REACTIONS WITH FURNACE SLAG

A low recovery in the steel of aluminium and silicon added to the ladle during furnace tapping and the phosphorus reversion to steel in the tap ladle have long been known to be the consequences of the reaction of liquid steel with the slag carried over from the furnace. However, no detailed study seems to have been made of this undesirable slag-metal reaction in the tap ladle. My deliberation on this subject here is based on the analysis of some Q-BOP heats (200 tonne) deoxidized with aluminium and/or silicon in the tap ladle.

The percentages of utilization of the ladle additions for low-carbon heats are 85 to 95% for Mn, 60 to 70% for Si and 35 to 65% for Al. The copious black fumes emitted at the time of the addition of ferromanganese to the tap stream indicate that the vaporization of manganese is the primary cause of some loss of manganese during the ladle addition. Losses in the ladle additions of aluminium and silicon are due to reactions with the furnace slag that is carried into the tap ladle. There is also the iron oxide-rich skull accumulating at the converter mouth, some of which falls into the ladle during furnace tapping and reacts with aluminium and silicon.

Material Balance for Aluminium:- The aluminium added to the tap ladle, $[\%Al]_a$, can be accounted for by the following summation.

$[\%Al]_a = [\%Al]_1$(dissolved residual) + $[\%Al]_2$(for deoxidation) + $[\%Al]_3$(as residual oxide inclusions) + $[\%Al]_4$(oxidized by furnace slag and fallen converter skull) + $[\%Al]_5$(oxidized by entrained air bubbles)

Numerous studies of air entrainment during furnace tapping indicate that less than 20 ppm O is introduced into the steel bath in the ladle. The total oxygen content of the Al-killed steel in the tap ladle is in the range 60 to 120 ppm. Therefore, for the sum $[\%Al]_3 + [\%Al]_5$ an average value of 0.01 is taken. The aluminium consumed in deoxidation of the steel is calculated from the oxygen content of the steel at turn down as measured by the oxygen sensor: $[\%Al]_2 = 1.125 \times [\%O]$. The amount of aluminium lost to the ladle slag is then given by the difference

$$[\%Al]_4 = [\%Al]_a - [\%Al]_t - [\%Al]_2 - 0.01 \qquad (1)$$

where $[\%Al]_t$ is the total aluminium content of the steel sample taken from the tap ladle.

Material Balance for Silicon:- For the Al-killed steel, the loss of silicon to the ladle slag, $[\%Si]_4$, is simplified to

$$[\%Si]_4 = [\%Si]_a + [\%Si]_f - [\%Si]_l \qquad (2)$$

where $[\%Si]_a$ = added to the tap ladle

$[\%Si]_f$ = content in steel at turn down

$[\%Si]_l$ = content in steel in tap ladle.

Mass of Slag Carry Over and Fallen Converter Skull:- The reaction of aluminium and silicon with the furnace slag and fallen converter skull is represented in a general form by the following equation

$$Fe(Mn)O_x + Al(Si) \rightarrow Fe(Mn) + Al(Si)O_x \qquad (3)$$

Using the average molecular masses and assuming 80% Fe_3O_4 for the converter skull, the following approximate relation is derived for the loss of aluminium and silicon to the ladle slag for a 200-tonne steel in the tap ladle.

$$[\%Al + \%Si]_4 \simeq 1.1\times10^{-6}\Delta(\%FeO_t + \%MnO)W_{fs} + 11\times10^{-5}W_{sk} \qquad (4)$$

where W_{fs} = mass of furnace slag carry over, kg

W_{sk} = mass of fallen converter skull, kg

$\Delta(\%FeO_t + \%MnO)$ = decrease in the oxide contents of the furnace slag during tapping.

For low-carbon Q-BOP heats, $\Delta(\%FeO_t + \%MnO)$ is about 20% for which equation (4) becomes

$$[\%Al + \%Si]_4 = 2.2\times10^{-5}W_{fs} + 11\times10^{-5}W_{sk} \qquad (5)$$

Phosphorus Reversion:- Another consequence of slag carry over is phosphorus reversion to the Al-killed steel in the tap ladle. For a 200-tonne heat, the phosphorus reversion Δ[ppm P] is represented by

$$\Delta[\text{ppm P}] = 0.05\Delta(\%P)W_{fs} \qquad (6)$$

where $\Delta(\%P)$ is the decrease in the phosphorus content of the furnace slag carried into the tap ladle. For low-carbon Q-BOP heats, $\Delta(\%P)$ is about 0.3% for which equation (6) becomes

$$\Delta[\text{ppm P}] \simeq 0.015W_{fs} \qquad (7)$$

The results of the analysis of Q-BOP heat logs, as described above, are plotted in Fig. 1 (the data points are scattered within the hatched area) which show the expected increase in the extent of phosphorus reversion with an increase in $[\%Al + \%Si]_4$ reacting with the ladle slag. For high phosphorus reversion of 45 ± 10 ppm, the mass of slag carry over estimated from equation (7) is W_{fs} = 3000 ± 670 kg, corresponding to a slag thickness (in a 200-tonne ladle) of about 117 ± 26 mm. In most cases of furnace tapping, the phosphorus reversion is about 20 ± 10 ppm for which the estimated slag carry over is about 1340 ± 670 kg. This estimated average slag carry over during furnace tapping corresponds to a slag thickness (in a 200-tonne ladle) of about 52 ± 26 mm which is in general accord with the plant observations, in the so-called minimum slag carry over practice.

When the heat is tapped open, with only the addition of ferro-manganese and a small amount of aluminium, the steel is not reduced enough to cause phosphorus reversion. In fact, in some cases of open-heat tapping with about 0.3 to 0.6% Mn addition, the phosphorus content of the steel decreases by about 10 ppm, due to the mixing of the carried over furnace slag with the steel during tapping.

For the average value of W_{fs} = 1340 ± 670 kg, giving Δ[ppm P] = 20 ± 10, the relations in Fig. 1 and equation (5) give W_{sk} = 470 ± 30 kg as the estimated amount of fallen converter skull.

SLAG-AIDED DEOXIDATION OF STEEL

Partial Deoxidation with Mn/Fe

The steel with low-nitrogen specification is tapped open with only the addition of ferro-manganese. The deoxidation with aluminium is done subsequently by either the wire-feed technique or other means of bulk addition, depending on the available ladle refining facilities.

An interesting aspect of deoxidation with manganese is that the iron also participates in the reaction.

$$[Mn] + [O] \rightarrow MnO$$

$$[Fe] + [O] \rightarrow FeO$$

The dissolution of the deoxidation product Mn(Fe)O in a neutral ladle slag, such as calcium silicate or calcium aluminate, will increase the extent of deoxidation. The prefused calcium aluminate is the preferred ladle slag, particularly for low-silicon, Al-killed steels.

The concept of slag-aided partial deoxidation of steel is by no means new. In the Perrin process developed in the early 1930's, the deoxidation of the open-hearth or Bessemer steel with ferro-manganese and ferro-silicon was enhanced by tapping the steel into a molten calcium (magnesium) aluminosilicate slag placed on the bottom of the tap ladle.

The plot in Fig. 2 depicts the progress of partial deoxidation of steel during tapping of a 200-tonne heat with the addition of 1800 kg lime-saturated calcium aluminate and ferro-manganese to the ladle at 1/8 ladle fillage. At the time of the ladle additions, the small quantity of steel in the ladle is almost completely deoxidized with the residual manganese in the steel being at a concentration of about 1.6% Mn. As the ladle is filled, the dissolved manganese is consumed by the deoxidation reaction and decreased to about 0.32% Mn when the ladle is full and the residual dissolved oxygen is reduced to about 300 from 650 ppm in the furnace.

The results obtained using this practice in the EAF and Q-BOP shops of U.S. Steel, Inc. (a Division of USX Corporation) are plotted in Fig. 3, for steels containing less than 0.003% Si and Al each. If no aluminate slag addition were made to the tap ladle, the deoxidation by Mn and Fe alone, with pure Mn(Fe)O as the deoxidation product, would have resulted in much higher levels of the residual dissolved oxygen in the steel as depicted by the dotted curve in Fig. 3.

A more detailed analysis of the EAF heats are given in Fig. 4, showing the expected linear relation between the ratio $\%MnO/\%FeO_t$ in the slag and %Mn in the steel for the tap samples from the EAF and the ladle. As known from the temperature dependence of the equilibrium constant for this deoxidation reaction, the slope of the line increases with a decrease in the steel temperature. These EAF plant data are consistent with the slag-metal equilibrium relation (broken line) as in BOF and Q-BOP for the slag basicity of $\%CaO/\%SiO_2 \simeq 3$.

As is seen from the data in Fig. 5, the deoxidation product [%Mn][ppm O] in the steel is related to the MnO content of the ladle slag. The plant data are in general accord with the equilibrium relation calculated from the thermochemical data by assuming $\gamma_{MnO} = 2$ for the thermodynamic activity coefficient of MnO (with respect to solid MnO) in lime-saturated calcium aluminate slags.

In these EAF trial heats, there was no argon stirring in the ladle during furnace tapping. Yet, we see from the plant data in Figs. 4 and 5 that the slag-aided partial deoxidation of steel achieved during furnace tapping is close to the levels determined by the slag-metal equilibrium. Apparently, there is sufficient mixing of slag and metal within the area of impact of the steel stream with the ladle slag for the slag-aided deoxidation reaction to progress at a relatively fast rate. Also, because of the off-centre entry of the steel stream into the ladle, there is slow rotation of the slag layer in the ladle. Thus, most of the ladle slag layer passes through the area of stream entry where there is the localized mixing of slag and metal.

In the EAF steelmaking, the nitrogen content of steel is about 35 to 45 ppm at tap. With partial deoxidation during furnace tapping from 600 or 700 ppm O to 280 to 320 ppm in the tap ladle, the nitrogen pick up was less than 10 ppm.

In these trial heats, the aluminium deoxidation was done with the wire-feeder, based on the level of the residual dissolved oxygen in the tap ladle. The efficiency of aluminium recovery was about 85 to 90 percent. Although the ladle slag contained 4 to 7% MnO and 7 to 12% FeO_t, there was little reduction of these oxides during the injection of the aluminium wire, accompanied by argon stirring at a moderate rate of 250 to 300 Nl/min.

<u>Effect of Slag Carry Over</u>

The carry over of furnace slag to the tap ladle is expected to affect adversely the extent of partial deoxidation of steel with manganese, iron and the aluminate ladle slag. This effect can be computed from the material balance and the relations in Figs. 4 and 5 for slag and metal compositions in the tap ladle.

$$\frac{(\%MnO)}{(\%FeO_t)} = 1.7\,[\%Mn] \tag{8}$$

$$(\%MnO) = 0.062[\%Mn][ppm\ O] \tag{9}$$

Definition of symbols:

W_{fs} = furnace slag carry over, kg

W_1 = amount of MnO in carry over slag, kg

W_2 = amount of FeO_t in carry over slag, kg

X = amount of MnO formed in deoxidation, kg

Y = amount of FeO formed in deoxidation, kg

The calculations are made for the following conditions: 200-tonne heat, 650 ppm O in steel at tap, ferro-manganese charged to aim at 0.4% Mn in steel and 1800 kg calcium aluminate ladle slag. From the material balance and equations (8) and (9), the following five equations are derived.

$$\frac{W_1 + X}{W_2 + Y} = 1.7\,\frac{800 - 0.775X}{2000} \tag{10a}$$

$$ppm\ O = \frac{130 - 0.225(X + Y)}{0.2} \tag{10b}$$

$$\%Mn = \frac{800 - 0.775X}{2000} \tag{10c}$$

$$\%MnO = \frac{(W_1 + X)100}{1800 + W_{fs} + X + Y} \tag{10d}$$

$$\%MnO = 0.062[\%Mn][ppm\ O] \tag{10e}$$

If there is extensive carry over of furnace slag, the iron will not contribute to deoxidation, instead some iron oxide in the slag will be reduced by Mn, hence Y will be a negative number. Calculations are also made for the case of no aluminate slag addition to the ladle. The results are plotted in Fig. 6.

For the limiting case of no slag carry over, the steel is deoxidized to a residual 274 ppm O. The extent of deoxidation becomes less and the residual %Mn decreases with an increase in the amount of the slag carry over.

When the calcium aluminate is not added to the tap ladle, deoxidation occurs to a small extent only when there is a 1000 to 1500 kg slag carry over as shown by the top curve in Fig. 6.

As is shown in Fig. 7, the participation of

iron in the deoxidation of steel with manganese decreases with an increase in the amount of the slag carry over. Above 1500 kg slag carry over, the iron no longer participates in the deoxidation reaction, instead, some iron oxide in the carry over furnace slag is reduced by manganese, as indicated by the negative values of ppm O reacting.

Mn/Si Semi-Killed Steel

In semi-killed steels deoxidized with silico-manganese and a small amount of aluminium, the deoxidation product is molten manganese aluminosilicate and the residual dissolved oxygen is about 50 ppm for the steel containing about 0.8% Mn and 0.2% Si. With the addition of 1000 kg prefused calcium aluminate to the ladle for a 200-tonne heat, the residual dissolved oxygen in the steel can be lowered to about 20 ppm with the Mn/Si deoxidation, because of the dissolution of the deoxidation products in the calcium aluminate ladle slag. Another advantage of the ladle slag is that it minimizes the adverse effect of the slag carry over on the silico-manganese deoxidation during furnace tapping.

In Mn/Si-killed, coarse grain steels of high drawability, as for instance the tyre-cord grade high-carbon steels, there is a narrow critical range for the total aluminium content. It should be within 15 to 25 ppm to ensure that the deoxidation product is molten manganese aluminosilicate of spessartite composition ($3\,MnO\cdot Al_2O_3\cdot 3\,SiO_2$), which has a high ductility required in cold drawing of the steel to a thin wire of about 0.2 mm diameter.

If the ladle slag used contains a high percentage of alumina, some of it will be reduced by silicon and subsequently precipitate as undesirable Al_2O_3 inclusions. These inclusions are nonductile and hence cause breakage during cold drawing to very thin sections. As noted from the equilibrium data in Fig. 8, the steel containing, for example, 0.2% Si may pick up an appreciable amount of aluminium by the reduction of alumina from the slag. In fact, experience has shown that when the Si-killed steel with 0.2% Si is mixed well with molten calcium aluminate by argon stirring, there is about 0.01% Al pickup. For Mn/Si-killed coarse grain steels, the ladle slag to be used is calcium silicate of wollastonite composition ($CaO\cdot SiO_2$), although it is less effective than the calcium aluminate ladle slag in enhancing the Mn/Si deoxidation.

Al-Killed Steel

Comments should be made first on the choice of the equilibrium constant for the solubility product of Al_2O_3 in the interpretation of the oxygen sensor readings in Al-killed steels.

Solubility Product Al_2O_3: Many careful measurements have been made of the solubility product of pure Al_2O_3 in high purity liquid iron using the emf technique. The reported values of the solubility product at 1600°C are in the range

$$[a_{Al}]^2[a_O]^3 \equiv [\%Al]^2[\%O]^3 = 2.4 \times 10^{-14} \text{ to } 1.2 \times 10^{-13}$$

The higher values are attributed to the interference with the emf readings caused by partial electronic conduction in the stabilized zirconia electrolyte of the emf cell, as in the commercial oxygen sensors used in the steel industry. In analysing the state of deoxidation in Al-killed steels it is, therefore, more appropriate to use 1.2×10^{-13} as the apparent solubility product at the ladle refining temperature of 1600 ± 10°C.

The oxygen sensor readings and analysis of steel samples from the tap ladle have indicated that the addition of lime and spar or prefused calcium aluminate to the ladle extends the deoxidation of steel with aluminium during furnace tapping. This is demonstrated by the plant data plotted in Fig. 9 where %Al is the total aluminium content of the steel. For an average total oxygen content of about 80 ppm in the steel, soon after furnace tapping, the concentration of dissolved aluminium is in fact about [%Al-0.009]. With this adjustment of %Al, the data points will be shifted towards the dotted curve for equilibrium with molten aluminate of composition 50:50 $CaO:Al_2O_3$. The entrained alumina inclusions in the melt are fluxed with the lime-rich ladle slag, i.e. the alumina activity is lowered, resulting in lower levels of residual dissolved oxygen.

In addition to its beneficial effect on the deoxidation reaction, the calcium aluminate ladle slag facilitates the removal of oxide inclusions and prevents the re-oxidation of the melt during argon rinsing for inclusion flotation.

In the plant trials with 30-tonne AOD heats, reported by Riley and Nusselt*, the oxygen sensor readings were taken after vigorous mixing of steel with the lime-rich aluminate slag with hard argon blowing. As is seen from their plant data in Fig. 10, the residual dissolved oxygen contents for 1650 and 1600°C are below the equilibrium curves for pure alumina, because of fluxing of the alumina inclusions with the calcium aluminate furnace slag with argon stirring. In fact, the oxide inclusions separated from the steel samples were found to have compositions similar to those of the calcium aluminate AOD slag.

STEEL DESULPHURIZATION DURING FURNACE TAPPING

The lime-saturated molten calcium aluminate has a high sulphide capacity and is a most appropriate medium for the desulphurization of aluminium-killed steels. However, no desulphurization occurs during furnace tapping onto a mixture of lime, prefused calcium aluminate and spar charged to the ladle before or during tapping. A vigorous mixing of molten slag and steel in the ladle is of course essential; this is achieved by the argon injection at the rate of 1.5 to 1.8 Nm^3/min at about 3-m depth in a 200-tonne ladle. For this high rate of argon injection, resulting in the required 100 to 150 W/t energy density of stirring for efficient desulphurization, without the injection of Ca–Si, there should be at least 700 mm height of freeboard above the slag layer to ensure minimum or no slag and metal splashing onto the ladle lid. In many steelmaking shops, the ladles are often over charged, leaving insufficient freeboard for hard argon stirring. In such cases, the Ca–Si or lime

---

* M.R. Riley and L.G. Nusselt, Jr. Fifth International Iron and Steel Congress, pp. 177-182. The Iron and Steel Society of AIME, 1986.

plus spar powder injection is made with argon flowing at about 1 Nm^3/min. With this practice, the extent of desulphurization with the calcium aluminate ladle slag and the injected powder mixture is usually less than 60 percent of that present in the steel at tap, as compared to 70 to 80 percent desulphurization with hard argon stirring and no Ca—Si powder injection.

In the conventional practice of desulphurization of steel in the ladle (with or without powder injection), the steel picks up nitrogen and hydrogen, even when the freeboard of the covered ladle is flushed with argon. Also, in this practice, the steel is tapped at relatively higher tap temperatures of 1650 to 1700°C, if there is no ladle furnace available for reheating.

Similar to the Perrin process of the 1930's for ladle dephosphorization or slag-aided deoxidation of steel, the desulphurization can be achieved by tapping the steel into a lime-saturated molten calcium aluminate and aluminium placed on the bottom of the ladle.

Exothermic Desulphurizing Slag

There are several patented disclosures and claims, and plant trials for the aluminothermic fusion of slag in the ladle for desulphurization during furnace tapping.

An exothermic slag mixture, EXOSLAG, was formulated and tested in 1984 at the Research Laboratory of USS, Inc. (a Division of USX Corporation). The composition of the EXOSLAG mixture is calculated from the heat and material balance consideration, including the additional Al_2O_3 formed during deoxidation of the steel. The following preferred mixture includes the additional lime needed to flux the alumina generated in the deoxidation reaction, to give a lime-saturated aluminate slag with the nominal ratio of $\%CaO/\%Al_2O_3 = 60/40$.

- 58% burnt lime
- 30% hematite ore (or iron ore concentrate)
- 12% aluminium powder (or finely shredded aluminium scrap)

The ingredients are in dry powder form, well mixed, and contained in moisture-proof bags of suitable size for easy handling. Because of the reduction of iron oxide in the aluminothermic reaction, 2000 kg EXOSLAG mixture gives about 1500 kg calcium aluminate and 500 kg iron which will increase the steel yield by about 0.25% in a 200-tonne heat.

The ignition temperature of the mixture has been experimentally determined to be in the range 870 to 890°C which is 60 to 40°C lower than the hot-face temperature of the lining in the preheated ladle. Experience in several plants has shown that with a well-preheated ladle, sufficient self ignition of the mixture occurs in about 5 minutes to permit tapping of the steel into the desired liquid calcium aluminate slag. It should be noted, however, that the ignited material is only partly liquid because of the presence of excess lime which will be fluxed subsequently with the alumina generated in the deoxidation of liquid steel with aluminium that was already added to the tap ladle together with the EXOSLAG mixture.

Several extensive plant trials have been conducted in U.S. Steel, Inc. to evaluate the performance of EXOSLAG in desulphurization of the Al-killed steel during tapping of EAF, BOF and Q-BOP. Similar results were obtained, irrespective of the steelmaking process. The data from 200-tonne BOF production heats of various grades of Al-killed steels are given in Fig. 11. For the EXOSLAG additions of 5 to 10 kg per tonne of steel, most of the data points are in the range of 25 to 50 percent sulphur removal during furnace tapping, the average being about 40 percent. A higher percentage of desulphurization was obtained in some heats using a 15 kg/tonne EXOSLAG addition. The scatter in the data is due, in part, to deviations from the recommended practice, and in part to variations in the carry over of the furnace slag, hence the variations in the residual aluminium content of steel in the tap ladle. More sulphur removal is usually achieved with a higher concentration of residual aluminium in steel in the tap ladle. The increase in the extent of desulphurization with an increase in the amount of EXOSLAG in the tap ladle is shown in Fig. 12 for a large number of production heats of various grades of low-and high-carbon steels, containing at tap 0.006 to 0.045% S.

Experience in plant trials has shown that argon bubbling through the porous plug of the ladle during tapping has no adverse or beneficial effect on the extent of desulphurization with EXOSLAG. However, if the steel is to be desulphurized to levels below 0.003% S, this can be achieved by hard argon stirring subsequent to furnace tapping. In the conventional method of desulphurization with the prefused calcium aluminate and lime and hard argon stirring (with or without the injection of Ca—Si), the sulphur content of the steel at tap should be 0.01% S or less to ensure that the steel is desulphurized to below 0.003% S. On the other hand, with the EXOSLAG practice, using 10 kg/tonne, the steel containing 0.015 to 0.020% S at tap can be desulphurized to about 0.007 and 0.01% S, respectively, during furnace tapping; then, subsequent hard argon stirring will lower the sulphur content to 0.003% or less.

Temperature Benefit in EXOSLAG Practice

An additional benefit realized from the use of EXOSLAG is that less energy is consumed from the steel to melt this slag than is required from the use of other types of synthetic slag mixtures. Consequently, with EXOSLAG, the furnace operator does not have to increase the tap temperature of the steel to accommodate the chill effect and melting of the ladle refining slag.

The temperature benefit realized with the EXOSLAG practice is derived from the difference between the overall heats of reactions in the tap ladle,

$$\Delta H = \Delta H_{EX}(\text{EXOSLAG practice}) - \Delta H_{CA}\ (\text{lime + aluminate practice}) \qquad (11)$$

for practices involving the same (i) amount of calcium aluminate ladle slag per tonne of steel, (ii) amount of aluminium addition for deoxidation, (iii) tap temperature, (iv) temperature of the preheated ladle, and (v) heat losses to the ladle lining.

The heat and material balance calculation,

using the thermochemical data, gives

$$\Delta H = -\ 18{,}670 \text{ kJ/tonne} \qquad (12)$$

Combining this with the heat capacity of liquid steel, dH/dT = 717 kJ/tonne, gives for the temperature benefit

$$\Delta T = \frac{18{,}670}{717} = 26°C \qquad (13)$$

That is, the heat losses incurred in the EXOSLAG practice will be 26°C less than that in the conventional practice using lime + prefused calcium aluminate in the tap ladle. Thus, the tap temperature can be decreased by 26°C when using the EXOSLAG practice as a replacement for the one generating the ladle refining slag from the addition of lime and prefused calcium aluminate.

The foregoing prediction of the temperature benefit has been well substantiated by the plant data shown in Fig. 13, where the temperature drop from the furnace to the ladle for heats made by the EXOSLAG practice is compared to those made with the ladle addition of lime plus prefused calcium aluminate. It is seen that the tap temperature of the EXOSLAG heats could have safely been lowered by about 24°C.

SUMMARY

I have cited various examples of reactions that occur in the ladle between liquid steel and slag during furnace tapping. The study of plant data has shown that losses of aluminium and silicon in the tap ladle are due mainly to their oxidation by the iron and manganese oxides in the furnace slag that is being carried into the ladle, and by the iron oxide-rich converter skull falling into the ladle during tapping. For 200-tonne Q-BOP heats, an average loss of %Al + %Si ≈ 0.08% to the ladle slag is attributed to a 1340 kg furnace slag carry over and 470 kg fallen converter skull. For this average amount of slag carry over, there is about 20 ppm phosphorus reversion in Q-BOP heats. The phosphorus reversion will be higher for BOF and EAF heats with higher phosphorus contents in the slag.

In the production of low-silicon and low-nitrogen steels with a narrow range for the residual aluminium, there are advantages to partial deoxidation of steel with ferro-manganese and calcium aluminate slag added to the tap ladle. For example, with the addition of 10 kg prefused calcium aluminate per tonne of steel and ferro-manganese to aim at 0.4% Mn, the steel at tap containing 650 ppm O is partially deoxidized to about 274 ppm O, in the limiting case of no furnace slag carry over. The extent of partial deoxidation of course decreases with an increase in the amount of slag carry over.

The analysis of plant data for Al-killed steels indicates that the concentration of residual dissolved oxygen in liquid steel is close to that in equilibrium with the calcium aluminate inclusions retained in the steel.

The desulphurization of the Al-killed steel by 40 to 50 percent during furnace tapping, even without argon stirring in the ladle, has now become a reality with the use of an exothermic slag mixture that is self ignited in the preheated ladle prior to furnace tapping. In the practice developed in U.S. Steel, Inc., the exothermic mixture, EXOSLAG, consists of 58% burnt lime, 30% hematite ore (or iron ore concentrate) and 12% aluminium powder (or finely shredded aluminium scrap). With the EXOSLAG practice, using 10 kg/tonne, the steel containing 0.015 to 0.020% S at tap can be desulphurized to about 0.007 and 0.01% S, respectively, during furnace tapping. Then, subsequent hard argon stirring (without Ca–Si injection) will lower the sulphur content to 0.003% or less. With this practice of desulphurization during furnace tapping, the pickup of nitrogen in steel is about 40 ± 10 ppm which is two-thirds to one-half of that occurring with hard argon stirring. Then, there is the temperature benefit of about 24°C in comparison to the conventional method of ladle desulphurization. That is, the tap temperature can be lowered by about 24°C with the EXOSLAG practice.

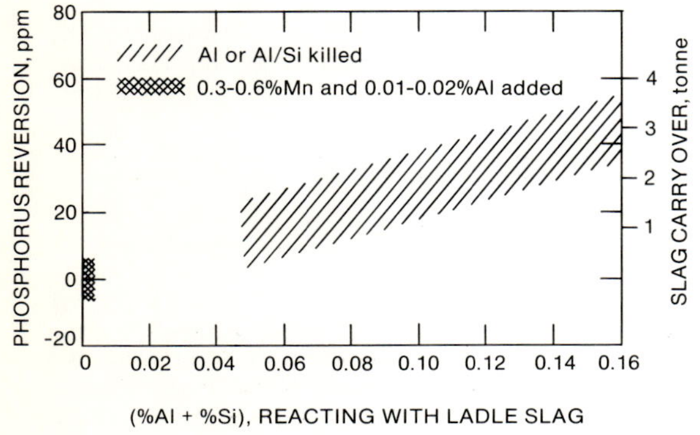

Fig. 1 Phosphorus reversion during tapping of 200-tonne Q-BOP heats is related to %Al + %Si reacting with the carry over furnace slag and fallen converter skull.

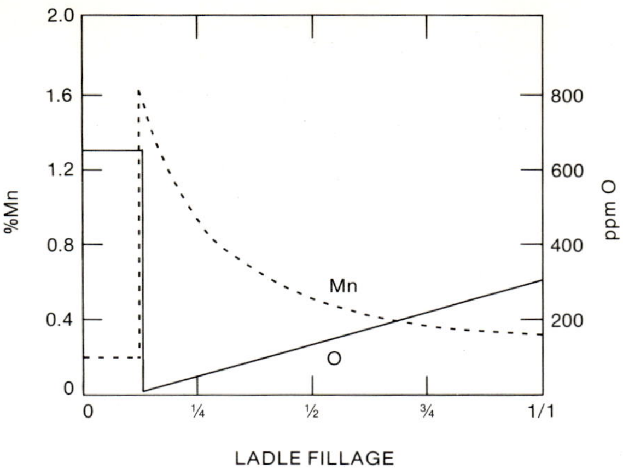

Fig. 2 Change in the dissolved contents of Mn and O in steel during furnace tapping of a 200-tonne heat; the ladle slag, 1800 kg lime-saturated calcium aluminate, and ferro-manganese charged at 1/8 ladle fillage to aim at 0.40% Mn in 200-t.

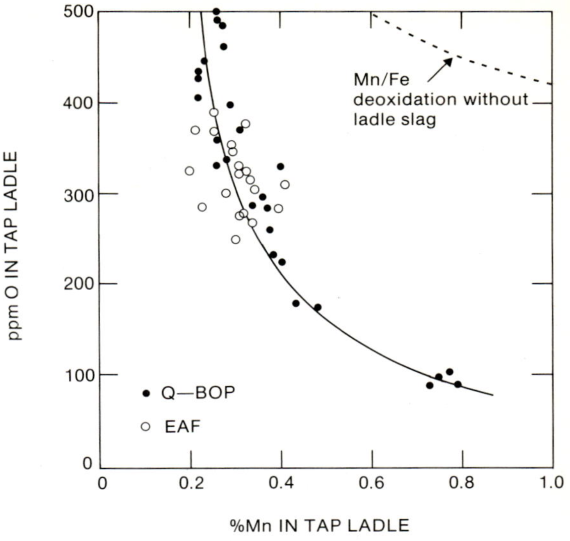

Fig. 3 Partial deoxidation of steel with ferro-manganese and calcium aluminate ladle slag during furnace tapping for steel in the ladle containing Si and Al each < 0.003%.

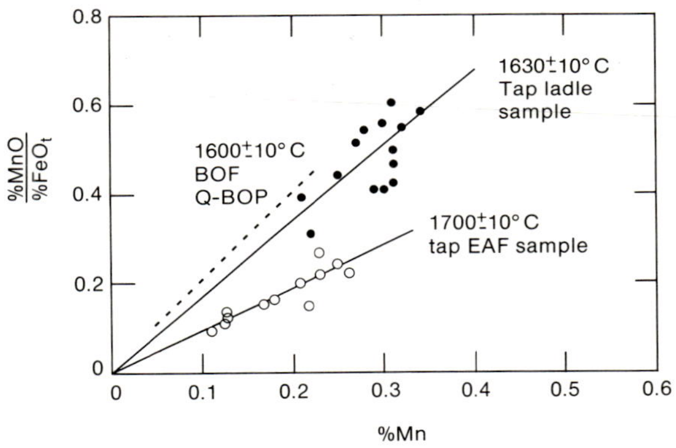

Fig. 4 The manganese content of steel in the EAF at tap and in the tap ladle is related to the ratio $\%MnO/\%FeO_t$ in the furnace slag and ladle slag and compared to that for the slag-metal equilibrium as in BOF and Q-BOP for $\%CaO/\%SiO_2 \simeq 3$.

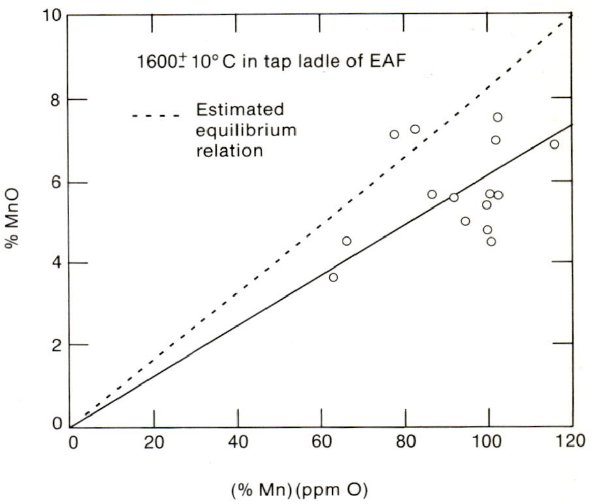

Fig. 5 Manganese deoxidation product is related to the MnO content of lime-saturated calcium aluminate ladle slag after EAF tapping.

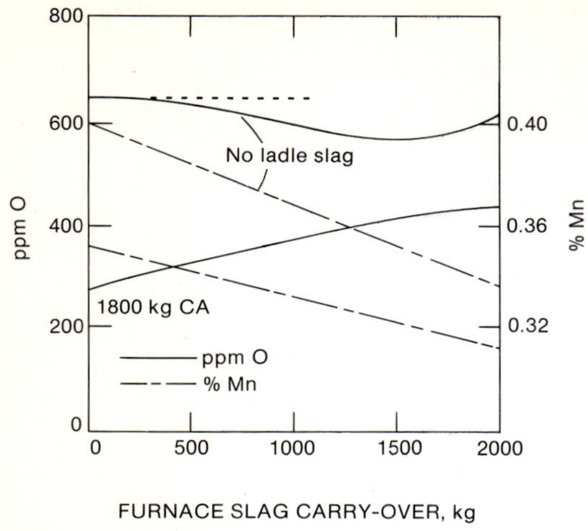

Fig. 6 Effect of furnace slag carry over on partial deoxidation of steel with manganese during furnace tapping for
(i) no addition of ladle slag and
(ii) addition of 1800 kg calcium aluminate (CA) in the tap ladle.

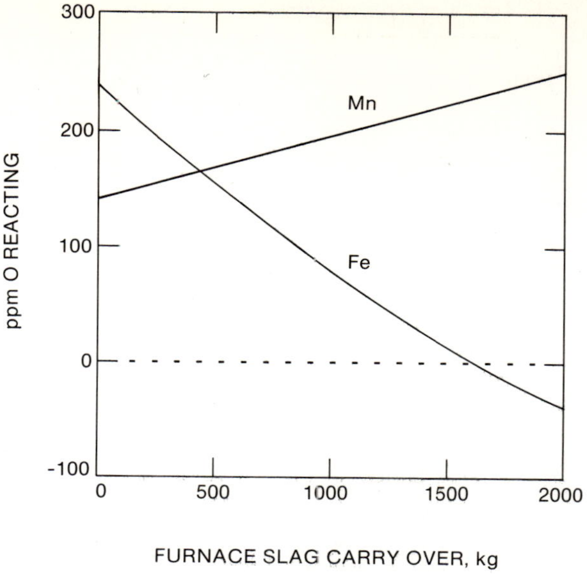

Fig. 7 Effect of furnace slag carry over on ppm oxygen reacting with Fe and Mn during furnace tapping with the addition of 1800 kg calcium aluminate and 0.4% Mn to the tap ladle.

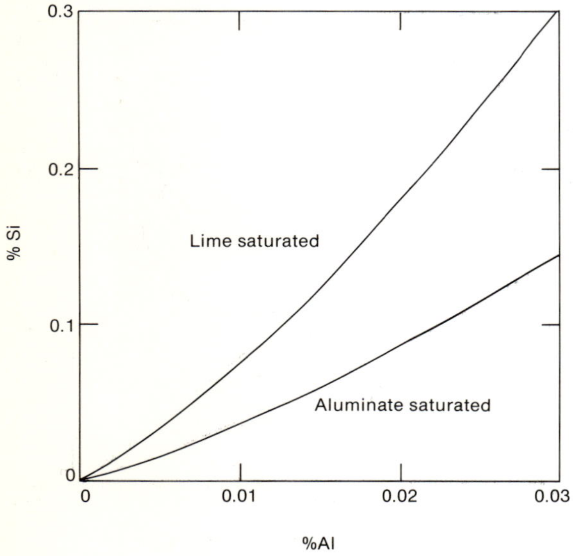

Fig. 8 Silicon and aluminium contents of steel in equilibrium with molten calcium aluminate slags containing ≃ 5% SiO_2 at 1600°C.

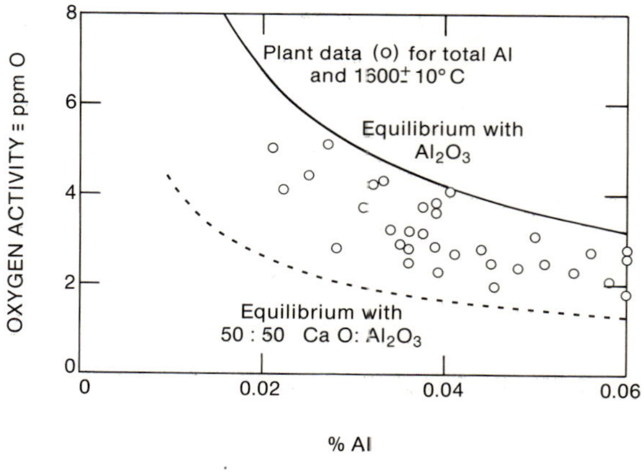

Fig. 9 Plant data for deoxidation with aluminium and calcium aluminate ladle slag during furnace tapping are compared with the equilibrium values for Al_2O_3 and molten calcium aluminate (50:50 $CaO:Al_2O_3$) inclusions.

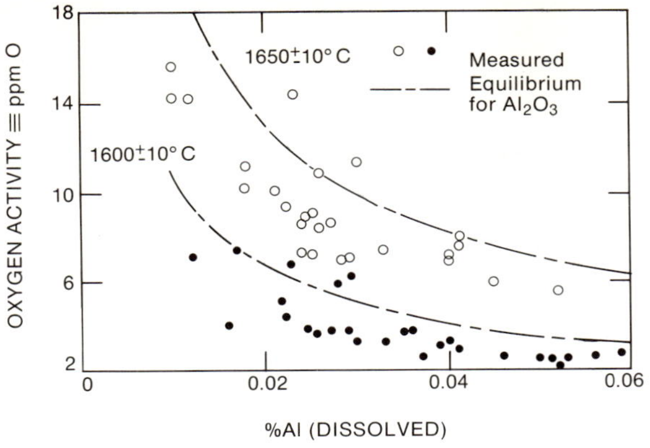

Fig. 10 Oxygen sensor measurements after Al deoxidation and hard argon stirring in 30-tonne AOD, reported by Riley and Nusselt.

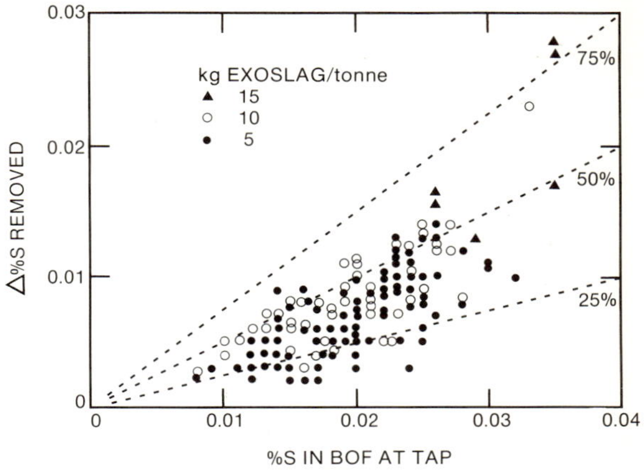

Fig. 11 Desulphurization of Al-killed BOF steels with EXOSLAG during furnace tapping.

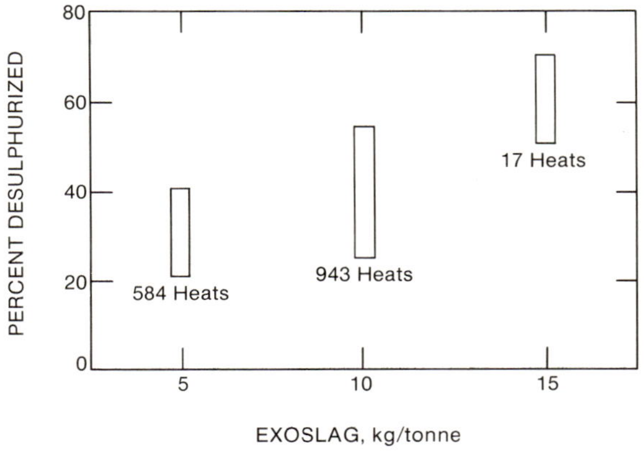

Fig. 12 Percent sulphur removed during furnace tapping increasing with an increase in the amount of EXOSLAG in the tap ladle for tap steel compositions in BOF in the range: < 0.1 to > 0.4% C and 0.006 to 0.045% S.

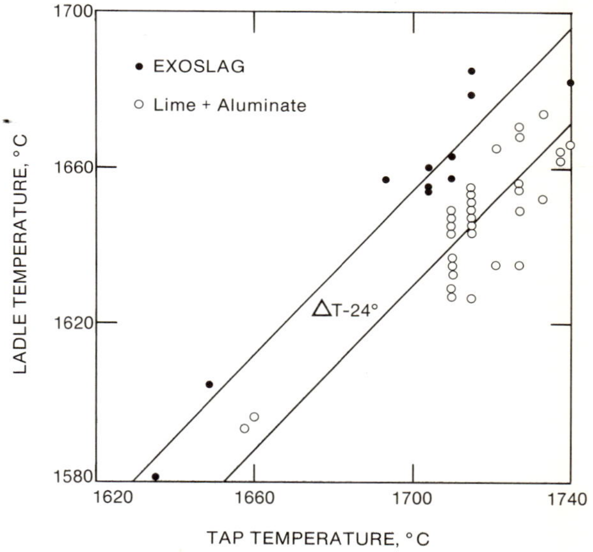

Fig. 13 Temperature loss during furnace tapping with EXOSLAG is compared with that for lime + prefused calcium aluminate added to the tap ladle.

Slag development in the BOS for the production of low phosphorus and sulphur steels

G J HASSALL and B C WELBOURN

Dr Hassall is in the Basic Studies Department at British Steel Technical, Teesside Laboratories; Mr Welbourn is in the Primary Processes Department at British Steel Technical, Teesside Laboratories.

SYNOPSIS

The process conditions necessary to consistently produce low S and P steels (<0·01 wt%) directly from low silicon hot metal (0.3 wt%) in the BOS process have been investigated. Fluxes similar to those used in hot metal pretreatment processes, in combination with lime and with or without dolomet have been used, with scrap or iron ore as coolant. Frequent sampling and analysis of metal and slag have indicated that the use of lime with 20 kg/tls (tonne of liquid steel) of a $CaO/Al_2O_3/CaF_2$ flux and with iron ore as coolant gives the slag compositional changes and metal temperature control required to optimise phosphorus removal during a blow. Because of the poor conditions generated in the BOS process for sulphur removal, it is necessary to limit the sulphur content of the hot metal to twice the sulphur specification in the final steel.

INTRODUCTION

Although the pretreatment of hot metal for sulphur removal prior to Basic Oxygen Steelmaking has become a well established practice, the removal of phosphorus at the same stage of operation has required a hot metal silicon content of approximately 0.2 wt%. This has resulted in the need for a desiliconisation process often carried out in the Blast Furnace runner.

Such process steps are costly in terms of flux consumption, time and temperature loss and in order to overcome these drawbacks, the possibility has been investigated of adding the same fluxes as used in the pretreatment processes to the BOS converter. The objective has been to try to produce steels with the same or improved final sulphur and phosphorus specifications that are possible via the hot metal pretreatment route to steel.

PILOT PLANT FACILITIES AND PRACTICE

A total of 53 experimental heats have been carried out on a 3 tonne magnesia lined pilot plant converter using a two hole oxygen lance operating at 660 m^3/h (S.T.P). Low silicon hot metal was supplied from an electric arc furnace and the average composition and temperature of the metal in the transfer ladle are given in Table 1. Scrap or iron ore was used as coolant and various fluxes in combination with lime were added at various times in the heats. Flux compositions are given in Table 2.

Metal samples were taken from the transfer ladle and during the course of and at the end of an experimental heat. Slag samples were recovered from frozen material on the rods carrying metal samplers and bath temperatures were also taken at regular intervals.

Flux additions were made to the converter either at floor level before moving the converter into the blowing position or during the experiments via a bunker/screen feed and water cooled chute system.

Bath stirring was carried out throughout each heat using air/nitrogen mixtures at flow rates varying from between 30 and 50 m^3/h and 4.5 and 8 m^3/h respectively.

Some changes were made in the lance height during the experiments, the objective being to maintain the ratio of oxygen reacting with carbon in the bath to total oxygen supplied at a value of between 0.85 and 1.0

EXPERIMENTAL CONDITIONS AND RESULTS

Three basic groups of experiments were carried out, each investigating the effects of different operating and flux practices on end point S and P contents in the steel.

Group 1

This group was split into 2 sub-groups both using conventional fluxes with granular fluorspar (3-6% of total lime feed). Dolomet, when used, was added on ignition.

In the first sub-group, fluxes which included a pre-sintered mixture of $CaO\text{-}Fe_2O_3$ were added either before or during the first 3 minutes of a blow. The aim CaO/SiO_2 ratio was 3.5 or 4.5 with an average total flux consumption of 34 kg/tls.

In the second group, fluxes were added in the first 3 minutes to give the same aim basicities as above with extra flux added 2-5 minutes before the end of the heat for an aim CaO/SiO_2 ratio of 7.0. Enhanced bath stirring took place during this period, and the average total flux for these heats was 48 kg/tls.

Typical slag compositional changes for these heats are shown in Figure 1 which is part of the pseudo - ternary in which all oxides, with the exception of CaO and SiO_2 are classified as RO. Generally, either the slag basicities were too low (shown as area A) or the 'oxide' content too high (shown as area B) not only during the heats but also at turn down. These factors, together with high tapping temperatures (>1650 °C) generally gave S and P contents in the final steel of >0.01 wt%. However, lower levels were obtained on a few occasions and the slag compositional changes for one heat (LPS3) are also seen in Figure 1 and show that there was a rapid move towards a reasonably high basicity (CaO/SiO_2 >2.0) whilst at the same time a reasonable oxidation state was maintained (45% (FeO_t + MnO)).

Group 2

In this group of experiments, again split into 2 sub-groups, the effect of carrying out pre-treatment on the hot metal or post- blow treatment on the steel was studied.

Hot metal was treated in the converter with CaO/CaF_2 or $CaO/Al_2O_3/CaF_2$ fluxes, with and without oxygen blowing and with gas stirring. The treated hot metal was not converted to steel.

Having identified the optimum flux mixture and practice from this first sub-group of experiments, treatment either before the main oxygen blow (with no slag/flux removal) or as a post blow stir, was carried out, both steps taking 5 minutes in total. All the dolomet and most of the lime was charged with the scrap before blowing commenced, with more lime being charged during the first three minutes. The total flux consumption ranged from 53 to 69 kg/tls in this group of heats.

The preferred method from this group of experiments for achieving low turn down S and P contents in the steel was identified as:

1. Pretreatment of hot metal in the converter with 20 kg/tls of a 70% CaO/20% Al_2O_3/10% CaF_2 flux, oxygen blowing for 2 minutes followed by gas stirring only for 3 minutes. The fluid slag/flux was retained within the converter.

2. Charging of scrap and dolomet followed by the main oxygen blow during which only lime was added as flux, the total flux consumption being 53 kg/tls.

Typical slag compositional changes as a result of adopting this practice are shown in Figure 2, with the transition from a high basicity, low 'oxide' state (region A) at the beginning of a blow to a reasonably high basicity, medium 'oxide' state (region B) taking place early in a blow.

Subsequently, as more lime was dissolved into the slag, a reduction in oxide content occurred (from region B to C) before increasing again by the end of a blow (region D).

Generally, despite pretreatment, low oxidation states and high temperatures (>1650 °C) at turn down gave final steels containing >0·01%P, although the sulphur contents averaged 0.008 wt%. Because of the pretreatment step, processing times were increased.

Group 3

In an attempt to optimise flux practice and process route to consistently obtain S and P contents below 0·01 wt% at the end of the blow and, at the same time, minimise process time, no pretreatment was carried out in this group of experiments, the flux mixture being added before blowing commenced. In several heats, iron ore replaced scrap as coolant with dolomet not being used in any blow. The lime charge was increased to compensate for no dolomet additions. Total flux consumption for this group of heats ranged from 60 to 90 kg/tls.

The slag compositional changes are typified by those shown in Figure 3, where some differences can be seen depending on the coolant used. A reasonably high basicity slag (CaO/SiO_2 between 2.0 and 3.0) was produced within 5 to 7 minutes when iron ore was used as coolant; this taking place at a time when iron ore additions were still being made to the converter. Similar slag basicities were only achieved between 8 and 10 minutes when scrap was used as coolant. Up to the time when a reasonably high basicity slag was present, the iron ore additions maintained the iron oxide content of the slag at high levels (>25%), whilst with scrap as coolant, the iron oxide content generally remained low (<20%). For blows having similar total flux consumption, end-of-blow basicities were generally higher when iron ore was used as coolant. The iron oxide content (Fe^{2+} and Fe^{3+} expressed as FeO, i.e. FeO_t) of the final slags were significantly higher; 35 to 45% FeO_t compared to 18 to 30% FeO_t when using scrap.

DISCUSSION

Considering first of all the changes achieved in the sulphur content of the metal during the pilot plant converter heats, approximately half the sulphur was removed whatever combination or weight of fluxes were used. Figure 4 shows an example of the actual and calculated sulphur partitions, (%S)/[%S], for a scrap based blow (LPS 3), the slag compositional changes associated with this blow having been referred to previously (Figure 1). Equilibrium partitions have been calculated using either the Flood model or a single line empirically derived equation. It can be seen in Figure 4 that actual partitions are higher than calculated, but both are low (4 to 12). Calculation of the equilibrium partition using optical basicity, however gives values ranging from 7 to 32, i.e. calculated higher than actual. In this latter model the equilibrium is dominated by the thermodynamics of the metal, whereas in the Flood model and the empirical equation, the slag dominates. Over the 14 minute blow, the sulphur content was reduced from 0.013 to 0.007 wt%.

These results suggest that changing the developing slag composition by suitable lime/flux additions will have little effect on the slag-metal partition of sulphur under the oxidising conditions found in the BOS converter process. As expected, the CaO/SiO_2 ratio was more important than the iron oxide content of the slag in influencing the partition under these conditions, but even the effect of basicity was not very great over the range covered by the pilot plant experiments (3.8 to 6.9 by the end of blow). End of blow actual partitions were generally higher than the calculated equilibrium values, although this trend was again reversed when using optical basicity to calculate the equilibrium partitions.

Only by reducing the oxidising conditions at the slag/ metal interface and either encouraging the transfer of sulphur from slag to the gas or increasing the slag bulk will sulphur removal be improved. Since such actions will adversely effect the partition of other elements such as phosphorus it will be necessary to pretreat the hot metal to a low sulphur content and use low sulphur-containing coolants and fluxes if ultra-low (<0.005 wt%) sulphur steel production is required. Since sulphur removal appears to be controlled by the thermochemistry, kinetic factors are only likely to have a secondary effect.

The partition of phosphorus between slag and metal in the pilot plant heats was found to be dependent on temperature, oxidation state (as measured by the FeO_t content of the slag) and basicity in decreasing order or significance. When lime assimilation took place slowly, generally when dolomet was added with scrap as coolant, actual phosphorus partitions were usually close to calculated equilibrium during the heat. With dolomet removed as a flux and extra lime added to compensate, conditions were developed in which big differences between the actual and calculated phosphorus partitions were evident. Despite this thermodynamic advantage, delayed or incomplete lime dissolution, high turn down temperatures and decreasing slag oxidation states acted against the early process benefits and only gave marginal improvements in turn- down phosphorus levels.

Figure 5 shows an example, and is again for the scrap- based blow LPS 3. After 9 minutes process time, the actual and equilibrium phosphorus partitions are very similar. From a start phosphorus content of 0.074 wt%, a final content of 0.008 wt% was achieved. Some phosphorus reversion was noticeable after 9 minutes.

As seen in the last group of heats, when scrap was replaced by iron ore as coolant, and again operating without dolomet as a flux, improvements in early lime dissolution were evident. Large differences between the actual and calculated equilibrium phosphorus partitions throughout the blows were observed. Iron ore also gave good oxidising conditions during the blows as well as better temperature control at turndown (<1650 °C), both factors contributing towards final steel phosphorus contents of less than 0.01 wt%.

The end of blow actual and calculated phosphorus partitions showed a close approach towards equilibrium according to an empirically derived equation, but the equilibrium calculated by the Flood model suggested that some of the final slags were still far from equilibrium. Figure 6 shows the distinct effect of bath temperature on the actual phosphorus partition at the end of the blows, a 25% increase in the partition taking place for a temperature decrease of 37 °C. This is very close to the temperature effect on the partition in the empirically derived equation.

The present experiments have shown that when conditions can be controlled such that there is always a large thermodynamic potential for transferring phosphorus from metal to slag, particularly during the early stages of a blow, then kinetic factors such as bath stirring will ultimately control the phosphorus content of the tapped steel.

CONCLUSIONS

Pilot plant BOS converter heats aimed at producing low (<0.01 wt%) sulphur and phosphorus steels have been carried out using a variety of fluxes and low silicon (0.3wt%) hot metal. From the results of the experimental heats, the process conditions required to consistently produce low S and P steels have been identified as:

(a) Limiting the sulphur content of the hot metal to approximately twice the aim specification of the steel

(b) Aiming for a turn down basicity (CaO/SiO_2) of 7 by charging up to 60 kg/tls of total flux, preferably excluding the additions of lump dolomet. Included in the 60 kg/tls would be 20 kg/tls of 70% CaO/20% Al_2O_3/10% CaF_2, all of which would be added prior to start of blow.

(c) Using ore coolant during the blow not only to control metal temperature (to a maximum turn down of 1650 °C) but also to maintain a high oxidation state (>25% FeO_t) and allow early dissolution of the lime.

ACKNOWLEDGEMENT

The authors would like to thank Dr. R. Baker, Director Research, British Steel plc for permission to publish this paper. Financial support by the ECSC, contract number 7219 C3/806, is also acknowledged.

TABLE 1
Average hot metal composition (wt%) and temperature °C

C	Si	Mn	S	P	Temp
4.00	0.31	0.73	.016	.079	1390

TABLE 2
Composition of flux materials used in trials

Material	Chemical Composition %										
	S	Mn	Fe_2O_3	FeO	SiO_2	CaO	MnO	MgO	Al_2O_3	P_2O_5	CoF_2
Lime	0.02	–	–	–	0.65	96.74	–	0.62	–	–	–
Iron Ore	–	–	96.1	0.11	2.5	0.2	0.02	0.2	–	–	–
Calcium Ferrate	–	0.028	50.9	–	1.0	44.82	0.2	2.4	0.52	0.076	–
Dolomet	–	–	–	1.1	0.8	58.7	–	39.0	–	–	–
Lime/Fluorspar Mixture	–	–	0.17	–	0.29	84.2	–	0.62	0.33	–	14.0
Lime/Fluorspar/ Alumina Mixtures	0.01	–	–	–	0.61	68.8	–	0.46	20.16	–	9.95
	0.01	–	–	–	0.69	68.8	–	0.48	10.09	–	19.91

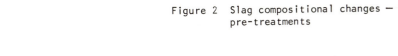

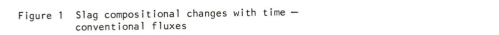

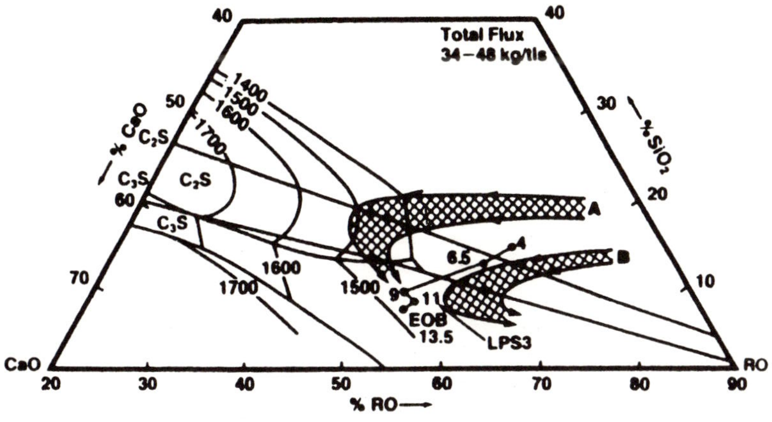

Figure 1 Slag compositional changes with time — conventional fluxes

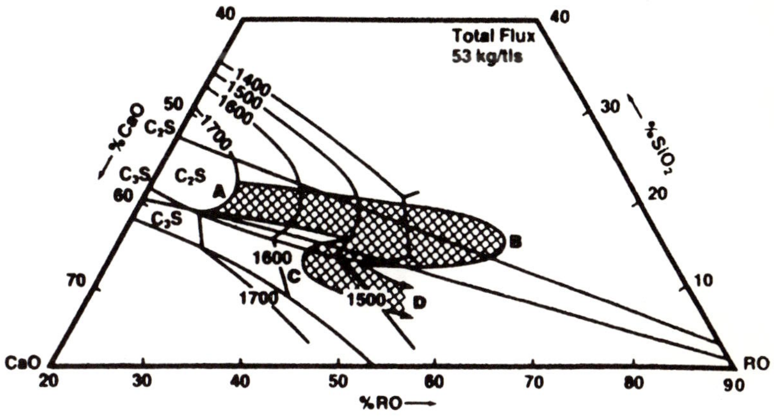

Figure 2 Slag compositional changes — pre-treatments

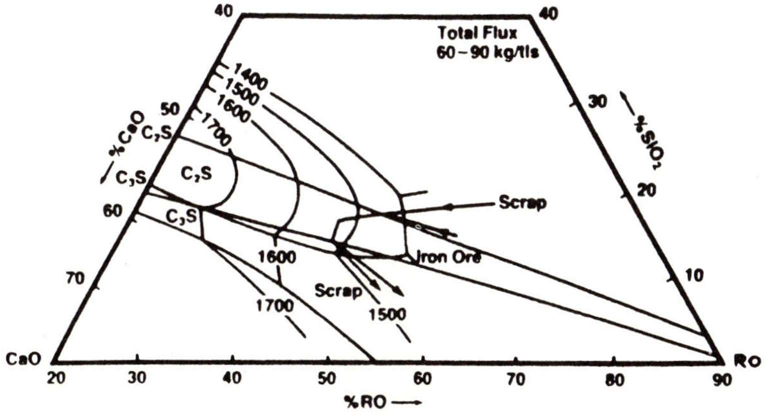

Figure 3 Slag compositional changes — no pre-treatment

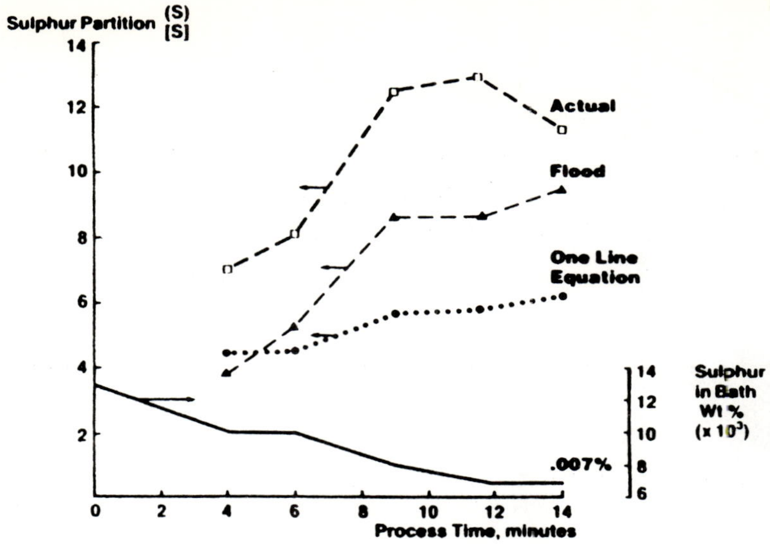

Figure 4 Bath sulphur changes and partitions heat LPS3

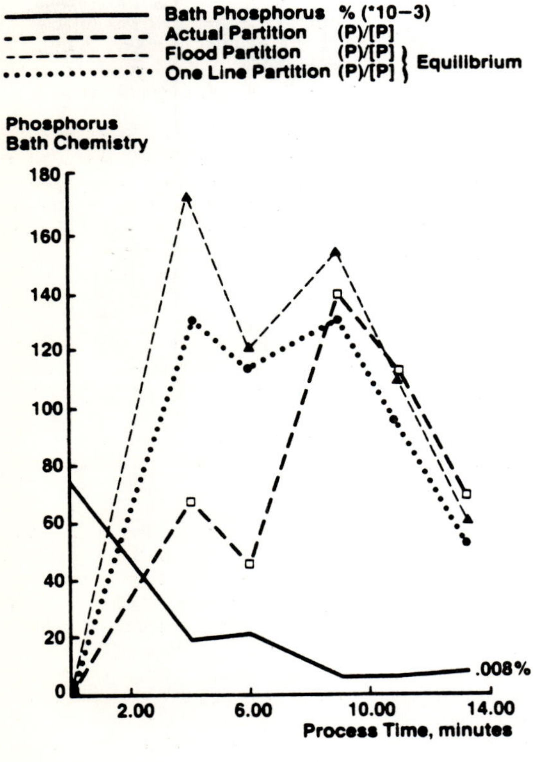

Figure 5 Bath phosphorus changes and partitions heat LPS3

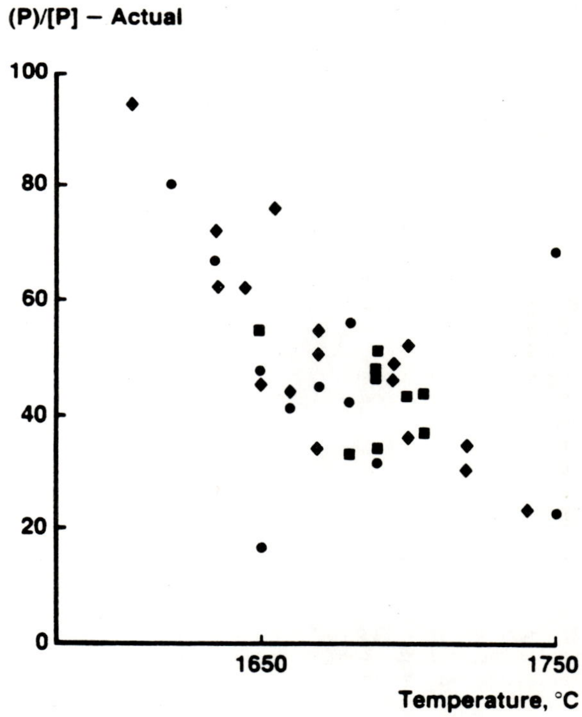

Figure 6 End of blow (P)/[P] actual vs. temperature

□

Use of slag characterisation techniques to optimise steelmaking flux and refractory practices

C B NUTT, P WILLIAMS and J QUIN

Mr Nutt is Export Sales Manager, Steetley Quarry Products, Hartlepool Works, PO Box 8, Hartlepool, Cleveland, TS24 0BY; Dr Williams is Research and Development Manager, Steetley Refractories Ltd., Steetley Works, Worksop, Nottinghamshire, S80 3EA; Dr Quin is Technical Sales Manager, Steetley Quarry Products, Southfield Lane, Whitwell, Nr. Worksop, Nottinghamshire, S80 3LJ.

SYNOPSIS

A technique of characterising industrial process slags has been developed over the past 7 years which uses both chemical analysis and the microstructural examination of cooled slags sampled during specific metallurgical processes, e.g. BOF, EAF, AOD, CLU etc. Slag characterisation studies are now considered as an additional aid in understanding complex slag and steelmaking metallurgical refinement during most steelmaking processes.

Slag chemistry viewed in isolation has been found to be misleading for defining slag basicities and slag bulk estimates. The achievement of the necessary metallurgical refinement during most steelmaking processes is determined by slag formation rates and slag compositional control, which in turn determine the performance of steel furnace refractories. To illustrate the slag characterisation technique a case study is presented for a basic oxygen steelmaking system which led ultimately to an optimised flux charge practice, improved slag formation rates, better dephosphorisation control and a two-fold increase in vessel refractory performance.

INTRODUCTION

The Steetley company holds a well established position within both the refractory and steelmaking industries as a supplier of refractory linings for all types of steelmaking applications including basic oxygen steelmaking (BOS) converters, and as a manufacturer and supplier of flux for basic steelmaking in the form of calcined dolomite.

It is therefore a natural progression for the two interests to be combined so that a service can be given to the steelmaker to optimise refractory life and flux practice. The approach discussed in this paper is restricted to BOS converter practices, although the techniques described can equally be applied to other primary and secondary steelmaking operations such as EAF, AOD and CLU[1].

The dolomitic lime flux is produced by heating Dolomite stone to a temperature of 1400°C in a rotary kiln. The resultant product has an apparent porosity of 50% to 55% and a bulk density of between 1500 and 1700 kg/m^3. The calcined dolomite contains 36% to 40% MgO and 57% to 60% CaO with small quantities of silica, alumina and iron oxide.

THE SLAG CHARACTERISATION TECHNIQUE

The achievement of the necessary metallurgical refinement during most steelmaking processes is largely determined by slag formation rates and slag compositional control which in turn determine the performance of steel furnace refractories.

In its most simple form a slag characterisation study involves chemical analysis and microstructural examination of cooled slag samples taken during the steelmaking operation. Experience has shown that slag chemistry alone can lead to misleading conclusions when attempting to interpret the steelmaking and slag making practice.

A chemical analysis shows only the relative amounts of slag oxide components, usually expressed in mass (weight) percent terms but it gives no information about the phase assemblages formed on cooling, or the degree of solution of the lime and magnesia components introduced in the flux addition.

This feature is illustrated in Figure 1 and shows a BOS converter first turndown slag with a low lime content (35.1% CaO) and a high iron oxide content (45.9% Fe_2O_3) which would normally have been sufficient to ensure that the lime component was fully fluxed (taken into solution). However, this particular example still shows evidence of "free" unreacted lime particles. Although the composition of this slag could be considered as a worst case example, it does emphasise that with this type of situation, slag basicity indices derived purely from a knowledge of the slag chemistry are meaningless unless the cooled slag phase assemblage is also taken into consideration. When this is combined with other cast data such as the quantities of fluxes charged, the timing of flux additions, hot metal and liquid steel analysis, mode of bath decarburisation and process times, a balanced interpretation of the slag making and steelmaking practice can be made.

Interpretation of the slag microstructure must also take into account the approximate slag sampling temperature and subsequent crystallisation path on cooling. It must be understood that the slag will usually crystallise under non-equilibrium conditions.

ESTIMATED SOLUBILITIES OF MAGNESIA IN STEELMAKING SLAGS

Within basic oxygen steelmaking the majority of converter slag compositions can be defined with the oxide phase equilibria system $CaO-MgO-SiO_2-FeOx$[2]. Analysis of slags by X-ray fluorescence techniques involves sample preparation using a fused glass bead technique, thus, the oxides of iron and manganese are arbitrarily reported as Fe_2O_3 and MnO. This does not necessarily relate to the oxidation state of these components in the slag. Manganese oxides are generally associated with the iron oxide containing phases in the slag system.

The compatibility of refractory linings with process slags and the achievement of required metallurgical refining can be significantly influenced by slag path trajectories [3,4]. For primary steelmaking units lined with magnesia based refractories, the solubility of magnesia in predominantly $CaO-FeO_x-SiO_2$ based slags at 1600°C is illustrated in Figure 2 with superimposed slag path trajectories. Using this information it is possible to estimate the relative solubility of magnesia in slags of various lime to silica ratios with respect to the corresponding slag path trajectory as shown in Figure 3.

Above a lime silica ratio of 3:1, end point slags have an estimated magnesia content (saturation) of between 4% and 6% depending on the type of slag path followed. Below a lime to silica ratio of 2:1 the slag system has a high capacity for magnesia solution. This feature has been shown by comparison of relative wear rates in laboratory rotary slag testing where higher wear rates are directly relatable to lower basicity slags.

It is not unusual for many BOS converters to reach temperatures of 1700°C at first turndown. At such temperatures the area of liquid demarcated in Figure 2 would be greatly increased and any refractory dissolution rates accelerated.

Adjustment of slag magnesia levels by adoption of a mixed lime/dolomitic lime flux practice can help to satisfy the early formed slags demand for magnesia with a magnesia source derived from the flux addition rather than from the vessel refractories. The amount of flux magnesia donated is dependent on the quantity of dolomitic flux addition used. Early formed process slags will then become chemically more compatible with a magnesia based refractory lining. Magnesia saturation levels in the first turndown slags are revealed in the slag microstructure with the appearance of magnesiowüstite precipitates. It is necessary to control the dolomitic lime addition to ensure magnesia saturation occurs in the turndown slags.

Dolomitic lime added as part of the flux charge also assists in the breakdown of dicalcium silicate shells that form around individual lime particles during the early stages of slag formation when iron silicate-rich slag compositions predominate [5]. This action has the effect of increasing the rate of solution of lime (CaO) into the slag bulk and as the slag basicity increases, the slag's capacity to take up magnesia is reduced. The dolomitic lime additions should be made at ignition or as early in the blow as possible to satisfy the early formed slag's magnesia requirements and to assist in the solution of lime.

The ultimate compatibility of the process slags with the refractory lining must be achieved whilst the necessary metallurgical refining characteristics of the slags are retained. In practice, it is often necessary for some compromise to be made between these two requirements. The best method of illustrating the slag characterisation technique is by a practical example which will also show how the refractory performance in a BOS converter was substantially improved.

CASE STUDY

This case study concerns a European BOS converter shop operating two from three vessels where low refractory lining life was a major factor controlling converter availability, and ultimately output of steel from the plant.

The first step was to establish the current practices with respect to the amount and timing of the flux charge (calcitic lime) addition, the lance practice, hot metal silicon levels and the physical appearance of the process slags and their compositional and cooled slag phase assemblages.

For slag characterisations it is normal to choose first turndown slags for examination since they largely represent the point at which the steelmaker hopes to have reached the desired metallurgical end point prior to tapping. Further reblowing or flux additions may lead to greater oxidation of the bath and occurrence of unfluxed slag component which may mask the main process trends which are intended to be studied.

The plant operating parameters for the standard flux practice are shown in Table 1. A typical first turndown slag analysis and cooled slag phase assemblage is illustrated in Figure 4. Within this cooled slag, the presence of free lime is readily discernible, (Figure 5). The other phases present were tricalcium silicate and dicalcium silicate within a calcium ferrite rich matrix. From slag characterisation studies and plant observations of operating parameters it was concluded that the slags were overlimed, the timing of the flux addition was far from ideal and the lance practice needed to be operated in a more controlled fashion to ensure that complete fluxing of the lime charge could occur. The true basicity of the slags sampled (reference Table 1) would be lower than the average $CaO:SiO_2$ ratio of 2.72:1.

The average slag magnesia level of 3.2% MgO reflects refractory magnesia pick-up from the converter lining and would account for the poor refractory performances recorded at the plant (350 to 550 heats).

The quantity of flux charged to the converter is a function of hot metal silicon and the level of dephosphorisation that has to be achieved. Using slag mass balance calculations, a graph indicating flux requirements for different hot metal silicon levels was produced on the assumption that all the lime charged ended up in solution. Figure 6 shows the graph for this plant based on a 210 tonne tap weight with flux additions designed to give as a first approximation an aim slag magnesia of 7% MgO i.e. 7% MgO in the slag from a flux source.

The operational parameters for the modified lime/dolomitic lime flux practice which applied for a full lining campaign are shown in Table 2. Minor modifications to the lance height control were required when the vessel "slopped" during the first few heats of its campaign. This slopping effect implied much faster rates of slag formation than had been achieved previously with the calcitic lime flux practice. The most significant effect of the modified flux practice were noted in the reduced CaO content in the slags, increased slag basicity and slag magnesia levels of 8% to 11% MgO. Physically, the slags appeared "creamy" and sustained a foam longer than slags associated with the calcitic lime practice which gave "flat", watery slags. The slags from the mixed flux practice had a more homogeneous slag microstructure with no evidence of "free" lime particles. Figure 7 illustrates the structure of a typical dolomitic lime containing slag. The slag consisted of primary phase tricalcium silicate laths with a granular form of dicalcium silicate. Magnesia saturation was identified by the appearance of magnesiowüstite clusters within the calcium ferrite rich matrix. Magnesiowüstite formation was only observed in the cooled slag phase assemblages at slag magnesia levels >6.5% MgO.

The modified dolomitic lime flux practice was operated to achieve an aim 8.0% MgO in the first turndown slag. The slag characterisation work enabled the lime flux charge to be reduced and the flux additions better timed. Free lime was rarely observed and the true basicity of the slag was realised. Phosphorus removal was improved.

The dramatic improvement achieved in converter refractory lining life once the dolomitic lime practice had been introduced to all vessels is illustrated in Figure 8.

CONCLUSIONS

The introduction of a dolomitic lime practice and the monitoring of slag formation using the slag characterisation approach has enabled the plant to achieve a 100% improvement in refractory lining performance and as a result has given steelplant personnel the confidence of converter availability.

ACKNOWLEDGEMENTS

The authors would like to acknowledge the boards of Steetley Quarry Operations Limited and Steetley Refractories Limited for permission to publish this paper.

REFERENCES

(1) P. WILLIAMS and J. QUIN

'The effective utilisation of raw materials, semis and consumables in the Iron and Steel Industry' SEAISI Malaysia Conference 21-23 May 1986

(2) A. MUAN and E. OSBORN

'Phase equilibria among oxides in steelmaking' 1965, Reading, Mass., Addison-Wesley

(3) K. L. FETTERS et al.

Trans AIME 1941, 145 P. 95

(4) R. BAKER

BSC internal report CAPL/SM/A/31/74, 1974

(5) P. WILLIAMS

PhD Thesis, Sheffield City Polytechnic, 1980

TABLE 1: Plant operating parameters for the standard flux practice.

Standard Flux Practice		
Hot Metal Silicon		0.5-0.8% Si
Total Lime Charge kg/t.s.		45 to 60
Total Dolomitic Lime Charge kg/t.s.		Nil
Lance Control		Uncontrolled
Flux Practice:	Retained slag plus 5 tonne lime.	
	Scrap and hot metal added.	
	Remainder of lime added.	
Slag Chemistry:	18-19%	SiO_2
	49-51%	CaO
	2.3-3.8%	MgO
	17-19%	Fe_2O_3

TABLE 2: Plant operating parameters for the modified flux practice.

Modified Flux Practice (March 1987)		
Hot Metal Silicon		0.5-0.8% Si
Total Lime Charge kg/t.s.		29 to 39
Total Dolomitic Lime Charge kg/t.s.		16 to 21
Lance Control		Limited Control
Flux Practice:	Retained slag plus 1 or 2 tonne lime.	
	Scrap and hot metal added.	
	All Dolomet added at ignition.	
	Remainder of lime added continuously over first 6 minutes.	
Slag Chemistry:	15-16%	SiO_2
	46-48%	CaO
	7.0-10.0%	MgO
	17-19%	Fe_2O_3

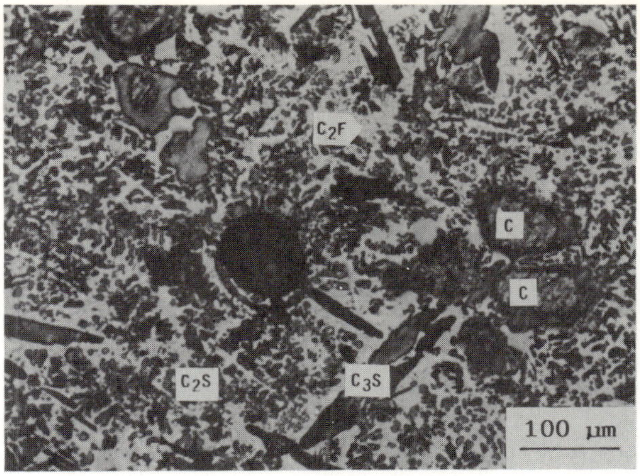

FIGURE 1: A poorly fluxed first turndown slag containing unreacted lime (C), tricalcium silicate (C_3S), dicalcium silicate (C_2S) and calcium ferrites (C_2F).

Slag composition (mass %) SiO_2 7.1, CaO 35.1, Fe_2O_3 45.9, MgO 6.7.

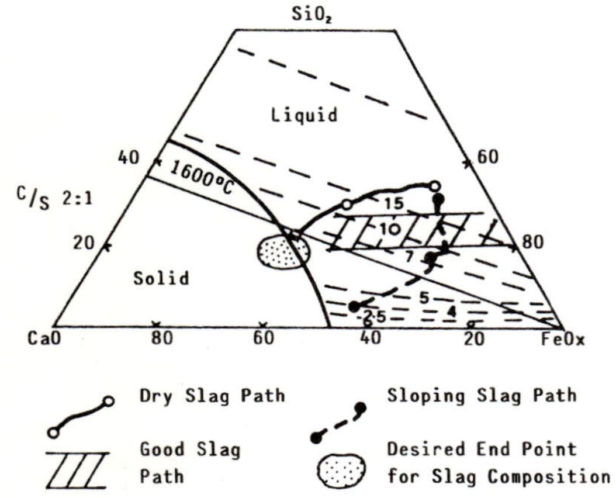

FIGURE 2: The solubility of magnesia in CaO-FeOx-SiO_2 slags at 1600°C superimposed on slag path trajectories.

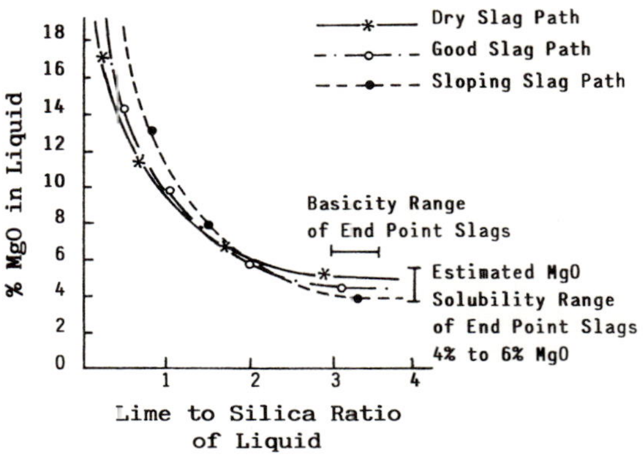

FIGURE 3: Estimated solubility of magnesia (MgO) in LD-type slags at 1600°C (from Figure 2).

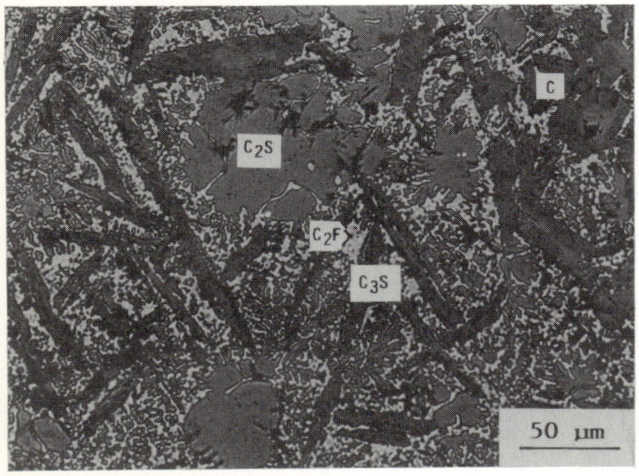

FIGURE 4: Typical first turndown slag from the standard flux practice. The slag microstructure shows "free" lime (C), tricalcium silicate laths (C_3S), dicalcium silicate (C_2S) and calcium ferrites (C_2F).

Slag composition (mass %) SiO_2 16.1, CaO 51.0, Fe_2O_3 22.5, MgO 3.7.

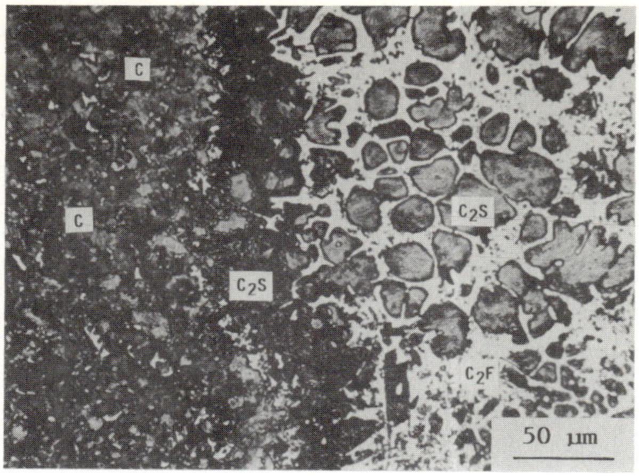

FIGURE 5: Details of partially reacted lime particle (C) from the same sample as Figure 4, showing the classic dicalcium silicate (C_2S) envelope which retards lime solution. The slag phase assemblages consist of dicalcium silicate (C_2S) and calcium ferrites (C_2F).

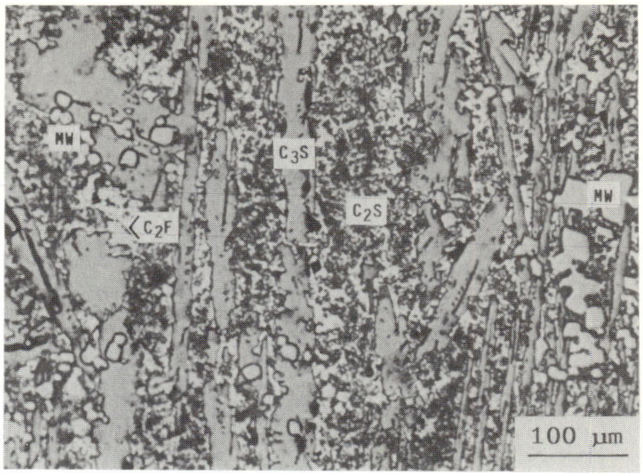

FIGURE 7: First turndown slag from the modified flux practice which shows magnesia saturation in the form of magnesio-wüstite (MW) associated with tri-calcium silicate laths (C_3S), dicalcium silicate (C_2S) and calcium ferrites (C_2F).

Slag composition (mass %) SiO_2 14.7, CaO 47.0, Fe_2O_3 21.9, MgO 9.9.

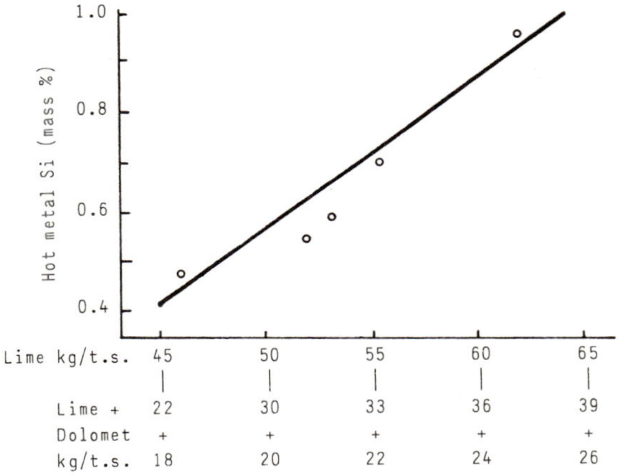

FIGURE 6: Lime and dolomet requirements for an aim 8% MgO content in the first turndown slag as a function of hot metal silicon level.

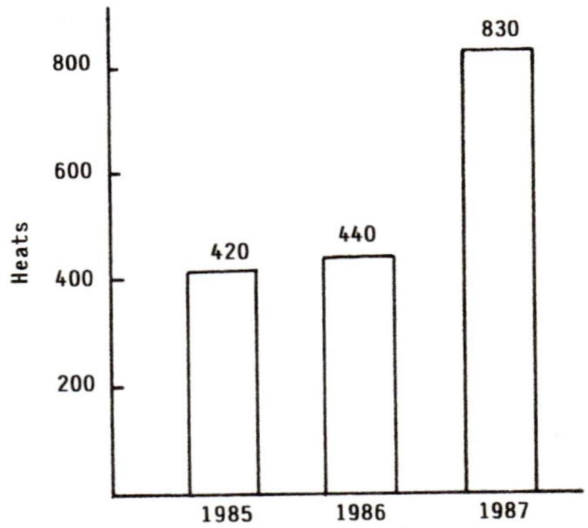

FIGURE 8: The improvement in refractory life during 1987 as a result of the modified flux practice.

Floating and settling behaviour of copper droplets in sodium silicate slag

T NAKAMURA, F NOGUCHI, Y UEDA, H TERASHIMA and T YANAGASE

Dr Nakamura, Dr Noguchi and Prof Ueda are in the Department of Metallurgical Engineering at the Kyushu Institute of Technology;
Mr Terashima is a graduate student at the Kyushu Institute of Technology;
Prof Yanagase is in the Department of Chemical Engineering at the Kyushu Sangyo University.

SYNOPSIS

Interfacial tensions between molten Cu and sodium silicate slag have been measured by the sessile drop method with X-ray radiophotography in order to study the floating and settling characteristics of Cu metal droplets in a sodium silicate slag. Interfacial tensions in the Cu - sodium silicate system decreased with increasing Cu content in the slag.

Flotation behaviour of Cu droplets with gas bubbles in the slag has been observed directly. The results were analysed using a concept of film coefficient and floating coefficient defined by Minto and Davenport. Flotation behaviour of Cu on the slag surface was also investigated quantitatively with a consideration of balance of density difference, surface tension and interfacial tension.

Settling velocities of Cu droplets in the slag have been measured. The results are discussed on a Stokes' law basis.

INTRODUCTION

Mechanical phase separation of Cu matte or metal droplets from Cu slags has been considered to be one of important problems in Cu smelting process. Minto and Davenport et al. studied the problem and proposed a model of entrapment and flotation of matte droplets in the Cu slag.(1)(2) Elliott and Mounier measured the interfacial tensions between Cu matte and iron silicate slag and discussed the flotation behaviour of the matte droplets in the slag.(3) Some of the present authors also investigated the influence of oxygen partial pressures on the interfacial tensions in the same system and succeeded in the direct observation of floating matte droplets with SO_2 gas bubbles in the iron silicate slag, (4)(5). To clarify the floating and settling behaviour of Cu droplets in slag, further experiments have been carried out using the Cu - sodium silicate slag system because it is easy to distinguish even a small Cu droplet in the slag due to a large density difference between sodium silicate slag and molten Cu.

EXPERIMENTAL

The slags used in the experiments were made from high quality quartz sand and extra grade sodium carbonate reagent. The required amounts were weighed out and melted in a Pt crucible at just above the melting point. After melting, slag was poured onto a water-cooled Pt dish. High purity copper droplets were used following hydrogen reduction.

The schematic diagram of the apparatus for high temperature X-ray radiophotography is shown in Fig.1. The furnace with an alumina reaction tube was heated by SiC rods and controlled by a PID system. An alumina support was able to be moved vertically and rotated from a control room when irradiated with X-rays. High alumina crucibles were used in the experiments. The sessile drop pictures of Cu and floating and settling behaviour were observed through films or image amplifier with a high quality video. Values of interfacial tension were calculated by a numerical method developed by Utigard.(6)

The slag compositions were analysed after the experiments. Analysis of silica was carried out by gravimetry and alumina and Cu were analysed by ICP and AA techniques respectively.

INTERFACIAL TENSION BETWEEN COPPER AND SODIUM SILICATE SLAG

Since alumina contents in sodium silicate slags after measuring the interfacial tension were always less than 2 mass%, the influence of alumina on the interfacial tension was ignored in this study. The molar ratios of sodium oxide to silica after the experiments were not changed from the starting compositions within the analytical errors. Values of interfacial tension between metal and slag are normally plotted against logarithmic values of partial oxygen pressures at the interface to calculate the value of interfacial excess quantity. However, it was very difficult to measure directly the partial oxygen pressures by zirconia O_2 sensor because it took a very long time to obtain equilibrium. Thus, interfacial tensions were plotted against Cu contents in the slag which would have a certain relationship to the oxygen partial pressures. The results obtained at 1473 K are shown in Fig. 2. An influence of the molar ratio of sodium oxide to silica on the interfacial tensions was not found.

On the other hand, Cu contents showed a large effect. For instance, the interfacial tensions were over 1 N/m below 0.1 mass % of Cu in the slag. The interfacial tension dropped to 0.57 N/m when the Cu content reached 6.8 mass %. The results indicate that oxygen works as an interfacial active element at the interface between Cu and slag.

FLOATING AND SETTLING BEHAVIOUR OF COPPER DROPLETS IN SODIUM SILICATE SLAG

Minto and Davenport discussed the flotation behaviour of Cu matte in Cu slag using two parameters, film coefficient(ϕ) and floating coefficient(Δ), defined by equations (1) and (2).

$$\phi = \gamma_s - \gamma_m - \gamma_{s/m} \quad (1)$$
$$\Delta = \gamma_s - \gamma_m + \gamma_{s/m} \quad (2)$$

They calculated both ϕ and Δ in the Cu slag and matte system, and reported that ϕ could be either positive or negative, depending on the matte and slag compositions but Δ was always positive under any conditions.(2) Elliott and Mounier discussed the same problem using their own data for interfacial tension etc. (3) However, neither Minto et al or Elliott et al have shown experimentally the floating behaviour of Cu matte or Cu metal droplets in the Cu slag at high temperature. Some of the present authors tried to observe directly the flotation of Cu matte droplets in iron silicate slag and succeeded in taking pictures of floating Cu matte droplets attached to gas bubbles at 1473 K. (5)

The same analysis was carried out in Cu - sodium silicate slag system. The values of ϕ and Δ in the system were calculated with data such as surface tensions etc. (7)(8) Partial oxygen pressures for the interfacial tension determinations were estimated from Cu contents in the slag using Yazawa's data which showed a relationship between partial oxygen pressures and Cu contents in sodium silicate slag.(9) While film coefficients were always negative, floating coefficients were positive in the system. Therefore, Cu droplets were expected to be floating attached to gas bubbles in the slag. Fig.3 shows the floating Cu droplet with CO_2 gas in a sodium silicate slag (Na_2O/SiO_2=1: molar ratio) at 1473 K. It is suggested that the floating mechanism proposed by Minto and Davenport(2) works well in Cu - sodium silicate system as in Cu matte - iron silicate slag.

The settling behaviour of Cu droplet in sodium silicate was investigated at 1473 K using the same apparatus. About 0.5 mm diameter of Cu droplets was the size limit observed using the X-ray video system. Settling velocities of Cu droplets were measured at 1473 K in the sodium silicate slag (Na_2O/SiO_2=4/6:molar ratio) for various diameters of Cu droplets.
The results are shown in Fig.4. Data were plotted as two curves; one was calculated from Stokes' law and the other from Hadamard-Rybczinski's law. (10)(11) The velocities, however, began to deviate due to the wall effect when the diameter was over 1 mm. (12) No more detailed discussion on this problem is presented in the paper because of limited space.

FLOATING BEHAVIOUR ON SODIUM SILICATE SLAG

The floating Cu droplets with attached gas bubbles eventually reach the surface of the slag. Then its stable floating behaviour of Cu on the slag surface was directly observed in the experiments.

Poggie et al. predicted this phenomenon and proposed a method for calculating the maximum size of metal droplet. (1) Deryabin et al also discussed the same matter in detail with respect to the contact angle. (13) Their treatments were considered to be sufficient as a first approximation. On the other hand, the present authors made an attempt to estimate a floating droplet shape using data for the surface tensions of metal and slag, the interfacial tension and the densities. (14) The immersed shape of the floating droplet in the slag was characterized using the contact angle, θ, and the interfacial tension as shown in Fig. 5. It is easy to calculate the interface shape between metal and slag using the same calculation technique which was used for calculating the interfacial tension value in this study. (7) Therefore, it is possible to estimate the maximum size of droplet on the slag surface by a numerical calculation checking the balance of density difference between metal and slag to the supporting force caused by the surface tension of slag at the interface of slag, metal and gas phase.

CONCLUSIONS

Floating behaviour of Cu droplets in a sodium silicate slag was investigated using two parameters, ϕ and Δ . The flotation coefficients at 1473 K in the system were positive and the flotation of Cu droplets was directly observed. The flotation of Cu droplets was also observed on the slag surface. A new method for the quantitative explanation of droplets floating on the surface was developed using data for interfacial properties in the system.

REFERENCES

(1)D.Poggi, R.Minto and W.G.Darvenport, J.Metal,21 Nov.,(1969),40.
(2)R.Minto and W.G.Darvenport, Trans.I.M.M.,81, (1972),C36.
(3)J.F.Elliott and M.Mounier, Can.Met.Quart.,21, (1982),415.
(4) and (5) T.Nakamura, F.Noguchi, Y.Ueda and A.Nakajyo, J.Min.Met.Tnst.Jap. in press.
(6)T.Utigard and J.M.Toguri, Met.Trans.B,16B, (1985),333.
(7)T.B.King, J.Soc.Glass.Tech.,35,(1951),241.
(8)Z.Morita and A.Kasama, J.Inst.Met.Jap.,40, (1976),787.
(9)G.Riveros, Y,Takeda and A.Yazawa, Preprint of MMIJ Fall Meeting,D5,(1983),17.
(10)W.Rybczinski, Bull.Acad.Cracovie (Ser A), (1911),40.
(11)M.J.Hadamard, Comp.Rend.Acad.Sci.,152,(1911), 40.
(12)R.Landenburg, Ann.Phys.,Paris,22,(1907),287.
(13)A.A.Deryabin, V.G.Baryshinikov and S.I.Popel, Izv.Akad.Nauk.SSSR Met.,(1971),No1,67.
(14)H.Terashima, T.Nakamura, F.Noguchi and Y.Ueda, Preprint of MMIJ Spring Meeting,(1988),479.

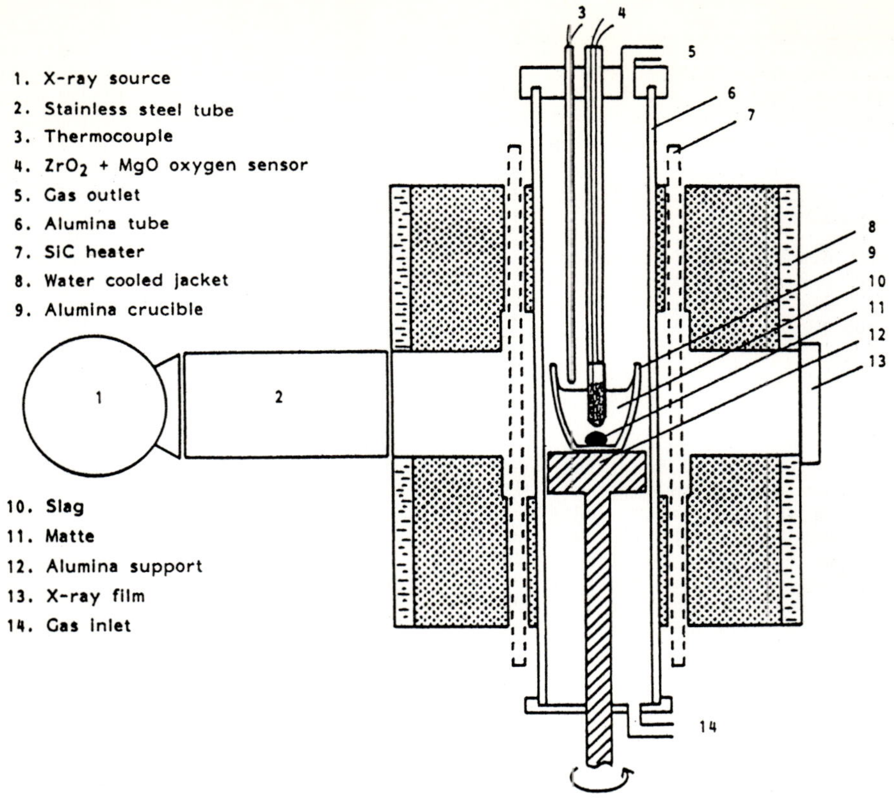

Fig.1 Schematic presentation of the apparatus for high temperature X-ray radiophotography.

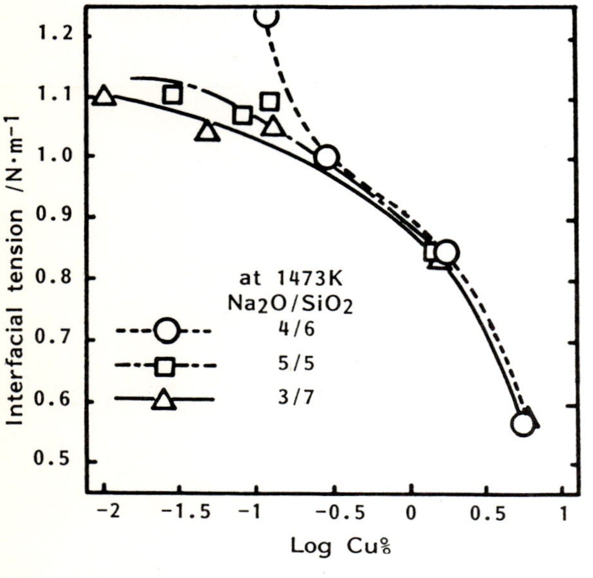

Fig.2 Plots of interfacial tensions in Cu - soda slag system against Cu contents at 1473 K.

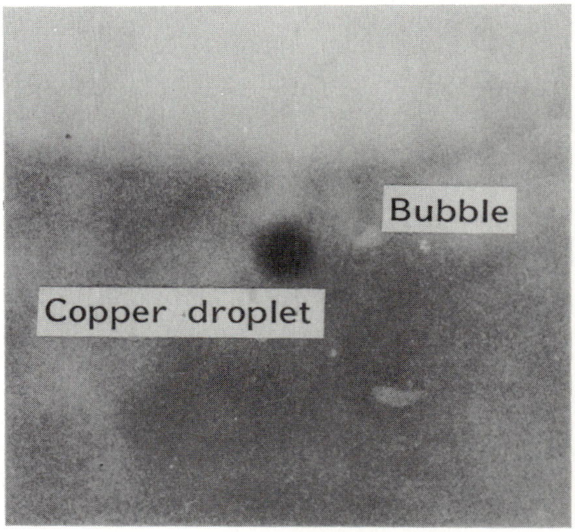

Fig.3 X-ray radiophotograph of Cu droplet with gas bubble in a sodium silicate slag.

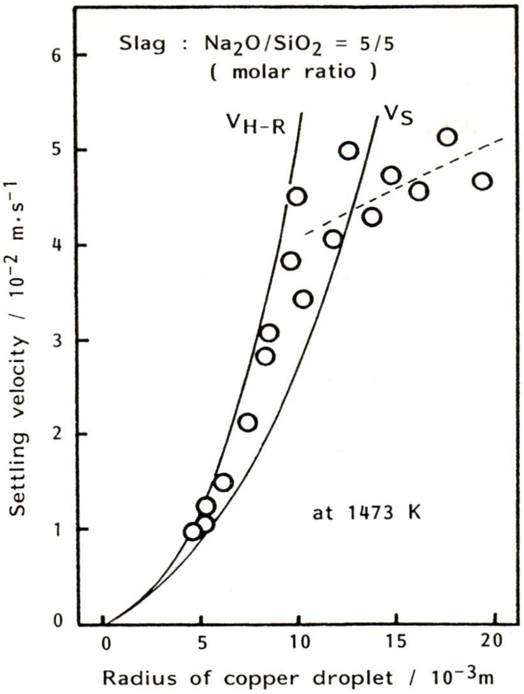

Fig.4 Relationship between settling velocities and the diameters of Cu droplets in a soda slag.

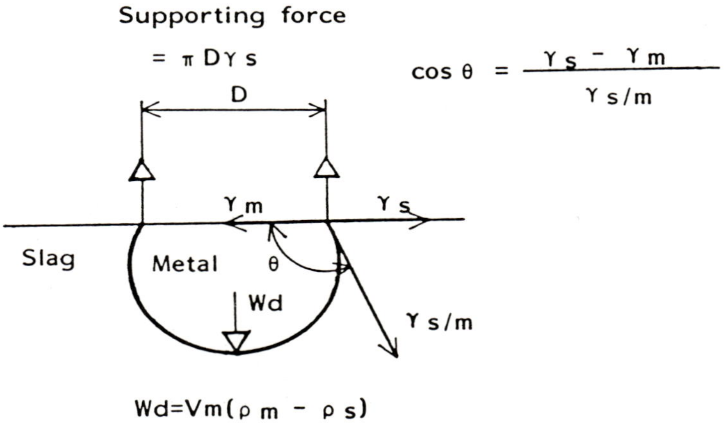

Fig.5 Model of metal droplet floating on the slag surface.

Mathematical expression of phosphorus distribution in steelmaking process by quadratic formalism

R NAGABAYASHI, M HINO and S BAN-YA

Drs Ban-ya and Hino are in the Department of Metallurgy at the Tohoku University, Sendai 980, Japan; Dr Nagabayashi, formerly at Tohoku University, is now in the Research and Development Lab, Shunan Works, Nisshin Steel Co. Ltd.

SYNOPSIS

The equilibrium of phosphorus distribution in steelmaking processes has been studied at temperatures from 1 300 to 1 680 °C for the slag systems of Fe_tO-P_2O_5-M_xO_y(M_xO_y = CaO, MgO, SiO_2) ternary and Fe_tO-P_2O_5-CaO-M_xO_y(M_xO_y = MgO, SiO_2) quaternary, to examine the potential validity of regular solution model. It was confirmed that the phosphorous and oxygen contents in liquid iron in equilibrium with slag can be estimated within an accuracy of ± 10 % by the quadratic formalism derived from an assumption of a regular solution for the slag.[1)]

INTRODUCTION

Many attempts have been made in the past to predict the phosphorus content in iron in slag-metal equilibrium by means of theoretical and empirical slag models. Some of the models were successful in evaluating the phosphorous distribution ratio for a restricted range of slag compositions, but the same modelwas inadequate for the data obtained from other sources for different compositions. In the present work, the potential validity of a regular solution model for slag was studied to formulate the equilibrium relation for phosphorus distribution for experimental results including the present work and other previous studies. This follows the successful application of the model for oxygen distribution between slag and metal in steelmaking process.[2,3)]

EXPERIMENTAL

The iron-phosphorus alloy (15 g) and premelted synthetic slag (6 - 8 g) were placed in a crucible and were melted under an argon atmosphere in a Keramax electric resistance furnace at a given temperature until the attainment of equilibrium. The slag systems studied in the present work were: Fe_tO-P_2O_5-CaO saturated with CaO, Fe_tO-P_2O_5-MgO saturated with (Mg,Fe)O, Fe_tO-P_2O_5-SiO_2 saturated with SiO_2, Fe_tO-P_2O_5-SiO_2-CaO saturated with $2CaO\cdot SiO_2$ and Fe_tO-P_2O_5-CaO-MgO saturated with (Mg,Fe)O, in order to test the model and to determine the values of the interaction energies between cations and phosphorus ions in slag. The previous data employed to check the applicability of the model were the experimental results by Winkler et al., Balajiva et al., Fischer et al., Peter et al. Knüppel et al., Trömel et al., Suito et al., and Shirota et al.

RESULTS

In the present work, the solubility limit of crucible materials in phosphate slags, ferric-ferrous iron equilibrium in the slag, oxygen distribution, and phosphorus distribution between slag and metal, were studied at the same time. However,it was difficult to show all of these experimental results due to the lack of space. As one example of the experimental results, Fig. 1 shows the phosphorus distribution ratio of Fe_tO-P_2O_5-M_xO_y (M_xO_y = CaO, MgO, SiO_2) ternary systems saturated with oxide M_xO_y, which are the crucible materials, at 1 600 °C. The phosphorous distribution ratio in Fe_tO-P_2O_5-CaO slag were the highest among the three ternary systems studied, and the maximum value of the distribution ratio was observed at the range of (wt%Fe_tO) = 10-20. The maximum phosphorous distribution ratios, L_P' = (%P)/[%P], in each slag system were $L_P' \simeq$ 1 000 in Fe_tO-P_2O_5-CaO slag, $L_P' \simeq$ 10 in Fe_tO-P_2O_5-MgO slag, and $L_P' < 1$ in Fe_tO-P_2O_5-SiO_2 slag respectively. The distribution ratio in Fe_tO-P_2O_5 binary slag was estimated to be $L_P' \simeq 1$ by extrapolation to 100 % Fe_tO in Fig. 1.

In our previous papers concerning oxygen distribution in steelmaking processes,[2, 3)] it was confirmed that the regular solution model was able to predict the oxygen content in iron within an accuracy of ± 10 % over a wide range of slag composition. In order to establish the same relation for dephosphorization in steelmaking processes, the experimental data for oxygen distribution were employed to test the validity of the model for phosphate slag, and to determine the interaction energies between cations and phosphorus ion. The equilibrium relation of oxygen in metal with iron oxide in slag can be written as follows:

$$Fe(l) + \underline{O} = (FeO)_{(in\ slag)} \quad (1)$$

$$\Delta G^\circ_1 = -30\ 615 + 13.86T \quad (cal) \quad (2)$$

$$RT\ln K_1 = RT\ln(a_{FeO}/a_O)$$
$$= RT\ln(X_{FeO}/a_O) + RT\ln\gamma_{FeO} \quad (3)$$

$$RT\ln\gamma_i = \sum_j \alpha_{ij}X_j^2 + \sum_{jk}\sum(\alpha_{ij} + \alpha_{ik} - \alpha_{jk})X_j \cdot X_k \quad (4)$$

where, X_j is the cation fraction of the component j in slag, and α_{ij} is the interaction energy between cation$_i$-O-cation$_j$. From the results obtained by the application of Eq. (1) to Eq. (4), it was confirmed that the regular solution model was satisfied for all the experimental results including the present work and previous studies by other investigators except for the extremely iron oxide-rich region. The interaction energies obtained were as follows:

$$\left.\begin{aligned} \alpha_{Ca^{2+}-P^{5+}} &= -60\ 000 \\ \alpha_{Mg^{2+}-P^{5+}} &= -\ \ 9\ 000 \\ \alpha_{Si^{4+}-P^{5+}} &= +20\ 000 \end{aligned}\right\} \text{(cal).} \quad (5)$$

DISCUSSION

From the above results, the equilibrium relation for dephosphorization with $FeO-Fe_2O_3-P_2O_5-SiO_2-CaO-MgO$ slags can be written as follows:

$$\underline{P} + 2.5\underline{O} = (PO_{2.5})_{(in\ slag)} \quad (6)$$

$$\log K_{P1} = 17\ 060/T - 8.510 \quad (\sigma = 0.4) \quad (7)$$

$$\Delta G^\circ_{P1} = -78\ 040 + 38.93T \quad (\sigma = 3\ 300) \quad (cal) \quad (8)$$

$$\underline{P} + 2.5(FeO)_{(in\ slag)}$$
$$= (PO_{2.5})_{(in\ slag)} + 2.5Fe(l) \quad (9)$$

$$\log K_{P2} = 328/T - 0.936 \quad (\sigma = 0.4) \quad (10)$$

$$\Delta G^\circ_{P2} = -1\ 500 + 4.28T \quad (\sigma = 3\ 300) \quad (cal) \quad (11)$$

$$RT\ln\gamma_{PO_{2.5}}$$
$$= -7\ 500X^2_{FeO} + 3\ 500X^2_{FeO_{1.5}} - 60\ 000X^2_{CaO}$$
$$- 9\ 000X^2_{MgO} + 20\ 000X^2_{SiO_2} + 460X_{FeO}\cdot X_{SiO_2}$$
$$- 60\ 000X_{FeO}\cdot X_{CaO} - 24\ 500X_{FeO}\cdot X_{MgO}$$
$$+ 22\ 500X_{FeO}\cdot X_{SiO_2} - 33\ 600X_{FeO_{1.5}}\cdot X_{CaO}$$
$$- 4\ 800X_{FeO_{1.5}}\cdot X_{MgO} + 15\ 700X_{FeO_{1.5}}\cdot X_{SiO_2}$$
$$- 45\ 000X_{CaO}\cdot X_{MgO} - 8\ 000X_{CaO}\cdot X_{SiO_2}$$
$$+ 27\ 000X_{MgO}\cdot X_{SiO_2} \quad (cal) \quad (12)$$

$$RT\ln\gamma_{FeO}$$
$$= -4\ 460X^2_{FeO_{1.5}} - 7\ 500X^2_{PO_{2.5}} - 10\ 000X^2_{SiO_2}$$
$$- 7\ 500X^2_{CaO} + 8\ 000X^2_{MgO}$$
$$- 15\ 460X_{FeO_{1.5}}\cdot X_{PO_{2.5}} + 10\ 940X_{FeO_{1.5}}\cdot X_{CaO}$$
$$+ 4\ 240X_{FeO_{1.5}}\cdot X_{MgO} - 22\ 260X_{FeO_{1.5}}\cdot X_{SiO_2}$$
$$+ 45\ 000X_{PO_{2.5}}\cdot X_{CaO} + 9\ 500X_{PO_{2.5}}\cdot X_{MgO}$$
$$- 37\ 500X_{PO_{2.5}}\cdot X_{SiO_2} + 24\ 500X_{CaO}\cdot X_{MgO}$$
$$+ 14\ 500X_{CaO}\cdot X_{SiO_2} + 14\ 000X_{MgO}\cdot X_{SiO_2} \quad (cal) \quad (13)$$

where the standard states for the activities of $PO_{2.5}$ and FeO are respectively hypothetical pure liquid $PO_{2.5}$ and FeO under an assumption of regular solution, and those of oxygen and phosphorus being 1 % by weight in iron.

The effect of temperature on the equilibrium constant of reaction (6) calculated by the model is shown in Fig. 2, in which the data reported by the investigators from five other groups are also shown by recalculation. The results obtained for different kinds of slag and by many different investigators are in fairly good agreement with each other. In Fig. 3, the phosphorous distribution ratio L'_P computed from the model is plotted against the measured L'_P-value of the data previously reported by other investigators. The agreement between both calculated and measured values was also good within the experimental error.

The conversion factors between the activities of conventional liquid P_2O_5 and Fe_tO and hypothetical liquid $PO_{2.5}$ and FeO were derived as follows;

$$P_2O_5(l) = 2PO_{2.5}(l,\ R.S.) \quad (14)$$

$$RT\ln a_{P_2O_5} = 2RT\ln a_{PO_{2.5}} + 12\ 600 - 55.14T \quad (cal) \quad (15)$$

$$Fe_tO(l) + (1-t)Fe(s\ or\ l) = FeO(l,\ R.S.) \quad (16)$$

$$RT\ln a_{Fe_tO} = RT\ln a_{FeO} - 2\ 040 + 1.707T \quad (cal) \quad (17)$$

CONCLUSIONS

It was confirmed that the quadratic formalism by an assumption of regular solution could be applied as an approximate expression of the activity coefficient of component i in slag, although the liquid slag is not a regular solution. The phosphorous and oxygen contents in metal in equilibrium with slag can be estimated within an accuracy of ± 10 % by the model calculation.

REFERENCES

1) J. Lumsden : Phys. Chem. of Process Metall. Part I, ed. by G. R. St. Pierre, Metall. Soc. Conf., Vol. 7(1961), p. 165, [Interscience], N. Y..
2) J.-D. Shim and S. Ban-ya : Tetsu-to-Hagané, 67 (1987), p. 1735 & 1745 ; Can. Metall. Quart., 23(1983), p. 319.
3) S. Ban-ya and M. Hino : Tetsu-to-Hagané : 73 (1987), p. 476.

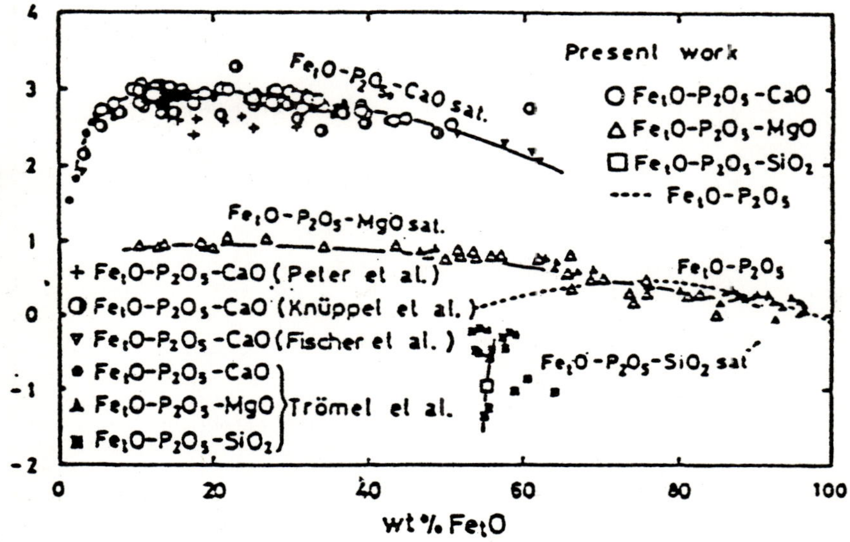

Fig. 1. Phosphorus distribution ratio between liquid iron and $Fe_tO-P_2O_5-M_xO_y$ (M_xO_y = CaO, MgO, SiO_2) slags saturated with solid M_xO_y at 1 600 °C.

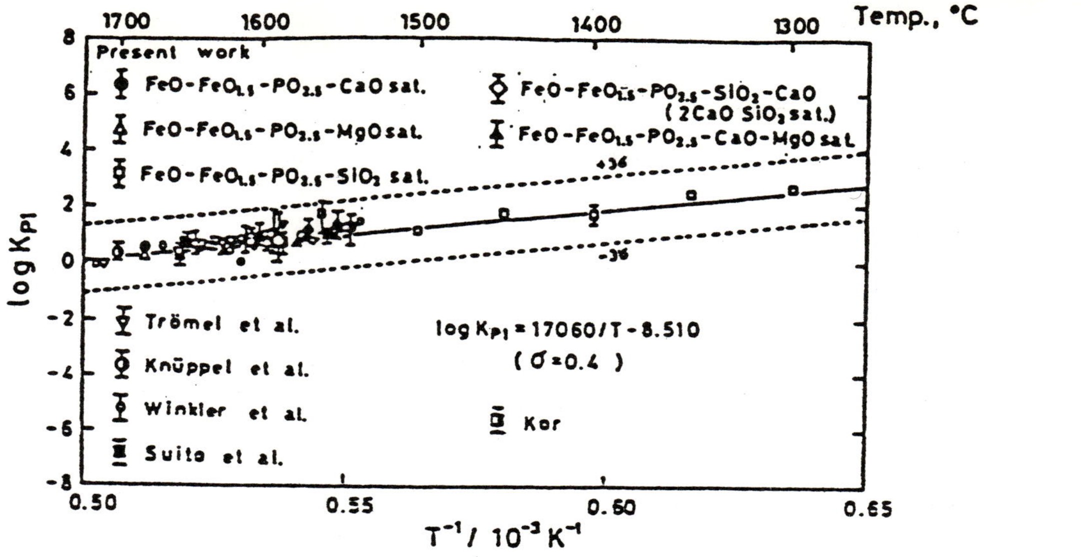

Fig. 2. Temperature dependence of $\log K_{P1}$.

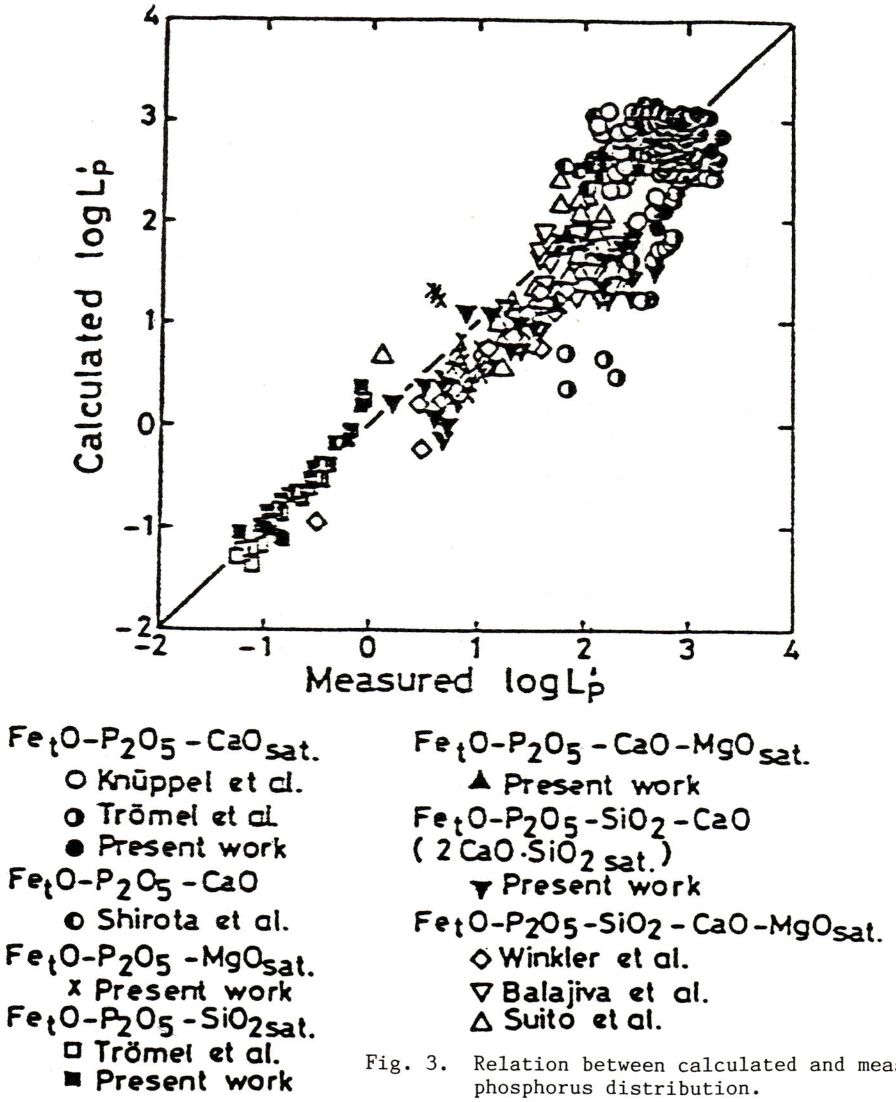

Fig. 3. Relation between calculated and measured phosphorus distribution.

Importance of slags on ladle treatment of steel

V PRESERN and A ROZMAN

Doc Dr Prešern, dipl ing met is Head of the Department for Steel Technology in the Slovenian Steelworks - Institut for Metallurgy in Ljubljana, Yugoslavia;
M Sc Rozman, dipl ing met is Head of Technology in Slovenian Steelworks - Steelworks Ravne, Yugoslavia.

SYNOPSIS

The influence of slags in ladle treat - ment of steel is presented regarding VOD and VAD process as well as calcium treat - ment in SLOVENIAN STEELWORKS. Details are given about preventing the carryover of furnace slags by tapping, and about the influence of slag composition on deoxidation, desulphurisation and steel cleanness.

INTRODUCTION

SLOVENIAN STEELWORKS consist of three steelworks with very different equipment for ladle treatment of steel, i.e, VOD process, VAD process, gas injection of powdered materials and cored wire injection.

To meet the demands for products of higher purity, with low oxygen and sulphur contents in aluminium killed steels; to be able to control the nitrogen pick-up; get sufficient steel cleanness (in terms of the elimination of inclusions and inclusion shape control) and required mechanical properties, the following conditions, regarding the ladle slag, have to be fullfilled[1-3] and are discussed from a technological point of view:

- high basicity of lime saturated slags with high sulphide capacity,
- low contents of FeO and MnO,
- good fluidity in operating range,
- proper quantity of slag,
- high energy of mixing for good metal-slag contact,

Our practice with different processes of treatment of steel in the ladle has shown that the desired level of refining the steel can only be achieved by providing the optimal slag-metal reaction conditions.

FORMATION OF LADLE SLAG

Steel in SLOVENIAN STEELWORKS is melted and oxided in electric arc furnaces and then treated in ladles. Efficient ladle refining of the steel to produce clean steel is attained only when the steel is treated under a basic and nonoxidising slag. One of the basic conditions for proper slag composition in the ladle, formed by the additions of lime, fluorspar or synthetic slags, is to minimise the carryover of the oxidising furnace slags into the ladles during tapping. Minimising its carryover will improve and make more predictable the ladle deoxidation and improve alloying efficiency.

Different systems are used for that reason in ladle metallurgy in SLOVENIAN STEELWORKS:

- eccentric bottom tapping[4,5] (in combination with VOD ladle treatment),
- reladling from the transport into the operating ladle[6,7] (in combination with SAFE-HEURTEY VAD treatment) by bottom pouring to another ladle,
- removing the furnace slag from the surface of the ladle (deslagging a tilted ladle),
- if phosphorous removal has been done successfully already in the melting furnace,it's possible to use furnace slag in the ladle with final deoxidation of steel and slag by tapping into the ladle,

One 85-tonnes UHP electric arc furnace with EBT is operating in STEELWORKS JESENICE and more than 90 % of the furnace slag is kept in the furnace. New slag is formed by additions of synthetic slag. The reladling method in STEELWORKS RAVNE to provide steel without furnace slag for two VAD units has also been giving good results regarding slag separation, but involves substantial temperature loss.

Typical slag compositions before and after treatment in different ladle metallurgy processes in SLOVENIAN STEELWORKS are shown in Table 1 and the concentration range of the used ladle slags in our conditions is presented in Fig. 1 (calculated for the system $CaO-Al_2O_3-SiO_2$).

To ensure low oxygen activity in the steel and to obtain high sulphur capacities, the most successful method is to use synthetic slags[8,9] which must meet the following general requirements:

- low oxygen potential,
- large solubility for alumina and sulphides,
- moderate fluidity,
- low melting point

To obtain good results such slags would be injected into the melt or added into the ladle and intensively mixed with the steel.

DESULPHURISATION

For good desulphurisation the oxygen activities in the steel bath have to be kept low by sufficient quantities of aluminium. The reaction product is alumina which forms a very suitable type of slag with lime. Investigations in the $CaO-Al_2O_3$ system show very high sulphur capacities for these slags[1,10,11]. As a third main component such slags also contain fluorspar or silica. Optimal compositions of the slags for desulphurisation are, on the basis of many reports, those near the lime saturated area.

The slags in the system $CaO-Al_2O_3$ are liquid by 1600^0 C from 37-58% CaO and with addition of fluorspar that range is even larger. It has been shown that a part of Al_2O_3 can be replaced by SiO_2 and the contents of silica up to 15% have no significant influence in decreasing the desulphurisation, because such slags can still be in the region of lime-saturation.

Experimental results on the thermo-dynamic properties for simple three or four component slags can be used for actual industrial slags with the help of different models. Such a model for sulphur distribution between slag and steel, based on optical basicity[2,12], is shown on Fig.2, where some results from our conditions are also presented.

The role of low oxygen for successful desulphurisation is known and theoretically explained; and Fig. 3 shows the influence of oxygen activity in the steel and slag composition (basicity) on the sulphur distribution.

The possibilities for producing low sulphur steel in our steelworks are shown in Table 2 and a favourable modified inclusion in low sulphur aluminium killed steel treated with CaSi gas injection can be seen on Fig. 4.

The use of proper slags in the quantity of $\sim$ 10 kg/t gives desulphurisation to more than 90 % and the distribution of sulphur between slag and steel of more than 250%.

Very suitable thermodynamic conditions for desulphurisation are presented in VOD or VAD units where beside a good slag a very important role is played by intensive mixing of steel.

METAL-SLAG CONTACT

The rates of metal-slag exchange reactions, provided that the slag is in the range of optimum composition, depend on interface area, diffusion coefficient of sulphur in liquid steel and gas stirring conditions.

Argon stirring through a porous plug, installed at the bottom or in the side of the ladle, is nowadays a common practice in the ladle refining of steel.

VOD and VAD units are equipped with vacuum systems and the very positive influence of lower pressures on the mixing energy is presented in Fig. 5[5,13].

The interdependence of the energy of mixing and the mass-transport coefficient for sulphur (k_S) can be calculated from the equation :

$$d(S)/dt = - A/V . k_S . ([S] - (S)/L_S)$$

where

A is reaction area
V is volume of steel
L is sulphur distribution

The calculated and some practical data[13] are presented in Fig. 6. The data for VAD in Steelworks Ravne are calculated for an argon stirring rate of 30 $l.min^{-1}$ and for 25 minutes of low vacuum (below 1 torr).

The mixing energy (E/A) for such conditions is 865 W/m^2, but without vacuum for the same conditions only 303 W/m^2.

Very efficient metal-slag contact is possible also by argon powder injection, where the effects of large contact interfaces during transitory reaction and metal-slag stirring are combined, while argon powder injection provides overall high stirring energy. Argon flow rates in our conditions are in the range 0,5-1,0 $Nm^3 min^{-1}$.

SLAG CONTROL BY CALCIUM TREATMENT

In the VOD or VAD process, much more care has to be taken for slag control in the ladle for calcium treatment of steel with gas injection of powdered CaSi or with cored-wire injection, where a proper slag is usually formed during tapping. Lime and fluorspar are used, as well as some aluminium granules to decrease the content of iron and manganese oxides in the slag. Typical ladle slag analysis before and after such treatments have already been shown in Table 1.

Typical aluminium content in calcium treated steel is 0.020 to 0.030 % and the activity of oxygen in the steel is mainly influenced by slag composition (and is in the range of 3 to 10 ppm after calcium treatment). It has been determined that the efficiency of steel treatment with calcium is sometimes more influenced by contents of iron and manganese oxides and basicity of slag, than by the quantity of used calcium.

CONCLUSIONS

The importance of slag control during treatment of steel in the ladle in SLOVENIAN STEELWORKS was presented. It has been shown that the desired level of refining of steel can only be achieved by providing the optimal slag metal reaction conditions and the most important factors to be controlled are:

- high basicity
- high sulphide capacity
- low content of FeO and MnO
- optimal quantity
- high energy of mixing

REFERENCES

1. Turkdogan E.:Archiv für Eisenhüttenwes. 54 (1983), Nr.2,Februar, P.45-52.

2. Riboud P.V., G.Gatellier:Ironmaking and Steelmaking 1985,Vol.12, No.2,P.79-86.

3. Prešern V., P.Bračun : Radex-Rundschau, Heft 1, 1986, P.45-56.

4. Baare R.D.,J.Overgaard,E.Rasmussen:1 st European Electric Steel Congress,Aachen Sept. 12 to 14, 1983, Paper C2.

5. Pellicani F., F.Villette, J.Dubois : SCANINJECT IV, Lulea, Sweden,June 11 to 13,1986, Paper 29.

6. Lamarque G. : Revue de Métallurgie-CIT, Octobre 1980, P.781-789.

7. Carmont F. et all:Revue de Métallurgie-CIT, Juin 1982, P.495-500.

8. Abratis H., H.J.Langhammer : Stahl und Eisen 100 (1980), P.1077-1078.

9. Koch K., D.Janke : Schlacken in der Metallurgie, Verlag Stahleisen mbh, Düsseldorf 1984, P.263-276.

10. Schürmann E., R.Bruder, K.Nürnberg, H. Richter:Arch.Eisenhüttenwes. 50 (1979), P.139-143.

11. Suzuki Y., T.Kuwabara : Ironmaking and Steelmaking, No. 2, 1978.

12. Duffy J.A., M.D.Ingram, D.Sommerville : J. Chem. Soc. Faraday Tran., 1978, 74, P.1410.

13. Kawakami K.et all : Stahl und Eisen 102 (1982), Nr. 5,P.227-231.

TABLE 1 : OPTIMAL SLAG COMPOSITIONS BEFORE AND AFTER TREATMENT WITH LADLE METALLURGY PROCESSES

PROCESS		CaO,%	MgO,%	SiO_2,%	Al_2O_3,%	FeO,%	MnO,%	CaF_2,%
VAD	before	52,5	10	15	12	1,5	1	8
	after	50	10	17,5	15	1,5	0,5	5,5
VOD	before	55	5	7	15	2	1	10
(Si-steel)	after	51	7	5	30	0,5	0	6,5
CaSi-gas	before	50	10	15	15	2,5	1,5	6
injection	after	49	12	17	16	1,0	0,5	4,5
CaSi-cored	before	50	10	15	15	2,5	1,5	6
wire inject	after	53,5	11	13	16	1,5	1	4

TABLE 2 : DESULPHURISATION POSSIBILITIES IN SLOVENIAN STEELWORKS FOR INITIAL SULPHUR CONTENTS FROM 0,020 TO 0,030 %

PROCESS	TYPE OF SLAG	QUANTITY (kg/t)	TYPICAL END SULPHUR (%)
VOD	new synthetic slag	8-12	<0,005%
VAD	new synthetic slag	8-12	<0,010%
Gas injection	deoxidised furnace slag	1,5-2(CaSi) 2 -4 (lime based powders)	<0,005% <0,005%

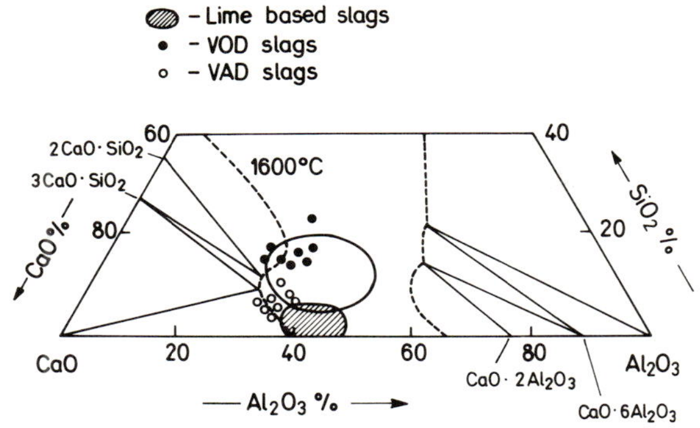

Fig. 1 : The concentration area of the used ladle slags

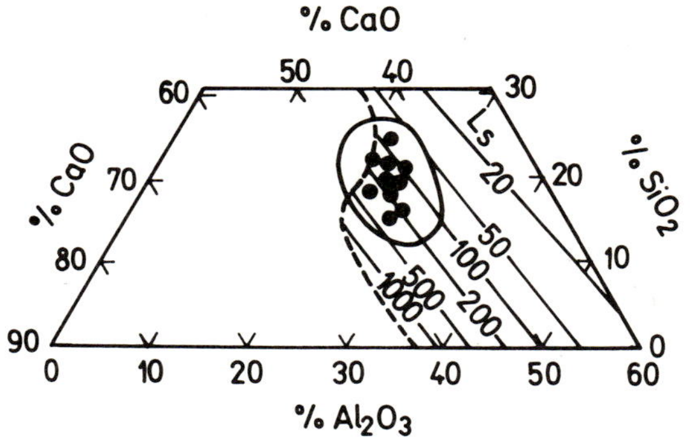

Fig. 2 : Sulphur distribution between slag and steel

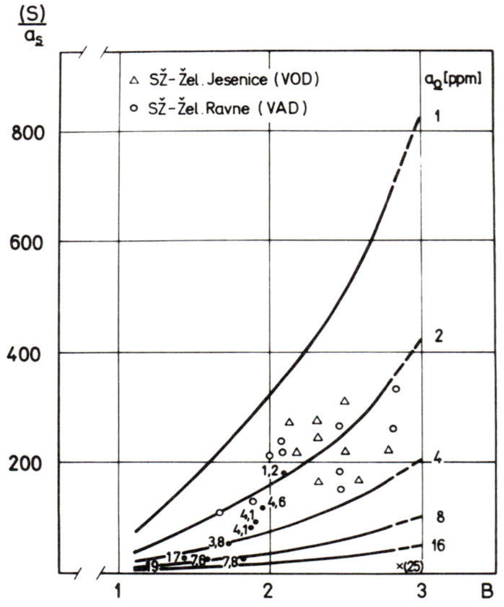

Fig. 3 : Influence of oxygen activity and slag composition on the sulphur distribution

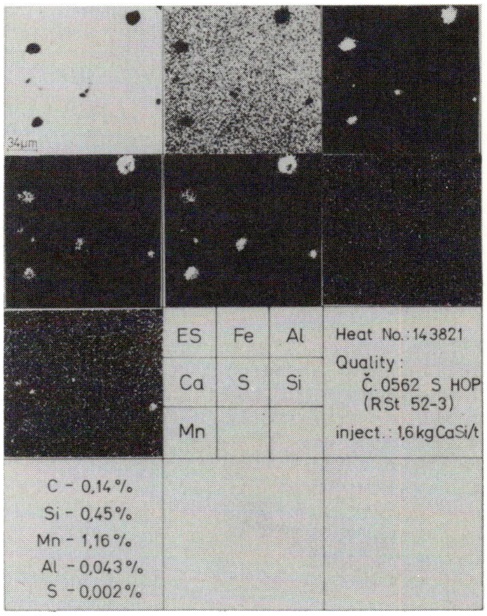

Fig. 4 : Modified inclusion in low sulphur aluminium killed steel treated with CaSi argon injection

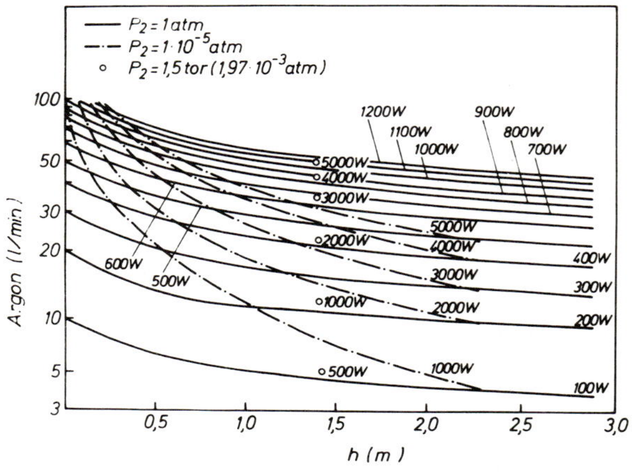

Fig. 5 : Influence of lower pressure on the mixing energy

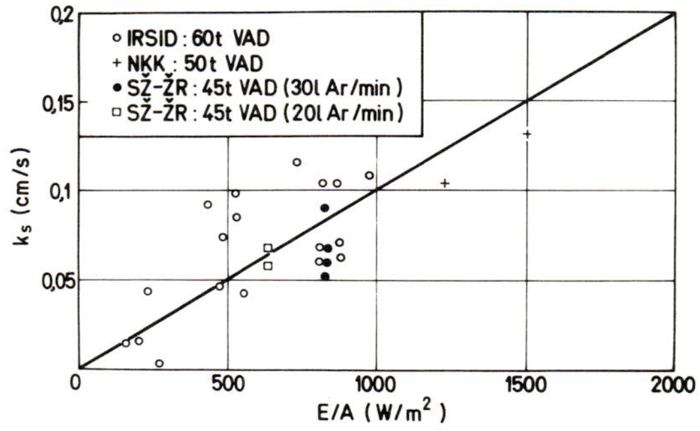

Fig. 6 : Dependence between the energy of mixing and the mass - transport coeeficient

Casting powder performance assessment in continuous casting by mould instrumentation

S G THORNTON, N HUNTER and B PATRICK

The authors are in the Casting and Reheating Department at the British Steel Research Laboratories, Middlesbrough, Cleveland, UK.

SYNOPSIS

Mould lubrication behaviour is of paramount importance in the continuous casting of steels. The formation of a uniform molten slag interface between the solidifying steel shell and the copper mould wall is essential for uniform heat transfer. Poor performance can lead to product surface defects and, in extreme cases, breakouts.

This paper describes a method of assessing casting powder performance, using thermocouples installed in the mould copper plates to generate information on heat transfer for the whole mould perimeter in real time. It also indicates how the information can be used to influence the casting operation, and aid in quality prediction.

INTRODUCTION

Casting powder is added to the metal surface during the continuous casting of steel and carries out several important functions. The powder layer insulates the meniscus preventing plating and oxidation and entraps inclusions from the melt aiding in the production of clean steels. The powder melts on the steel surface and the resulting molten slag infiltrates the gap between the steel shell and the mould copper plates, facilitated by the action of mould oscillation. This mechanism provides lubrication, and most importantly forms an interface with a high thermal impedance thus giving rise to a smooth, uniform transfer of heat from the steel through to the mould cooling water. This uniformity and stability is fundamental for the production of good surface quality and high security of operation. The operation of the casting powder is shown schematically in Figure 1. The molten slag solidifies on the cold mould wall but remains liquid adjacent to the steel shell. The slag is carried down the length of the mould and out at the bottom. Casting powder is added continuously during casting (usually manually), to ensure uniform cover and consumption rate.

It is important that a stable interface is established around the whole perimeter of the mould. Any local starvation of slag can lead to localised uneven heat transfer and in the extreme, contact between the steel and copper, leading to cracking of the steel or sticking. Poor performance of the casting powder can be caused by many factors. The chemical and physical properties of a given powder may be unsuited to specific casting operations, or may become so due to contamination, for example by moisture pickup in storage or alumina absorption during casting. The operation may also be affected by machine operating conditions such as mould level control or mould oscillation conditions. The performance of the Submerged Entry Nozzle (SEN) is also important as blockages, misalignment or incorrect immersion depths can significantly alter steel flow patterns in the mould leading to cold areas of the meniscus.

This paper describes mould instrumentation designed to give a means of assessing mould performance during casting thus enabling the development of optimum casting practice, including the choice of casting powders and operational parameters. The information generated by such instrumentation can be used in retrospect to assess periods of operation. It can also be used in real time as part of an on-line grading/quality prediction system, and to enable changes to be made to improve mould performance. The increasing need to achieve high levels of hot charging and direct rolling is reducing the possibility of off-line inspection and means that methods of assessing mould performance during casting are becoming increasingly important.

INSTRUMENTATION AND METHODS TO ASSESS MOULD PERFORMANCE

Existing methods of assessing mould operation include monitoring of mould friction and average mould heat transfer. These methods however, give only a macroscopic picture of mould performance and any localised effects are masked. Using these methods standard ranges can be defined within which casting conditions should lie, for specific steel grades, casting powders and casting speeds. A great deal of data is necessary to achieve this and the method cannot be used to give rapid feedback on casting conditions, or detect breakouts.

A more useful way of gaining information is by installing thermocouples in the copper plates and monitoring temperature behaviour during casting (1). Thermocouples installed in this way can be used to detect sticking of steel to the copper thus allowing sticker breakouts to be prevented. For this purpose two or more thermocouples are installed at different distances below the meniscus, in several columns around the mould perimeter. Systems of this type are well reported (2) and some are commercially available. However, experience has shown that it is also very useful to monitor the behaviour of copper temperatures with time, and in particular using thermocouples installed as indicated in Figure 1. The thermocouple is located far enough below the meniscus so that variation in the steel level does not affect the measurement, but high enough to correspond to a region of high heat flux, where the initial shell is still forming. The distance from the hot face should be as low as practically possible in order to increase sensitivity. Variability of the copper temperatures with time indicates the level and stability of heat transfer in this region and can identify problems related to the operation of the casting powder as described previously. The following Sections illustrate this with examples and show how an on-line system has been developed to enable analysis of mould conditions to be performed in real time.

RELATIONSHIP BETWEEN COPPER TEMPERATURE VARIABILITY AND SURFACE DEFECTS

In one trial six thermocouples were installed across the central 800 mm of the inner radius broad plate of a continuous slab casting mould, at positions detailed in Figures 1 & 2. A sample of results when casting 0.083% carbon steel is shown in Figure 2. The diagrams show the temperatures recorded over a thirty minute period. Variability of the temperatures is greater for positions C and D (either side of centreline). This type of behaviour occurred consistently on over 50 casts of similar steel grades when casting slab widths ranging from 1 175 mm to 1 850 mm and 223 mm thick. Slabs cast with high temperature variability in the central region were invariably found to have longitudinal cracks in this area. During the trials many casting parameters were altered in a controlled manner and different casting powders were tried, with little effect on the temperature behaviour or cracking performance. All the trials were carried out using the same design SEN, this being constructed from graphitised alumina with a zirconia-bearing insert to reduce wear at the slag line. Later studies revealed that the cracking was greatly reduced using a similar design SEN, but with an insert of higher thermal conductivity. Extended trials were carried out to confirm this using instrumentation as described above, and typical results are shown when casting 0.077% carbon steel in Figure 3. This time copper temperature variability in the central region is very low and the slabs corresponding to this period were found to be crack-free. This Figure also illustrates two other effects. In the middle of the extract there was a ladle change to a very similar steel grade accompanied soon after by a step rise in casting speed of 0.1 m/min, shown in the bottom diagram. At the speed change the temperatures rose gradually in the central region whereas in the previous trials such abrupt changes caused rapid fluctuation in temperatures and an increase in the probability of cracking. The ability of a powder to respond in this way to rapid changes in casting conditions is an important quality, as in practical use such changes are frequent. It can also be seen from the Figure that the temperatures are variable at one side of the mould and then, after the ladle change the variability switches to the other side. This probably indicates a change in steel flow patterns in the mould which can be a particular problem when partial blockage of one or both SEN ports occurs.

Figure 4 shows two examples when casting 0.25% carbon steel on a round bloom caster. Three thermocouples are shown, these being at three different levels in the copper at the front of the mould, (i.e. the "loose side"). The steel level is about 140 mm below the top of the copper. In Figure 4(a) over thirty minutes pass before the copper temperatures stabilise indicating the formation of a stable slag film. Where this type of temperature variability persisted, longitudinal cracks were found to occur, some of which resulted in breakouts. Stable heat transfer conditions finally result at about 43 minutes, after the characteristic dip in the lower thermocouple temperature. Following this discovery a new practice was adopted where a fast melting start powder was added to the metal surface for the first few minutes of casting, before commencing addition of the normal powder. The results are shown in Figure 4(b) and it can be seen that temperature variability is less and stability is attained much more quickly. This modification to casting practice resulted in a rapid improvement in surface quality. A further feature can be seen after 55 minutes in Figure 4(a). Here the introduction of temperature variability corresponds to the argon gas flow down the stopper being turned off. On a round bloom using a straight through SEN, argon plays an important part in facilitating the flow of fresh steel to the meniscus. Without this there is an increased risk of meniscus cooling and plating of the metal surface leading to, amongst other effects, impairment of the casting powder performance.

REAL TIME ANALYSIS OF COPPER TEMPERATURES

The examples discussed so far are based on a retrospective analysis of previously logged data. This is a useful method of investigation when a particular mould-related problem is being studied. Information of this nature however, would be very useful if it were to be included as part of an on-line quality prediction system and to aid in the control of casting conditions. It is necessary therefore for information on thermal variability to be generated in real time and to be used by the grading system and displayed to casting controllers. A system has been developed to achieve this and installed on a round bloom caster (3). It is also being installed on slab casters within British Steel.

The system is based on a PC which is capable of scanning over 100 analogue signals every second whilst carrying out statistical analysis and updating VDU displays in the control room. Temperature variability indices are calculated as indicated in Figure 5 where a statistical representation is given for the temperatures at position A in Figure 3. The thermocouples and certain plant signals such as casting speed and mould level are scanned for a period of about 2 minutes (this period can be varied to represent a specified cast length, such as one slab). For each of these periods a variability index is calculated for each thermocouple in the mould and analysis of the results is carried out using methods developed through experience. The results are then displayed for this period to the casting controllers who may take action if predetermined levels are exceeded. The nature of the displays used and actions employed must be determined for each situation separately. In the case of the current system operating on a round bloom caster, the most useful information is gained from 4 thermocouples equally spaced around the tube at the same level just below the meniscus. The variability index is calculated for each of these thermocouples and the average is displayed on a VDU which updates at the end of each period. Levels have been determined, above which deterioration in surface quality is likely. If these levels are consistently exceeded then certain actions are taken which have been found to affect powder performance and so heat transfer conditions. For example, argon flow rate may be adjusted. As well as identifying adverse casting conditions, the system is also useful in giving a rapid assessment of different casting powders. This is illustrated in Figure 6 where heat transfer conditions are shown for two powders with different viscosities when casting the same steel grade. The Figure is similar to the VDU display with a clear picture of recent history being shown, as well as current conditions. In the first half of the extract a high viscosity powder (24.5 poise) is used and it is clear that heat transfer conditions are very unstable. A change is made to a lower viscosity powder (4.2 poise) in the middle of a ladle and stable conditions are quickly established. Later surface inspection confirmed a corresponding improvement in surface quality.

As stated previously the information generated by the system could be used in conjunction with other parameters by a grading computer, or could be stored for later assessment of casting practice. An example of such an analysis is shown in Figure 7. Here 2 powders with similar properties were used during two campaigns casting very similar steel grades. The Figure shows the distribution of temperature variability for the period over which the powders were used. Figure 7(a) shows an assessment of Powder "A" which corresponded to good surface quality and resulted in low variability for most of the campaign. Powder "B" however gave opposite results as can be seen in Figure 7(b), and resulted in poor surface quality.

APPLICATION OF REAL TIME THERMAL MONITORING SYSTEMS TO SLAB MOULDS

Systems based on the principles outlined in this paper are currently being developed for use on slab moulds. The thermocouple array in this case consists of about 40 per mould with the majority being situated just below meniscus level. There is a concentration of thermocouples in the centre of the mould which is a particularly sensitive area for longitudinal facial cracking. Heat transfer conditions for the whole mould are assessed through consideration of variability, absolute levels, rates of change and symmetry of measured temperatures. Some lower thermocouples are installed to monitor slag development down the length of the mould plate, local starvations being indicated by high variability for a lower thermocouple. These thermocouples are also utilised to give the facility for detection of sticking type behaviour.

CONCLUSION

The use of this instrumentation provides a means of "looking" into the mould during casting to monitor casting conditions and to identify and optimise those parameters which influence operation and quality.

REFERENCES

1. Irving, W. R., Dewar, W. A. G., and Perkins, A: Thermal Control Requirements for Continuous Casting. International Iron and Steel Congress, Dusseldorf, 1974.

2. Bardet, P., Leclercq, A., Mangin, M., Sarter, HB.: Control of Continuous Slab Casting at SOLLAC. Journees Siderurgiques, ATS, Paris, December 1982.

3. Byrne, A., Powell, J., Perkins, A., and Hunter, N.: The Commissioning and Work-up of the 3-Strand Bloom Caster at BSC Clydesdale Works. 4th International Conference on Continuous Casting, Brussels, May 1988.

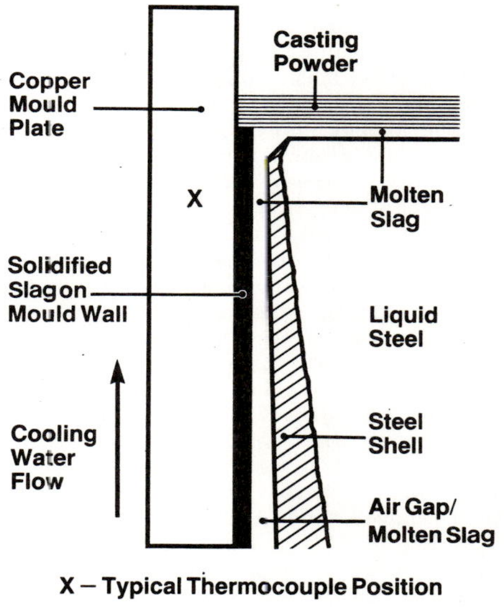

Figure 1 - Schematic Diagram of Conditions in the Interface between Steel and Copper.

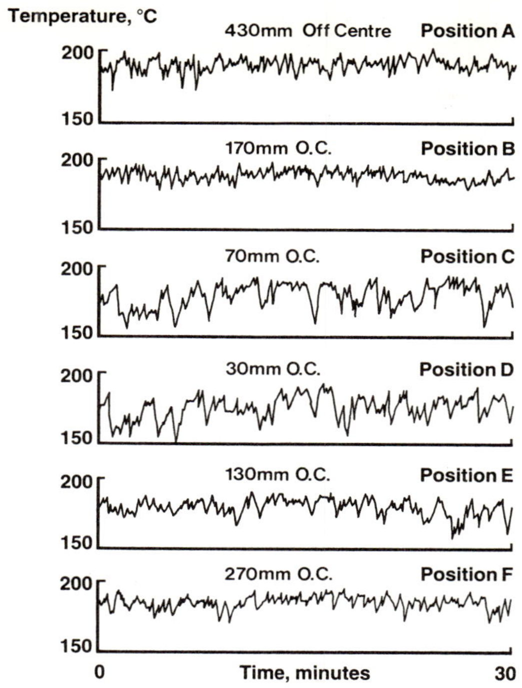

Figure 2 - Copper Temperatures Across the Central 800 mm of a Slab Mould Broad Face Plate - 0.083% Carbon Steel.

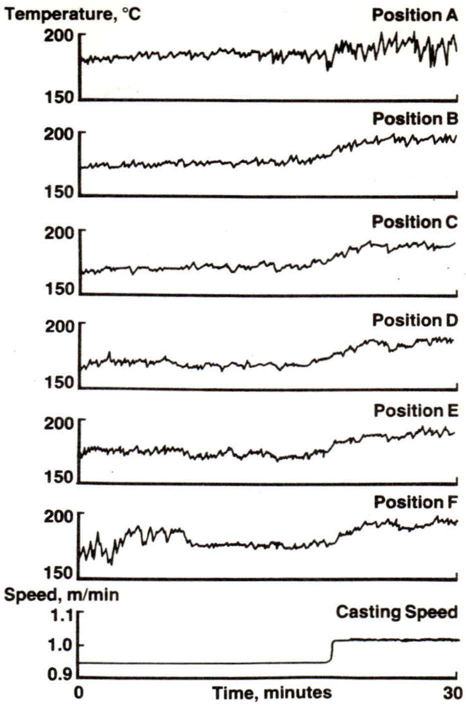

Figure 3 - Copper Temperature Across the Central 800 mm of a Slab Mould Broad Face Plate, Together with the Corresponding Casting Speed - 0.077% Carbon Steel.

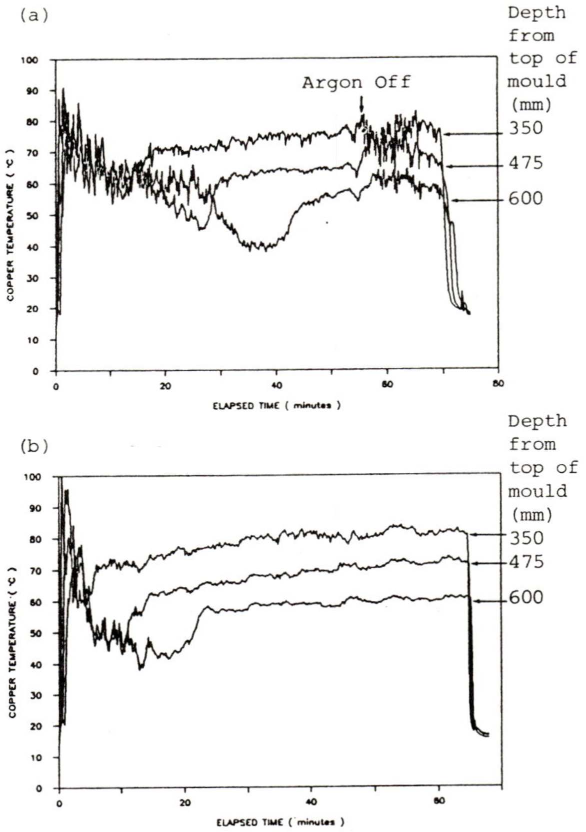

Figure 4 - Copper Temperatures at Different Distances down the Mould When Casting 0.25% Carbon Steel, With and Without Starter Powder.
(a) Without Starter Powder.
(b) With Starter Powder.

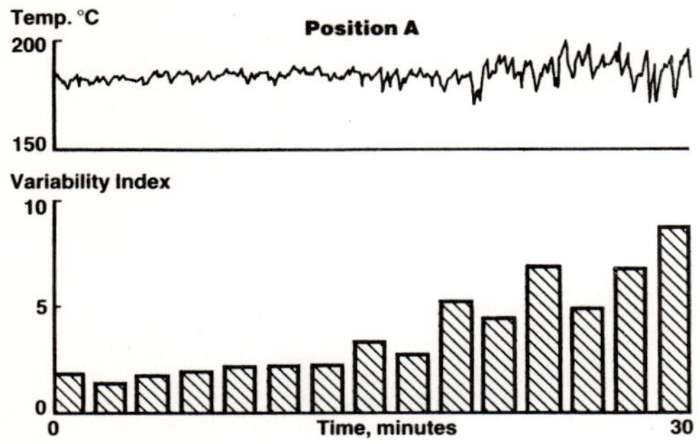

Figure 5 - Statistical Representation of temperature Variability.

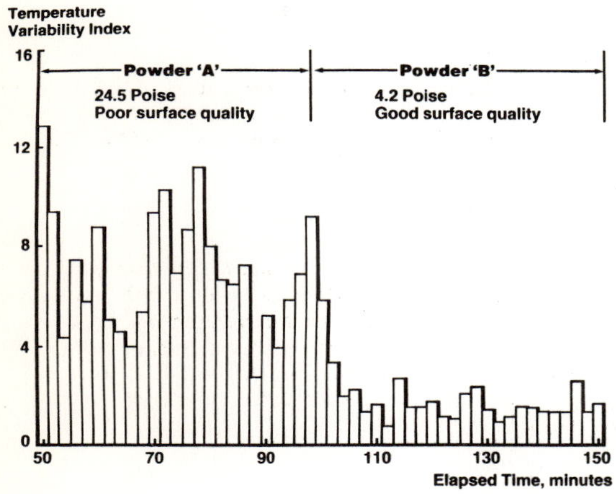

Figure 6 - Comparison of Resulting Thermal Variability for High and Low Viscosity Powders When Casting Round Blooms.

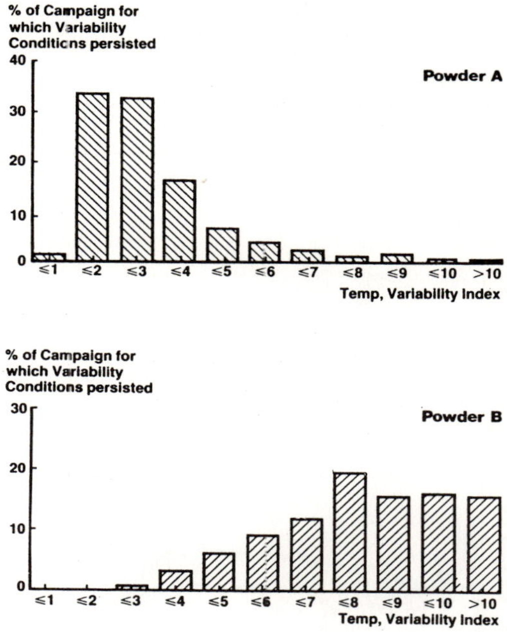

Figure 7 - Statistical assessment of Casting Powders Resulting in Good and Poor Surface Quality.
(a) Good Surface Quality.
(b) Poor Surface Quality.

Influence of casting condition on the frictional force between mould and shell in high speed continuous casting practice

M WASHIO, K HAMAGAMI and T KOSHIKAWA

M Washio is in the Steelmaking Department at Kawasaki Steel Corporation; K Hamagami is Deputy Manager of the Steelmaking Department at Kawasaki Steel Corporation; T Koshikawa is General Manager of the Steelmaking Department at Kawasaki Steel Corporation.

Synopsis

The frictional force between mould and shell is measured to clarify the mechanism of mould lubrication and prevent the occurrence of breakouts during high speed casting practice.

1. Introduction

In the continuous casting of slabs, high speed casting is important for increasing the productivity. However, breakouts are more likely to occur under high speed casting. To clarify the mechanism of mould lubrication, the frictional force between mould and shell was measured by means of load cells. Based on these experimental results, the optimum condition of mould lubrication is discussed for the purpose of preventing the occurrence of breakout from high speed casting practice.

2. Apparatus and experimental condition

As shown in figure 1, load cells were fitted onto the supporting pins of the oscillating table of the No.3 continuous casting machine at Chiba Works of Kawasaki Steel Corporation.[1] Low carbon Al-killed steels were cast for measuring the drag force; and the frictional forces between mould and shell can be calculated from the drag force. The experimental conditions are shown in table 1, where the negative stripping ratio, N, is defined in equations (1) and (2):

$$V_r(1 + N/100) = 2f_s \quad (1)$$

$$x = s/2(\sin 2\pi ft) \quad (2)$$

Where x : displacement of mould oscillation, s : stroke of mould oscillation, f : frequency of mould oscillation, t : time, V_r : casting speed, N : negative stripping ratio.

Chemical compositions and specifications of four kinds of casting powder in this experiment are shown in table 2.

3. Results

The value ΔF is defined as the amplitude of the frictional force per unit area of the inner surface of the mould. Figure 2 shows the effect of casting speed on the value of ΔF. As the casting speed increases, ΔF increases, independent of the oscillation condition.

The effect of negative stripping ratio on the frictional force is shown in figure 3 for three kinds of casting powders. ΔF decreases apparently with decreasing negative stripping ratio.

The effect of the viscosity of the casting powders on the frictional force is shown in Fig. 4 at a constant casting speed. Viscosity can be determined by sampling from the mould at the same time as measuring the frictional force and subsequent measurement using a viscometer. As shown in Fig. 4, the frictional force was decreased with lowering viscosity with all kinds of mould powders.

The relation between the frictional force and index of frequency of breakouts is shown in figure 5. As the frictional force decreases, the frequency of breakouts tends to be reduced and breakouts can be eliminated when the frictional force is controlled to be under 200 gf/cm^2.

4. Discussion

The frictional force between the shell and the mould F(t) is composed of the solid friction, $F_s(t)$, and liquid state lubrication, $F_l(t)$, in the following equation.[1,2]

$$F(t) = F_s(t) + F_l(t) \quad (3)$$

$F_s(t)$ is caused by the contact between the shell and the mould or solid slag and the mould. $F_l(t)$ is composed of the shear stress in liquid slag film cased by relative motion between the shell and the mould. $F_l(t)$ is calculated by using the oscillation condition and viscosity of the mould powder as follows:

$$F_l(t) = -\eta_l\,(Vr + Vm)/d_l \quad (4)$$

Where η_l : apparent viscosity of the lubricating liquid between the shell and the mould, d_l : mean thickness of lubrication liquid, Vm : oscillating velocity of the mould,

$$Vm = \pi f_s(\cos 2\pi ft) \quad (5)$$

By substituting equation (1) and (5) into (4), the maximum value of $F_l(t)$ is given as follows.

$$F_{l\,MAX} = -\eta_l(Vr/d_l) \times (\pi/200N + \pi/2+1) \quad (6)$$

Under the conditions in Table 3, $F_{l\,MAX}$ is calculated from equation (6) by assuming that the thickness of the lubricating liquid is constant in the mould; and it is estimated by consumption of mould powder, though the thickness of the lubricating liquid depends on the interface temperature between the shell and the mould.

The relationship between calculated $F_{l\,MAX}$ and distance from menisucus is shown in figure 6. Under conditions I and II which have a stronger shear stress than σ_{TS},[4] there is a possibility of breakout. This is the same tendency as in figure 5 which shows that there is a possibility of breakout occurring when using powders A and B at a high casting speed, 1.75 m/min. From the above findings, proper casting powder and proper oscillation condition should be chosen to prevent the occurrence of breakout during high speed casting practice.

5. Conclusions

The frictional force between the shell and the mould was measured by means of a load cell fitted on the supporting pin of the oscillating table. Results obtained were as follows:

1) Frictional force increases with higher casting speed.
2) Frictional force decreases with lower negative stripping ratio at the same casting speed.
3) Frictional force decreases when using powder having a lower viscosity for any kind of mould powder
4) Mould powders having low viscosity and the choice of a low negative stripping ratio are suitable for preventing breakouts at a high casting speed.

References

1) H.Nakato,et.al ; Journal of Metals, March (1984), 44
2) S.Omiya,et.al ; Tetsu-to-Hagane 68(1982)S926
3) H.Mizukami,et.al ; Tetsu-to-Hagane 70(1984)S151
4) H.Kitaoka,et.al ; Kawasaki Tech. Report 12 (1980) 497

Table I Experimental Conditions

Mould oscillation:	Stroke	S = 7.8 mm
	: Frequency	f = 100～150 cpm
	: Negative stripping ratio of averaged velocity	N = 5～30 %
Casting speed		Vr= 1.65～2.04 m/min
Slab size		230 × 900～1100 mm

Table II Chemical compositions of mould powder(wt%)

Powder	CaO/SiO_2	Al_2O_3	F	Li	Ba	Mg	Viscosity (poise at 1300°C)	Softening point (°C)
A	1.05	6.4%	8.8%	0%	4.8%	0%	0.9 P	960
B	0.85	6.0	8.8	0.5	7.0	0	1.3	890
C	0.95	5.7	8.8	0.5	0	2.8	0.8	910
D	0.95	5.1	8.8	0.5	0	4.3	0.9	910

Table III Casting condition for calculation
(Mould size : 230mm × 1000mm)

	I	II	III	IV	V
V_R (m/min)	1.75	1.75	2.00	1.75	2.00
N (%)	30	10	5	10	5
Mould powder	A	B	C	C	D
Consumption of mould powder (kg/t)	0.33	0.41	0.40	0.45	0.40
dl (cm)	0.0089	0.011	0.011	0.012	0.011

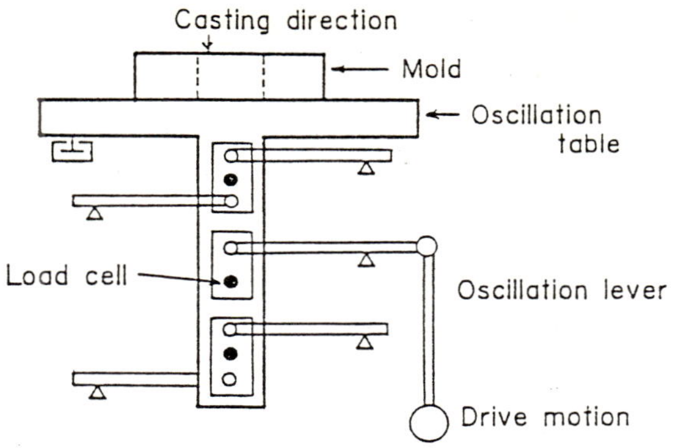

Fig.1 Schematic view of oscillation generator.

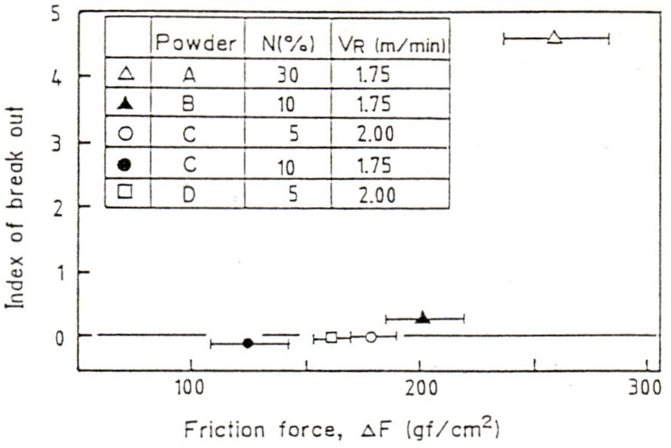

Fig.4 Relationship between frictional force and viscosity of mould slag.

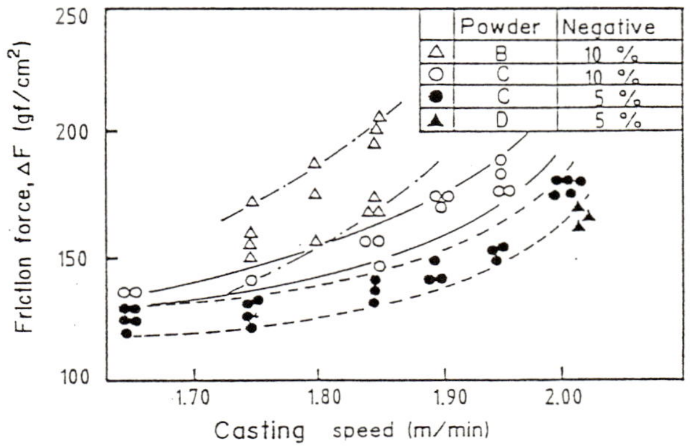

Fig.2 Influence of casting speed on frictional force between mould and the shell measured by load cell.

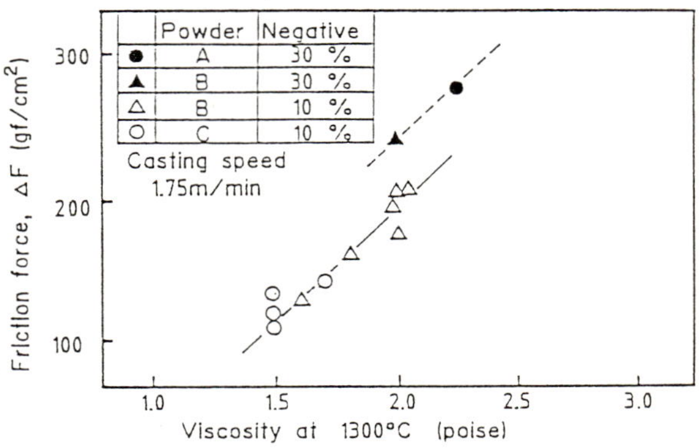

Fig.5 Influence of frictional force by index on occurrence of breakout.

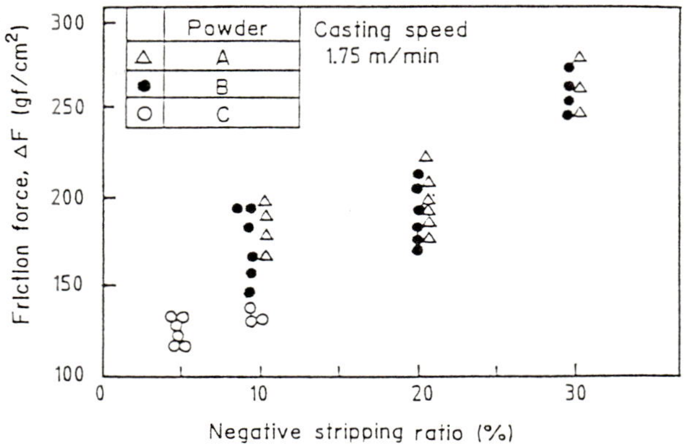

Fig.3 Effect of negative stripping ratio on frictional force.

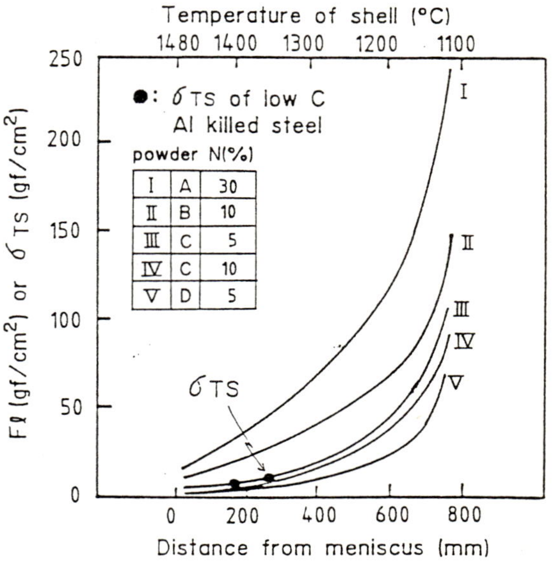

Fig.6 Variation of F_l calculated and σ_{TS} of low C Al killed steel with distance from meniscus.

Investigations concerning structure and properties of steel slags

J GEISELER and R SCHLOSSER

Dr Geiseler is the Representative Manager of the Research Institute of the Forschungsgemeinschaft Eisenhüttenschlacken (FEhS), Duisburg;
Dr Schlösser is a Scientific Collaborator in the Research Institute of the FEhS, Duisburg.

SYNOPSIS

Steelmaking slags are a very suitable material for road construction provided their volume stability is adequate. This can be jeopardized mainly by free lime and free magnesia. These mineral phases react with moisture forming hydroxides which are much greater in volume than the oxides. Consequently the slag can crack and can even disintegrate completely. Therefore, measures are presented to limit the content of both free lime and free magnesia.

Examinations concerning steel slags low in MgO

Linz-Donawitz(LD) slags which are formed using lime as a flux contain less than 3 % MgO. The volume stability of such slags can be jeopardized mainly by free lime.

From the microscopic investigations it is evident that the free lime can be divided into residual lime and lime precipitated during the solidification and subsequent cooling (Fig. 1). In both categories forms exist which are different in appearance and particle size. So the residual lime can be subdivided into grainy lime with particle sizes mainly between 3 and 10 µm (Fig. 2a) and spongy lime with particle sizes mainly between 6 and 50 µm (Fig. 2b). The precipitated lime can exist on C_2F-grain boundaries (Fig. 2c) or within C_3S crystals (Fig. 2d) and has particle sizes mainly of less than 4 µm. All kinds of free lime can hydrate but most significant is the so-called spongy free lime with grain sizes up to 50 µm and which is a form of residual free lime. Moreover, the slags can contain macroscopically visible fragments of undissolved lime which can hydrate very easily. Besides the content of total free lime the volume stability also depends on the porosity of the slag. So the aim is to reduce total free lime content, its grain size, and porosity.

Investigations concerning steel slags rich in MgO

Steel slags rich in MgO include electric arc furnace slags, Open-Hearth(OH) slags and LD slags which are produced using dolomite or dolomitic lime as a flux. Microscopic studies showed that in LD slags magnesium oxide appears mainly in solid solution with FeO and MnO, so-called magnesiowustites, while unbound MgO (periclase) was very seldom found. Fig. 3a shows a typical form of magnesiowustites with a core rich in MgO surrounded by an area rich in FeO and MnO. In arc furnace slags and OH slags the MgO was also observed to form spinels of various compositions (Fig. 3b) and sometimes silicates like merwinite, melilite etc. Spinels and silicates cannot hydrate, as is known from the behaviour of these mineral phases in nature. On the other hand unbound MgO and some magnesiowustites can hydrate. Since FeO and MnO can delay or even prevent the hydration of MgO the question of which magnesiowustites can hydrate was examined. For this purpose the hydration reactions of synthetic magnesiowustites with variable composition were tested under different conditions of temperature and pressure. It was borne in mind that the laboratory results obtained at 100 °C/ 1 bar are best comparable to reactions taking place under natural conditions. So free magnesia was defined as the content of unbound MgO and magnesiowustites with more than 70 % MgO. The unbound MgO also includes the macroscopically visible fragments of undissolved refractory lining or dolomite which were found in some slags. Fig. 4 shows an LD slag with undissolved dolomite inclusions.

Measures to reduce the content of hydratable mineral phases in steel slags

Experience from the use of LD slags in experimental roads has shown that slags with contents of free lime less than 7 % are causing no damage. There are several possibilities to reduce the content of free lime:

1. Treatment of the molten slag in the furnace before tapping
2. Treatment of the molten slag after tapping in the slag ladle
3. Treatment of the solidified slag

First of all the addition of lime, which depends on the silicon content of the hot metal, has to be optimized by means of computerized process control.

Holding the slag in the slag ladle for a fixed time has no effect on the solution of the residual lime.

The addition of filter dust to the molten slag after blowing lowers the free lime content somewhat . The best results are obtained by mixing the molten LD slag with molten blast furnace slag. Fig. 5 shows the mixing process. The decrease of free lime - related to the initial free lime content - can amount to nearly 100 % in relation to the mixing proportion of LD slag and blast furnace slag as Fig. 6 demonstrates. This method, however, may delay the operating sequence in the plant and involve some capital investment. Thus it is proposed to blow filter dust into the molten slag with oxygen as a carrier gas.

Finally the free lime content can be reduced in the solidified slag by weathering the slag under natural or artificial conditions. An alternative method to obtain slags with sufficient volume stability is to separate the slags with regard to their free lime content.

Investigations concerning the free magnesia content of steel slags have shown that good solution of the added dolomite is most important. For this purpose the dolomite should be added at an early stage and grain size should be as small as possible. Then the MgO reacts with other components forming mineral phases which will not hydrate. In any case one must avoid large fragments of dolomite or refractory lining entering the molten slag at an advanced stage when they can no longer be dissolved.

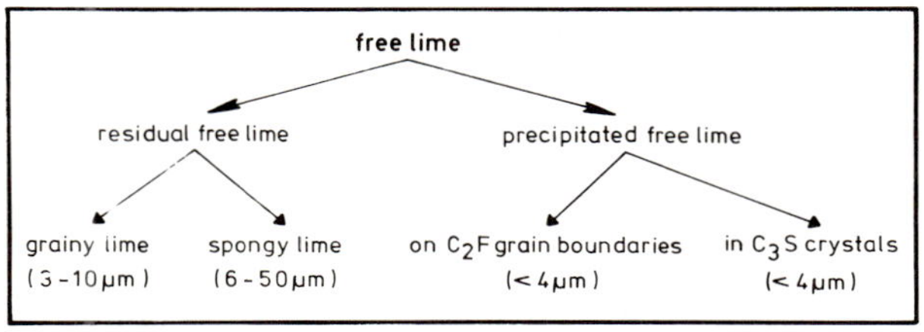

Fig. 1 Different kinds of free lime

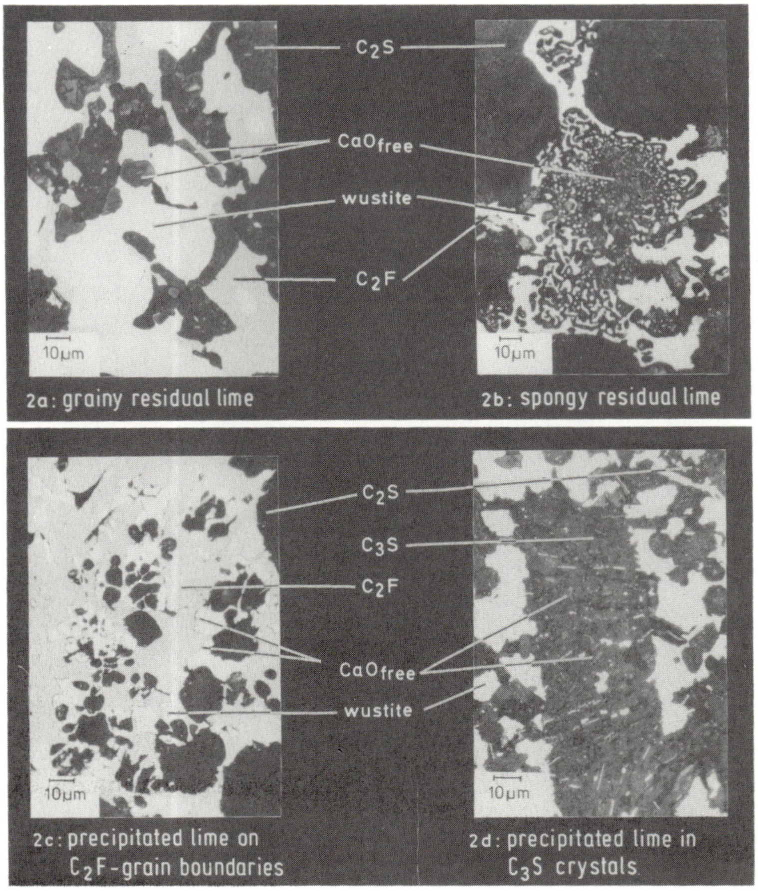

Fig. 2 a - 2 d Micrographs showing different kinds of free lime

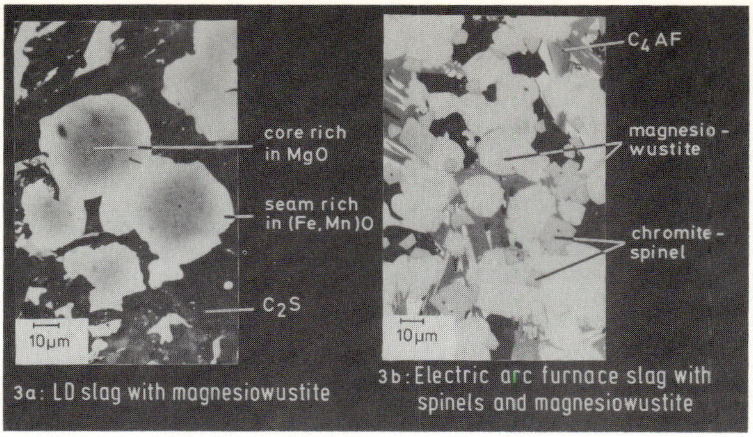

Fig. 3 a - 3 b   Micrographs showing different kinds of MgO-phases

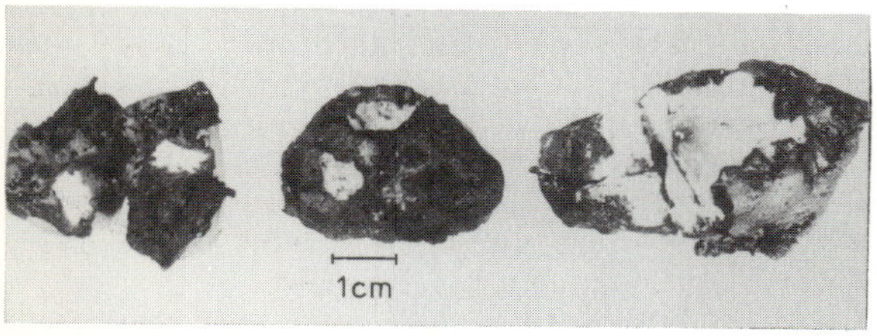

Fig. 4

LD slag with undissolved dolomite inclusions

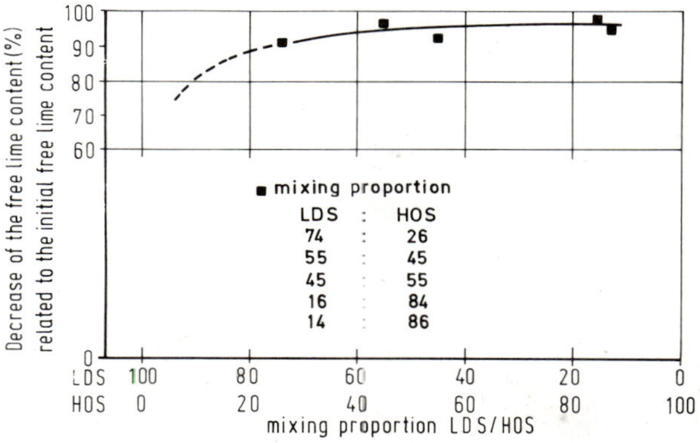

Fig. 6

Percentage decrease of the free lime content in relation to the mixing proportion of LD slag and blast furnace slag

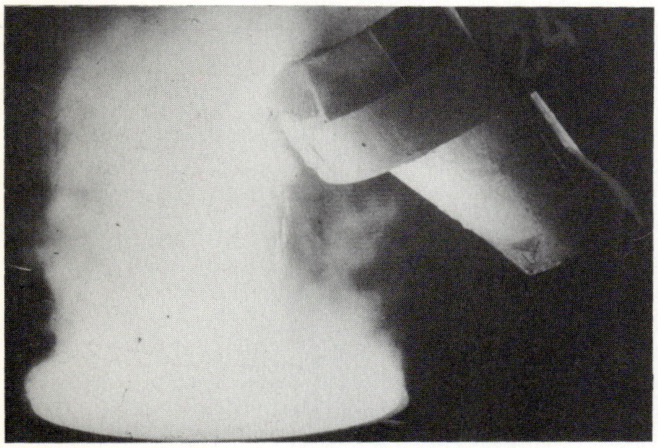

Fig. 5

Mixing of molten LD slag with molten blast furnace slag

Solubility of $MgO-Cr_2O_3$ in $MgO-Al_2O_3-SiO_2-CaO$ melts

K MORITA and N SANO

Dr Morita and Prof Sano are in the Department of Metallurgy at the University of Tokyo.

Synopsis

The solubility of $MgO{\cdot}Cr_2O_3$ in $MgO-Al_2O_3-SiO_2-CaO$ melts was measured at 1600°C as a function of oxygen partial pressure and slag composition. When CaO is added to $MgO-SiO_2$ melts in air, the solubility of $MgO{\cdot}Cr_2O_3$ increases on account of the presence of Cr^{6+}, particularly for the melts containing a large amount of CaO. On the contrary, under reducing conditions, Cr^{6+} is not stable and the addition of CaO leads to a decrease in the solubility of $MgO{\cdot}Cr_2O_3$ through a decrease in Cr^{2+} content. For $MgO-SiO_2$ melts, the solubility increased significantly with lowering oxygen partial pressure from 0.21 atm. to 3.6×10^{-13} atm.

Introduction

Recently a new smelting reduction process to make stainless steel has been developed especially in Japan. The combustion of coke or coal can be used instead of electric power as a heat source in this process. The presence of undissolved chromite particles in the slag, which are identified as $MgO{\cdot}Cr_2O_3$, has been observed to decrease the chromium yield during smelting reduction of chromium ore. However, when CaO or SiO_2 is added to the slag, undissolved chromite disappears and chromium yield increases up to 96%.[1] Thus, the control of slag composition to enhance the solubility of chromite is one of the most important factors in order to promote the performance of smelting reduction of chromium ore.

In this study, the solubility of $MgO{\cdot}Cr_2O_3$ in $MgO-Al_2O_3-SiO_2-CaO$ melts was measured for the purpose of optimizing the slag composition for smelting reduction.

Experimental

Synthesized $MgO{\cdot}Cr_2O_3$ was mainly used as chromite, and slags were prepared by mixing reagent grade chemicals for respective components. Platinum, magnesia and molibdenum crucibles were used as containers which held 4 grams of slags together with a piece of $MgO{\cdot}Cr_2O_3$.

After they were kept in a Tammann furnace at 1600±2°C for 18 or 21 hours under various atmospheres for equilibration, the slag samples were separated from chromite. All components of the slag sample were analysed by wet chemical methods, and some of the slag samples were identified by X-ray diffraction. The solubility of $MgO{\cdot}Cr_2O_3$ was determined as total chromium content. Contents of Cr^{2+}, Cr^{3+} and Cr^{6+} were also separately analysed for some samples.

Results and Discussion

Solubility in air

The solubility was found to be 1.1 to 4.5wt% Cr_2O_3 for $MgO-SiO_2$ melts (broken line in Fig.1) which is compatible with the $MgO-SiO_2-Cr_2O_3$ phase diagram reported by Keith.[2] When Al_2O_3 was added to the melts, the solubility decreased to less than 1wt% Cr_2O_3, and Cr in the coexisting $MgO{\cdot}Cr_2O_3$ was replaced by Al in proportion to Al_2O_3 content in the slag.

For MgO-CaO melts the solubility substantially rose to as much as 40 to 55wt% CrO_x as shown in Fig.2, although more than 50% of chromium was Cr^{6+}. When the oxygen partial pressure was lowered from in air to 10^{-5} atm. however, liquid phase could not be observed at any compositions. This fact indicates that the existence of Cr^{6+} promoted the solubility in air. The reason would be that hexavalent chromium oxide CrO_3 has low melting point, 196°C, and it is likely to act as an effective solvent for $MgO{\cdot}Cr_2O_3$.

For $MgO-Al_2O_3-SiO_2-CaO$ melts, which are typical for smelting reduction of chromium ore, the solubility decreased with increasing MgO and Al_2O_3 contents as shown in Fig.3. The effect of CaO/SiO_2 ratio on the solubility in the melts was found to be unexpectedly small as long as CaO was contained. When 22.1wt% of BaO was added to 20wt%MgO-20wt%Al_2O_3-30wt%SiO_2-30wt%CaO melts, an increase in the solubility from 1.1wt% to 4.4wt% Cr_2O_3 was observed.

Solubility under reducing conditions

For $MgO-SiO_2$ melts, the solubility of $MgO \cdot Cr_2O_3$ at $Po_2=2.73 \times 10^{-11}$ atm. was demonstrated in Fig. 1 (solid line), compared with that in air. At the $2MgO \cdot SiO_2$ saturation for the same system, the solubility was observed to increase from 0.7 to 8.2wt% Cr with lowering the oxygen partial pressure from 0.21 to 3.6×10^{-13} atm. as shown in Fig.4. According to the

$$CrO(\text{in slag}) + 1/4O_2 = CrO_{1.5}(\text{in slag}) \quad (1)$$

reaction:(1), the value of Cr^{2+}/Cr^{3+} reasonably increased with lowering oxygen partial pressure to 3.6×10^{-13} atm. where 25% of chromium existed as Cr^{2+}. The increase in solubility can be interpreted by the decreases in both $\gamma_{CrO1.5}$ and γ_{CrO}, which were simultaneously measured by equilibrating Ni-Cr melts with the slags and $MgO \cdot Cr_2O_3$.

As shown in Fig.4 and Fig.5, contrary to the results in air, the CaO addition to the above slag gave rise to a decrease in the solubility and also in the value of Cr^{2+}/Cr^{3+}. The latter is due to a larger increase in γ_{CrO} than $\gamma_{CrO1.5}$ by the CaO addition (Fig.6). For $MgO-SiO_2$ melts saturated with SiO_2, however, the CaO addition had no effect on the solubility of $MgO \cdot Cr_2O_3$ at $Po_2 = 2.73 \times 10^{-11}$ atm. being around 11wt% Cr which is greater than the maximum value for the $2MgO \cdot SiO_2$ saturated melts at $Po_2 = 3.6 \times 10^{-13}$ atm.

For 20wt%$MgO-Al_2O_3-SiO_2-CaO$ ($CaO/SiO_2 = 1$) melts, the solubility decreased with increasing Al_2O_3 content at $Po_2 = 2.73 \times 10^{-10}$ atm. This trend is consistent with that in air, though the solubility was twice as much as that in air.

Conclusions

Careful adjustment of slag composition, mainly by lowering MgO and Al_2O_3 contents, is recommended to enhance the dissolution of $MgO \cdot Cr_2O_3$ for promoting the reduction of chromium ore.

The strongly reducing condition on smelting reduction also functions favourably for the dissolution of chromium ore into molten slags through the appearance of CrO.

References

1) S.KOUROKI K.MORITA and N.SANO : Proc. Int. Conf. Recent Advances in Mineral Science and Technology, (1984), p.935 [MINTEK]
2) M.L.KEITH : J. Am. Ceram. Soc., 37(1954), p.490

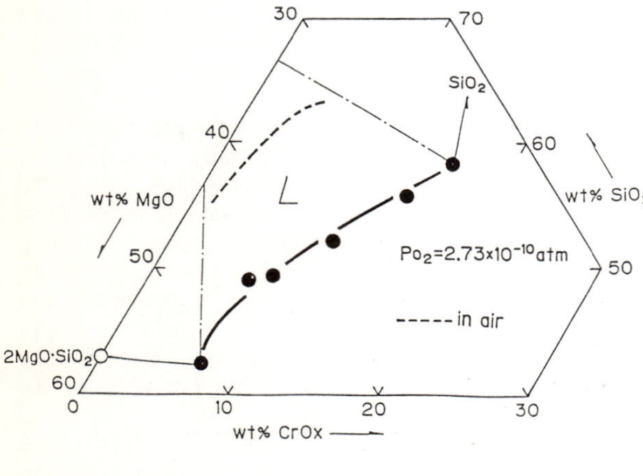

Fig.1 Comparison of the solubility of $MgO \cdot Cr_2O_3$ in the $MgO-SiO_2(-CrOx)$ system at 1600°C under different oxygen partial pressures.

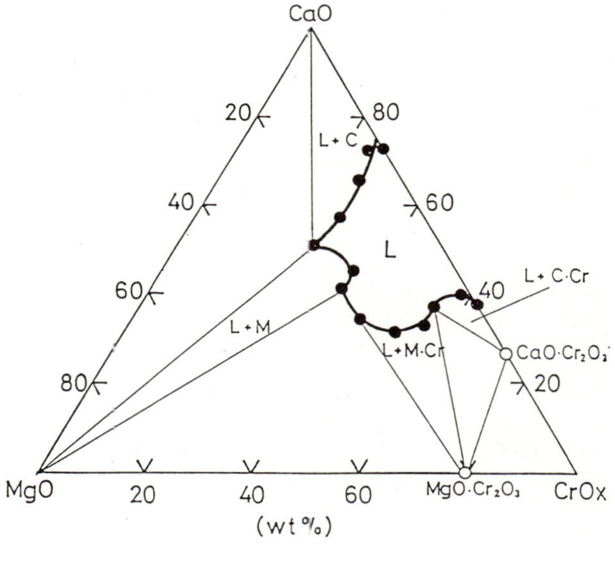

Fig.2 Liquidus in the MgO-CaO-CrOx system at 1600°C in air.
(L:liquid, M:MgO, C:CaO, Cr:Cr_2O_3)

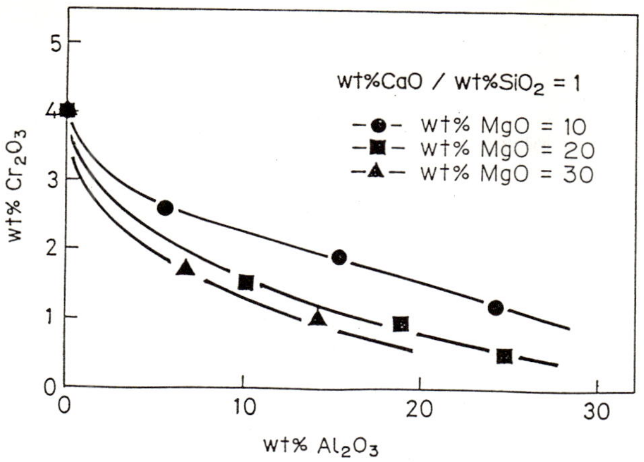

Fig.3 Effect of Al_2O_3 content on the solubility of $MgO \cdot Cr_2O_3$ in the $MgO-Al_2O_3-SiO_2-CaO$ system at 1600°C in air.

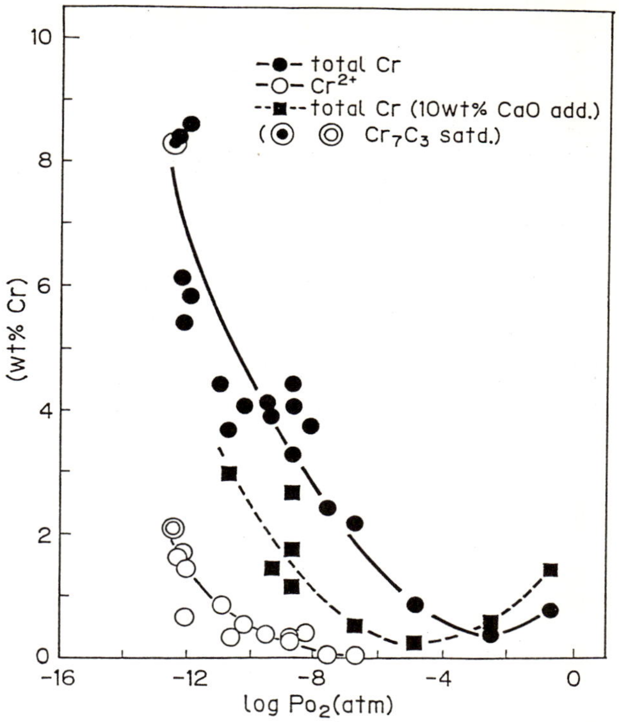

Fig.4 Oxygen partial pressure dependence of the solubility of $MgO \cdot Cr_2O_3$ in the $MgO-SiO_2(-CaO)$ system ($2MgO \cdot SiO_2$ satd) at 1600°C.

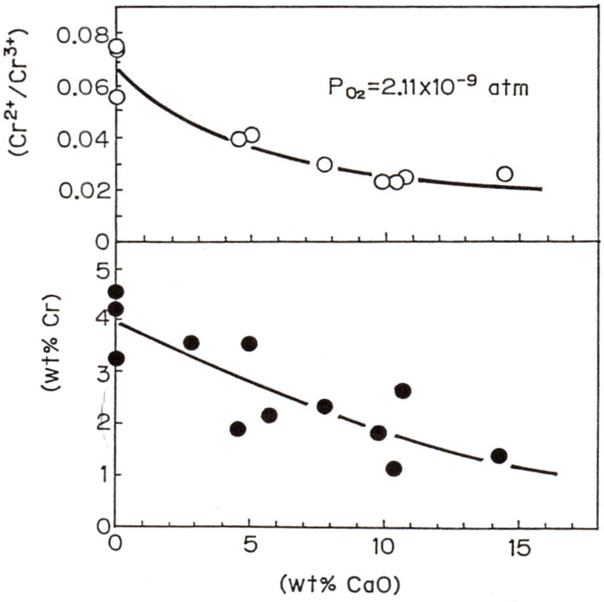

Fig.5 Effect of CaO content on the solubility of $MgO \cdot Cr_2O_3$ and Cr^{2+}/Cr^{3+} ratio in the $MgO-SiO_2-CaO$ system ($2MgO \cdot SiO_2$ satd) at 1600°C.

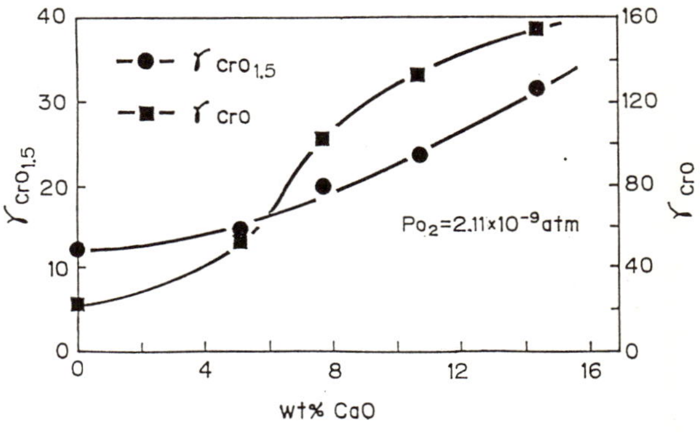

Fig.6 Effect of CaO content on the activity coefficients of $CrO_{1.5}$ and CrO in the $MgO-SiO_2-CaO-CrOx$ system ($2MgO \cdot SiO_2$, $MgO \cdot Cr_2O_3$ satd) at 1600°C.

□

Activities of cobalt and copper oxides in silicate and ferrite slags

A KATYAL and J H E JEFFES

Dr Katyal, formerly studying at Imperial College, is now with Cookson Group PLC; Prof Jeffes is Emeritus Professor of Extraction Metallurgy, Department of Materials, Imperial College, London.

ABSTRACT

The distribution of cobalt and copper between a copper-cobalt alloy and a variety of slags was determined by a levitation melting technique at oxygen potentials fixed by CO/CO_2 gas mixtures and at temperatures from 1250 to 1350°C. The activities of CoO and $CuO_{0.5}$ in the slags were calculated from these data.

SUMMARY

The activities of CoO and Cu_2O (as $CuO_{0.5}$) in slags were determined by equilibrating Co-Cu alloys with melts of varying compositions in a levitation melting apparatus in which the atmosphere was controlled to produce known oxygen potentials at temperatures from 1250-1350°C. The slags used were of compositions similar to those encountered in industrial furnaces, based on 'FeO'-SiO_2, 'FeO'-CaO-SiO_2, 'FeO'-CaO-Al_2O_3-SiO_2, and 'FeO'-CaO melts.

The melts and slags were analysed for their principal elements after equilibration: the iron in the slags was analysed for Fe^{2+} and Fe^{3+} but the small samples available made this differentiation difficult.

The activity coefficient of CoO at high dilution in 'FeO'-SiO_2 slags was found to be close to unity. Some slight decrease in its values was observed as the 'FeO' content of the slag was increased.

The activity of Cu_2O, expressed as $CuO_{0.5}$, was in good agreement with previous studies in 'FeO'-SiO_2 melts.

The activities of both CoO and $CuO_{0.5}$ were found to increase with the addition of CaO to 'FeO'-SiO_2 slags, the effects being slight and marked, respectively.

The activity of CoO in CaO-'FeO' slags was found to be somewhat greater than in 'FeO'-SiO_2 slags.

These results are discussed in terms of the slag/metal and slag/matte distributions encountered in industrial practice.

INTRODUCTION

The principal source of primary cobalt is as a by-product of copper and nickel smelting, as cobalt is found in copper and nickel deposits. The copper and nickel concentrates are first treated pyrometallurgically to enrich the cobalt content in either a matte, speiss or a metal phase. The bulk of concentrates are thus reduced for further treatment by hydrometallurgical methods. A high retention of cobalt in the matte, speiss or metal phases is therefore advantageous for a good recovery of cobalt by subsequent processes.

Copper concentrates are smelted to produce two immiscible liquid phases, i.e. slag and a copper-rich matte. One purpose of the matte smelting is to ensure a complete sulphidization of all the copper and other recoverable elements such as Co, Ni etc. During smelting the conditions are more reducing in order to produce a discard slag containing low percentages of copper, cobalt, nickel etc. Matte is then treated in a converter where it is blown with air or oxygen enriched air. The slag produced from the converting is high in copper and other valuable recoverable metals such as cobalt, nickel etc. which is either returned to the matte smelting (first stage) or is smelted separately in an electric furnace to recover a copper-cobalt-iron alloy[1,2].

Although an Ellingham diagram (Figure 1) is useful in interpreting smelting processes as described above, it relates only to pure phases as standard states. A knowledge of activity coefficients of the components in various phases would be advantageous in order to control the process conditions for a minimum loss of cobalt in the discard slag. At the time of commencement of this work very little data was available for the activity coefficient of cobalt oxide in the slag and speiss phases[3,4]. The present work was undertaken to study the slag/metal/gas equilibria between liquid copper-cobalt alloys and various slags encountered in the smelting processes which would assist the elucidation of the distribution of cobalt between slag/matte, slag/metal and slag/speiss phases encountered in the industrial processes.

EXPERIMENTAL

The gas/slag/metal equilibrium studies were carried out using a levitation melting technique. A sketch of the levitation apparatus is shown in Figure 2. The levitation cell consisted of a 15mm o.d., 150mm long transparent silica tube with both ends open. The tube was mounted on a fixed brass platform and was sealed with O-rings and brass couplings at both ends. Inlet and outlet holes for the incoming and outgoing gases were located in the brass couplings. The top end of the brass coupling had a viewing glass through which temperature was recorded by means of a two colour pyrometer. A full account of the levitation apparatus has been given elsewhere[5].

Preliminary levitation experiments showed that lower temperatures in the range 1250-1400°C could only be obtained with an alloy containing less than 5% cobalt. The design of levitation coils was as recommended by Jahanshahi and Jeffes[6] which gave stable and lower temperatures. The temperature of the levitated droplet was further controlled by varying the power supply of the RF generator set and by varying the helium gas flow rate through the levitation cell. To reduce the temperature of the droplet, the power supply was increased which lifted the droplet in the upper coil, the upper coil acting mainly as a magnetic "stopper". Temperature of the droplet was further controlled by passing up to 700 ml/min of helium gas through the levitation cell. At the temperature of investigations and up to 700 ml/min of helium flowrate, the enhanced vaporization of copper-cobalt alloy was insignificant and no copper fumes were observed in the gas stream.

Stocks of copper-cobalt alloys and slags used for the equilibration studies were prepared prior to the experiments. Slag was encapsulated in the alloy to produce samples weighing approximately 1.0±0.1g of alloy and 85±15mg of slag made by ramming the slag into a drilled copper-cobalt alloy cylinder . The compositions of stock slags are given in Table 1.

The oxygen potential at the slag/metal interface was controlled by using a mixture of carbon monoxide and carbon dioxide. All the gases viz CO, CO_2 and He were dried and freed of oxygen by passing them through drying agents and the helium passed through a 'getter' furnace. Further details of drying agents and oxygen removal are given in reference 5. The gases were metered and the oxygen potentials of the incoming and outgoing gases through the levitation cell were measured with a solid electrolyte probe of lime stabilized zirconia. The levitation cell was found to be 'leak free' as the oxygen potential of the incoming and outgoing gases was the same.

A high impedance electrometer was used for obtaining the emf signals from the solid electrolyte probe. The CO/CO_2 ratio was not determined analytically but the values calculated from the emf measurements were in good agreement with the values of CO/CO_2 measured by orifice flowmeters.

The emissivities of metal droplets were fairly constant at up to 5% cobalt content in the copper; the temperature recorded by the two colour pyrometer agreed to within ±3°C with that of a Pt-Pt/13%Rh thermocouple.

Previous investigators[7,8] have shown that equilibrium between gas/metal/slag phase was attained within one minute. A series of experiments were conducted by varying the equilibration time from 5 minutes to 20 minutes. As there was no significant difference in the results obtained, an equilibration time of ten minutes was chosen for most of the experiments.

The slag sampling technique was similar to that used by Taylor[9]. The slag was sampled by touching the levitated droplet with an alumina rod and the metal droplet was subsequently quenched in a copper mould. Previous work[9] has shown that this slag sampling technique gave a minimum disturbance of the equilibrium during sampling as it allowed the slag to cool out of contact with the metal phase.

The slag sampled was weighed and was generally of 20 to 30 mg. Sometimes a second sample of slag was taken if the quantity of slag collected was small, before quenching the droplet.

Both the slag and metal samples were analysed for all the components (viz Cu, Co, Fe, Si, Ca, Al) after equilibration. A Perkin Elmer atomic absorption spectrophotometer was used for analyses. Ca, Si and Al were not detected in the metal samples. The mutual interference of the elements in the analytical instrument used was checked and was found to be negligible. Approximately 50% of the metal samples were analysed in duplicate and the reproducibility was within ±2% of the determined value. A complete bead of slag sample was taken for analysis in order to avoid any errors due to inhomogeneity on cooling.

The ferric analyses were carried out on several slag samples using a colorimetric method[10]. As the ferric iron was determined by taking a difference between total iron and ferrous iron, the accuracy of ferric iron was low and is estimated to be ±20%.

CALCULATION OF RESULTS

The results of the levitation experiments are summarised in Table 2. Full details are available in reference 5.

The metal analyses for Cu, Co and Fe usually totalled 100±1%. The iron content of the metal was always less than 1% and its effect on the activity of Co in the alloy was considered to be small. The oxygen contents of the alloys were not determined. Estimation of oxygen contents from thermodynamic calculations indicated that its effect on the activity of Co and of Cu would be negligible.

The activities of Co and Cu in the metal were calculated from the data of Taskinen[11]. Copper contents in the alloys lay in the region where the Co activities are a linear function of its atom fraction so that Cu behaves ideally.

The oxygen pressures at equilibrium with CO/CO_2 gas mixtures at various temperatures were calculated from the relationship:

$$CO + \tfrac{1}{2}O_2 = CO_2$$

$$\Delta G^{o} = -66950 + 20.2T \text{ cal/mol} \qquad [1]$$

Slag mol fractions were calculated on the assumption that the species were CoO, CuO(0.5), FeO, CaO and SiO_2. Analysis of the slags for FeO(1.5) caused serious problems because of the small amounts of slag available and of preventing changes in Fe(2+)/Fe(3+) ratios of the samples during preparation for analysis. In most cases the Fe(3+) contents of the slags varied from 2-6%, with an error margin of ±20%. In such a slag, the presence of Fe(3+) would cause an underestimate of N(CoO) and $N(CuO_{0.5})$ of less than 1%. In the case of the calcium ferrite slag (slag E) the presence of Fe(3+) was taken into account in calculating mol fractions.

The slag analyses totalled to within 2.5% of 100%. In the case of slags C and D the lime to silica ratio was adjusted to that of the initial slag composition. This was calculated from:

$$100-(\text{wt\% } CuO_{0.5} + \text{wt\% } CoO + \text{wt\% } FeO) = \text{wt\%}(CaO + SiO_2)$$

$$CaO/SiO_2 = R$$

where R is the original ratio of the slag.

The activities of copper and cobalt oxides are expressed in terms of $CuO_{0.5}$ and CoO. The former was chosen rather than Cu_2O since two separate Cu^+ ions are formed in slag solutions as shown by the fact that the copper content of slags is proportional to $pO_2^{1/4}$ and to a_{Cu} [13,14,15]. Using $a(CuO_{0.5})$ gives constant values of $\gamma(CuO_{0.5})$ as $N(CuO_{0.5})$ approaches zero whereas $\gamma(Cu_2O)$ approaches zero as $N(Cu_2O)$ approaches zero.

Activities of $CuO_{0.5}$ and CoO were calculated relative to the pure liquid states using the equations:

$$Co_{(1)} + \tfrac{1}{2}O_2 = CoO_{(1)}$$

$$\Delta G^o = -53{,}219 + 16.49T \text{ cal/mol} \qquad [2]$$

based on [16] and [17]. The entropy of fusion of CoO was estimated as 4.50 cal/mol deg in agreement with the value used by Wang, Santander and Toguri[4] leading to

$$CoO_{(s)} = CoO_{(1)}$$

$$\Delta G^o = 9351 - 4.50\ T \text{ cal/mol*} \qquad [3]$$

$$2Cu_{(1)} + \tfrac{1}{2}O_2 = 2CuO_{0.5(1)}$$

$$\Delta G^o = -29260 + 10.488T \text{ cal/mol} \qquad [4]$$

This equation is that used by Altman and Kellog[18] based on the data of Mah et al.[19]

RESULTS

All activities are calculated relative to the liquid standard state.

FeO-SiO$_2$ slags

$a(CuO_{0.5})$ is plotted as a function of $N(CuO_{0.5})$ in slag A (FeO/SiO_2 ratio = 3.0) and slag B (FeO/SiO_2 ratio = 1.85) in Figure 3. No systematic difference between the two slag compositions could be observed. The limiting slope at infinite dilution gives a value of $\gamma(CuO_{0.5}) = 2.89$, there being some evidence that at higher copper concentrations it falls appreciably below this value.

a(CoO) is plotted against N(CoO) in Figure 4. As with $CuO_{0.5}$ there is little obvious difference between the results in slag A at slag B. The value of $\gamma(CoO)$ = 0.91 is obtained for slag A and 1.00 for slag B.

FeO-SiO$_2$-CaO slags

The results for slags C and D are given in Figs. 5 and 6. In these slags FeO in slag B was replaced by 28-29% CaO and $\gamma(CuO_{0.5})$ was more than doubled to 7.81 and $\gamma(CoO)$ to 2.08.

The effects of smaller additions of CaO to $FeO-SiO_2$ slags of FeO/SiO_2 weight ratios of 1.20 and 1.85 are shown in Figures 9, 10, 11, and 12. The resulting values of $\gamma(CoO)$ and $\gamma(CuO_{0.5})$ are given in Table 1. It will be noted that the scatter in these experimental results produces some anomalies but the effect of increasing the CaO content of the slags is, in general, to increase the activity coefficients of both $CuO_{0.5}$ and CoO.

***NOTE:**

Since writing this, some evidence has indicated that an entropy of fusion of 6.0cal/mol deg may be more appropriate[20].

FeO-CaO slags

The results for a 'calcium ferrite' slag, E, are shown in Figures 7 and 8. The values of $\gamma(CoO)$ are slightly higher than in iron silicate slags but $\gamma(CuO_{0.5})$ is lower than in iron silicate slags.

FeO-SiO$_2$-Al$_2$O$_3$ slag

The effect of replacing half of the CaO in slag G by Al_2O_3 (slag K) is shown in Figures 11 and 12. This appears to have the opposite effect to CaO and to cause a slight lowering of $\gamma(CoO)$ and $\gamma(CuO_{0.5})$.

DISCUSSION

The above results are summarised in Table 1. In this table are also given the initial slag compositions, calculated in terms of FeO, SiO_2, CaO and Al_2O_3 species. As stated above, analysis for Fe_2O_3 caused serious difficulties because of the small slag samples available. The figures for Fe_2O_3 in Table 1 represent the approximate Fe_2O_3 contents of the slags after equilibration experiments. Only in the case of slag E, calcium ferrite, is the Fe_2O_3 content of the slag greater than 5 wt %.

It is clear from Figures 3 and 4 that the effect of varying the FeO/SiO_2 ratio of an iron silicate slag has little effect on the activities of $CuO_{0.5}$ or CoO although a slight trend to lower activity coefficients in higher FeO slags may be detected. This is in accord with the values obtained by Taylor and Jeffes[7,21] on $CuO_{0.5}$ and NiO in such slags.

The effect of additions of CaO are, however, quite striking in increasing $\gamma(CuO_{0.5})$ but the effect on $\gamma(CoO)$ appears to be positive but much less so than for $\gamma(CoO_{0.5})$. These are illustrated in Figure 13.

It is interesting to speculate on the reason for these observed effects and to relate them to the stabilities of the compounds formed between the slag components.

In Table 3 are listed the heats of formation from the oxides of some silicates and ferrites. It has been noted[7] that the slight but definite lowering of $\gamma(NiO)$ in iron silicate slags by increased 'FeO' content is due to the stability of $NiFe_2O_4$ compared with Ni_2SiO_4 and because Fe_2O_3 becomes progressively more important as the FeO content of the slag increases. The slight effect observed on $\gamma(CoO)$ may be due to the smaller difference in the stabilities of Co_2SiO_4 and $CoFe_2O_4$.

The addition of lime to an iron silicate slag would strongly lower the activity of SiO_2 and correspondingly increase the activity of FeO and promote the formation of Fe_2O_3. If this were the case, the activity coefficient of $CuO_{0.5}$ would **decrease** but in fact it **increases**. This effect is well documented in other work by Elliot et al.[22] and Altmann[23] and presents something of a conundrum in terms of the stabilities of the binary compounds. Turkdogan[24] points out that this is analogous to increase in activity coefficients of FeO and MnO in slags, but in these cases the silicates involved are much more stable than the non-existent copper silicates.

The authors suggest that this could be fruitful field for study by theoreticians.

Comparison with other results

Wang, Santander and Toguri[4] measured the solubility of NiO and CoO in iron silicate slags. Their results indicate a value of $\gamma(CoO)$ of 1.14 in silica saturated slag, which has an FeO/SiO_2 ratio of 1.3. This is in fair accord with the present values given in Table 1.

Yazawa[25] has calculated values of $\gamma(CoO)$ of 1.7 and 2.0 in calcium ferrite and iron silicate slags. These values are relative to the solid CoO standard state, and recalculated to the liquid standard state these values are changed to 0.85 and 0.98 respectively. The value for iron silicate slags is in good agreement with the present work but the value for calcium ferrite is appreciably lower than ours.

Matte-Metal distribution of cobalt

Imris[2] has given an account of the distribution of cobalt between matte and slag in the Rokana smelter in which a matte containing 59.6% Cu and 0.99% Co was in co-existence in a reverberatory furnace with a slag containing 0.46% Co. Using values of $p(O_2) = 1.2 \times 10^{-4}$ and $p(S_2) = 1 \times 10^4$ atm. to define conditions at the slag/matte interface, assuming $\gamma(CoO) = 1.2$ and ideal behaviour of cobalt sulphide in matte. The predicted slag/matte distribution ratio of 0.88 w/w comes very close to the distribution observed at Rokana with a 60% copper matte.

Slag-Metal distribution of cobalt

The behaviour of cobalt in partitioning between metal and slag during the fire refining of copper was discussed by Jeffes and Jacob[26] who concluded that it could be explained if CoO(s) had an activity coefficient of 0.94 in the slags involved. Converting this to the liquid standard state would give a value of $\gamma_{CoO(l)} = 0.46$. The higher values obtained in the present work means that cobalt will be retained in the metal about twice as well as was concluded by Jeffes and Jacob.

REFERENCES

1) Young R.S., 'Cobalt' Monograph series 108, Published by Reinhold Corporation, N.Y., 1948.
2) Imris I., Trans. Inst. Min. Metall., 1982, 91, C153.
3) Wang S.S., Kurtis A.J., and Toguri J.M., Canadian Met. Quart., 1973, 12, No. 4, p383
4) Wang S.S., Santander N.H., and Toguri J.M., Met. Trans., 1974, 5, Jan., p261
5) Katyal A., 'The Physical Chemistry of Cobalt in Slags and Speiss phases'. Ph.D. Thesis, London University, 1983
6) Jahanshahi S., and Jeffes J.H.E., Trans. Inst. Min. Met., 1981, 90, C138
7) Taylor J.R., and Jeffes J.H.E., Trans., Inst. Min. Met., 84, 1975, C136-147
8) Distin P.A., Hallet G.D., and Richardson F.D., J.I.S.I., 1968, 206, p821-833
9) Taylor J.R., "The activities of Cu_2O and NiO in silicate slags". Ph.D. Thesis, University of London, 1973
10) "Methods for Chemical Analysis of Metals", p328, Published by ASTM, 1953
11) Taskinen P., Zeitschrift fur Metallkunde, 1982, 73 (7), 445
12) Stull D.R. and Prophet H., JANAF Thermodynamics Tables, NS RDS NB537, U.S. Department of Commerce, Washington D.C., 1971
13) Richardson F.D., and Billington J.C., Trans. Inst. Min. Met., 1955-6, 65, 7
14) Toguri J.M. and Santander, N.H., Can. Met. Quart., 1969, 8, 167
15) Toguri J.M. and Santander, N.H., Met. Trans., 1972, 3, 586
16) Barin I., and Knacke O., "Thermochemical properties of inorganic substances", Springer Verlag Berlin + New York, 1973 and Supplement 1977
17) Hultgren R., et al. "Selected values of thermodyanmic properties of the elements", American Society for Metals, 1973
18) Altman R., and Kellogg H.H., Trans. Inst. Min. Met. 1972, 81, C163
19) Mah A.D., et al, Report of investigation, U.S. Bureau of Mines No 7026, 1967, p20
20) Jeffes J.H.E., Proceedings of Terkel Rosenqvist Symposium, Trondheim, 1988, The Norwegian Institute of Technology, Trondheim, p225
21) Taylor J., and Jeffes J.H.E., Trans. Inst. Min. Met., 1975, 84, C18
22) Elliot B.J., See J.B., and Rankin W.J., Trans. Inst. Min. Met. 1978, 87, C204
23) Altman R., Trans. Inst. Min. Met., 1978, 87, C23
24) Turkdogan E.T., 'Physicochemical properties of molten slags and glasses" Metals Society, London, 1983, p116
25) Yazawa A., "Extractive Metallurgical Chemistry with special reference to copper smelting", 28 Congress of IUPAC, Vancouver, 1981
26) Jeffes J.H.E. and Jacob K.T., "The physical chemistry of copper fire refining", Symposium on Metallurgical Chemistry - National Physical Laboratory, HMSO, 1971.

TABLE 1

S. No	Type of Slag	Temp °C	$-\log p_{O_2}$	a_{Cu} x10	a_{Co} x10	a_{Fe} $x10^2$	N $CuO_{0.5}$ $x10^2$	N CoO $x10^2$	N FeO $x10^2$	N SiO_2 $x10^2$	N CaO	% Cu Slag	a $CuO_{0.5}$ $x10^2$	% Co Slag	a CoO $x10^2$	% Fe Slag	a FeO	γ $CuO_{0.5}$	γ CoO
1	Slag	1310	9.0582	9.48	5.09	7.60	1.26	9.63	63.46	25.64		1.16	3.90	8.26	8.26	51.39	0.4836	3.10	0.86
2	'A'	1360	8.4920	9.70	2.84	6.90	1.83	5.63	65.06	27.48		1.70	4.77	4.87	5.48	53.24	0.4781	2.61	0.97
3	I	1380	8.2750	9.73	2.62	5.09	1.62	5.93	66.69	25.75		1.49	5.14	5.07	5.44	53.87	0.3634	3.17	0.92
4	R	1255	9.7238	9.74	3.37	8.41	1.01	5.44	66.05	27.50		0.92	3.22	4.63	4.74	53.12	0.4843	3.18	0.87
5	O	1270	9.5376	9.76	2.66	8.92	1.43	4.46	66.41	27.69		1.32	3.42	3.83	3.90	53.87	0.5265	2.39	0.88
6	N	1260	9.6614	9.76	3.49	8.85	1.17	5.33	66.62	26.87		1.09	3.28	4.61	4.98	54.46	0.5127	2.80	0.93
7		1342	8.1117	9.69	3.04	2.43	2.30	11.83	62.46	23.41		2.13	6.24	10.19	10.60	50.86	0.3175	2.71	0.90
8		1368	7.8244	9.39	4.22	3.92	2.34	18.49	56.37	22.79		2.10	6.75	15.46	15.79	44.55	0.5346	2.88	0.86
9	S	1305	8.5366	9.75	2.62	3.32	2.50	8.61	62.05	26.84		2.28	5.36	7.38	8.82	50.23	0.4069	2.14	0.96
10	I	1325	8.3044	9.59	3.56	2.55	2.17	15.26	55.18	27.28		1.98	5.84	12.92	11.89	44.18	0.3242	2.68	0.78
11	L	1270	8.9574	9.78	2.50	3.75	2.26	6.43	64.94	26.37		2.04	4.77	5.40	7.16	51.58	0.4318	2.11	1.11
12	I	1315	8.4198	9.65	4.28	3.48	2.07	16.95	59.43	21.54		1.84	5.62	14.06	13.91	46.61	0.4349	2.72	0.82
13	C	1395	7.5356	-	3.68	2.80	2.69	15.40	56.00	25.91		2.47	7.45	13.14	14.74	45.83	0.3994	2.77	0.96
14	A	1325	7.9584	9.74	3.82	1.81	2.85	20.40	51.98	24.73		2.61	7.18	17.37	19.01	41.77	0.3430	2.52	0.93
15	T	1335	7.8445	9.60	3.98	1.60	2.88	22.32	49.81	24.99		2.65	7.35	19.06	20.35	40.21	0.3087	2.55	0.91
16	E	1370	7.4566	9.67	3.45	1.76	4.91	23.97	49.34	21.77		4.38	8.36	19.78	19.33	38.61	0.3601	1.70	0.81
17		1285	8.4287	9.59	3.92	2.33	2.80	20.46	51.06	25.67		2.59	6.12	17.64	17.46	41.59	0.4109	2.19	0.85
18		1355	7.6208	9.65	3.33	2.34	3.23	20.84	48.97	26.95		2.92	7.96	17.51	17.95	38.90	0.4666	2.46	0.86
1	Slag	1350	8.6024	9.45	4.62	6.16	1.52	8.58	53.05	36.84		1.43	4.51	7.50	8.46	43.80	0.4188	2.96	0.99
2	'B'	1378	8.2966	9.33	4.36	5.79	1.56	9.24	52.33	36.86		1.42	4.10	7.80	8.58	41.79	0.4254	2.62	0.93
3		1320	8.9422	9.57	4.78	8.21	1.22	8.56	53.71	36.51		1.12	4.04	7.37	7.79	43.69	0.5302	3.31	0.91
4	I	1308	9.8016	-	4.77	7.16	1.55	8.27	52.75	37.42		1.42	3.88	7.04	7.80	42.47	0.4526	2.50	0.94
5	R	1345	8.0782	9.54	4.07	2.45	2.58	16.85	47.38	33.20		2.39	6.22	14.49	14.34	38.52	0.3221	2.41	0.85
6	O	1370	7.8028	9.63	3.47	1.68	2.81	13.22	47.56	36.44		2.62	6.06	11.48	13.05		0.2300	2.45	0.99
7	N S	1325	8.3046	9.56	3.86	2.95	2.07	14.59	52.22	31.12		1.89	5.82	12.38	12.89	41.89	0.3748	2.81	0.88
8	I	1320	8.3620	9.57	3.17	3.78	2.62	11.51	51.65	34.21		2.41	5.76	9.83	10.45	41.71	0.4760	2.20	0.91
9	L	1325	8.3046	9.78	2.43	2.66	2.38	7.76	52.81	37.04		2.16	5.90	6.54	8.12	42.03	0.3376	2.48	1.05
10	I	1285	8.7750	9.67	3.53	3.47	2.31	10.82	54.01	32.86		2.11	5.04	9.16	10.55	43.24	0.4109	2.18	0.98
11	C	1280	8.8354	9.70	2.49	3.29	1.79	7.42	57.46	33.32		1.68	4.98	6.47	7.34	47.32	0.3854	2.78	0.99
12	A	1330	7.9012	9.50	3.12	-	2.93	22.27	44.92	29.88		2.71	7.29	19.13	15.74	36.44	-	2.49	0.71
13	T	1370	7.4566	9.70	3.41	1.61	3.22	19.68	47.85	29.24		2.97	8.38	16.81	19.11	-	0.3392	2.60	0.97
14	E	1320	8.0159	9.65	3.67	2.07	2.61	19.99	46.09	31.40		2.41	6.99	17.08	18.02	37.41	0.4903	2.68	0.90
1	Slag	1290	9.2160	9.60	4.46	8.26	1.69	7.06	66.83		24.43	1.55	3.81	6.00	7.06	53.73	0.5543	2.26	1.08
2	'E'	1330	8.7487	9.49	3.99	9.29	1.89	6.39	66.95		24.77	1.74	4.44	5.47	7.59	54.16	0.6685	2.35	1.19
3	C	1300	9.0970	9.68	3.76	9.28	2.31	6.07	68.82		22.79	2.12	3.98	5.17	6.59	55.47	0.6335	1.72	1.09
4	A	1270	9.4588	9.51	4.13	9.15	1.68	5.82	67.62		24.88	1.53	3.55	4.92	6.64	53.99	0.5914	2.11	1.14
5	L	1250	9.7080	9.55	4.22	10.01	1.79	5.51	68.54		24.15	1.63	3.27	4.67	6.40	54.84	0.6231	1.82	1.16
6	C F	1350	8.5236	9.61	3.35	6.14	2.05	5.52	68.46		23.97	1.86	4.81	4.66	6.72	54.61	0.4574	2.34	1.21
7	I E	1315	8.0201	9.71	3.05	3.09	2.57	13.76	63.39		20.28	2.34	7.13	11.67	15.70	50.80	0.6106	2.77	1.14
8	U R	1255	8.7440	9.79	3.62	2.81	2.36	15.11	63.18		19.36	2.10	5.63	12.48	15.70	49.32	0.4991	2.39	1.04
9	M R	1275	8.4964	9.83	3.18	2.95	2.79	12.11	65.44		19.67	2.45	6.11	9.88	14.63	50.52	0.5425	2.19	1.21
10	I	1355	7.5672	-	3.30	2.32	3.16	15.45	63.11		18.24	2.83	8.24	12.85	18.91	49.61	0.4921	2.61	1.22
11	T	1360	7.1736	9.73	2.54	1.29	4.09	19.72	58.21		17.98	3.63	10.62	16.28	21.76	45.42	0.4074	2.60	1.10
12	E	1300	7.8574	9.78	3.18	2.01	3.12	24.35	55.03		17.50	2.80	8.18	20.36	23.22	43.50	0.5712	2.62	0.96

TABLE 1 (continued)

S. No.	Type of Slag	Temp °C	$-\log p_{O_2}$	a_{Cu} x10	a_{Co} x10	a_{Fe} x10²	N $CuO_{0.5}$ x 10	N CoO x10²	N FeO x10²	N SiO_2 x10²	N CaO x10²	% Cu Slag	a $CuO_{0.5}$	% Co Slag	a CoO x10²	% Fe Slag	a FeO	γ $CuO_{0.5}$	γ CoO
1	Slag	1305	9.1614	9.29	4.52	10.06	4.94	3.09	23.88	36.80	35.74	0.51	3.75	2.94	6.97	21.50	0.6016	7.60	2.25
2	'C'	1355	8.5916	9.61	3.44	8.68	6.10	2.97	26.48	35.49	34.45	0.62	4.54	2.81	6.06	23.71	0.5659	7.45	2.05
3		1310	9.1029	9.49	4.41	9.98	5.54	3.18	23.04	37.15	36.07	0.57	3.83	3.04	6.89	20.77	0.6020	6.92	2.16
4	CaO	1340	8.7590	9.55	4.03	9.23	5.25	3.03	27.98	34.74	33.72	0.53	4.29	2.86	6.83	24.96	0.5862	8.17	2.25
5	FeO	1295	9.2797	9.60	4.37	8.46	5.21	3.42	24.68	36.21	35.16	0.53	3.62	3.25	6.54	22.16	0.4968	6.95	1.91
6	SiO_2	1295	9.2797	9.59	3.83	10.48	4.91	3.12	28.81	34.29	33.29	0.50	3.65	2.93	5.73	25.65	0.6156	7.43	1.84
7		1365	7.5827	9.84	2.88	-	18.20	7.33	30.44	30.65	29.76	1.81	7.96	6.78	14.68	26.63	-	4.37	2.00
8		1320	8.0875	9.56	3.67	3.74	11.03	7.54	29.63	31.32	30.41	1.10	6.73	6.99	16.61	25.98	0.6469	6.10	2.20
9		1375	7.3758	9.64	3.73	2.42	15.07	9.66	30.59	29.55	28.68	1.49	8.16	8.89	19.50	26.60	0.4584	5.42	2.02
10		1330	8.5916	9.65	4.20	4.31	14.24	11.43	30.65	28.66	27.84	1.40	6.93	10.46	19.52	26.52	0.7569	4.86	1.71
11		1260	8.8068	9.48	4.91	3.42	9.59	10.20	29.91	29.89	29.04	0.95	5.29	9.39	18.74	26.03	0.5300	5.51	1.84
12		1325	8.0300	9.52	4.36	2.93	10.65	10.72	32.29	28.37	27.55	1.05	6.77	9.80	19.99	27.93	0.5120	6.36	1.87
13		1280	8.5608	9.57	4.87	3.44	17.90	10.44	30.68	28.97	28.12	1.77	5.72	9.57	19.67	26.60	0.5537	3.20	1.88
14		1295	8.3805	9.66	4.43	1.92	9.65	10.10	28.57	30.62	29.74	0.96	6.07	9.33	18.71	24.98	0.3170	6.29	1.85
15		1315	8.1454	9.43	4.67	2.49	13.33	10.93	27.95	30.33	29.46	1.32	6.51	10.08	20.84	24.37	0.4259	4.89	1.91
16	Slag	1375	8.2865	9.46	4.75	5.90	6.42	4.65	22.14	39.58	32.98	0.66	5.03	4.42	9.74	19.88	0.4389	7.83	2.09
17	'D'	1360	8.4496	9.62	4.19	7.20	5.49	3.55	24.94	38.70	32.26	0.56	4.80	3.36	8.28	22.34	0.5223	8.74	2.33
18		1310	9.0159	9.76	3.92	8.53	5.20	2.90	22.94	40.16	33.47	0.53	4.06	2.76	6.77	20.67	0.5687	7.81	2.33
19	CaO	1330	8.7851	9.81	3.01	8.29	5.47	2.87	24.38	39.38	32.83	0.56	4.38	2.72	5.48	21.89	0.5721	8.01	1.91
20	FeO	1310	9.0159	9.60	3.48	7.04	4.85	2.97	22.98	40.12	33.44	0.50	4.06	2.83	5.99	20.70	0.4690	8.37	2.01
21	SiO_2	1335	8.7283	9.43	4.90	7.46	5.45	4.73	20.65	40.41	33.66	0.56	4.37	4.51	9.06	18.62	0.5086	8.00	1.92
22		1375	7.3461	-	4.81	2.08	26.08	20.20	25.37	28.27	23.56	2.53	8.63	18.21	29.14	21.60	0.4561	3.31	1.44
23		1365	7.4546	9.65	4.24	2.61	13.79	14.66	26.22	31.50	26.24	1.36	8.42	13.42	25.03	22.68	0.5637	6.10	1.71
24		1325	7.9019	9.53	4.34	2.45	11.81	14.37	26.59	31.56	26.27	1.16	7.30	13.16	23.03	23.02	0.4940	6.18	1.60
25		1300	8.1931	9.78	3.38	4.32	10.61	9.97	27.29	33.64	28.04	1.06	6.73	9.23	16.76	23.87	-	6.34	1.68
26		1345	7.6755	9.54	4.40	2.26	12.24	15.29	24.97	31.92	26.60	1.21	7.81	14.02	24.63	21.65	0.4720	6.28	1.61
27		1350	7.6198	9.53	4.23	2.72	13.83	15.78	26.25	30.87	25.72	1.36	7.96	14.40	24.00	22.66	0.5738	5.75	1.52
1	Slag	1420	7.7334	9.02	5.59	4.93	12.80	12.81	34.63	42.72	8.78	1.22	5.80	11.57	14.03	29.68	0.4325	4.52	1.10
2	'F'	1370	8.2595	9.13	5.77	6.53	10.40	10.20	35.22	41.44	12.10	0.98	5.00	8.93	12.80	29.28	0.5291	4.82	1.26
3	FeO/SiO_2	1430	7.6319	8.79	6.39	5.88	13.80	13.99	32.20	43.61	8.81	1.31	5.86	12.35	16.45	26.95	0.5238	4.23	1.18
4	1.2	1320	8.8187	-	6.10	4.67	13.90	18.55	29.37	41.32	9.36	1.32	4.19	16.39	11.86	24.24	0.3478	3.01	0.64
5	CaO 10%	1320	8.8187	9.40	5.85	3.58	10.40	18.52	34.83	41.46	4.15	1.00	4.22	16.60	11.37	29.63	0.2666	4.05	0.62
1	Slag 'G'	1370	8.2595	9.13	5.57	5.02	10.10	14.31	31.00	34.43	19.24	0.98	5.05	12.92	12.35	26.55	0.4068	5.03	0.87
2	FeO/SiO_2	1330	8.7041	8.98	6.76	6.56	10.30	12.22	29.70	35.90	21.15	1.00	4.28	11.05	13.50	25.49	0.4971	4.16	1.10
3	1.2	1350	8.5336	9.58	3.81	6.13	8.90	5.64	32.87	38.19	22.40	0.87	4.71	5.11	7.54	28.24	0.4513	5.27	1.33
4	CaO 20%	1335	8.7018	9.62	3.30	6.55	9.50	4.68	31.14	39.47	23.76	0.93	4.50	4.23	6.27	26.66	0.4701	4.74	1.34
1	Slag 'K'	1380	8.1518	9.32	7.07	6.47	13.70	18.62	31.72	34.87	10.01	1.37	4.84	15.20	16.10	33.56	0.5329	3.55	0.87
2	F/S1.2 CaO	1300	9.0523	9.57	6.26	6.38	11.10	16.95	31.61	35.44	9.46	1.29	3.91	5.57	11.52	38.06	0.4855	3.51	0.68
3	10% Al_2O_3	1370	8.3141	9.81	2.90	7.61	14.20	5.53	32.81	42.20	12.49	1.24	5.13	2.52	6.04	38.39	0.5509	3.63	1.10
4	10%	1340	8.6454	9.48	3.30		12.90	5.83	31.90	42.42	13.11	1.08	4.59	6.49	6.35	36.22		3.57	1.09
1	Slag 'H'	1360	8.3686	9.93	5.92	6.06	14.50	17.40	40.50	29.48	10.79	1.55	4.85	0.00	12.79	38.33	0.4830	3.33	0.74
2	F/S 1.85	1350	8.5336	8.44	3.47	6.36	13.50	6.27	45.10	35.76	11.52	1.33	4.76	14.28	6.86	30.16	0.4682	5.53	1.10
3	CaO 10%	1360	8.4232	9.02	1.29	6.21	13.30	2.88	46.36	35.99	13.43	1.51	5.03	12.62	2.63	31.85	0.4648	3.80	0.91
4		1365	8.3685	8.99	3.79	5.49	11.40	7.41	43.57	35.68	12.18	1.18	4.99	11.40	7.80	34.56	0.4144	4.38	1.05
1	Slag 'J'	1320	8.9347	8.39	0.00	7.50	15.60	0.00	44.09	32.22	22.13	1.24	4.53	15.76	0.00	25.49	0.5586	2.90	0.00
2	F/S 1.85	1410	7.8362	9.31	6.88	7.56	14.10	16.27	36.20	29.94	16.17	1.02	5.36	14.27	16.86	25.22	0.6529	3.80	1.04
3	CaO 20%	1360	8.3686	-	6.29	5.43	16.00	14.43	38.38	28.05	17.53	1.29	4.80	4.68	13.59	26.32	0.4328	2.99	0.94
4		1320	8.8187	9.55	6.40	6.36	12.70	13.22	42.26	25.89	17.34	1.17	4.17	4.94	12.42	25.65	0.4737	3.27	0.94

TABLE 2

Activity coefficients of CuO(0.5) and CoO in dilute solution in various slags

No	FeO/SiO_2 w/w ratio	Initial Slag Compositions (w %) FeO	SiO_2	CaO	Al_2O_3	Fe_2O_3	γ°_{CoO}	$\gamma^{\circ}_{CuO_{0.5}}$
A	3.0	75	25	-	-	4.52	0.91	2.89
B	1.85	65	35	-	-	1.35	1.00	2.89
C	1.20	39	32	29	-	1.92	2.08	7.81
D	1.0	35	36	28	-	NA (1.9)	2.08	7.81
E	-	79	-	19	-	14.4	1.16	2.32
F	1.20	49	41	10	-	2.3	1.49	4.03
G	1.20	43.6	36.3	20	-	2.69	1.29	4.54^x
H	1.85	58.5	31.5	10	-	(2.3)	1.05	3.94
J	1.85	51.9	28.1	20	-	(2.7)	1.00	3.04^x
K	1.20	43.6	36.3	10	10	(2.3)	1.14	3.62

FeO 71.85

SiO_2 80.09

TABLE 3

ΔH_{298} for orthosilicates and ferrites/mole basic oxide (Kcal)

	SiO_2	Fe_2O_3
NiO	-1.8	-4.1
$CuO_{0.5}$	N/A	-8.7
FeO	-4.3	-5.3
CoO	-4.8	-5.9
CaO	-16.8	-4.8

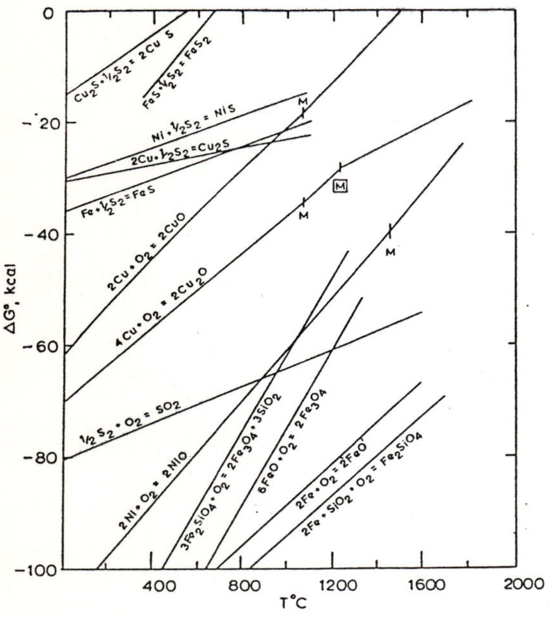

1) Standard Gibbs free energies of formation of oxides and sulphides of iron, copper, nickel and cobalt

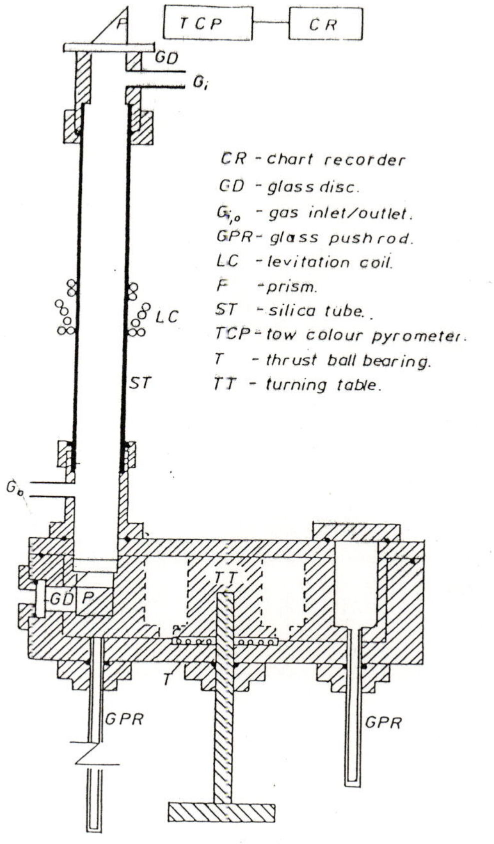

2) The levitation apparatus

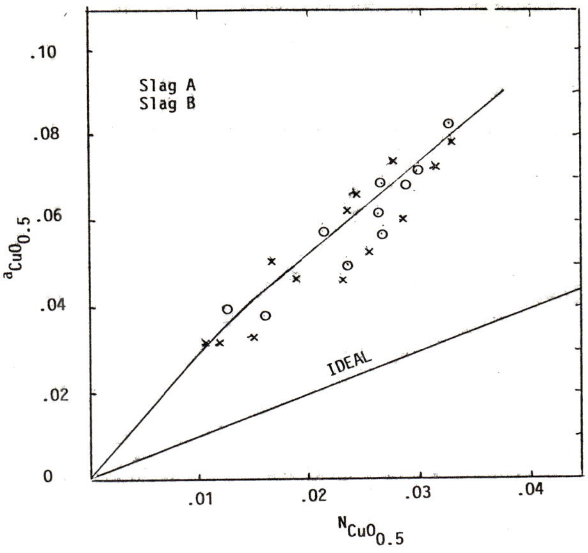

3) Activity of $CuO_{0.5}$ in slags A and B (Iron silicate)

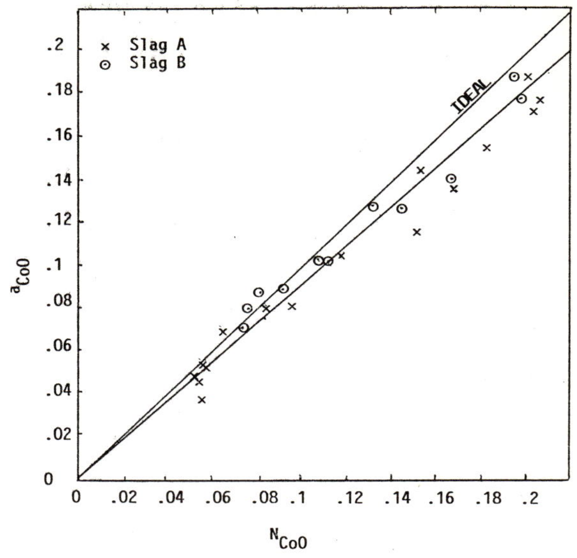

4) Activity of CoO in slags A and B (Iron silicate)

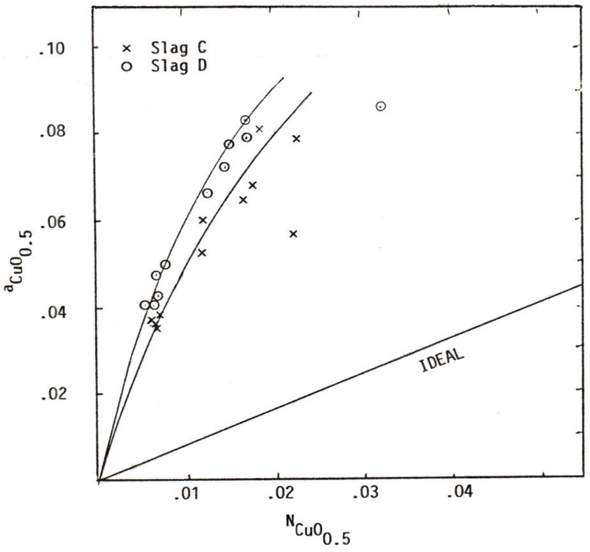

5) Activity of $CuO_{0.5}$ in slags C and D (Lime iron silicate)

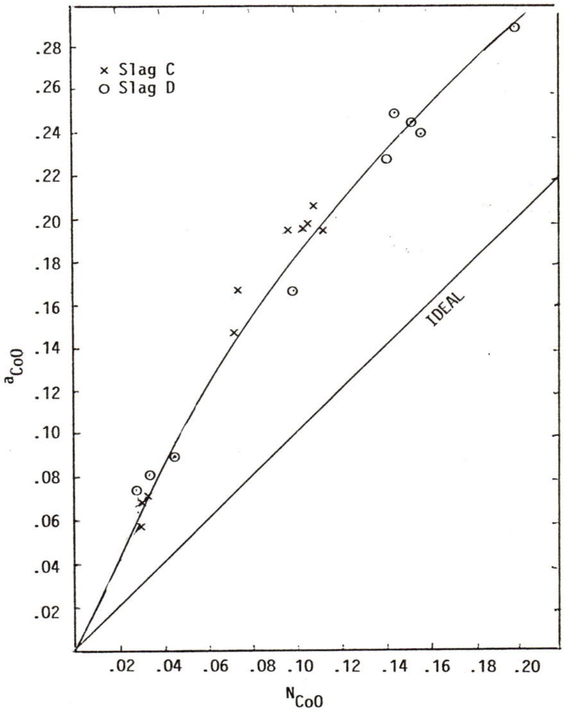

6) Activity of CoO in slags C and D (Lime iron silicate)

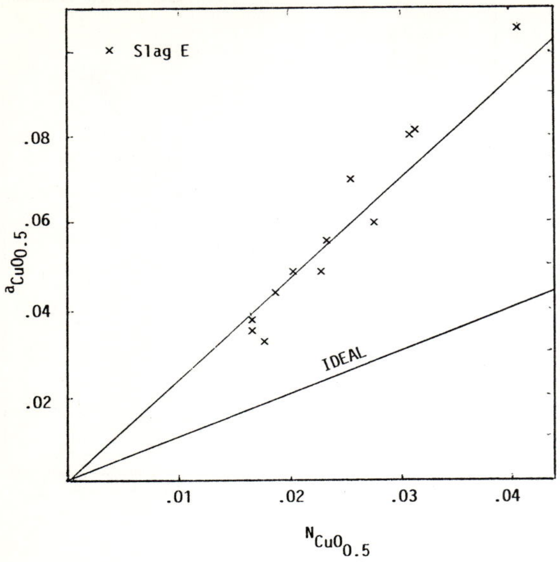

7) Activity of $CuO_{0.5}$ in slag E (Calcium ferrite)

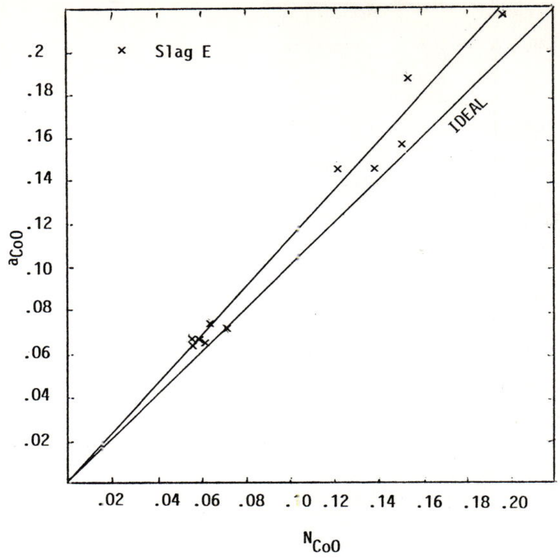

8) Activity of CoO in slag E (Calcium ferrite)

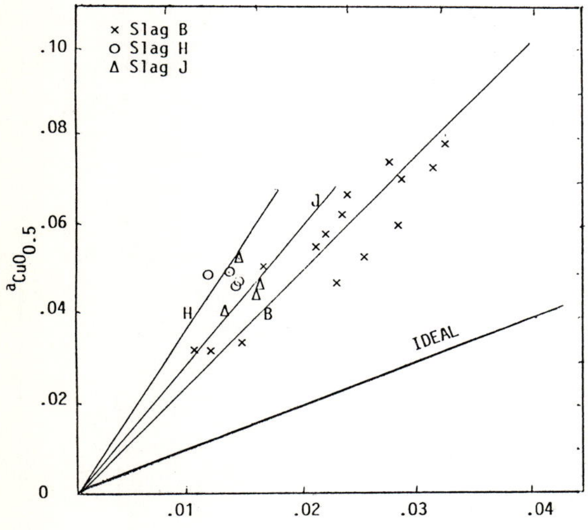

9) Activity of $CuO_{0.5}$ in slags B, H and J (Varying CaO contents)

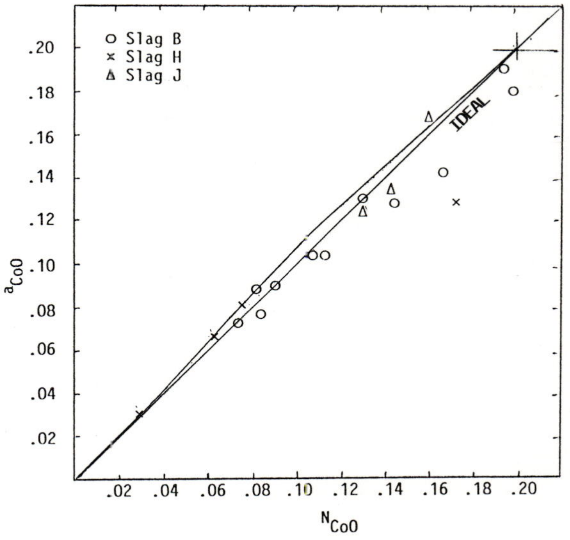

10) Activity of CoO in slags B, H and J (Varying CaO contents)

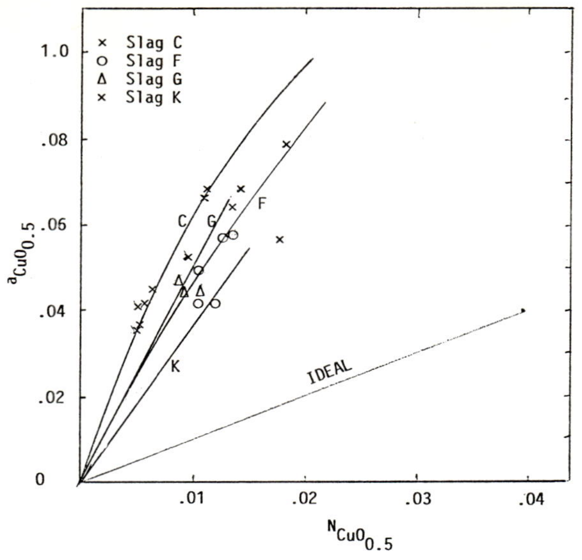

11) Activity of $CuO_{0.5}$ in slags C, F, G, K (Varying CaO and alumina contents)

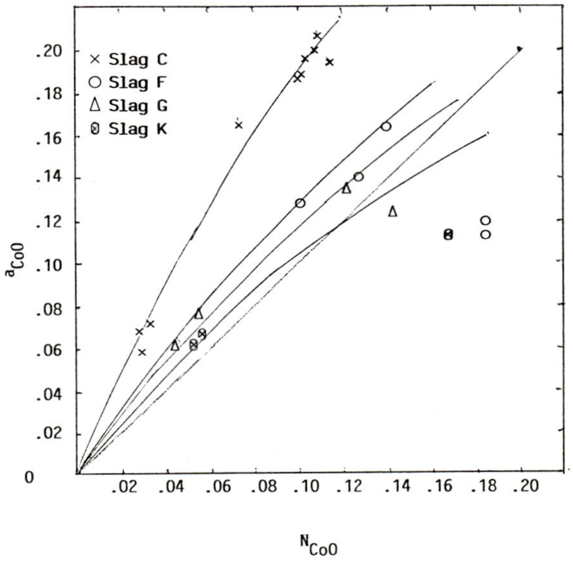

12) Activity of CoO in slags C, F, G, K (Varying CaO and alumina contents)

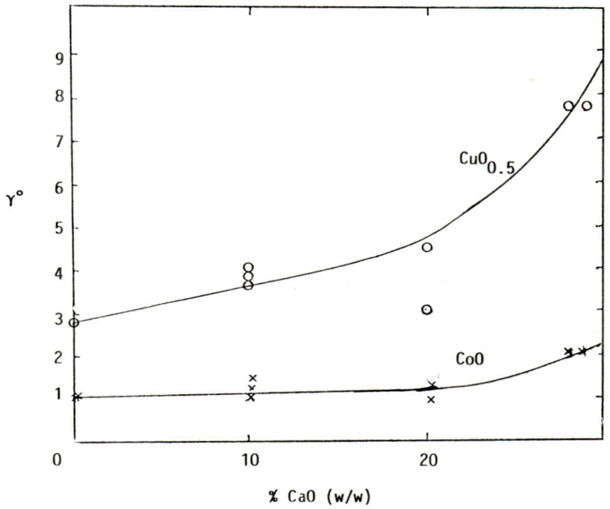

13) Effect of CaO content of slags on activity coefficients of $CuO_{0.5}$ and CoO

Distribution equilibria of Sb between Na_2CO_3 melt and molten copper - effects of partial pressure of CO_2 and activity of Na_2O

T FUJISAWA, CH YAMAUCHI and H SAKAO

Dr Fujisawa and Prof Yamauchi are in the Department of Metallurgy, Faculty of Engineering at Nagoya University, Nagoya, Japan; Dr Sakao, formerly Professor in the Department of Metallurgy, is now Professor Emeritus at Nagoya University.

SYNOPSIS

Distribution equilibria of antimony between sodium carbonate slag and molten copper was determined at 1523K as functions of partial pressures of oxygen and CO_2. Activities of Na_2O and Na_2CO_3 in the Na_2O-CO_2-Sb_2O_5 system were measured by EMF method for various partial pressures of CO_2 at 1423 to 1523K. Solubility of CO_2 in the melt was also measured at 1523K. The distribution ratio of antimony between the slag and molten copper was discussed in terms of the slag composition and partial pressures of oxygen and CO_2.

INTRODUCTION

There has been a growing emphasis on using special slags in the fire-refining of crude copper. At the same time, the interest is also increased in the production of ultra high pure copper low in residual elements. It was known that the soda slag treatment of molten copper is an effective method for removing Sb and As[1)-6)].

The purpose of the present work is to determine the thermodynamics of the removal of Sb from molten copper by using sodium carbonate slag.

THERMODYNAMIC CONSIDERATIONS

The distribution reaction of Sb between slag and metallic copper may be represented as[1)5)]

$$Sb + \nu/2O_2 = SbO_\nu \tag{1}$$

However, one of the aims of this work is to examine the effect of the partial pressure of CO_2 and hence the activity of Na_2O in slags on the distribution ratio of Sb. In this case, it is more convenient to use an equation which contains the term of the activity of Na_2O. Such an equation may be expressed as

$$Sb + \eta/2Na_2O + \theta/4O_2 = Na_\eta SbO_m \tag{2}$$

$$m = (\eta + \theta)/2 \tag{3}$$

and the distribution ratio of Sb defined by (%Sb in slag)/[%Sb in copper] can be represented by

$$L_{Sb} = \frac{(\%Sb)}{[\%Sb]} = \frac{K \cdot M_{Cu} \cdot n_T \cdot \gamma_{Sb}}{100 \cdot \gamma_{Na_\eta SbO_m}} \cdot a_{Na_2O}^{\eta/2} \cdot P_{O_2}^{\theta/4} \tag{4}$$

where K is the equilibrium constant of Eq.(2), M_{Cu} the molar mass of copper, n_T the molar amount of 100 g slag, a and γ are, respectively, Raoultian activity and activity coefficient, and θ corresponds to the valence of Sb.

Equation (4) indicates that the distribution ratio is affected by the activity of Na_2O in the refining slag and the partial pressure of oxygen, provided that the activity coefficients of Sb and $Na_\eta SbO_m$ are kept constant. In addition, under a condition of constant activity of Na_2O, the plot of log L_{Sb} against log P_{O_2} could give a linear relationship with a slope of $\theta/4$ which suggests the dissolved species of Sb in the slag.

EXPERIMENT

The distribution equilibria of Sb between pure Na_2CO_3 slag and metallic copper containing various amounts of oxygen was measured at 1523K. CO_2-CO-Ar gas mixture was blown into the melt to control partial pressures of CO_2 and O_2 in the system. The Sb content in the Na_2CO_3 slag after the equilibration was always adjusted to rd. 1wt%. After the experiments slag and metallic copper were quenched and analysed. $3Na_2O$-Sb_2O_5 premelted slag was also used for the experiment.

Activities of Na_2O and Na_2CO_3 in Na_2O-CO_2-Sb_2O_5 melt were determined by EMF method, using beta"-alumina as a solid electrolyte, for various partial pressures of CO_2 at 1423 to 1523K. The electrochemical cell used in the present work may be represented as

Pt,O_2(g,P_{O_2}*),Na(g,P_{Na}*) (Na_2O-SiO_2-Fe_2O_3 ref.melt) | Na^+ beta"-Al_2O_3 | Na(g,P_{Na}),CO_2(g,P_{CO_2}),O_2(g,P_{O_2}),Pt (Na_2O-CO_2-Sb_2O_5 sample melt)

The reversible EMF is related to the activities of Na_2O and Na_2CO_3. The activity of Na_2O in the reference melt was measured in advance by using pure Na_2CO_3 as a sample melt.

The solubility of CO_2 in Na_2O-Sb_2O_5 melt was measured by equilibrating samples with CO_2-Ar gas mixture at 1523K. After the experiment, the CO_2

as well as other components in samples were chemically analysed.

RESULTS AND DISCUSSION

The results of distribution experiment with pure Na_2CO_3 are shown in Fig.1 where log L_{Sb} are plotted against log P_{O_2} for various CO_2 pressures. The value of P_{O_2} was calculated from the analysed oxygen content in the metallic copper by using appropriate thermodynamic data. The partial pressure of CO_2 has great influence on the distribution ratio, and Sb can be easily removed from oxygen bearing copper, especially at low partial pressure of CO_2. The slopes of the lines connecting the experimental points in the figure are near to 5/4, suggesting that the pentavalent oxide of Sb is predominant in the Na_2CO_3 slag under the present experimental conditions.

The results of the solubility measurements of CO_2 in Na_2O-Sb_2O_5 melts imply that Na_2CO_3 slag is decomposed according to the stoichiometric reaction corresponding to the formation of Na_3SbO_4 species in the slag.

Therefore, the distribution equilibria can be represented by the following reaction:

$$Sb + 3/2Na_2CO_3 + 5/4O_2 = Na_3SbO_4 + 3/2CO_2 \quad (5)$$

or

$$Sb + 3/2Na_2O + 5/4O_2 = Na_3SbO_4 \quad (2.1)$$

The activity of Na_2O in the Na_2O-CO_2-Sb_2O_5 system at 1523K is shown in Fig.2, in which the slag composition is represented by $N = n_{Na_2O}/(n_{Na_2O}+n_{Sb_2O_5})$ where n is the number of moles. Since in the distribution experiment the Sb content in the Na_2CO_3 slag after the equilibration was always adjusted to rd. 1wt% ($N \fallingdotseq 0.996$), the corresponding values of a_{Na_2O} under various partial pressures of CO_2 were substituted into Eq.(4). Then the relationship between log L_{Sb} and log P_{O_2} for various P_{CO_2} can be well represented as shown by straight lines in Fig.1.

The activity of Sb_2O_5 in Na_2O-CO_2-Sb_2O_5 melt may be calculated from the activity of Na_2O by integrating the Gibbs-Duhem relation. Under the condition of constant P_{CO_2}, the Gibbs-Duhem relation for the Na_2O-CO_2-Sb_2O_5 system is given as:

$$N \, d\ln a_{Na_2O} + (1-N) \, d\ln a_{Sb_2O_5} = 0 \quad (6)$$

The activity of Sb_2O_5 at N=0.75 was determined from the distribution experiment of Sb between $3Na_2O$-Sb_2O_5 slag and metallic copper, and used as the starting point of the integration. The calculated activity of Sb_2O_5 is shown in Fig.3 for three different partial pressures of CO_2. The partial pressure of CO_2 has great influence on the activities of the components in the Na_2O-CO_2-Sb_2O_5 system. The existence of a liquid state miscibility gap was suggested at the Na_2CO_3-rich composition region in the system.

From Fig.3, the distribution ratio L_{Sb} at various slag compositions can be estimated by using the following reaction equilibria[7]:

$$2Sb(l) + 5/2O_2(g) = Sb_2O_5(l) \quad (7)$$

$$: \Delta G°/J = -924,775 + 430.75T$$

The L_{Sb} estimated are plotted in Fig.4 against the slag composition N by using P_{O_2} as a parameter for P_{CO_2}=0.08 MPa at 1523K.

CONCLUSIONS

The thermodynamics of the removal of Sb from molten copper by using sodium carbonate slag were investigated. The distribution equilibria can be represented by the following reaction: $Sb + 3/2Na_2O + 5/4O_2 = Na_3SbO_4$. The higher P_{O_2} and the lower P_{CO_2} are preferable for the removal of Sb. With increasing the concentration of antimony oxide in the slag, the distribution ratio passes through a maximum value. It is found that this is the most effective slag composition for removing antimony.

REFERENCES

1) A.Yazawa: Proceedings of the Second International Symposium on Metallurgical Slags and Fluxes, Lake Tahoe, Nevada, U.S.A., (1984), TMS-AIME, p.701.
2) P.Taskinen: Scand. J. Metall., 11(1982), p.150.
3) I.V.Kojo, P.Taskinen and K.Lilius: Erzmetall, 37(1984), p.21.
4) I.V.Kojo: Acta Polytechnica Scandinavica, Chemical Technology and Metallurgy Series No.161, (1985), p.1.
5) G.Riveros, Y.J.Park, Y.Takeda and A.Yazawa: J. Min. Metall. Inst. Japan, 102(1986), p.415.
6) Ch.Yamauchi, K.Ohtsuki, T.Fujisawa and H.Sakao: J. Japan Inst. Metals, 52(1988), p.561.
7) Kohza-Gendai-no-Kinzoku-Gaku, Seiren-Hen 2, Hitetsu-Kinzoku-Seiren, ed. by Japan Inst. Metals, (1980), p.320.

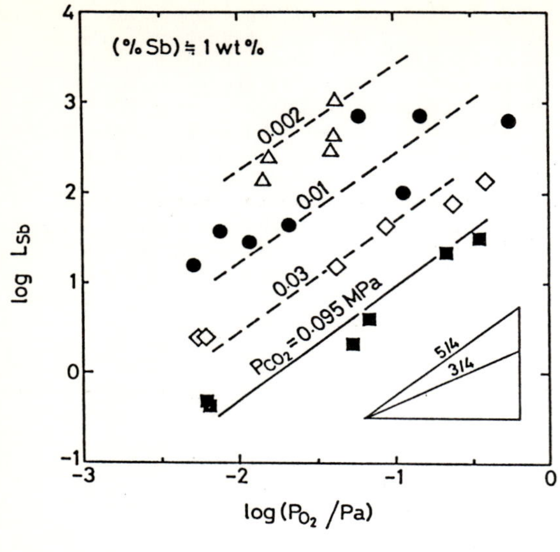

Fig.1 Relationship between log L_{Sb} and log P_{O_2} for various partial pressures of CO_2 at 1523K.

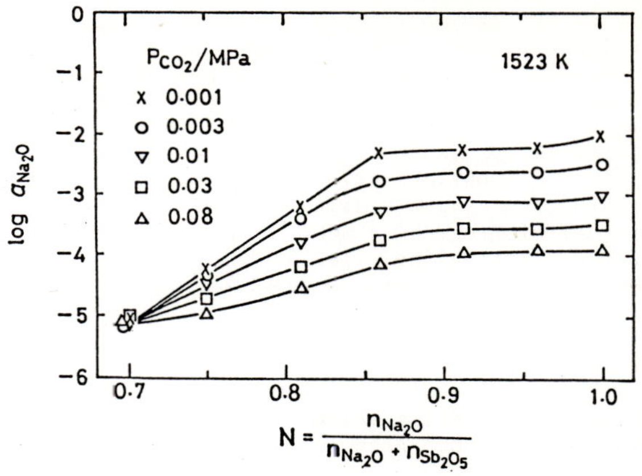

Fig.2 Activity of Na_2O in the Na_2O-CO_2-Sb_2O_5 system at 1523K.

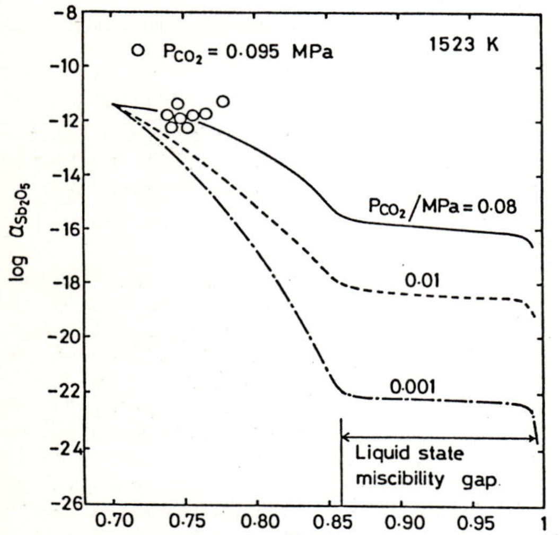

Fig.3 Activity of Sb_2O_5 in the Na_2O-CO_2-Sb_2O_5 system at 1523K.

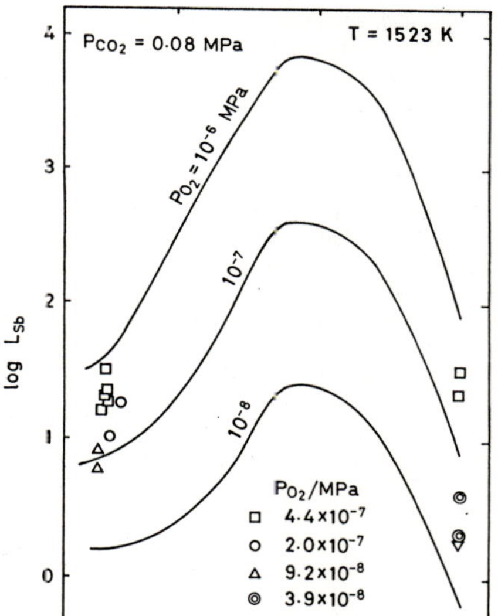

Fig.4 Relationship between estimated distribution ratio of Sb and slag composition N for P_{CO_2}=0.08 MPa at 1523K.

Thermodynamics of P, S and Na_2O in CaO based slags containing Na_2O

R J FRUEHAN, J J PAK and A H CHAN

RJF: Carnegie Mellon University, MEMS Department
JJP: Oak Ridge National Laboratory (formerly at Carnegie Mellon University)
AHC: Linde Division-Union Carbide (formerly at Carnegie Mellon University).

SYNOPSIS

The thermodynamics of P, S, and Na_2O in CaO based slags was investigated using a variety of experimental techniques including slag-metal and gas-slag equilibration and electrochemical measurements. The addition of Na_2O to CaO-SiO_2 slags significantly increases the phosphate ($C_{PO_4^{-3}}$) and sulfide (C_S) capacities. For ladle dephosphorization slags (CaO-FeO-SiO_2) the addition of 20% Na_2O increases $C_{PO_4^{-3}}$ by a factor of 25, and, 2% Na_2O in reducing ladle slags (CaO-Al_2O_3-SiO_2) increases $C_{PO_4^{-3}}$ by a factor of 200. These results indicate Na_2SiO_3 slag additions will benefit actual processes. The activity of Na_2O in Na_2O-CaO-SiO_2-(Al_2O_3) slags is increased by CaO; this is one reason why small additions of Na_2O improved desulfurization and dephosphorization significantly.

Introduction

Previous studies by the authors[1,2], and others, have shown that Na_2O based slags have higher sulfide and phosphate capacities than CaO based slags. However the use of Na_2O slags, usually achieved by the addition of Na_2CO_3, or Na_2SiO_3, in iron and steelmaking, has been limited because of excessive cost, vaporization of Na_2O and refractory wear. It has been shown that the addition of Na_2O to CaO based slags increase phosphorus removal from hot metal[3] and steel[4]. In this paper, recent results on the effect of Na_2O on the sulfide and phosphate capacities of CaO based slags, and, on the activity of Na_2O in these slags is presented.

The phosphate and sulfide capacities are defined by equations (1) and (2) respectively.

$$C_{PO_4^{-3}} = \frac{(\%\ PO_4^{-3})}{p_{P_2}\ p_{O_2}^{5/4}} \qquad (1)$$

$$C_S = (\%S)\left(\frac{p_{O_2}}{p_{S_2}}\right)^{1/2} \qquad (2)$$

where p_{P_2}, p_{O_2}, and p_{S_2} are the phosphorus, oxygen and sulfur pressures respectively, and ($\%PO_4^{-3}$) and (%S) are the phosphate and sulfide concentrations in the slag. It has been shown that the phosphorus (L_P) and sulfur (L_S) distribution ratio between the slag and iron can be related to the respective capacities if the oxygen pressure in the system is defined.

$$L_P = \frac{(\%P)}{[\%P]} \qquad (3)$$

$$L_S = \frac{(\%S)}{[\%S]} \qquad (4)$$

It is possible to relate L_P and L_S to their respective capacities if the oxygen pressure in the system is known.

$$\log C_S = \log L_S - \log f_S + 1/2 \log p_{O_2} + \log K_S \qquad (5)$$

$$\log C_{PO_4^{-3}} = \log L_P - 5/4 \log p_{O_2} - \log f_P - \log \frac{m_{PO_4^{-3}}}{m_P} + \log K_P \qquad (6)$$

where f_i is the activity coefficient, M_i is the molecular weight of specie i and K_S and K_P are the equilibrium constants for reactions (7) and (8) respectively

$$1/2\ S_2 = \underline{S}\ (1\ wt\ \%\ in\ Fe) \qquad (7)$$

$$1/2\ P_2 + \underline{P}\ (1\ wt\ \%\ in\ Fe) \qquad (8)$$

In most cases the oxygen pressure is controlled by equilibrium between Fe in the metal and FeO in the slag. For the phosphorus distribution ratio measurements for CaO-SiO_2-Na_2O slags C-CO equilibrium was also obtained.

Experimental

Several different experimental techniques were employed depending on the slag and thermodynamic quantity being measured. Details of the experimental techniques are given in recent publications[1,2,5-7]. For CaO-SiO_2-Na_2O blast furnace type slags and CaO-Al_2O_3-SiO_2-Na_2O ladle type slags the sulfur and phosphorus partition ratios between the slag (2 grams) and carbon saturated iron (2 grams) contained in graphite crucibles were measured at 1400 and 1500°C in a CO atmosphere. For CaO saturated slags relevant to ladle dephosphorization (CaO-FeO-SiO_2-Na_2O) the ratio between slag (8 grams) and iron (30 grams) contained in CaO crucibles was measured at 1600°C in Ar. For steelmaking slags, saturated with $2CaO$-SiO_2, the partition ratio between the slag and solid iron foil contained in iron crucibles was measured at 1500°C in Ar.

Sulfide capacity measurements were performed at 1400°C by gas (CO-CO_2-SO_2) equilibration for CaO-SiO_2-Na_2O slags contained in Pt crucibles. Sulfur distribution measurements between slag and carbon saturated iron were also performed. The activity of Na_2O in CaO-SiO_2-Na_2O and CaO-Al_2O_3-SiO_2-Ba_2O slags was measured using an electrochemical technique. The electrolyte was β alumina which contained the reference electrode of a slag of known Na_2O activity. The cell can be represented by

Air, Pt | Na_2O (sample) | β alumina | Na_2O (reference) | Pt, air

Results

Phosphorus: The addition of Na_2O to CaO-SiO_2 slags significantly increases the phosphorus distribution between the slag and carbon saturated iron and the phosphate capacity. In computing the $C_{PO_4^{-3}}$ it was assumed that the oxygen pressure was controlled by C-CO equilibrium, as was found in previous work. As shown in Figure (1) at a constant SiO_2 content of 50%, $C_{PO_4^{-3}}$ and L_P increase approximately linearly with Na_2O content. The substitution of 20% Na_2O for CaO at a constant basicity (B) of one increases the phosphate capacity by about a factor of 100 at 1400°C.

$$B = \frac{\%Na_2O + \%CaO}{\%SiO_2}$$

As shown in Figure (2) for CaO saturated slags containing about 50% CaO - 40% FeO - 8% SiO_2 the addition of 5% Na_2O increases the phosphate capacity by a factor of 25 at 1600°C. For slags saturated with $2CaO$-SiO_2 containing about 45% CaO - 20% FeO - 25% SiO_2 - 10% P_2O_5 the addition of up to 15% Na_2O increases the phosphate capacity by only about a factor of 3. In the calculation it was assumed that the oxygen pressure was controlled by Fe-FeO equilibrium. The addition of Na_2O to reducing ladle slags significantly increases the phosphate capacity. As shown in Figure (3) the addition of only 2% Na_2O increases $C_{PO_4^{-3}}$ by a factor of 200.

The results indicate that the addition of Na_2O to ladle oxidizing dephosphorization slags saturated with CaO would be expected to improve the process. In fact, Na_2SiO_3 additions are commonly used in such processes. The results also indicate that Na_2O additions to reducing ladle slags (CaO-Al_2O_3-SiO_2) may help retard phosphorus reversion. Calculations indicate that for a steel containing 0.01% Al, a 2% Na_2O addition will increase $C_{PO_4^{-3}}$ by over a factor of 200 and will stop rephosphorization in many cases. For a steel with 0.03% Al, 2% Na_2O will significantly retard rephosphorization.

Sulfur: The addition of Na_2O to CaO-SiO_2 slags increased the sulfur partition ratio with carbon saturated iron (L_S) as indicated in Figure (4). By comparing the sulfide capacity from the gas equilibration experiments to that calculated from L_S it was concluded that the oxygen potential in the slag-metal experiments was controlled by Fe-FeO equilibrium. The activity of FeO was calculated using a regular solution model. The sulfide capacity correlated to the optical basicity (Λ) but the correlation was different than that for CaO based slag[8] and is given by

$$\log C_S = 11.86\Lambda - 11.33 \qquad (9)$$

The addition of up to 3% Na_2O to a typical ladle slag, containing about 47% CaO - 42% Al_2O_3 - 7% SiO_2, did not significantly increase the sulfur partition ratio. However, these slags, even without Na_2O, gave very low sulfur levels. (0.003 - 0.005%). Consequently, further improvement in desulfurization may have been difficult to detect.

Activity of Na_2O: In preliminary experiments, the activity of Na_2O in binary Na_2O-SiO_2 was measured and good agreement with previous work[9] was obtained confirming the experimental technique. The results for Na_2O-SiO_2-CaO slags at 1400°C are shown in Figure (5) as a function of CaO content at constant Na_2O/SiO_2 ratios. The activity of Na_2O increases even though it's concentration is decreasing. As shown in Figure (6) the activity of Na_2O increases with the CaO/SiO_2 ratio.

Richardson's solution model, in which the ternary solution is treated as an ideal mixing of the two binaries, fits the system very well. Therefore, the activity of CaO in the ternary was calculated using the model and the results are shown in Figure (7).

The activity of Na_2O in CaO-Al_2O_3-SiO_2 ladle type slags is very high as shown in Figure (8). The activity for a slag containing only 8% Na_2O is as high as for a 50% Na_2O-SiO_2 slag. Again Richardson's model represents the data reasonably well.

It is evident that for both systems CaO significantly increases the activity of Na_2O. This is one reason why even small additions of Na_2O to CaO based slags significantly improves dephosphorization and desulfurization. Reactions (10) and (11) are enhanced by high Na_2O activities,

$$\frac{3}{2}(Na_2O) + \underline{P} + \frac{5}{4}O_2 = (Na_3\,PO_4) \qquad (10)$$

$$Na_2O + \underline{S} = (Na_2S) + 1/2\ O_2 \qquad (11)$$

Summary and Conclusions

The addition of Na_2O to CaO based slags significantly improves the phosphate and sulfide capacity. Of particular interest, and possible commercial value, is that 5% Na_2O addition to CaO saturated - 40% FeO slags increases the phosphorus distribution ratio by about a factor of 5. Therefore, these additions will help in ladle dephosphorization. The addition of 2% to reducing ladle slags increases L_P by about a factor of 50 indicating these additions may stop phosphorus reversion. The activity of Na_2O is increased by CaO in slags. For example, for activity of Na_2O in a 49% CaO - 39% Al_2O_3 - 10% SiO_2 - 2% Na_2O is as high as in a 50% Na_2O-SiO_2 slag. The high Na_2O activity is one reason these slags have improved phosphate and sulfide capacities.

Acknowledgements: The authors wish to thank the Center for Iron and Steelmaking Research at CMU and the National Science Foundation Grant 8421112 for support of the research.

REFERENCES

1. A.H. Chan and R.J. Fruehan: Metall. Trans. B, 1986, vol 17B, p. 491.
2. J.J. Pak and R.J. Fruehan: Metall. Trans. B, vol 17B, 1986, p. 797.
3. M. Muraki, H. Fukushima and N. Sano: Trans. ISIJ, vol 25, 1985, p. 1025.
4. Y. Mujawaki, Y. Matsuda, H. Tanake and H. Funanokawa: European Steelmaking Conference, Vol 2, 1984, Strasburg, France.
5. J.J. Pak and R.J. Fruehan: submitted to Metall. Trans. B.
6. A.H. Chan and R.J. Fruehan: submitted to Metall. Trans. B.
7. J.J. Pak and R.J. Fruehan: submitted to Metall. Trans. B.
8. D.J. Sosinsky and I.D. Sommerville: Metall. Trans. B, vol 17B, 1986, p. 331.
9. S. Yamaguchi, A. Imai, and K.S. Goto: Scand. J. Metall. Vol 11, 1982, p. 263.
10. F.D. Richardson: Trans. Faraday Soc., Vol 52, 1956, p. 1312.

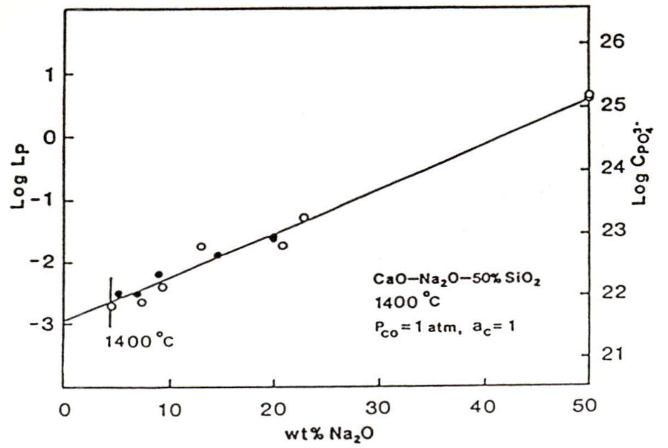

Figure 1. Phosphorus partition ratio and phosphate capacity of CaO-Na_2O-SiO_2 slag as a function of Na_2O content at a fixed SiO_2 content (50 wt%) at 1400°C and P_{CO} = 1 atm.

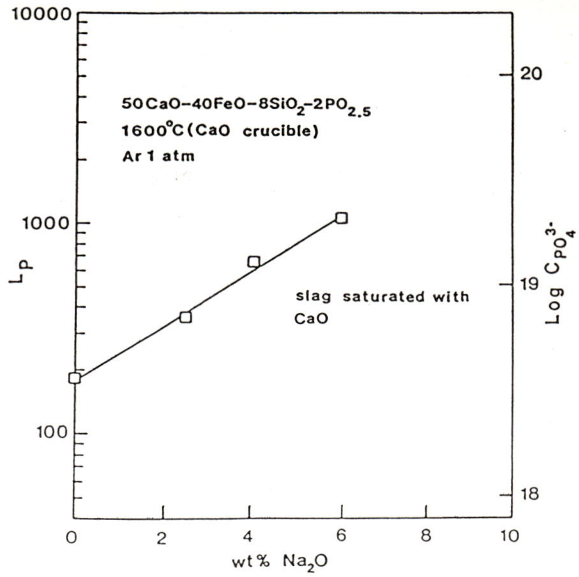

Figure 2. The effect of Na_2O addition on phosphorus partition between CaO Saturated CaO-FeO_T-SiO_2 slag and liquid iron at 1600°C.

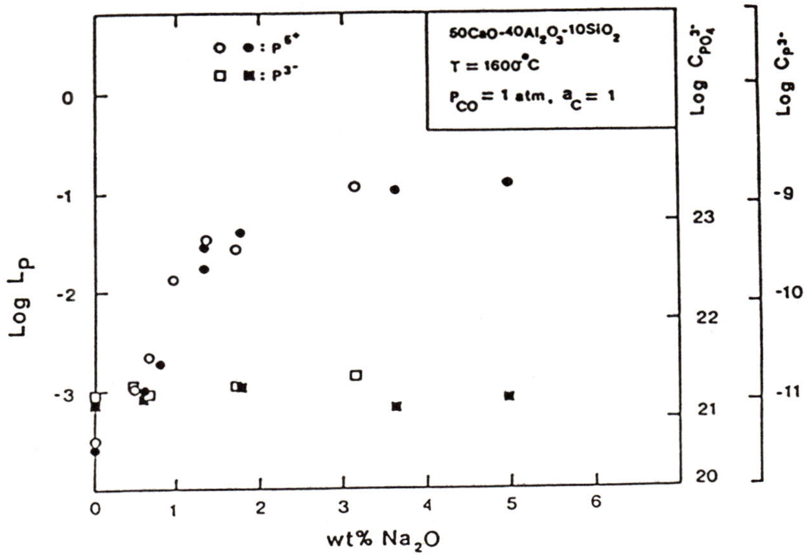

Figure 3.

The effect of Na_2O on the phosphorus distribution between ladle slag and carbon saturated iron at 1600°C (solid: phosphorus from slag to metal, open: metal to slag.

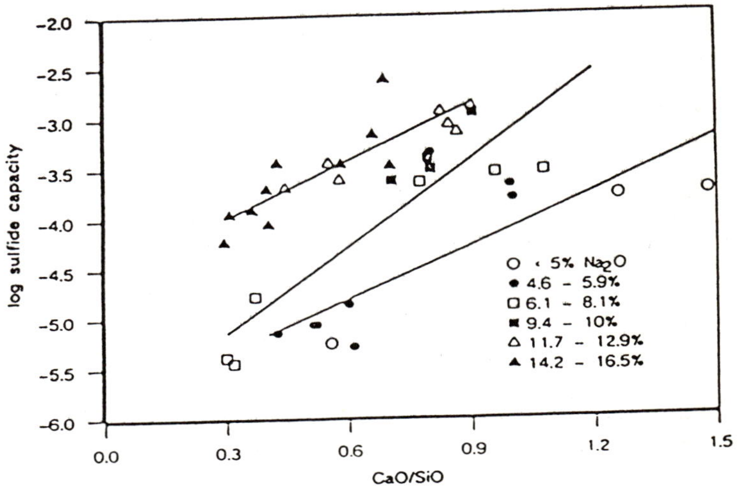

Figure 4.

The sulfur partition ratio versus CaO/SiO_2 (mole fraction) at several different Na_2O contents at 1400°C.

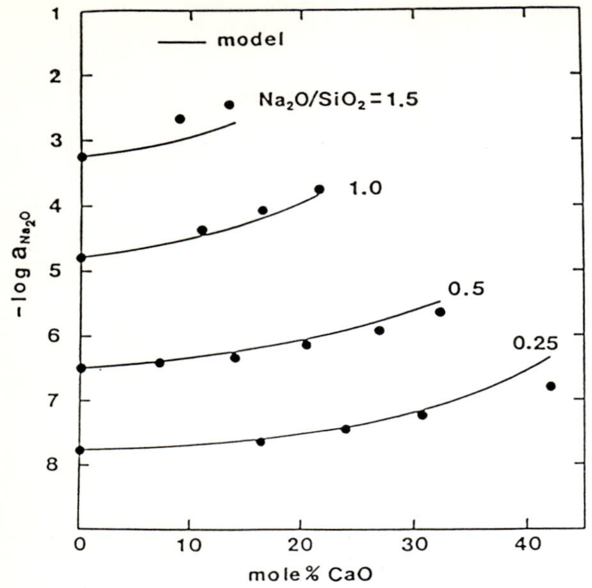

Figure 5. Effect of additions of CaO to Na_2O-SiO_2 melts on the activity of Na_2O at 1400°C.

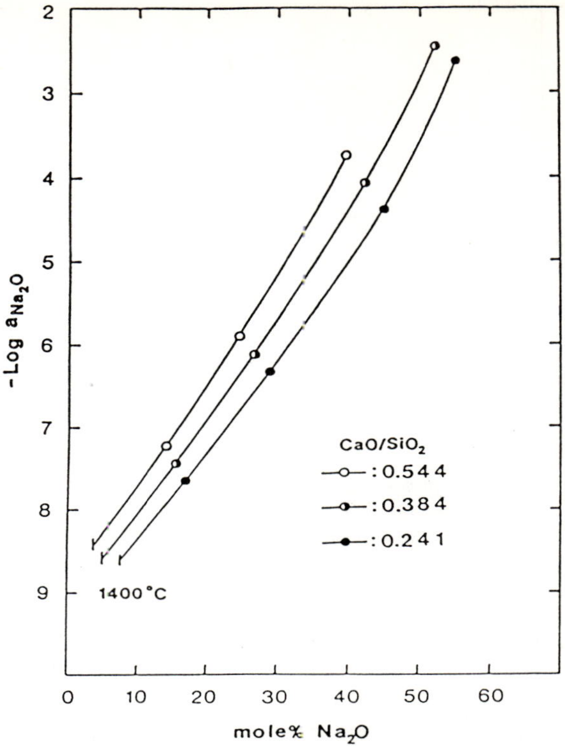

Figure 6. The activity of Na_2O as a function of Na_2O content at fixed CaO-/SiO_2 ratios at 1400°C in CaO-SiO_2-Na_2O melts.

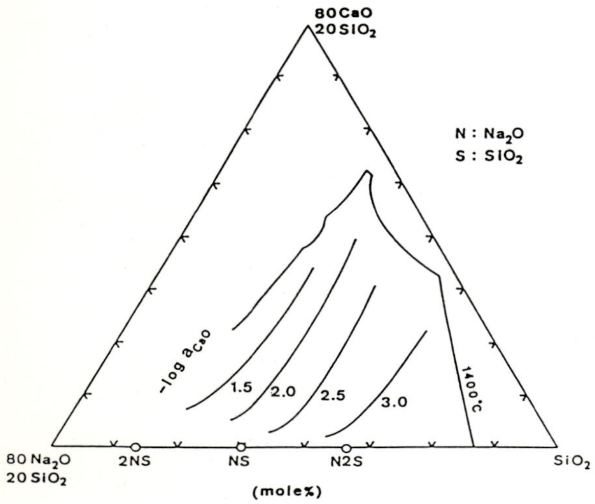

Figure 7. Iso-activity lines for CaO in the CaO-Na_2O-SiO_2 system at 1400°C.

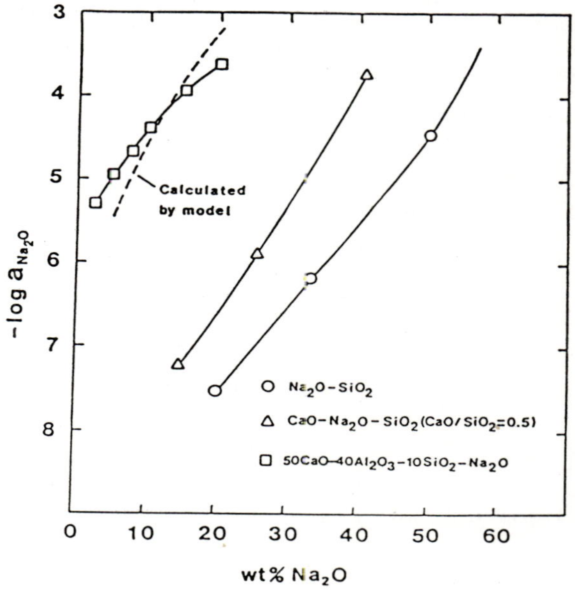

Figure 8. The Na_2O activities in various types of slags at 1500°:C.

Vaporization and condensation of polymeric melts of Li_2O-B_2O_3 and other binary systems

K S GOTO, L-W ZHANG and M SUSA

The authors are with the Tokyo Institute of Technology.

SYNOPSIS

The vapor species from the network-former oxide melts are estimated by means of thermodynamics. The results of the subsequent experiments are consistent with the above estimation and suggest that the vapor species is $LiBO_2$ for Li_2O-B_2O_3 systems. By measuring the rates of evaporation and condensation for the same system, the evaporation and the sticking coefficients are estimated.

INTRODUCTION

Network-former oxides play a very important role in iron- and steelmaking processes. A great many studies on such oxide melts have been reported with respect to thermodynamic properties[1], structures[2] and so on. However, there are very few reports on their vaporization and condensation[3-5] in spite of the importance in industry, which is especially recognized in the field of thin-film manufacturing processes. In this paper, we study the vaporization and condensation of polymeric melts and estimate the vapor species, the evaporation and the sticking coefficients. For the present purpose, first, the vapor species from the melts are estimated thermodynamically. Secondly, to verify the validity of the thermodynamic estimation, thin films of Li_2O-B_2O_3 are synthesized with the use of the vacuum deposition method. The experiment contains X-ray diffraction and IMA studies and the measurements of the rates of vaporization and condensation. Finally, on the basis of the above results we discuss the vapor species and the coefficients of vaporization and sticking.

THERMODYNAMIC ESTIMATION OF VAPOR SPECIES

For the determination of vapor species from melts, mass-spectroscopy is usually employed. But we estimate the vapor species from the Li_2O-B_2O_3 , Na_2O-SiO_2 and Na_2O-P_2O_5 oxide melts by use of thermodynamic consideration.

For Li_2O-B_2O_3 system, the main vapor species are $LiBO_2$, Li, LiO, Li_2O, B_2O_3 and so on. Their vapor pressures can be estimated by considering the chemical reactions producing the vapor species. The procedure will be demonstrated for the partial pressure of $LiBO_2$.

$$1/2Li_2O(l) + 1/2B_2O_3(l) = LiBO_2(g) \qquad (1)$$

$$\Delta G^\circ_{LiBO_2} = -(1/2)RT\cdot\ln(P^2_{LiBO_2}/a_{Li_2O}\cdot a_{B_2O_3}) \qquad (2)$$

where P_{LiBO_2} is the partial pressure of $LiBO_2$ and a_{Li_2O} and $a_{B_2O_3}$ are the activities of Li_2O and B_2O_3 in the liquid Li_2O-B_2O_3 system. When the values of $\Delta G^\circ_{LiBO_2}$, a_{Li_2O} and $a_{B_2O_3}$ can be obtained as functions of temperature and composition, P_{LiBO_2} can be calculated at different temperatures and compositions. Vapor pressures of other important species over fused Li_2O-B_2O_3 systems can be calculated in the same way.

The above consideration can give Figure 1. The same method has been applied to $Na_2O\cdot SiO_2$ and $Na_2O\cdot P_2O_5$ systems. In this thermodynamic calculation, the data of Gibbs free energy of formation are obtained from JANAF Tables[6] and the activity data are taken from Ref.7 for $Li_2O\cdot B_2O_3$ and Ref.8 for $Na_2O\cdot SiO_2$ and $Na_2O\cdot P_2O_5$.

EXPERIMENTAL

The conventional vacuum deposition method was employed for synthesizing thin films. Figure 2 shows the schematic view of the vacuum deposition apparatus. $3Li_2O\cdot 2B_2O_3$, $Li_2O\cdot B_2O_3$ and $Li_2O\cdot 2B_2O_3$ were used as the source materials. The source and the substrate temperatures were varied from 850 to 1050°C and 50 to 450°C, respectively. The pressure in the chamber was about 5×10^{-5} torr during the depositions.

Glass plate with low alkali content and silicon wafers were used as the substrates.

X-ray diffraction was used to identify the structure of the thin films. The compositions were determined by the IMA study.

The vaporization rate was calculated from the weight loss of the source materials. The deposition rate was determined by measuring the thickness of the thin films using ellipsometry, which was corrected by the stylus measurement.

RESULTS

(a) Structure of thin films

All deposited thin films were colorless and transparent.

X-ray diffraction analyses revealed the following. The as-deposited film under the substrate temperature of 100°C is in the amorphous state. But by annealing it is crystallized to $Li_2B_2O_4$. The bulk sample used as the source is a crystal of $Li_2B_2O_4$.

(b) Compositions of thin films

Figure 3 shows the relationship between the mole ratio of Li/B in the deposited films and that of the source materials for three kinds of Li_2O-B_2O_3 systems. The source temperature was fixed to

be 850°C but the substrate temperatures were varied from 100 to 400°C. This figure indicates the following:
(1)Under the condition of Li/B<1 for the sources, the composition of the films is independent of that of the source. It is strongly dependent on the substrate temperature.
(2)Under the condition of Li/B>1 for the sources, Li/B of the films increases with increasing Li/B of the sources.
(3)Li/B of the films increases with decreasing the substrate temperature.
In the figure, the experimental results reported by Levasseur et al[9] and Ito et al[10] are also shown for comparison.

(c) Rates of vaporization and condensation

Figure 4 shows the relation between the film thickness and the deposition time. The deposition conditions were as follows: the source material was $Li_2O{\cdot}B_2O_3$, the source temperature 850°C, and the substrate temperature was 40 to 70°C. The deposition rate increased with time and after 100 seconds became nearly constant.

Under the same condition, the vaporization rate (J) was measured to be $2.50x10^{-6}g/cm^2s$ by using the weight loss of the source material.

DISCUSSION

(a) Estimation of vapor species

On the basis of the results of X-ray diffraction and Figure 3, it can be estimated that the predominant vapor species is $LiBO_2$ for $Li_2O-B_2O_3$ system. This estimation is consistent with the result of the previous thermodynamic consideration.

Since the $LiBO_2$ species is condensed on the substrate as it is for the low substrate temperature, the mole ratio of Li/B of the film is 1. 1.But when the substrate temperature is high, the ratio is not 1.
very In order to explain this very complicated phenomenon, we must consider the decomposition reaction of the condensed matter on the substrate surface. But this is a future subject.

(b) Evaporation and sticking coefficients

Under the condition of Fig.4, it can be estimated that the predominant vapor species, $LiBO_2$, deposits on the substrate as it is, from the above discussion. The above vaporization rate (J) could be converted to be $3.03x10^{16}/cm^2s$ by using the molecular weight of $LiBO_2$.On the other hand, the theoretical evaporation rate(Je) under the same condition can be calculated to be 1.13x $10^{17}/cm^2s$, by using Herz-Knudsen's equation[11]. The evaporation coefficient α, defined by α=J/Je, is estimated to be 0.268. This value is almost the same as those of Al_2O_3, Ga_2O_3, and In_2O_3[11], 0.3.

Figure 5 shows the dependence of the sticking coefficient s, defined by s=(dN/dt)/J, on the film thickness. The deposition rate dN/dt was calculated from Fig. 4 and the film density, which was estimated with the use of Lorentz-Lorenz equation[12] and the refractive index obtained by ellipsometry. The sticking coefficient increases with increasing thickness or deposition time, and becomes constant, 0.858, at around 500 nm in thickness.

CONCLUSION

The vapor species from the network-former oxide melts were estimated by means of thermodynamics. The results of the subsequent experiments, which consist of synthesizing and analysing thin films of $Li_2O-B_2O_3$ systems, are consistent with the above estimation and suggest that the vapor species is $LiBO_2$. By measuring the rates of evaporation and condensation for the same system, the evaporation and the sticking coefficients are estimated to be 0.268 and 0.858 (the maximum value), respectively.

REFERENCES

1.J.F.Elliott, M.Gleiser and V.Ramarkrisha: "Thermochemistry for steelmaking, vol.2", (1963), Addison-Wesley Press.
2.E.T.Turkdogan: "Physicochemical properties of molten slags and glasses", (1983), The Metals Society.
3.L.I.Maissel and R.Glang: "Handbook of thin film technology", (1970) p.1-65, McGraw-Hill.
4.L.W.Zhang, M.Kobayashi and K.S.Goto: Solid State Ionics, 18&19 (1986) p.741-746.
5.L.W.Zhang, M.Yahagi and K.S.Goto: Solid State Ionics, 18&19 (1986) p.1163-1169.
6.D.R.Stall and H.Prokhet: JANAF Thermodynamical Table, Nat.Stand. (1971).
7.M.Ito, S.Sato and T.Yokokawa: J.Chem.Thermodynamics, 8 (1976) p.339.
8.S.Yamaguchi, A.Imai and K.S.Goto: J.Japan Inst. of Metals, 47 (1983) p.739 and S.Yamaguchi and K.S.Goto: ibid, 48(1984) p.43.
9.A.Levasseur, M.Kbala and P.Hagenmuller: Solid State Ionics, 9&10 (1983) p.1439.
10.Y.Ito, K.Miyauchi and T.Oi: J.Non-crystalline Solids, 57 (1983) p.398.
11.Japan Society for the Promotion of Science, 131st Committee: "Thin film handbook", (1983) OHM-sha.
12.E.S.Larsen and H.Berman: "The microscopical determination of nonopaque minerals, 2nd ed.", Washington D.C.U.S. Gedogical Survey, Bull. (1934).

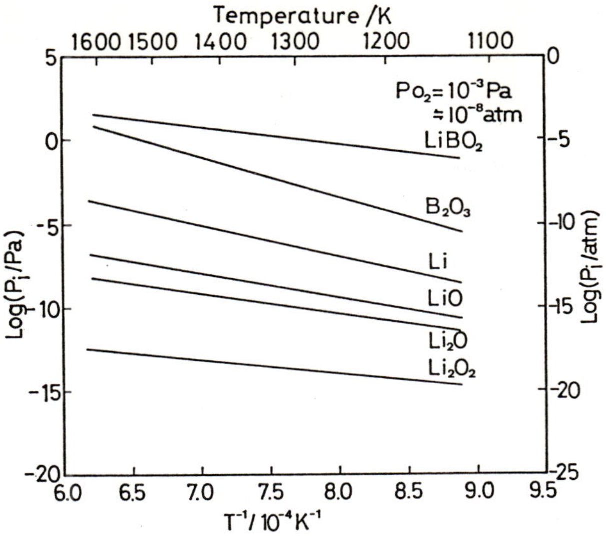

Fig.1 Temperature dependence of partial pressures of vapor species over molten $0.5Li_2O\cdot0.5B_2O_3$.

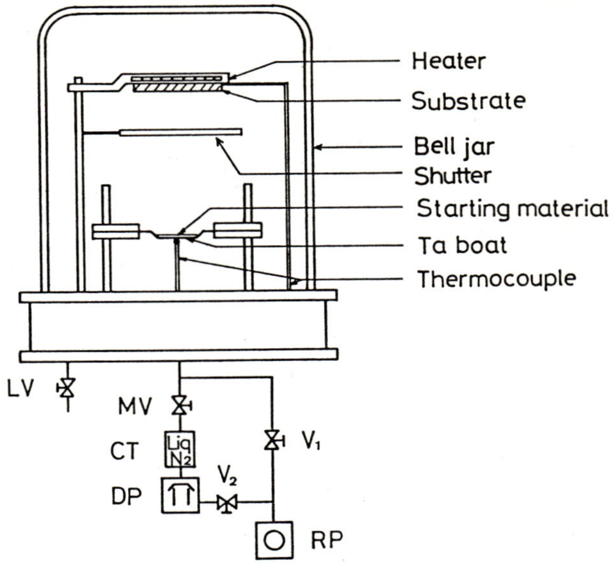

Fig.2 Schematic view of experimental apparatus.

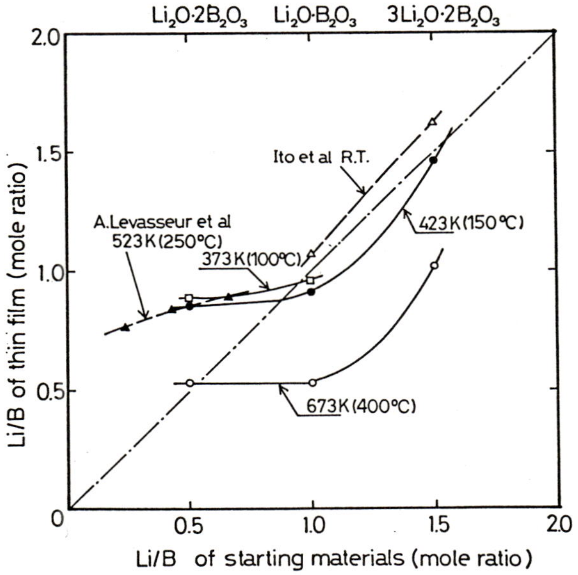

Fig.3 Relationship between thin-film composition and source composition (source temperature; present work,1123K; Levasseur et al, 873K; Ito et al,unknown).

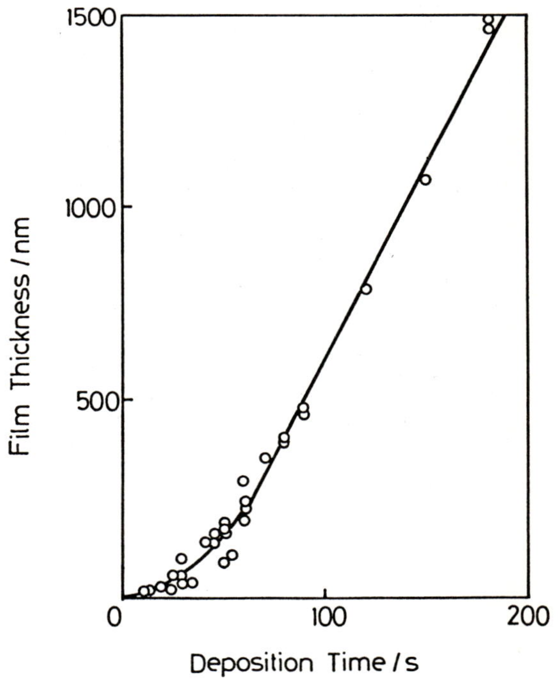

Fig.4 Relation between the film thickness of $Li_2O\cdot B_2O_3$ oxide film and the deposition time.

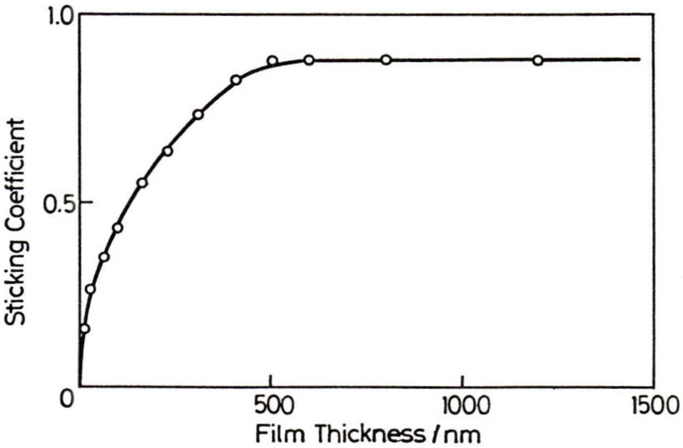

Fig.5 Dependence of the sticking coefficient on the thickness of $Li_2O\cdot B_2O_3$ oxide film.

Quasichemical model for thermodynamic properties of multicomponent slags

A D PELTON, G ERIKSSON and M BLANDER

Prof Pelton is Professor of Metallurgy at Ecole Polytechnique de Montreal, Montreal, Canada; Dr Eriksson is a Senior Research Associate; Dr Blander is a Senior Scientist and Group Leader in the Chemical Technology Division/Materials Science Program at the Argonne National Laboratory, Argonne, IL, USA.

SYNOPSIS

A modified quasichemical model has been shown to represent thermodynamic properties of binary molten slags, with a small number (≤ 7) of parameters, and leads to good predictions of ternary and probably higher order systems, based solely on data for the subsidiary binaries. Analyses of the MnO-SiO_2 binary system provides a good representation of all reliable thermodynamic data on this system with only three parameters. When combined with analyses of the other pertinent binaries, predictions of activities of components in the MnO-FeO-SiO_2 and MnO-CaO-SiO_2 ternaries are in very good agreement with measurements.

INTRODUCTION

A modified quasichemical model has been developed for the representation and analysis of the thermodynamic properties of molten slags and glasses.[1–5] The parameters of the model for a binary system are obtained by a simultaneous least-squares optimization of all available thermodynamic and phase equilibrium data.[6] Quantitative representations at all compositions have been obtained in all binary systems involving SiO_2, CaO, MgO, MnO, Na_2O, and FeO. For all ternary subsystems studied, quantitative representations of the thermodynamic properties and phase diagrams at all compositions were obtained by using only the parameters from the binary subsystems.

DISCUSSION

In order to illustrate the power of the method, we discuss some of our results for MnO containing systems. For the MnO-SiO_2 system, we used an optimization procedure coupled to the model[6] to simultaneously fit data on the liquidus temperatures and activities at 1400°C, 1500°C, and 1600°C,[7] the phase diagram below 1400°C,[8] the melting point of 1785°C, the Gibbs energy of fusion of MnO[9] and SiO_2,[9] and the Gibbs energies of the compounds at their melting points.[10] The best fit led to an equation for the energy parameters in the model that was independent of temperature and contained only three constants. The self-consistency of the procedure was illustrated by the excellent representation of all of the input data by the equations deduced from the quasichemical model. For example, we exhibit the calculated phase diagram in Fig. 1, which is consistent with the data of Rao and Gaskell,[7] at or above 1400°C and with the published phase diagram[8] below 1400°C. The measured and calculated activities of MnO exhibited in Fig. 2 are very well represented by the calculations at all three temperatures. In addition, we have performed similar analyses of two other binary systems, MnO-CaO and MnO-FeO, based on much less data.[10–13] However, since these binaries are simpler than the MnO-SiO_2 system, the analyses are relatively reliable.

The above results on the three binaries, when combined with prior results on the FeO-SiO_2 or CaO-SiO_2 binary systems, were used to calculate the thermodynamic properties of the two ternaries, MnO-FeO-SiO_2 and MnO-CaO-SiO_2, using the asymmetric approximation discussed earlier.[1–5] This approximation is consistent with the best available theory and has been previously used successfully to predict the properties of ternaries based solely on the properties of the binary subsystems.[1–5] Calculations and measurements of the activities of SiO_2 and FeO in the MnO-FeO-SiO_2 system at 1550°C are exhibited in Figs. 3 and 4, and of the activities of MnO in the MnO-CaO-SiO_2 system at 1600°C are exhibited in Fig. 5. These calculations are in excellent agreement with experimental data.[14,15]

CONCLUSION

Our success in predicting properties of ternaries from the results of data analyses of subsidiary binaries lends confidence in our methods and indicates that they can be used to predict properties of multicomponent systems at all compositions and temperatures from the evaluated binary parameters. We have developed interactive programs that rapidly calculate equilibria involving the six component slags (SiO_2, CaO, MgO, MnO, Na_2O, and FeO) with other phases, such as solid oxides, molten steel, gases, etc. For example, we have performed calculations for a reported steelmaking reaction between a slag and a steel alloy[16] with 0.6 wt. % Si, 0.7% Mn, and 4.1% C with about 3.1 tons of CaO, 0.9 tons of dolomite, and 7.0 tons of O_2 added per 100 tons of metal. Our calculated final compositions of the metal and slag given in Table I compare very well with the measured values. Work is in progress to include more components.

ACKNOWLEDGMENTS

This work was supported, in part, by the U.S. Department of Energy, Division of Materials Science, Office of Basic Energy Sciences, under Contract W-31-109-ENG-38 with the Argonne National Laboratory.

REFERENCES

1. M. Blander and A. D. Pelton, "Computer-Assisted Analyses of the Thermodynamic Properties of Slags in Coal Combustion Systems," ANL/FE-83-19, Argonne National Laboratory, Argonne, IL (1983). Available from NTIS, U.S. Dept. of Commerce, Washington, D.C.

2. A. Pelton and M. Blander, "Computer Assisted Analyses of the Thermodynamic Properties and Phase Diagrams of Slags," Proc. of the Second Interml. Symposium on Metallurgical Slags and Fluxes, H. A. Fine and D. R. Gaskell, Eds., TMS-AIME, Warrendale, PA (1984) pp. 281-294.

3. M. Blander and A. Pelton, "Analyses and Predictions of the Thermodynamic Properties of Multicomponent Silicates," Proc. of the Second Internl. Symposium on Metallurgical Slags and Fluxes, H. A. Fine and D. R. Gaskell, Eds., TMS-AIME, Warrendale, PA (1984) pp. 295-304.

4. A. D. Pelton and M. Blander, Metall. Trans.. **17B**, 805-815 (1986).

5. M. Blander and A. D. Pelton, Geochim. Cosmochim. Acta **51** 85-95 (1987).

6. A. D. Pelton and M. Blander, CALPHAD **12**(1), 97-108 (1988).

7. K. D. P. Rao and D. R. Gaskell, Metall. Trans. **12B**, 311 (1981).

8. F. P. Glasser, Am. J. Sci. **256**, 405 (1958). See also Phase Diagrams for Ceramists, No. 101.

9. I. Barin, O. Knacke, and O. Kubaschewski, "Thermochem. Props. of Inorg. Substances" (and supplement), Springer-Verlag, NY (1973, 1977).

10. H. Schenk, M. G. Frohberg, and R. Nünninghof, Arch. Eisen. **4**, 269 (1964).

11. S. Raghavan, G. N. K. Iyengar, and K. P. Abraham, Trans. Indian Inst. Met. **33**, 51 (1980).

12. W. A. Fischer and H. J. Fleischer, Arch. Eisenh. **32**, 6 (1961).

13. A. D. Pelton, Ber. Bunsenges. Phys. Chem. **84**, 212 (1980).

14. H. Gaye and D. Coulombet, IRSID Report PCM-RE.1064, "Données Thermochimiques et Cinétiques Relatives à certains matériaux sidérurgigues," Maizières-Les-Metz, France (1984).

15. H. Gaye and P. Riboud, "Données Expérimentales sur les activités des constituants de laitiers," Rep. CECA #7210-CA/3/303 (November 1981).

16. W. T. Lankford, Jr., N. L. Samways, R. F. Craven, and H. E. McGannon (eds.) "The Making, Shaping, and Treating of Steel," 10th Edition, Assn. of Iron and Steel Engineers, Pittsburgh, PA (1985) pps. 468-470.

Table I. Calculated and Measured[16] Values of the Compositions (wt %) of Slag and Steel Blown With O_2 at 1600°C

	Calculated	Measured†
Steel		
Mn	0.176	0.2
C	0.067	<0.1
Si	0.119×10^{-5}	0.0
Slag		
CaO	42.43	42.6
FeO	30.81	29.5
SiO_2	14.62	14.6
MnO	7.86	6.3
MgO	4.29	4.5

†Reference 16: These values were taken from Figure 13 on p. 469. The total of 97.5% for the slag analyses indicates that there is some error in the analyses or the plot.

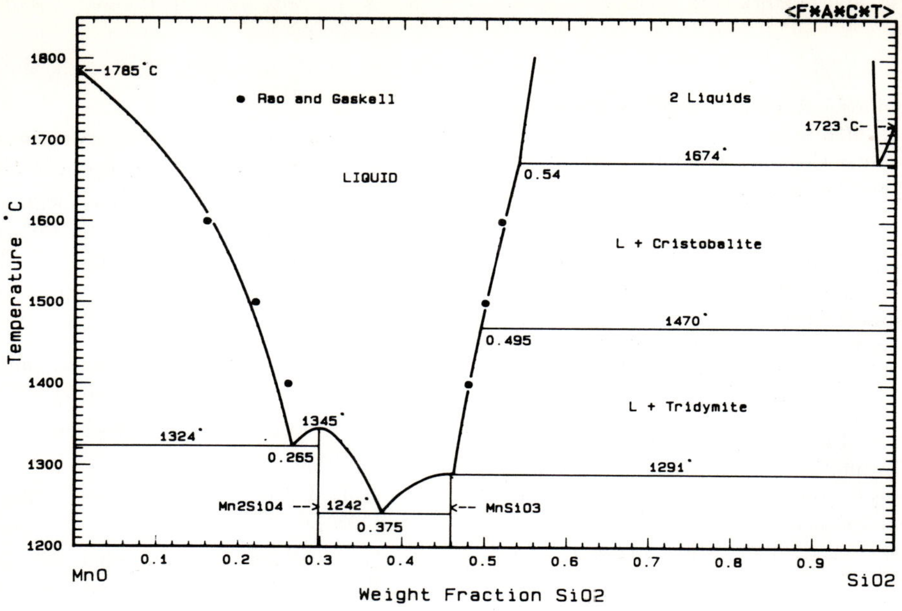

Fig. 1. Calculated MnO-SiO_2 Phase Diagram.

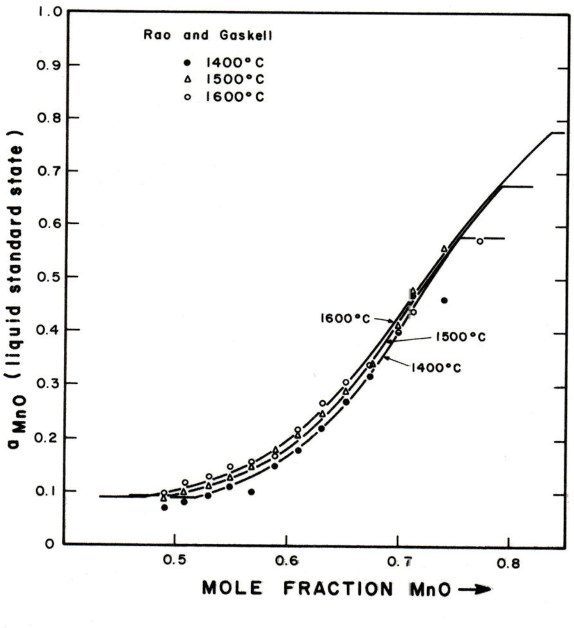

Fig. 2. Activities of MnO in MnO-SiO_2 Slags at 1400, 1500, and 1600°C. Points are from Rao and Gaskell.[7] Lines are Calculated.

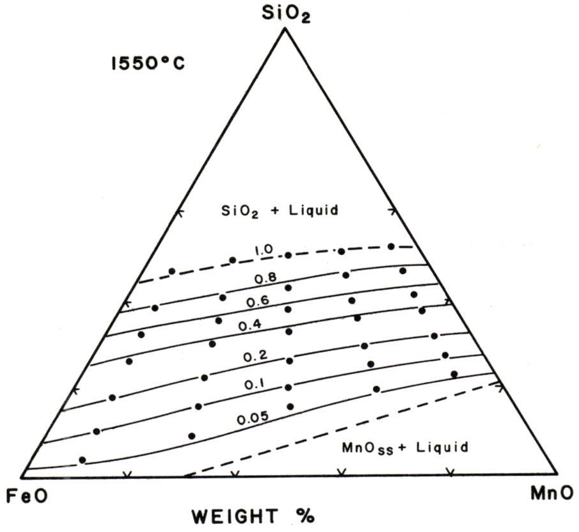

Fig. 3. Experimental[14,15] (Lines) and Calculated (Points) SiO_2 Activities in SiO_2-FeO-MnO Slags at 1550°C.

Fig. 4. Experimental[14,15] (Lines) and Calculated (Points) FeO Activities (Liquid Standard State) in SiO_2-FeO-MnO Slags at 1550°C.

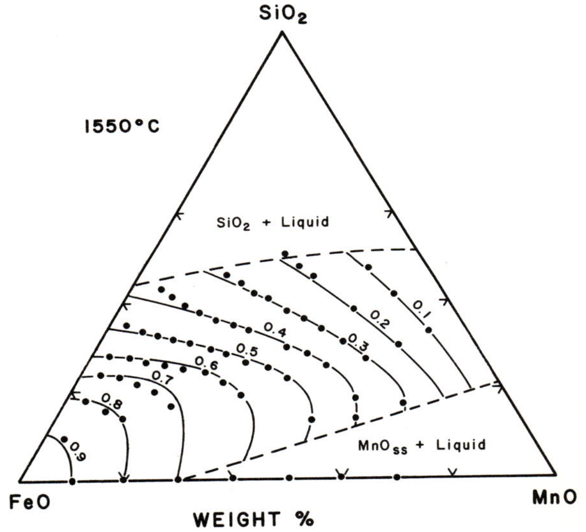

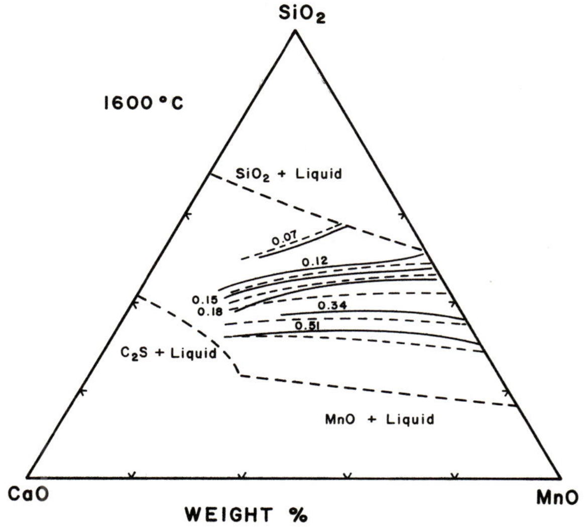

Fig. 5. Experimental[14,15] (Solid Lines) and Calculated (Dashed Lines) MnO Activities (Solid Standard State) in SiO_2-CaO-MnO Slags at 1600°C.

Mass spectrometric determination of activity of P_2O_5 dissolved in Fe_tO and interaction parameters of CaO, MgO, MnO and SiO_2 on P_2O_5

N OHHARA, S KAMBAYASHI, S NUNOUE and E KATO

Mr Ohhara, Mr Nunoue and Dr Kato are in the Department of Materials Science and Engineering, Waseda University; Dr Kambayashi, formerly in the same department, is now with Toshiba Corp.

SYNOPSIS

The Henrian and Raoultian activities of P_2O_5 in the dilute solution of P_2O_5 in Fe_tO in the temperature range 1370-1390°C have been determined by the use of a Knudsen cell-mass spectrometer combination and the chemical equilibrium

$$P_2O_5(\text{in slag})+PO(g)=3PO_2(g)$$

The effects of CaO, MgO, MnO and SiO_2 on the activity coefficient of P_2O_5 in the dilute solution of P_2O_5 dissolved in Fe_tO have been determined by the use of the same technique.

INTRODUCTION

The activity of Fe_tO in the Fe_tO-P_2O_5 system was determined by Ban-ya and Watanabe(1) and this is the only determination which has been performed with this system. In the gas phase above liquid Fe_tO-P_2O_5 solutions P_2, PO and PO_2 exist, and in this study the activity of P_2O_5 in the Fe_tO-P_2O_5 system has been determined mass spectrometrically by the use of chemical equilibrium

$$P_2O_5(\text{in slag})+PO(g)=3PO_2(g) \qquad (1)$$

and the iron vapor in equilibrium with solid iron was used as the internal standard.

The activities in the PbO-P_2O_5 system were determined by the use of the same technique(2). In the gas phase above this system, PbO(g) also exists, and both of the activities of PbO and P_2O_5 can be determined mass spectrometrically. The determined values of the activity of P_2O_5 is in good agreement with the calculated values from the activity of PbO by means of application of the Gibbs-Duhem equation. These results indicate that this technique can be applied to the determination of the activity of P_2O_5 in slags.

The effects of added oxides to that activity have also been determined by the use of the same technique.

EXPERIMENTAL METHOD

The Knudsen cell-mass spectrometer combination (Hitachi RM-6E) employed for this study was similar to that used for the study of the MgO-SiO_2 system(3), and a picoammeter (Keithley model 417) was used for the measurement of ion current instead of the pulse counting apparatus which was used in the study of the MgO-SiO_2 system. The Knudsen cell, made of high purity alumina was 15mm inside diameter with an 0.5mm orifice. The temperature of the cell was measured by two W26%Re-W5%Re thermocouples. A pure iron crucible containing the sample was placed in the Knudsen cell. The vapor of iron was used as the internal standard for the measurement of ion currents. The samples were made from reagent grade FeO(99.9% purity) and $3FeO\cdot P_2O_5\cdot 8H_2O$(>98% purity).

The activity of P_2O_5 in the sample is related to the partial pressure of PO and PO_2 through the chemical equilibrium indicated by Eq.(1). The equilibrium constant K of the reaction (1) is given by

$$K=P^3_{PO_2}/a_{P_2O_5}\cdot P_{PO} \qquad (2)$$

and when the partial pressures of these gases in equilibrium with the standard state of P_2O_5 are indicated as P^o_{PO} and $P^o_{PO_2}$, $a_{P_2O_5}$ is expressed as

$$a_{P_2O_5}=(P_{PO_2}/P^o{}_{PO_2})^3/(P_{PO}/P^o{}_{PO}) \qquad (3)$$

Therefore, $a_{P_2O_5}$ is related to the ion currents of PO and PO_2 when vapor of Fe is used as the internal standard.

$$a_{P_2O_5}=\frac{\left\{(I^+_{PO_2}/I^+_{Fe})/(I^{+o}_{PO_2}/I^+_{Fe})\right\}^3}{(I^+_{PO}/I^+_{Fe})/(I^{+o}_{PO}/I^+_{Fe})} \qquad (4)$$

The concentration of P in the iron crucible was always less than 1wt pct, and $a_{Fe}=1$ is used for the calculation of $a_{P_2O_5}$.

The activity of P_2O_5 in the Fe_tO-P_2O_5 system when another oxide was added was determined by the same use of the technique. The concentrations of the added oxides were about 5 mol pct, and the concentration range of P_2O_5 was from 1.5 to 9.5 mol pct.

EXPERIMENTAL RESULTS

For the Fe_tO-P_2O_5 system, the determinations were performed in the temperature range 1370 - 1390°C and in the concentration range 0.75 - 9.91 mol pct P_2O_5. Figure 1 shows the dependence of the ion intensity ratios on time and on temperature. As shown in this figure, the ratios were stable for long hours and reproducible against the change in temperature. The compositions of the samples are indicated in Table 1 and the determined ion intensities ratios are shown in Table 2. The obtained Henrian activity of P_2O_5 plotted against $X_{P_2O_5}$ is shown in Fig. 2. The ion current ratio at $X_{P_2O_5}=1$ is obtained by extrapolating the determined activity values in the dilute solution range of P_2O_5 concentrations to $X_{P_2O_5}=1$. The Raoultian activity of Fe_tO is calculated from the activity of P_2O_5 by means of application of the Gibbs-Duhem equation, and the calculated results are shown in Fig. 3.

The effects of CaO, MgO, MnO and SiO_2 on the activity coefficient of P_2O_5 are shown in Fig. 4.

DISCUSSION

The ions of PO^+ and PO_2^+ are not fragment ions and the precursors of these ions are PO and PO_2, respectively(2).

Thermodynamic property of $P_2O_5(g)$ is not given in the literature, and the vapor pressure of P_4O_{10} is used for obtaining the Raoultian activity of P_2O_5 for which solid P_2O_5 is chosen as the standard state. The relation between the activity of P_2O_5 and the partial pressure of P_4O_{10} is given by the following equation,

$$a_{P_2O_5}=(P_{P_4O_{10}}/P^o_{P_4O_{10}})^{1/2} \qquad (5)$$

where $P^o_{P_4O_{10}}$ is the partial pressure of P_4O_{10} in equilibrium with $P_2O_5(s)$. The partial pressure of P_4O_{10} in equilibrium with PO(g) and $PO_2(g)$ through the following chemical reaction

$$P_4O_{10}(g)+2PO(g)=6PO_2(g) \qquad (6)$$

is given by

$$P_{P_4O_{10}}=K'P^6_{PO_2}/P^2_{PO} \qquad (7)$$

The standard free energies for the formation of $P_4O_{10}(s)$, $P_4O_{10}(g)$, PO(g), $PO_2(g)$ given in JANAF Tables(4) were used for the calculation. The free energies of $P_4O_{10}(g)$ in equilibrium with $P_2O_5(s)$ in the temperatures higher than 1227°C are not given in the table and these values were calculated by extrapolating the values for the temperature range lower than 1227°C to higher temperatures, and the vapor pressures of P_4O_{10} at 1370°C, 1380°C and 1390°C were calculated as 700atm, 709atm and 719atm, respectively. The calculated Raoultian activities of P_2O_5 are shown in Table 1, and the activity coefficients for the dilute solution range of P_2O_5 concentrations are

$$\gamma_{P_2O_5}=(2.2\pm0.8)\times10^{-15} \qquad (1370^oC)$$
$$=(2.5\pm0.8)\times10^{-15} \qquad (1380^oC)$$
$$=(2.8\pm 1\ )\times10^{-15} \qquad (1390^oC)$$

The errors indicated in the above equations are equal to twice the standard deviations obtained for the determinations at each temperature.

The Raoultian activity coefficients when liquid P_2O_5 is chosen as the standard state can be calculated from the above-mentioned activity coefficients by the use of the standard free energy change for the transformation $P_2O_5(s)=P_2O_5(l)$ given by Turkdogan and Peason(5), and the following values are obtained

$$\gamma_{P_2O_5}=(3.4\pm1.2)\times10^{-13} \qquad (1370^oC)$$
$$=(4.0\pm1.3)\times10^{-13} \qquad (1380^oC)$$
$$=(4.7\pm1.7)\times10^{-13} \qquad (1390^oC)$$

As indicated in Fig. 2, the activity of P_2O_5 in the Fe_tO-P_2O_5 system obeys Henry's law up to 8 mol pct of P_2O_5. The concentrations of P_2O_5 in the almost all ternary systems used in this study were less than this concentration, and as indicated in Table 3, the difference in the ratio Fe^{2+}/Fe^{3+} in the samples was small. Therefore, the change in the structure of the solvent Fe_tO is considered to have been small throughout this study, and the interaction parameters which indicate the effect of the added oxide to the activity coefficient of P_2O_5 can be calculated from the experimental results. The obtained interaction parameters are

$$\epsilon^{CaO}_{P_2O_5}=-23\pm3 \qquad \epsilon^{MgO}_{P_2O_5}=-20\pm2$$
$$\epsilon^{MnO}_{P_2O_5}=-13\pm1 \qquad \epsilon^{SiO_2}_{P_2O_5}=-4\pm1$$

CONCLUSION

The Henrian and Raoultian activities of P_2O_5 in the dilute solution of P_2O_5 in Fe_tO have been determined at 1370-1390°C. The activity of P_2O_5 obeys Henry's law up to the 8 mol pct P_2O_5. The Raoultian activities of Fe_tO have been calculated from the activity of P_2O_5.

The interaction parameters of CaO, MgO, MnO and SiO_2 on P_2O_5 have been determined.

ACKNOWLEDGMENT

The authors are grateful to Kawasaki Steel Co. Ltd. for the chemical analysis of the samples and also to Showa Denko Co. Ltd. for supplying pure iron crucible.

REFERENCES

(1) S. Ban-ya and T. Watanabe: Tetsu-to-Hagané, vol. 63 (1977), p. 1809

(2) S. Kambayashi, and H. Awaka and E. Kato: Tetsu-to-Hagané, vol. 71 (1985), p. 1911

(3) S. Kambayashi and E. Kato: J. Chem. Thermodynamics, vol. 15 (1983), p. 701

(4) JANAF Thermochemical Tables. NSRDS-NBS. (1971), supply. (1974)

(5) E. T. Turkdogan and J. Peason: JISI, vol. 175 (1953), p. 398

Table 1. Chemical compositions of samples for $Fe_tO-P_2O_5$ and Raoultian activity of P_2O_5

No.	$X_{P_2O_5}$	t in Fe_tO	$a_{P_2O_5} \times 10^{17}$		
			1370°C	1380°C	1390°C
1	0.0075	0.955	1.2	1.4	1.5
2	0.0202	0.949	3.8	4.5	5.4
3	0.0271	0.947	7.2	8.7	9.2
4	0.0427	0.953	10.0	11.4	14.0
5	0.0499	0.955	12.4	----	----
6	0.0691	0.958	14.8	18.1	20.8
7	0.0764	0.959	15.6	18.0	20.0
8	0.0991	0.953	26.3	32.3	47.6

Table 2. Experimental values of ion current ratios for $Fe_tO-P_2O_5$ systems

No.	$I^+_{PO}/I^+_{Fe} \times 10$			$I^+_{PO_2}/I^+_{Fe} \times 10^2$		
	1370°C	1380°C	1390°C	1370°C	1380°C	1390°C
1	1.51	1.54	1.68	1.21	1.22	1.25
2	3.02	3.11	3.40	2.22	2.28	2.41
3	4.03	4.34	4.49	3.03	3.18	3.15
4	4.47	4.77	4.99	3.49	3.59	3.75
5	5.71	----	----	4.07	----	----
6	6.23	6.57	6.94	4.45	4.66	4.78
7	6.37	6.76	7.26	4.56	4.70	4.79
8	8.06	8.51	8.68	5.87	6.16	6.79

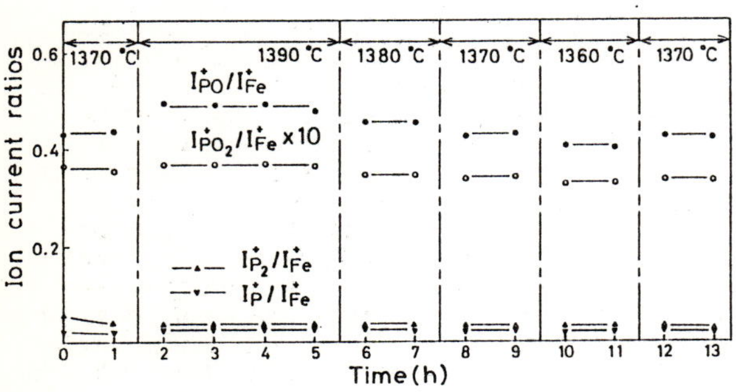

Fig. 1 Time dependence of ion current ratios for the $Fe_tO-P_2O_5$ system at $X_{P_2O_5}=0.0427$

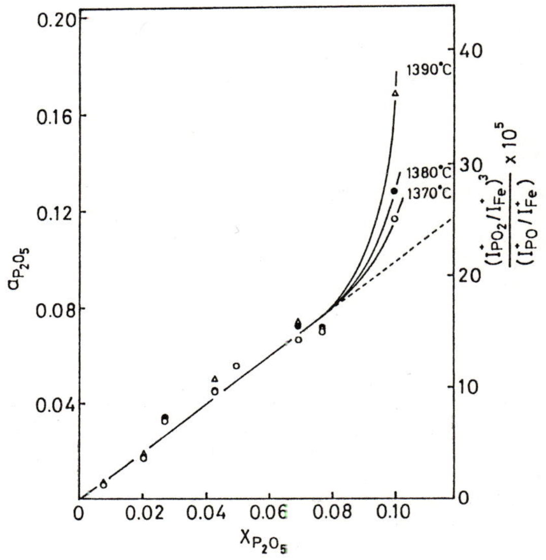

Fig. 2 Henrian activity of P_2O_5 for $Fe_tO-P_2O_5$ systems.

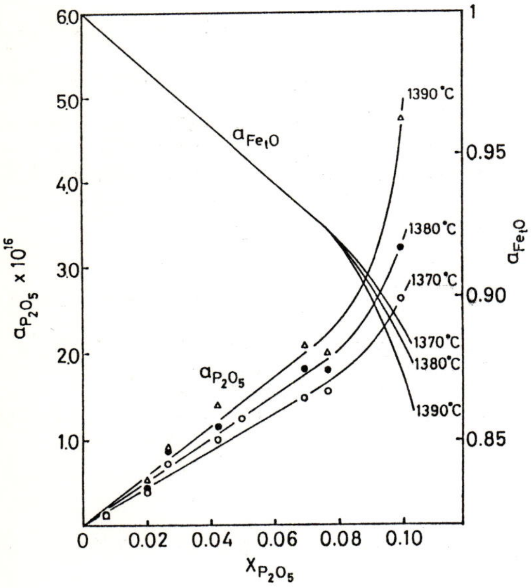

Fig. 3 Raoultian activities for $Fe_tO-P_2O_5$ systems.

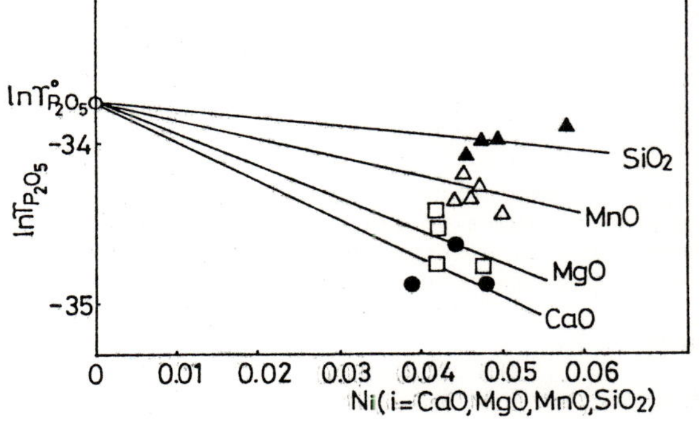

Fig. 4 Effect of added oxides to the activity coefficient of P_2O_5.

Thermodynamics of oxide + chloride + phosphate melts

H FUJIWARA and M IWASE

Mr Fujiwara, PhD Student, and Dr Iwase, Associate Professor, are with the Department of Metallurgy, Kyoto University, Kyoto, Japan.

SYNOPSIS

Based on experimental data for the activities of P_2O_5, an empirical equation was derived for expressing the activities of components in MO + MCl_2 + P_2O_5 melts.(M represents alkaline earth elements).

INTRODUCTION

In physicochemical studies of extraction metallurgy, a long-standing challenge has been the search for an appropriate formula expressing the activities of components in molten slags. By employing the regular solution model, Lumsden[1] derived an expression, which has been successfully applied to oxide melts[2,3]. Application of this model to MO + MCl_2 + P_2O_5 fluxes, however, does not seem to be promising.

In the present study, firstly, emf measurements were conducted to obtain the activities of P_2O_5 in MO + MCl_2 + P_2O_5 melts. By using the experimental results, an empirical equation was then derived to express analytically the activities of components in MO + MCl_2 + P_2O_5 melts.

EXPERIMENTAL

The details of the experimental apparatus and procedures have been reported elsewhere[4]. About 200 g of Cu + P alloy and 25 to 35 g of slag were melted in a magnesia crucibel under a stream of purified argon inside a SiC resistance furnace, which was equipped with a mullite reaction tube.

The experimental procedure consisted of measuring the open-circuit emfs of solid-state galvanic cell, and subsequently sampling the liquid copper and the molten slag for chemical analysis. The electrochemical cell can be expressed as

$$Mo/Mo + MoO_2//ZrO_2(MgO)//(Cu + P) + (P_2O_5)/Mo.$$

The electrochemical half cell, Mo/Mo + MoO_2// $ZrO_2(MgO)$, consisted of a zirconia tube and a two-phase mixture of Mo + MoO_2. The zirconia tube, closed at one end and stabilized by 9 mol pct of MgO, had an i.d. of 4 mm, an o.d. of 6 mm and a length of 50 mm. A molybdenum rod of 3 mm diameter was used as an electrical lead to the reference electrode, which consisted of 4 parts Mo and 1 part MoO_2 by weight. The electrical contact to the slag electrode was also made by a molybdenum rod of 3 mm diameter. Dissolution of Mo in liquid copper was negligibly small. In order to change the concentrations of P_2O_5 in the slag, granules of pre-melted MO + MCl_2 + P_2O_5 ternary slags were added and stirred into the melt. The molar ratio of MO/MCl_2 in the ternary slag was adjusted such that the MO/MCl_2 molar ratio within the molten fluxes was kept constant during the experimental runs. The concentration of phosphorus in the slag was also changed by the addition of MO + MCl_2 binary slags; emf measurements were conducted with increasing and decreasing P_2O_5 contents.

The equilibrium reaction underlying the emf measurements can be formulated as

$$2\ P(\text{in Cu}) + (5/2)\ O_2(g) = P_2O_5(\text{in slag}) \quad ......(1)$$

$$\Delta G°(1) = 2\ R\ T\ \ln h(P) + (5/2)\ R\ T\ \ln P(O_2)(\text{slag}) - R\ T\ \ln a(P_2O_5) \quad ......(2)$$

By using the available thermochemical data for the free energy change for the dissolution of gaseous diatomic phosphorus in liquid copper[5], and the standard free energy of formation of pure liquid P_2O_5[6], values for $\Delta G°(1)$ can be expressed as

$$\Delta G°(1) = -\ 1{,}284{,}750 + 505.2\ T \quad \text{Joule} \quad ......(3)$$

The open-circuit emf of the cell is given by[7]

$$E = \frac{R\ T}{F} \ln \frac{P(O_2)(\text{ref.})^{1/4} + P_e^{1/4}}{P(O_2)(\text{slag})^{1/4} + P_e^{1/2}} \quad ......(4)$$

where

$$\log (P_e\ /\ \text{atm}) = 20.40 - 6.45 \times 10^4\ /\ T[8] \quad .....(5)$$

The oxygen partial pressure at the reference electrode, Mo + MoO_2, was calculated by using the authors' previous results[9]. The activities of P_2O_5 referred to pure super-cooled liquid P_2O_5 and were then obtained by using equation(2).

EXPERIMENTAL RESULTS AND DISCUSSION

The activity-composition relationship for BaO + $BaCl_2$ + P_2O_5 and CaO + $CaCl_2$ + P_2O_5 ternary slags at 1473 K is shown in figures 1 and 2, respectively; the activities of P_2O_5 show a significant dependence on the MO/MCl_2 molar ratio; substitution of MCl_2 for MO increases the activities of P_2O_5.

With an assumption that the anions in MO + MCl_2 + P_2O_5 consist of merely PO_4^{3-}, O^{2-} and Cl^-, the anion fractions for these ions can be calculated by using the Temkin model. The oxygen anion fractions for fixed MO/MCl_2 mole ratios can be read on the auxiliary scale above the diagrams in figures 1 and 2; the activities of P_2O_5 showed a sharp increase at $X(O^{2-}) = 0$. This behaviour can be interpreted by taking account of the ionic equilibrium within the flux;

$$P_2O_5 + 3\,O^{2-} = 2\,PO_4{}^{3-} \quad\ldots\ldots\ldots\ldots\ldots\ldots\ldots\ldots(6)$$

$$a(P_2O_5) = X(PO_4{}^{3-})^2 / X(O^{2-})^3\,K'(6) \quad\ldots\ldots\ldots(7)$$

Equation(7) implies that the P_2O_5 activities increase sharply when $X(O^{2-})$ is close to zero, in conforming to the experimental results.

Figure 3 shows the logarithm of the activity coefficients of P_2O_5 as the function of $X(O^{2-})$ for BaO + $BaCl_2$ + P_2O_5 melts. At $X(O^{2-}) >$ 0.05, the logarithm of the activity coefficient could be well expressed as a linear function of $X(O^{2-})$, being independent of the $BaO/BaCl_2$ mole ratio. At oxygen anion fractions smaller than 0.05, however, significant deviation from the straight line is observed. This would presumably be due to the polymerization of phosphate ions;

$$2\,PO_4{}^{3-} = P_2O_7{}^{4-} + O^{2-} \quad\ldots\ldots\ldots\ldots\ldots\ldots\ldots(8)$$

The calculation of oxygen anion fraction based on the Temkin model does not accout for such a polymerization as expressed by equation(8). Hence, the departure from the straight line is not surprising.

Figure 4 shows the effect of temperature on the activities of P_2O_5 in BaO + $BaCl_2$ + P_2O_5 and CaO + $CaCl_2$ + P_2O_5 melts. Except for the concentration ranges corresponding to $X(O^{2-}) = 0$, a decrease in temperature by a magnitude of 50 K result in a decrease in the P_2O_5 activities by an order of about 1.5. The sharp increase in the activities is still observed near $X(O^{2-}) = 0$, indicating that the ionic structure or distribution within MO + MCl_2 + P_2O_5 melts would be fairly independent of temperature.

Figure 5 illustrate the relation between $RT\ln\gamma(P_2O_5)$ and $X(O^{2-})$ for BaO + $BaCl_2$ + P_2O_5 melts, As shown in this figure, at relatively high oxygen anion fraction, the relation could be well expressed by a linear function of $X(O^{2-})$, being independent of the MO/MCl_2 molar ratio;

$$RT\ln\gamma(P_2O_5) = m\;X(O^{2-}) + C \quad\ldots\ldots\ldots(9)$$

where m and C denote constants, respectively. As discussed elsewhere[4], a plot of iso-oxygen-anion fraction lines drawn on a ternary composition triangle for the MO + MCl_2 + P_2O_5 system consist of straight lines. Hence, by using equation(9), iso-activity curves for P_2O_5 can readily be drawn on a composition triangle.

Equation(9) is also advantageous in the ternary Gibbs-Duhem integration for the calculation of the activities of other components. The present authors have shown that the activities of MO and MCl_2 can be given by equations(10) and (11), respectively[4];

$$RT\ln\gamma(MO) = RT\ln[(r+1)/(r+2)]$$
$$-\,m\,[\,X(O^{2-}) + 2\ln(3 - X(O^{2-})) - r/(r+2)$$
$$-\,2\ln(2(r+3)/(r+2))\,] \quad\ldots\ldots\ldots\ldots\ldots\ldots(10)$$

$$RT\ln\gamma(MCl_2) = RT\ln[4(1+r)/(r+2)^2]$$
$$-\,2m\,[\,X(O^{2-}) + 3\ln(3 - X(O^{2-})) - r/(r+2)$$
$$-\,3\ln(2(r+3)/(r+2))\,] \quad\ldots\ldots\ldots\ldots\ldots\ldots(11)$$

where r denotes the molar ratio of MO/MCl_2.

Additional information drawn from the present results pertains to the phosphate capacity. Figure 6 shows the relation between phosphate capacity, C(P), as defined below, and $X(O^{2-})$ for BaO + $BaCl_2$ + P_2O_5 slags.

$$C(P) = X(PO_4{}^{3-}) / P(P_2)^{1/2}\,P(O_2)^{5/4} \quad\ldots\ldots\ldots(12)$$

As shown in figure 6, the phosphate capacity depends significantly upon $X(P_2O_5)$. It has been postulated that the phosphate capacities would be independent of $X(P_2O_5)$. However, for MO + MCl_2 + P_2O_5 melts, this is not the case.

$a(P_2O_5)$;	Activity of P_2O_5 referred to pure liquid P_2O_5.
h(P) ;	Henrian activity of phosphorus in molten copper.
K'(6) ;	Apparent equilibrium constant for reaction(6).
$P(O_2)$(slag);	Equilibrium oxygen partila pressure at the slag electrode.
$P(O_2)$(ref.);	Equilibrium oxygen partial pressure at the reference electrode.
P_e ;	Oxygen partial pressure at which the ionic and n-type electronic conduc conductivities are equal.
R ;	The gas constant.
T ;	Temparture(K).
γ (i);	Activity coefficient of component i.

ACKNOWLEDGEMENTS

Helpful comments, suggestions, discussion and encouragements were given by Professor Alex McLean, Department of Metallurgy, University of Toronto, and these are gratefully acknowledged. One of the authors (M.I.) also appreciates financial support from Nippon Kokan (NKK).

REFERENCES

[1] J. Lumsden, "Thermodynamics of Molten Salt Mixture", 1966, Academic Press, London.
[2] I. D. Sommerville, I. Ivanchev and H. B. Bell, in "Chemical Metallurgy of Iron and Steel", 1973, London, Iron and Steel Inst.

[3] S. Ban-ya, M. Hino and H. Takezoe, in "Proceedings of the 2nd International Symposium on Metallurgical Slags and Fluxes" pp.395/416, H. A. Fine and D. R. Gaskell, eds., TMS/AIME.
[4] M. Iwase et al., Iron and Steelmaker, vol.15, 1988, no.5, pp.69/80.
[5] M. Iwase et al., Steel Research, vol.56, 1985, pp.319/26.
[6] E. T. Turkdogan and J. Pearson, JISI, vol. vol.175, 1953, pp.398/401.
[7] H. Schmalzried, Z. Elektrochem., vol.66, 1962, no.7, pp.572/76.
[8] M. Iwase et al., Trans. Jpn. Inst. Met., vol.25, 1984, no.2, pp.43/52.
[9] M. Iwase et al., Electrochimica Acta, vol.19, 1979, no.3, pp.261/66.

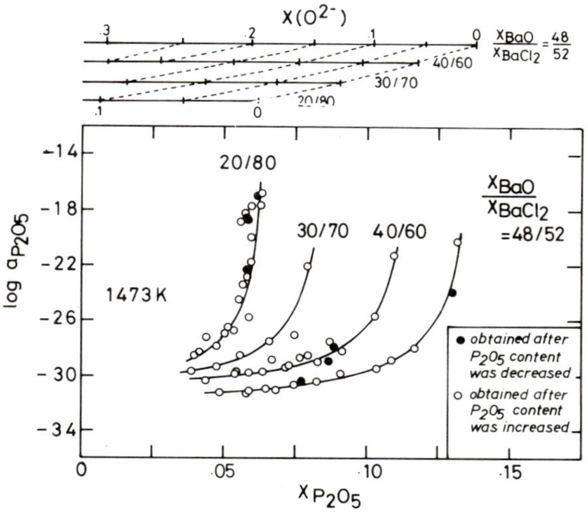

Figure 1 Activities of P_2O_5 as the function of $X(P_2O_5)$ at 14573 K for BaO + $BaCl_2$ + P_2O_5 melts.

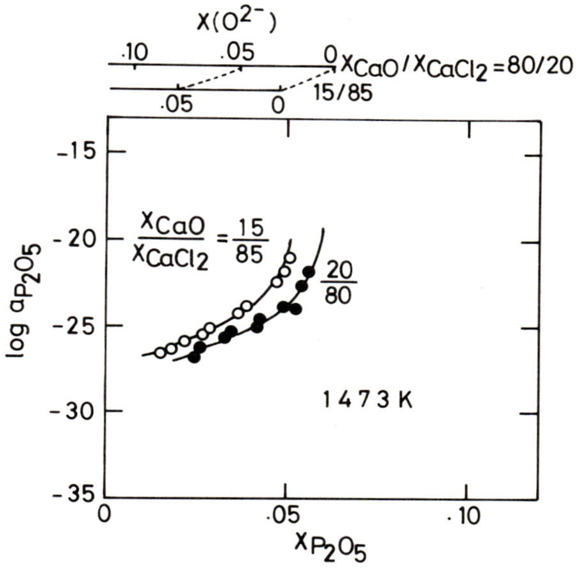

Figure 2 Activities of P_2O_5 as the function of $X(P_2O_5)$ at 1473 K for CaO + $CaCl_2$ + P_2O_5 melts.

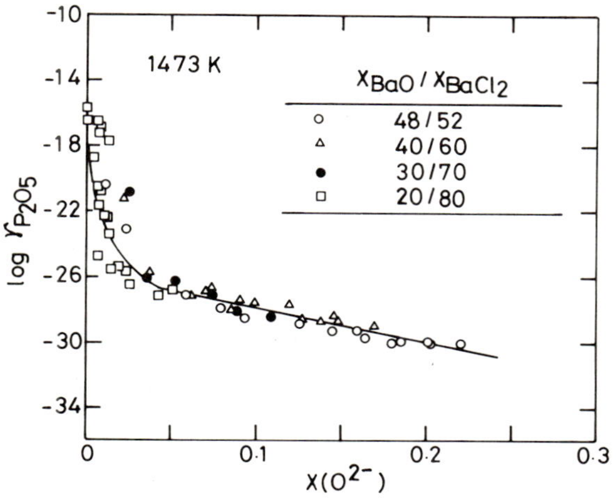

Figure 3 Activity coefficients of P_2O_5 in BaO + $BaCl_2$ + P_2O_5 as the function of $X(O^{2-})$ at 1473 K.

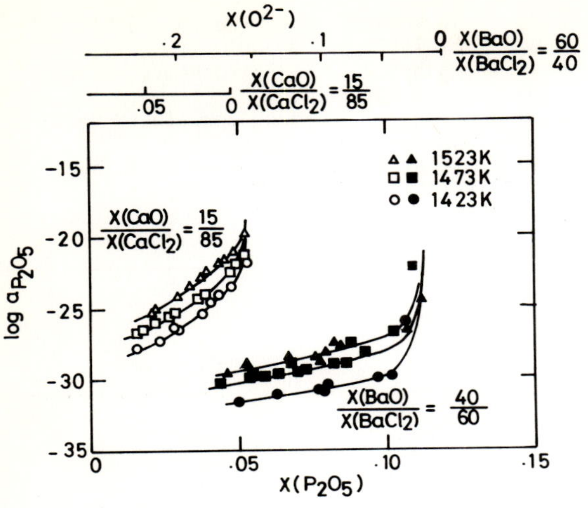

Figure 4 Effect of temperature on the activities of P_2O_5 in BaO + $BaCl_2$ + P_2O_5 and CaO + $CaCl_2$ + P_2O_5 melts.

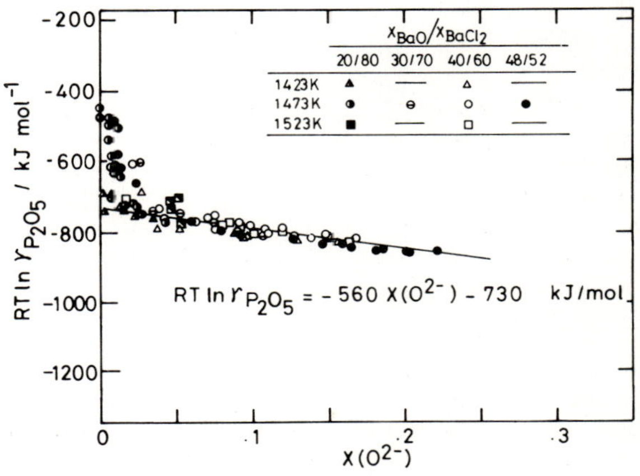

Figure 5 Relation between R T ln $\gamma(P_2O_5)$ and $X(O^{2-})$ for EaO + $BaCl_2$ + P_2O_5.

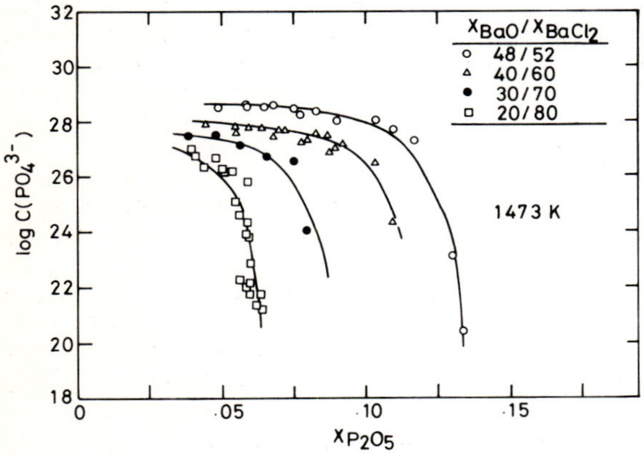

Figure 6 Phosphate capacities as the function of $X(P_2O_5)$ for BaO + $BaCl_2$ + P_2O_5 melts at 14573 K.

Thermodynamic studies on steel denitrization by flux

R INOUE and H SUITO

The authors are with the Research Institute of Mineral Dressing and Metallurgy, Tohoku University, Katahira Sendai, Japan 980.

Synopsis

Thermodynamic study on steel denitrization has been carried out in order to find out the best flux components. Based on the measured nitrogen distribution ratios between liquid iron and various slags at 1550 ~ 1650°C, the feasibility of steel denitrization has been discussed.

Introduction

Despite the fact that nitrogen control at low level is one of the most important refining processes in steelmaking, not much work has been reported, particularly, the study on steel denitrization by using flux.

It is the purpose of this investigation to determine the nitrogen distribution between slag and liquid iron, and slag and gas for various slag compositions and to offer the discussion based on the thermodynamic considerations.

Thermodynamic Considerations of Denitrization of Steel using Flux

The denitrization reaction between slag and metal in the absence of carbon can be written as

$$\underline{N} + 3/2(\,O^{2-}\,) = (\,N^{3-}\,) + 3/2\,\underline{O} \qquad [1]$$

The nitride capacity is defined by

$$C_N = (\,\%N\,)\cdot a_{\underline{O}}^{3/2} / a_{\underline{N}} = K\cdot a_{O^{2-}}^{3/2} / f_{N^{3-}} \qquad [2]$$

and the nitrogen distribution ratio L_N is given by

$$L_N = (\,\%N\,) / [\,\%N\,] = C_N\cdot f_{\underline{N}} / a_{\underline{O}}^{3/2} \qquad [3]$$

Similarly to the desulfurization reaction, the high value for C_N and low value for $a_{\underline{O}}$ are required to obtain higher value of L_N.

In the present work, the oxygen potential is controlled by the following $\underline{Al}/(Al_2O_3)$ equilibrium.

$$\underline{Al} + 3/2\,\underline{O} = 1/2\,(\,Al_2O_3\,) \qquad [4]$$

From Eqs.[1] and [4], we have

$$\underline{N} + \underline{Al} + 3/2\,O^{2-} = N^{3-} + 1/2\,(\,Al_2O_3\,) \qquad [5]$$

The standard free-energy change[1-3] in Eq. [5] has been calculated at 1600°C for the various oxides such as CaO, BaO, TiO_x, etc. As a result, the order of the ΔG^o value is as follows : $TiO_x < SiO_2 < ZrO_2 < Al_2O_3 < Ce_2O_3 < O < MgO < BaO < Li_2O < CaO$, suggesting that the oxide such as TiO_x, ZrO_2, etc. is more favorable to denitrization than the basic oxides. In order to estimate the relative magnitude of $a_{Al_2O_3}$, the ΔG^o value[1,2] for the following reaction is calculated at 1600°C.

$$MO + 2\,\underline{Al} + 3\,\underline{O} = MO\cdot Al_2O_3(s) \qquad [6]$$

The results are as follows : $BaO < Li_2O < CaO < MgO < TiO_2 < Al_2O_3$, indicating that as a result of lowering the oxygen potential with increase in basic oxide level, the denitrization becomes more favorable. The following three components thus have to take into account the denitrization flux ; that is, Al_2O_3 which controls the oxygen potential, the component which reduces $a_{Al_2O_3}$ and the component which increases nitride capacity.

Experimental

A $LaCrO_3$ vertical resistance furnace was used under deoxidized Ar atmosphere for the experiments on nitrogen distribution between slag and metal in the temperature range of 1550 to 1650°C. Twenty-five g of Fe-0.001 ~ 2%Al(-0.01%N) alloy and 8 g of slag containing 0 ~ 0.5% AlN were melted in Al_2O_3, MgO or CaO (Yoshizawa Lime Industry, Co., Ltd.) crucible for 1 ~ 5 hrs.

A SiC horizontal furnace was used for the slag-gas equilibrium experiment. The mixture of Ar-N_2-H_2 gas was circulated through the zone filled by Mn-MnO powder mixture by pump. Three g of slag on the Al_2O_3 boat was equilibrated at 1450°C for 8 ~ 13 hrs. The oxygen potential was measured by ZrO_2(+MgO) solid electrolyte.

The slag systems used in this work are as follows : $CaO-Al_2O_3$(-10 ~ 20% CaF_2), $CaO-SiO_2$, $MO-SiO_2-Al_2O_3$ (MO= CaO, BaO, Li_2O), $CaO-Al_2O_3$-MgO (-TiO_2, ZrO_2, CeO_2) and $MO-TiO_2-Al_2O_3$.

Results and Discussion

1. Nitrogen distribution between slag and liquid iron

Figure 1 shows the nitrogen distribution ratio L_N plotted against Al, Si contents in metal for the $CaO-Al_2O_3$(CA), $CaO-SiO_2$(CS) and $CaO-Al_2O_3-10\%CaF_2$(CAF) systems. It is clear in the $CaO-Al_2O_3$ system that the result obtained by CaO crucible is about seven times greater than that by Al_2O_3 crucible at the constant [%Al]. A slight effect of CaF_2 on L_N was observed. The log C_N values of $CaO-Al_2O_3$ slag at 1550 ~ 1650°C calculated from the value of $a_{Al_2O_3(s)}$[4] and ΔG^o[1] for $2\underline{Al}+3\underline{O}=Al_2O_3(s)$ reaction are given along with the previous results[5,6] in Fig.2, indicating that the log C_N values in the Al_2O_3 crucible is slightly greater than those in the CaO crucible at 1600°C. The addition of TiO_2, CeO_2 and ZrO_2 to the MgO-saturated $CaO-Al_2O_3$ flux ((%CaO)/(%Al_2O_3) =1) increases the L_N value in the increasing order of $ZrO_2 < CeO_2 < TiO_2$ as shown in Fig.3.

Figure 4 shows the results for the $MO-SiO_2-Al_2O_3$ (MO= CaO, BaO, Li_2O) systems, in which the order of the L_N value at the constant [%Al] and MO/SiO_2 mole ratio = 1 is as follows : Li_2O<CaO< BaO. The results for the $MO-TiO_2-Al_2O_3$(MO= CaO, BaO, Li_2O) system are also shown in Fig.4. It is seen that the flux containing BaO and TiO_2 is one of the best denitrization flux.

2. Slag-gas equilibrium

The temperature dependence of nitride capacity $C_N^{3-}(=(\%N)\ P_{O_2}^{3/4}/P_{N_2}^{1/2})$ for the $CaO-Al_2O_3$ system is obtained from the results of the slag-gas experiments and those of the slag-metal experiments shown in Fig.2. The results are shown in Fig.5, where the C_N^{3-} values in the $CaO-Al_2O_3$ system calculated from the previous workers' data[5,6] and those of the $BaO-SiO_2-Al_2O_3$ system are also included.

3. Steel denitrization by flux

Refining limit of [%N] has been estimated from the mass balance equation given by $[\%N]_f = [\%N]_i/[\ 1 + L_N \cdot W_{slag} \times 10^{-3}\ ]$ for the $BaO-TiO_2-Al_2O_3$ and $BaO-SiO_2-Al_2O_3$ systems. $[\%N]_f$ and $[\%N]_i$ represent the final and initial contents of nitrogen in metal, respectively and W_{slag} denotes the slag weight (Kg/ton steel). The calculated results are shown in Fig.6, in which the slag consumption 2 Kg/ton steel and 30 Kg/ton steel are examined for the case of initial 40 ppm N. It can be seen that by use of the $BaO-SiO_2-Al_2O_3$ slag with 30 Kg/ton steel, the nitrogen content can be lowered to 20 ppm at the 0.05% Al content. In the same Al content, the use of the $BaO-TiO_2-Al_2O_3$ slags with only 2 Kg/ton steel makes it possible to decrease nitrogen to 16 ppm.

Conclusion

The $BaO-TiO_2-Al_2O_3$, $Li_2O-TiO_2-Al_2O_3$ and $BaO-SiO_2-Al_2O_3$ systems were found to have excellent facility for liquid steel denitrization. The potential use of the flux denitrization will be expected for other metals and alloys such as super alloy.

References

1) D.Janke : 'Clean Steel', The Metals Society, London, (1983), p.202
2) E.T.Turkdogan : Physical Chemistry of High Temperature Technology, (1980) [Academic Press]
3) G.K.Sigworth and J.F.Elliott : Metal Science, 8(1974), p.298
4) R.H.Rein and J.Chipman : Trans.AIME, 33(1965), p.415
5) A.N.Morozov, A.G.Ponomarenko and Yu.E.Kozlov : Izv.Acad. nauk USSR, Metally, (1971), No.6, p.53; (1974), No.3, p.64
6) K.Schwerdtfeger and H.G.Schubert : Met.Trans. B., 8B(1977), p.535

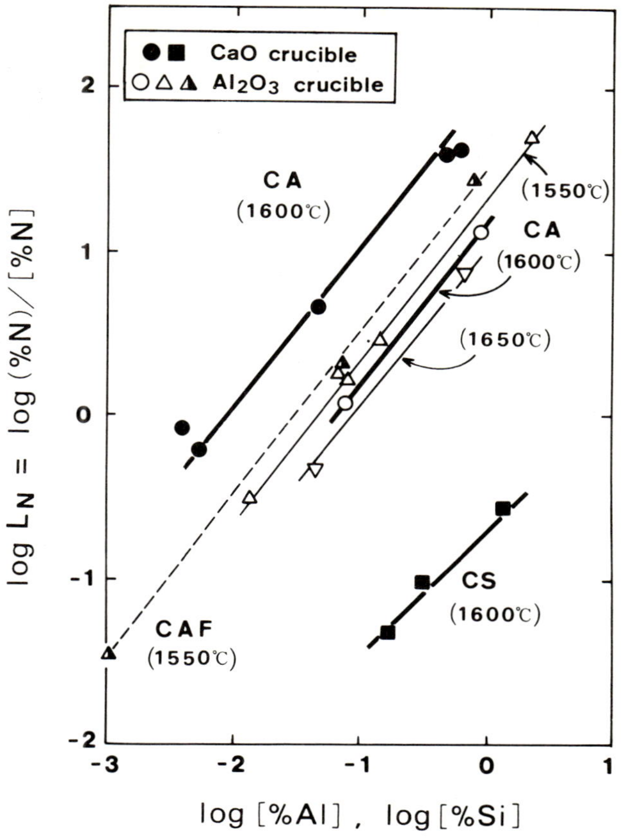

Fig.1 Logarithms of L_N in $CaO-Al_2O_3(-CaF_2)$ and $CaO-SiO_2$ systems plotted against log [%Al] and log [%Si].

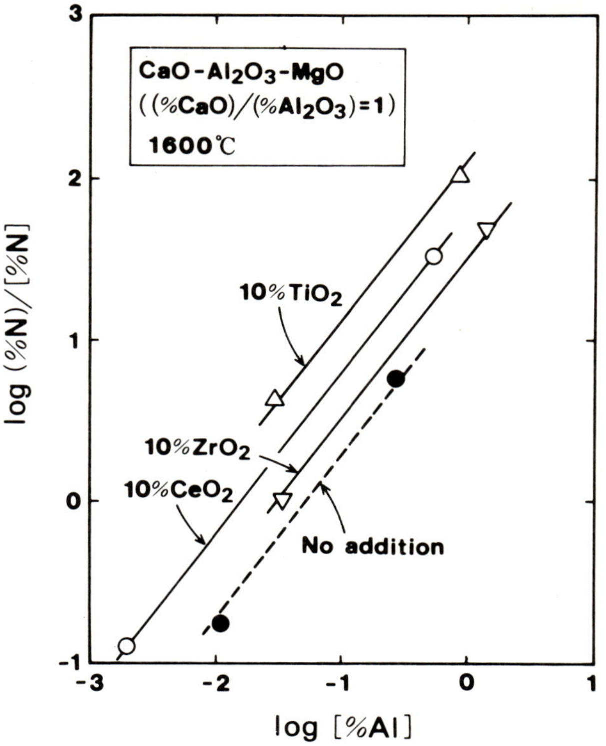

Fig.3 Effect of TiO_2, CeO_2 and ZrO_2 addition on log L_N.

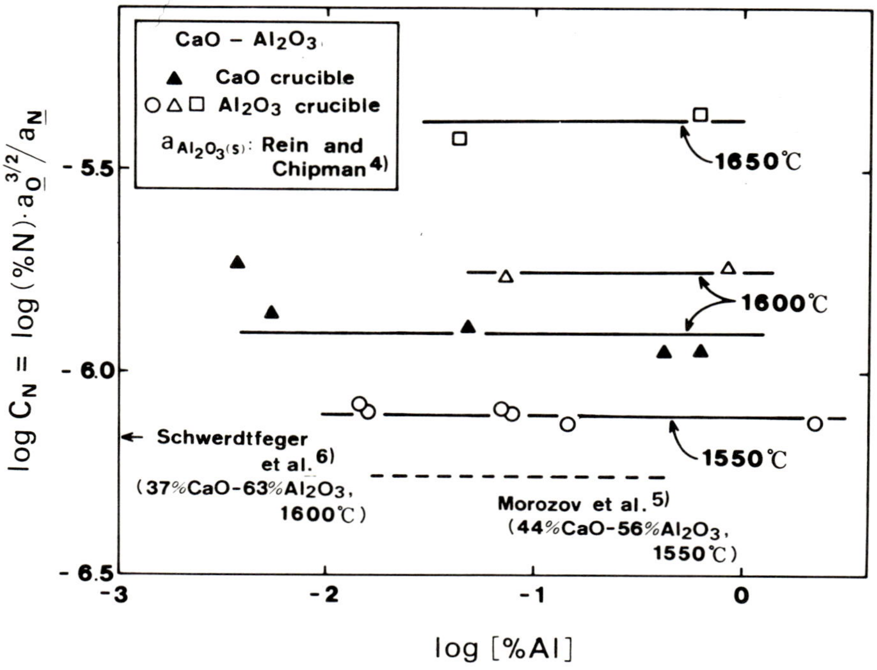

Fig.2 Logarithms of C_N in $CaO-Al_2O_3$ system at 1550 ~ 1650°C.

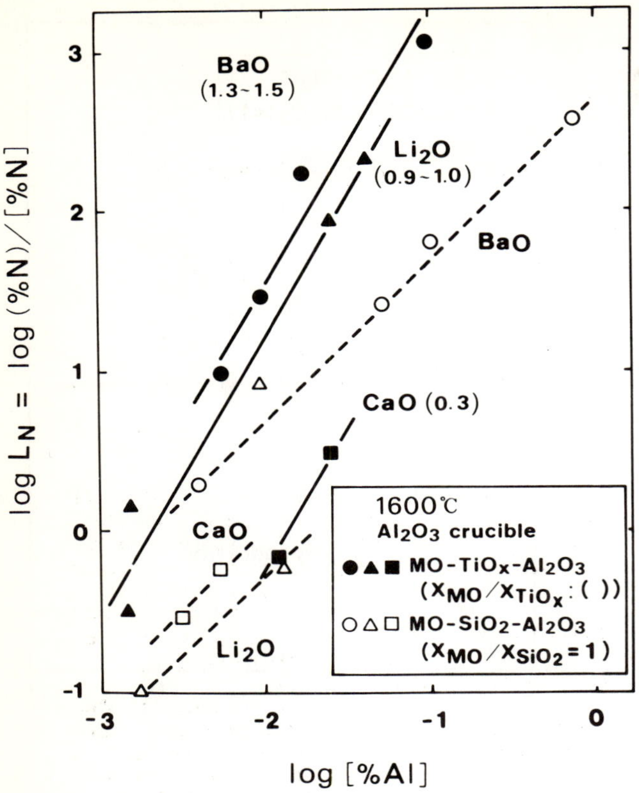

Fig.4 Comparison of log L_N between MO-SiO_2-Al_2O_3 (MO= BaO, Li_2O, CaO) and MO-TiO_2-Al_2O_3 systems.

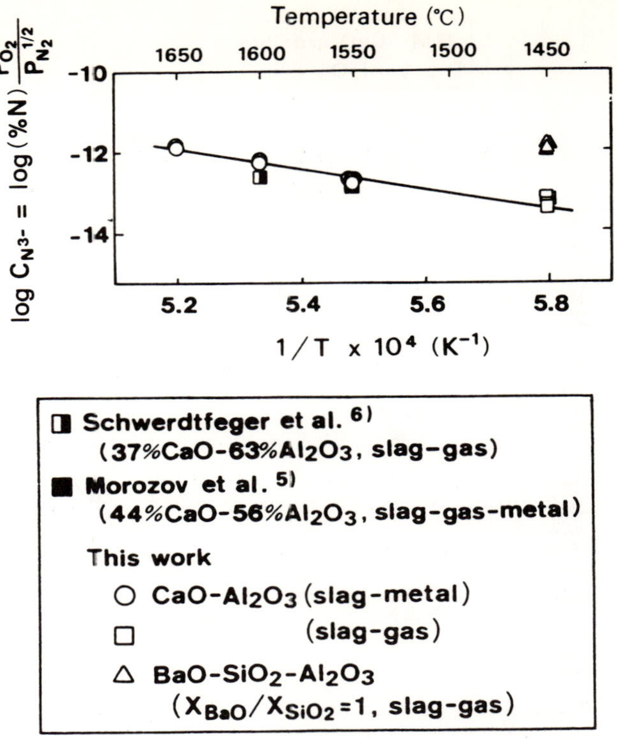

Fig.5 Temperature dependence of $\log C_{N^{3-}}$.

$$[N]_f = [N]_i / (1 + L_N \cdot W_{Flux} / 1000)$$

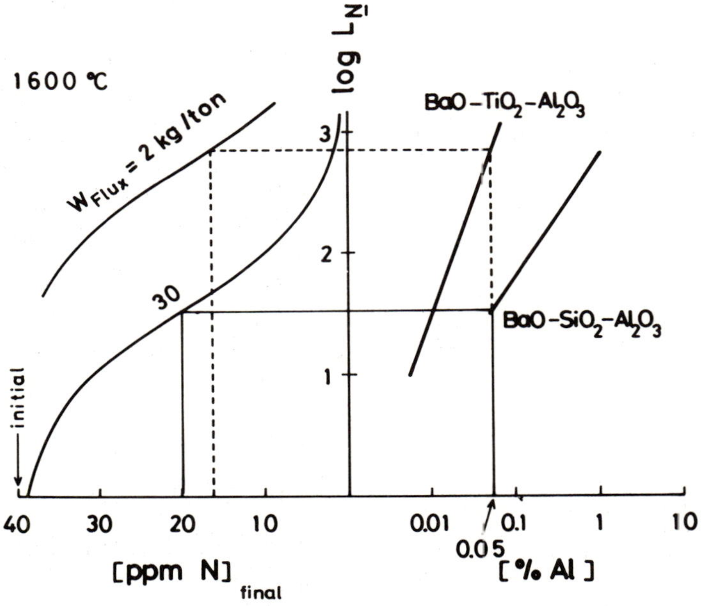

Fig.6 Refining limit of N in steel in BaO-based fluxes.

Heats of formation of liquid Fe_tO-CaO-SiO_2 and Fe_tO-MnO-SiO_2 slags

S BAN-YA, Y IGUCHI, H ISHIZUKA, T SHIBATA and M ARAI

Profs Ban-ya and Iguchi are in the Department of Metallurgy, Tohoku University; Messrs Ishizuka, Shibata and Arai, formerly graduate students at Tohoku University, are now with Kawasaki Steel Corporation, The Japan Steel Works Ltd., and Nippon Steel Corporation, respectively.

SYNOPSIS

The heats of mixing of solid SiO_2, CaO, or MnO with a solvent melted in an iron vessel were directly measured at 1420° or 1450 °C by the modified isoperibol calorimeter. The solvent was Fe_tO, Fe_tO-SiO_2 or Fe_tO-CaO in equilibrium with solid iron. The heats of formation of liquid ternary slags (Fe_tO-CaO-SiO_2, Fe_tO-MnO-SiO_2) in the liquid state of each slag component were calculated from the direct measured values and heats of fusion of SiO_2, CaO and MnO. These results are discussed in comparison with the estimated values by the thermochemical data and by the regular solution model proposed by J. Lumsden.

INTRODUCTION

Heats of formation of liquid slags are important thermochemical data for metallurgical reactions, and the thermodynamic properties of liquid slags have been studied both from chemical equilibrium and hypothetical calculation of available data. However, only a few directly measured heats of formation of liquid slags are available. Our previous studies have reported the heat of mixing of liquid Fe_tO-SiO_2 slags, determined directly by the modified isoperibol calorimeter at 1420 °C.[1]

In this study, heats of mixing of liquid Fe_tO-CaO-SiO_2 slags at 1420°C and Fe_tO-MnO-SiO_2 slags at 1450°C have been measured, and the contours of iso-heat of mixing have been obtained.

EXPERIMENTALS

Calorimeter

An isoperibol calorimeter(iso-thermal jacket type) used in this experiment consists of a calorimeter block made of iron and an outer alumina jacket kept at a constant temperature, which are set in an alumina reaction tube heated by a molybdenum wire resistance furnace as shown in Fig. 1. The calorimeter block made of low carbon steel has a hole for the thermocouple to detect a temperature change and another hole for the inner heater to determine the heat capacity of the calorimeter block. The solvent, pure Fe_tO, Fe_tO-SiO_2 or Fe_tO-CaO, was premelted in an iron crucible. The solute, solid silica, lime or MnO, has a cylindrical shape, which is connected with a molybdenum rod at the upper position. Mixing is made by moving up and down, and rotating the rod. Silica samples were made from a transparent quartz tube, but lime and MnO samples were sintered and prepared from each pure oxide powder by ourselves.

Experimental procedure

A certain time (1~2 hrs.) was required for the whole assembly to reach thermal equilibrium at the experimental temperature and for the solvent to reach chemical equilibrium with solid iron. Mixing was started by lowering and dipping the solute sample. Equilibrium attainment of Fe^{3+}/Fe^{2+} in the quenched solvent before mixing was confirmed by chemical analysis, in comparison with the results by Ban-ya et al.[2] The atmosphere is a purified argon. The temperature change of the calorimeter block reached a maximum within 30 ~ 70 sec in most cases. It might mean that mixing is very fast in these systems. After mixing, the determination of heat capacity of the calorimeter block was repeated by supplying a given amount of electric power to the Pt-Rh inner heater. The enthalpy change of the system on mixing was obtained by analysing the temperature-time curves drawn on a recorder chart on the basis of Newton's law of cooling. The obtained enthalpy change(ΔH) consists of the heat of fusion(ΔH_m^T) of the solute and the heat of mixing(ΔH_b^M) of the solvent, if it was the binary slag. Therefore, the heat of mixing of ternary slags in the liquid standard state(ΔH_t^M) was evaluated by combining the heat of fusion of solute (for SiO_2 and CaO from the literature but for MnO for this study) and the heat of mixing of the liquid binary slags obtained in the previous and present studies. When SiO_2 is dissolved in the liquid Fe_tO-CaO slag, the heat of mixing(ΔH_t^M) for the liquid Fe_tO-CaO-SiO_2 ternary slag is expressed by the following equation:

$$\Delta H_t^M = \Delta H - X_{SiO_2} \cdot \Delta H_m^T (SiO_2) + (1 - X_{SiO_2}) \cdot \Delta H_b^M \quad (1)$$

EXPERIMENTAL RESULTS AND DISCUSSIONS

Heat of mixing of Fe_tO-CaO-SiO_2 slags at 1420 °C

At first the heat of mixing for liquid Fe_tO-CaO binary slags was obtained by dissolution of solid lime into liquid Fe_tO. The value by Chang and Howald[3] was used as the heat of fusion of CaO. The results are shown in Fig. 2. Mixing of the Fe_tO-CaO system is more exothermic than that of the

Fe_tO-SiO_2 system reported in our previous study.[1] No direct measured data were available for comparison, so we tried to evaluate the heat of mixing from activity measurements. Several investigators, including Ban-ya, have reported the activities in this system, but clear temperature dependence (which is useful for calculation of heats of mixing) has not been obtained.[2, 4~7] α -function is almost constant over the composition range of $X_{CaO}>0.15$. Therefore, the heat of mixing was evaluated on the assumption of the regular solution. As shown in Fig. 2, the present results agree well with the calculated value except the range of $X_{CaO}<0.1$. From the present data, partial molar heats of mixing were also evaluated.

For the ternary slag, mostly solid CaO was dissolved into the liquid Fe_tO-SiO_2 binary slag and a few measurements were carried out by dissolution of SiO_2 into Fe_tO-CaO slag. The heats of mixing obtained by the different mixing sequence, however, were consistent with each other within the experimental error, when the final slag compositions were the same. This means that the heats of mixing of binary Fe_tO-SiO_2 in the previous study and Fe_tO-CaO in this study are reliable. From the results shown in Fig. 3, mixing of this system is exothermic and generated heats rise with increases in the concentration of CaO and SiO_2. The present result was compared with the value calculated from the activities reported in the literature on the assumption of the regular solution, and also compared with the value calculated with the Lumsden's model by using interaction energies determined by Lumsden,[8] and Ban-ya and Hino.[9] The measured and calculated values became close with each other away from pure Fe_tO and Fe_tO-CaO binary slag.

Heat of mixing of $Fe_tO-MnO-SiO_2$ slags at 1450°C

In this system, solid MnO was dissolved in liquid Fe_tO or Fe_tO-SiO_2 slag and the heat of mixing was evaluated by Eq.(1). So, the accuracy of the heat of fusion of MnO quite affects that of the heat of mixing. However, the heat of fusion published for MnO is scattered,[10~13] while liquid Fe_tO-MnO binary slag could be assumed to be an ideal solution from the activity measurements.[12, 14~16] Therefore, in this study the heat of fusion of MnO was evaluated from the enthalpy change when MnO was dissolved into liquid Fe_tO. It is 10000 ± 1500 cal/mol. By combining this heat of fusion of MnO and the heat of mixing of Fe_tO-SiO_2 binary in the previous study, the heats of mixing of $Fe_tO-MnO-SiO_2$ slags were obtained and shown in Fig. 4. The contours of iso-heat of mixing were estimated for this ternary slag. In this system, mixing is more exothermic with increasing SiO_2 and MnO.

For comparison, the Lumsden's model was applied and stable phases at steelmaking temperatures were chosen as standard states, such as liquid Fe_tO, solid SiO_2 and solid MnO. From Fig. 5, it was confirmed that the regular solution model gave a good approximation to evaluate the heat of mixing except for high Fe_tO concentration.

CONCLUSION

The heats of mixing of liquid slags have been directly measured at 1420° or 1450°C by the modified isoperibol calorimeter.
(1) Mixing of the Fe_tO-CaO slag is exothermic. This slag could be approximately treated as a regular solution except the range of $X_{CaO} < 0.1$.
(2) Mixing of the liquid ternary $Fe_tO-CaO-SiO_2$ and $Fe_tO-MnO-SiO_2$ slags are more exothermic with increasing CaO, SiO_2 and MnO contents.
(3) Obtained results are compared with the estimated values by the thermochemical data and by the regular solution model proposed by J. Lumsden.

ACKNOWLEDGMENTS

The authors are grateful to Drs. M. Hino and F. Ishii (Department of Metallurgy, Tohoku University) for their valuable contributions, and are also thankful to Messrs. K. Okada, M. Uemura, T. Igarashi and M. Fujiwara for assistance in the experiment. Part of this work was supported by the Ministry of Education, Science and Culture of Japan under a Grant-in-Aid for Scientific Research No.57550411(1982~1983).

REFERENCES

1) S.Ban-ya, Y.Iguchi, H.Honda and H.Ishizuka: Tetsu-to-Hagane,71(1985),p.846
2) S.Ban-ya, A.Chiba and A.Hikosaka:Tetsu-to-Hagane,66(1980),p.1484
3) D.R.Chang and R.A.Howaeld:High Temp.Sci., 15(1982),p.209
4) K.L.Fetters and J.Chipman:Trans.AIME, 145(1941),p.95
5) C.R.Taylor and J.Chipman:Trans.AIME, 154(1943),p.228
6) H.L.Bishop, N.J.Grant, and J.Chipman: Trans.AIME,212(1958),p.185
7) H.Fujita,Y.Iritani and S.Maruhashi:Tetsu-to-Hagane,54(1968),p.359
8) J.Lumsden:Phys. Chem. of Process Metall. Part 1 (1961),p.165 (Intersci. Pub., New York)
9) S.Ban-ya and M.Hino:Tetsu-to-Hagane, 73(1987),p.476
10)J.F.Elliott and M.Gleiser:"Thermochemical Properties of Inorganic Substances"(1960), Addison Wesely Pub.
11)K.K.Kelly:"Contribution to the Data on Theoretical Metallurgy",XIII, 1960, U.S.Bureau of Mines
12)W.A.Fischer and H.J.Fleischer:Arch.Eisenhüttenw.,32(1961),p.1
13)H.Fujita and S.Maruhashi:Tetsu-to-Hagane, 56(1970),p.830
14)J.Chipman,J.B.Gero and T.B.Winkler:Trans. AIME,188(1950),p.341
15)W.A.Fischer and P.W.Bardenheuer:Arch.Eisenhüttenw.,39(1968),p.559,p.637
16)S.Ban-ya,M.Hino and I.Kikuchi:Tetsu-to-Hagane, 72(1986),p.s936

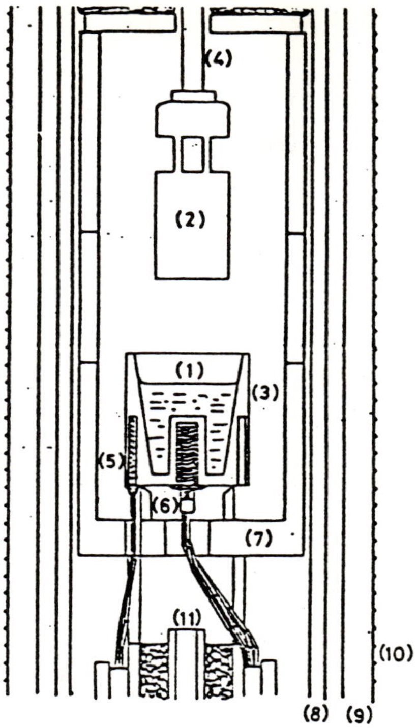

(1) Solvent
(2) Solute
(3) Iron crucible for solvent(calorimeter)
(4) Molybdenum rod
(5) Thermocouple for measurement (Pt·10%Rh-Pt)
(6) Internal heater for calibration (Pt·10%Rh wire)
(7) Alumina isothermal jacket
(8) Alumina reaction tube
(9) Alumina furnace tube
(10)Molybdenum heater

Fig. 1 Calorimeter assembly

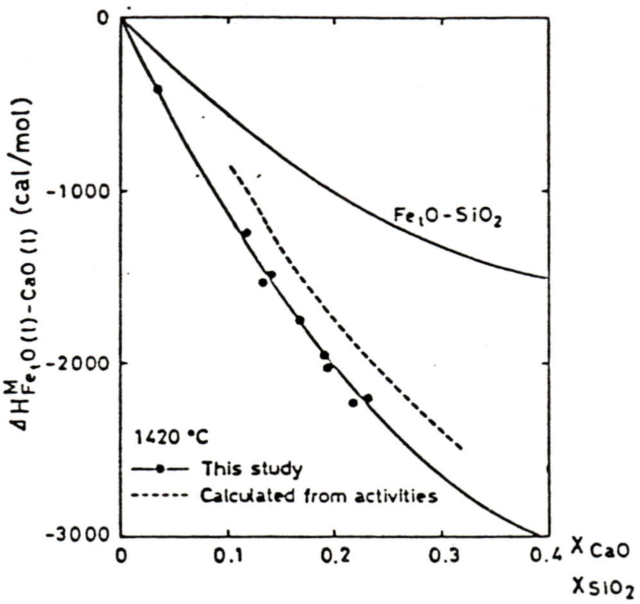

Fig. 2 Heats of mixing for liquid binary Fe_tO-CaO slags in comparison with those for Fe_tO-SiO_2 slags.

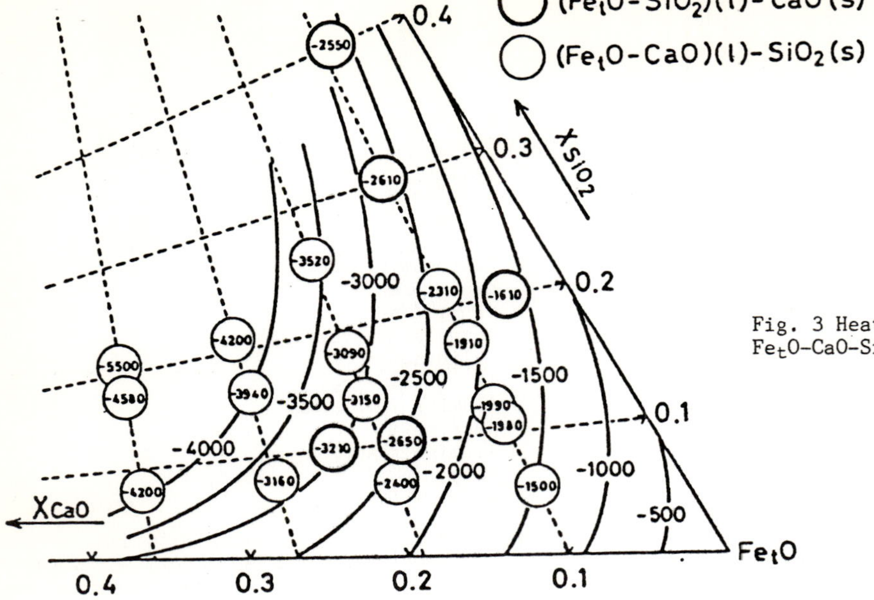

Fig. 3 Heats of mixing for liquid ternary Fe_tO-CaO-SiO_2 slags at 1420 ℃ (cal/mol).

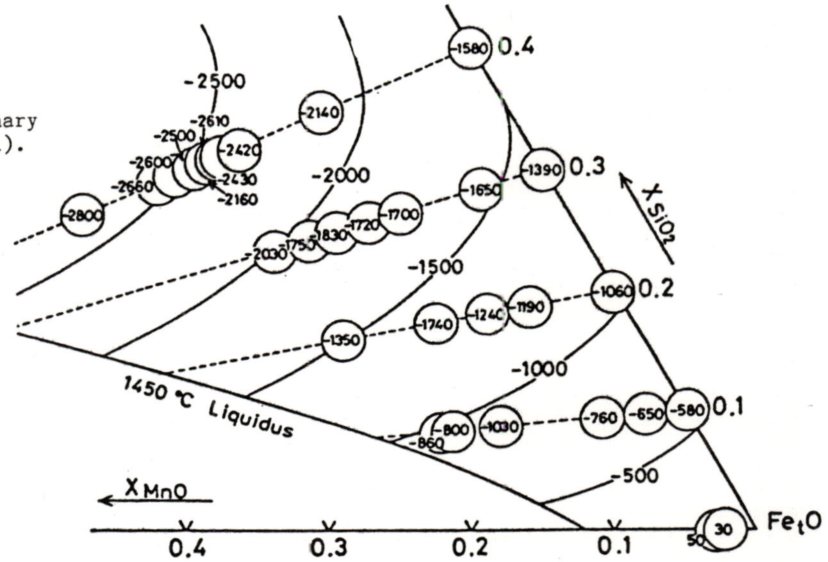

Fig. 4 Heats of mixing for liquid ternary Fe_tO-MnO-SiO_2 slags at 1450°C (cal/mol).

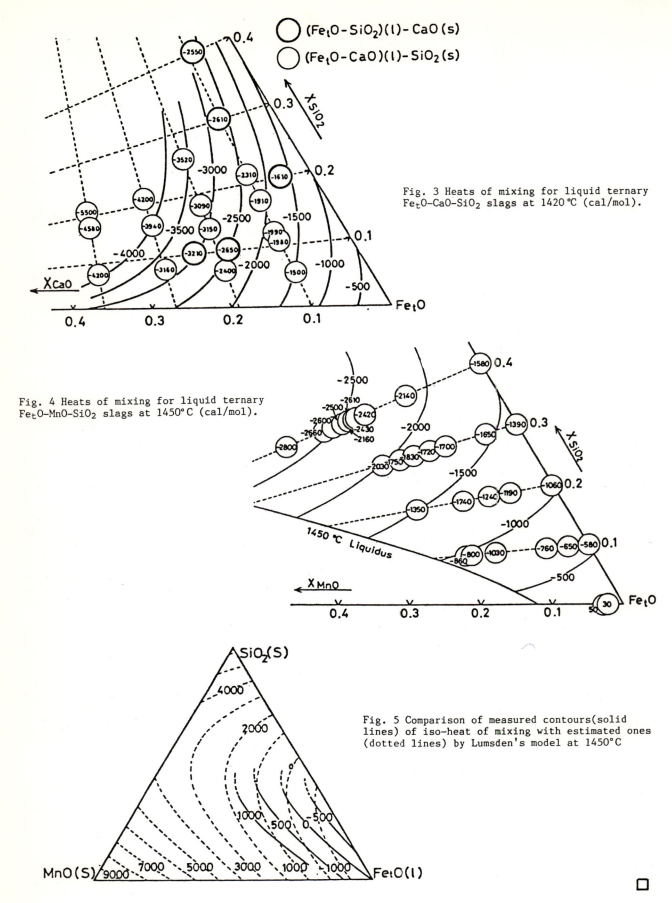

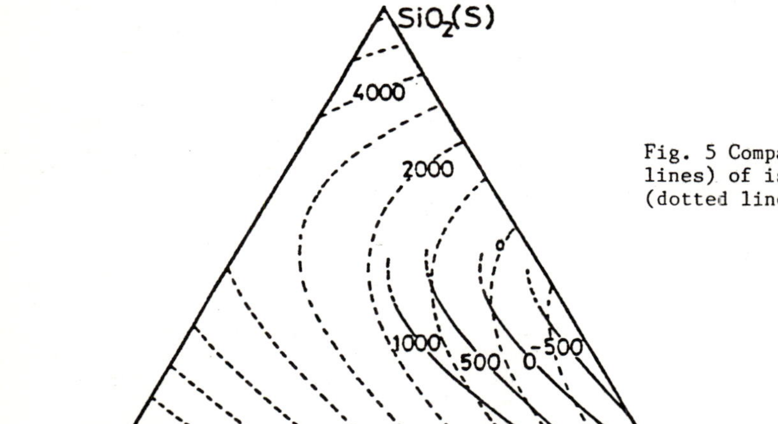

Fig. 5 Comparison of measured contours(solid lines) of iso-heat of mixing with estimated ones (dotted lines) by Lumsden's model at 1450°C

Sulphide capacity measurements in CaO-FeO-SiO_2 melts at 1400 and 1480°C

C SAINT-JOURS and M ALLIBERT

The authors are in the Laboratoire de Thermodynamique et Physico-Chimie Métallugiques, ENSEEG, BP.75, 38402 St Martin d'Hères, France.

SYNOPSIS

Sulphide capacity measurements were carried out by analysis of the sulphur content of samples exposed to CO-CO_2-SO_2 atmospheres for about 15h. Results were obtained mainly in the FeO-poor domain using Pt crucibles, but some experiments were carried out with silica-saturated samples in silica crucibles at high FeO contents. Conversely to the behaviour expected from literature data on FeO-SiO_2 the sulphide capacity does not change very markedly from one binary system to the other. Optical basicity or thermodynamic models are in agreement with measurements in the FeO-poor domain but deviate from them close to the FeO-SiO_2 binary.

INTRODUCTION

The prediction of sulphur or sulphide thermodynamics behaviour in oxide melts is of interest for iron and steelmaking as well as for sulphide deposit geology. The necessary thermodynamic modelling of these complex melts can be achieved only on the basis of experimental data on simpler binary and ternary systems. Among these basic systems the CaO-FeO-SiO_2 melts have not been studied in detail for sulphide capacity. Moreover the large variation of sulphide capacity measured by FINCHAM and RICHARDSON (1) from CaO-SiO_2 to FeO-SiO_2 is not reflected in the data obtained by BRONSON and ST-PIERRE (2) on ternary mixtures. Sulphide capacity measurements were thus undertaken to provide a larger data base for the thermodynamic modelling of this system for which oxide activity values have already been determined by TIMUCIN and MORRIS (3) and by the present author (4).

Sulphur behaviour in oxide melts was characterised by FINCHAM and RICHARDSON (1) by a measurable quantity called sulphide capacity Cs:

$$Cs = (wt\%S) \times Po_2^{1/2}/Ps_2^{1/2} \qquad (1)$$

where Po_2 and Ps_2 are the partial pressures of oxygen and sulphur in a gaseous phase at equilibrium with the oxide melts in which sulphur is dissolved. This relationship is based on the interchange equilibrium:

$$O^{=} + ½\, S_2 \rightarrow S^{=} + ½\, O_2 \qquad (2)$$

For thermodynamic calculations it is preferable to present this equilibrium as a competition between Fe and Ca ions for oxygen and sulphur ion surroundings:

$$CaO + ½\, S_2 \rightarrow CaS + ½\, O_2 \qquad (3)$$

$$FeO + ½\, S_2 \rightarrow FeS + ½\, O_2 \qquad (4)$$

The melt may then be described as a solution at infinite dilution of the thermodynamic species CaS and FeS, the respective amounts of which depend on CaO and FeO activities.

EXPERIMENTAL PROCEDURE

The measurements were carried out according to the procedure proposed by FINCHAM and RICHARDSON (1): samples of oxide mixtures are heated in a furnace and exposed to a gaseous mixture of CO, CO_2 and SO_2, the composition of which is known and maintained constant. When an equilibrium is reached, the samples are removed from the furnace, quenched and analysed for sulphur content. The knowledge of temperature and gas composition allows the calculation of oxygen and sulphur partial pressure and leads to the determination of Cs for each sample.

Samples

The studied mixtures were prepared by mixing CaO, SiO_2 and Fe_2SiO_4. CaO is obtained by roasting $CaCO_3$ (98% Rectapur-Prolabo) at 1000°C for 12 h. SiO_2 is simply dried in air at 300°C (99.9%. Fluka). Fe_2SiO_4 is prepared by sintering a mixture of Fe powder, Fe_2O_3 (99%) Rectapur-Prolabo) and SiO_2 for about 12 h at 1100°C in a quartz ampoule sealed under vacuum. X-Ray analysis confirmed the formation of Fe_2SiO_4 within its sensitivity limit (∿ 5%).

Containers

Platinum boats were used for samples containing less than 20wt%FeO. Five silica-saturated mixtures, including FeO-rich samples, were placed in silica crucibles. For these samples the liquid composition is calculated from the phase diagram (5).

Gaseous mixtures

The gaseous mixtures were obtained by mixing gases from three cylinders containing pure CO, pure CO_2 and pure SO_2. The mixtures were obtained by controlling the gas flow on the three lines by an automatic regulation set. The total gas flow is about 65cc/min with respective typical flows of about 15.30 and 20cc/min with an individual error of ± 1cc/min. The fluxes are mixed in a tube outside the furnace and then brought to the central zone, close to the samples, by a 4/6mm diameter 40cm long alumina tube.

Temperature

Temperature is measured in the centre of the furnace, close to the samples, by a PtRh 6%/PtRh 30% thermocouple placed as shown in figure 1. Absolute error on temperature, due to calibration and thermal gradient between samples and thermocouple, is estimated at about ± 10°C.

The furnace temperature is controlled to within ±1°C.

Equilibrium and procedure

The samples, made of unsintered mixed powders placed in platinum boats, are introduced in the furnace at 700-800°C. The gas flow is turned on and the temperature is raised to a predetermined value and maintained for about 15h. The samples are then removed in air and water quenched in their platinum boats.

A preliminary study on the CaO-MgO-SiO_2 system showed that equilibrium is reached within 5 h after melting of the oxide mixture. This duration, for a 8 mm thick bath, agrees with the studies reported by Turkdogan (6) in similar cases.

Chemical analysis

The total weight of a sample is between 2 and 3 g and allows two or three samples to be taken for analysis. The method (7) comprises dissolution of the sulphur-containing oxide in a reducing solution followed by transfer of H_2S into a separate silver-containing solution. A silver sensitive electrochemical probe is used to monitor the silver ion concentration and permits back titration of the precipitated silver sulphide. The error on sulphur content is estimated at ± 0.02wt% on values of about 0.1% corresponding to sulphide capacities ranging from $3x10^{-5}$ to $5x10^{-4}$.

The overall error on sulphide capacities is estimated at ± 35%, and more than ± 50% for the lower values of Cs ($\sim 10^{-5}$).

RESULTS

The oxygen and sulphur partial pressures were set at $7.38x10^{-8}$atm and $3.55x10^{-2}$atm respectively at 1480°C, and $1.77x10^{-3}$atm and $4.49x10^{-2}$atm at 1400°C. Oxygen partial pressure is chosen to avoid reduction of FeO and dissolution of iron in the platinum containers. Sulphur potential is selected to obtain a low but measurable sulphur content. A higher sulphur potential cannot be obtained with the gaseous sources used because they would correspond to very low CO_2 flows i.e. a high uncertainty on oxygen and sulphur potential.

The results are reported in Table 1 and figures 2 and 3 along with measurements carried out in silica crucibles for high-FeO silica-saturated samples. Experimental data given by BRONSON and ST-PIERRE at 1500°C for FeO contents of 14 and 18 wt% are significantly higher. This discrepancy could originate in a high oxygen pressure in their experimental set up (sealed container). For instance, for a sample mean composition of 0.45 SiO_2, 0.43 CaO and 0.12 FeO (molar fractions) and 0.01 to 0.1 mole of sulphur S the oxygen partial pressure inside their vessel was estimated from thermodynamic calculation to range from 10^{-6} to 10^{-4} atm, pressures under which sulphate may form beside sulphide.

Modelling of the sulphide capacity behaviour

The variation in sulphide capacity with composition can be predicted with the model proposed by SOSINSKY and SOMMERVILLE (8,9), based on correlations with measured optical basicities, when possible. The application of this model gives the iso-Cs lines presented on Figure 4.

The present authors extended to polyanionic mixtures a thermodynamic model based on the models proposed by KAPOOR and FROHBERG (10) and developed for steelmaking slags by H. GAYE (11). This model is based on the description of melts with associated species containing a single anion (O,S,F). This description is aimed at translating the effect of anionic bindings into thermodynamic properties using the following basic assumptions:

- the electronic state of each anion depends primarily on its cationic surroundings and characterises the energy of the whole solution (including cations).
- the energy levels that can be occupied by an anion are limited to a few typical configurations identified by anion-cation associations called "cells".

The stoichiometry of associates or "cells" is presented in Table 2 together with the various parameters of the model. The free energy of mixing of this solution is expressed as follows:

$$G = \sum_i X_i(\Delta G°_i + RT \ln x_i) + \sum_v \sum_{ij} \varepsilon^v_{ij} x_i x_j (x_i - x_j)^v \quad (5)$$

where i and ij designate a cell and a couple of cells, x_i the molar fraction of cell i, $\Delta G°_i$ the free energy of formation of cell i, ε^v_{ij} an interaction coefficient for a pair ij and v the order corresponding to a Redlich-Kister polynomial. The sulphide behaviour is represented only by considering the associations CaS and FeS and by neglecting the variation of their free energy of formation with the oxide composition just as if S interacted only with O and its first cation neighbours. This S-O interaction is assumed to be unaffected by the oxygen ion electronic state, which means that the sulphide activity coefficient at infinite dilution is independent of oxide composition as suggested earlier by FINCHAM and RICHARDSON.

In the present case, sulphide capacity in the CaO-SiO_2 system was used to adjust the free energy of formation of the CaS cell as shown in Fig. 5. The corresponding parameter for FeS was adjusted to fit the Cs values in the ternary system because the FeO-SiO_2 sulphide capacities were determined only for a single composition (1). Moreover this value seems very high when compared to values estimated from the optical basicity model.

Indeed both this latter model and our thermodynamic model satisfactorily represent the ternary data (Table 3) but do not extrapolate to the Cs value measured on the FeO-SiO_2 binary. ($\sim 1x10^{-4}$ instead of $55x10^{-4}$ as measured). It should be noted that, although the thermodynamic model is slightly better (Fig. 6) in the silica-rich domain than the optical basicity model, due to its greater number of parameters, neither model predicts Cs values measured for mixtures having a FeO/SiO_2 weight ratio greater than 0.3. However the discrepancy is limited to a factor of 3 instead of 50 and may be attributed to the assumption of constant activity coefficient for FeS.

CONCLUSION

Sulphide capacities were measured at two temperatures in a portion of the liquid domain of the $CaO-FeO-SiO_2$ system. The examined composition range was limited by the availability of FeO resistant containers at high FeO activity. Despite this limitation, it seems that the published Cs value in the $FeO-SiO_2$ binary is much too high (by a factor of 10) so that substitution of CaO by FeO in a silicate would not greatly facilitate sulphur solubility.

Models are available to represent or predict the variation of sulphide capacities with oxide composition up to a FeO contents of about 20-25 wt%. For larger amounts, new developments associated with further experimental investigations are needed.

REFERENCES

(1) C.J.B. FINCHAM, F.D. RICHARDSON
Proc. Roy. Soc. (London), 1954, vol.A 223, p.40.

(2) A. BRONSON, G.R. ST PIERRE
Met. Trans. B, 1981, Vol.12B, p.729.

(3) M. TIMUCIN, A.E. MORRIS
Met. Trans., 1970, 1, P.3193.

(4) A. DHIMA, B. STAFA, M. ALLIBERT
High Temp. Science, 1986, Vol.21, p.143.

(5) Slag Atlas - Verlag Stahleisen m.b.h, 1981, Düsseldorf.

(6) E.T. TURKDOGAN, M.L. PEARCE
Trans. Met. Soc. AIME, 1963, Vol.224, p.316.

(7) H. BOZON, S. BOZON
Analysis, 1978, 6, p.243.

(8) I.D. SOMMERVILLE, D.J. SOSINSKY
Proc. Second Intern. Sympos. Metall. Slags and Fluxes, The Metall. Soc. AIME, 1984, p.1015.

(9) D.J. SOSINSKY, I.D. SOMMERVILLE
Met. Trans. B, 1986, Vol.12B, p.729.

(10) M.L. KAPOOR, M.G. FROHBERG
Chem. Met. of Iron and Steel Symposium, Sheffield, 1971, p.17.

(11) H. GAYE, D. COULOMBET
"Données thermochimiques et cinétiques relatives à certains matériaux sidérurgiques". Rapport CECA N° 7210-CF/301, Mars 1984.

Table 1 : Experimental results for sulphide capacities in $CaO-FeO-SiO_2$ at 1400°C and 1480°C.

N° Sample	Compositions %SiO_2	%CaO	%FeO	1480°C $Cs.10^4$	1400°C $Cs.10^4$	ΔH_r (KJ)
1	35	44	21	4.92		
2	40	37	23	1.80	0.96	257.5
3	41	39	20	1.58	0.65	164
4	41.5	46.5	12	1.85		
5	42	50	8	1.90		
6	43	40	17	2.05	0.42	461.3
7	44	30	26	1.70	0.63	258.2
8	44	42	14	1.70		
9	45	50	5	1.70		
10	46	4	50	4.00		
11	46	34	20	1.06	0.57	176.3
12	46	36	18	0.64		
13	47	38	15	0.69	0.44	136
14	47.5	44	8.5	0.47		
15	49	40	11	0.36		
16	49	29	22	0.75	0.30	285.7
17	50	34	16	0.42	0.20	227.2
18	53	12	35	2.26		
19	54.5	36	9.5	0.45		
20	55	15	30	1.39		
21	55	23	22	0.39	0.28	100
22	56	39	5	0.26		
23	57	28	15	0.40	0.29	91.2
24	58	22	20	0.50		
25	59	34	7	0.26		
26	60	30	10	0.26	0.10	241
27	62.5	32	5.5	0.16		
28	63	27	10	0.26		

Table 2 : Parameters used in the thermodynamic modelling of the binary and ternary boundary systems.

Cells i stoich.	$\Delta G°_i$ (joule)	Interaction parameters equat (5) (Joule) Binary	0	1	2
1 CaO	0	1-2	-58600	-29300	0
2 FeO	0	1-4	-67000	0	0
3 Si½O	0	2-3	12560	-6280	31400
4 Fe2/3O	0	2-4	7120	0	0
5 Ca½Si1/4O	-47100	2-5	25120	0	0
6 CaS	155700 -125.6T *	3-4	20930	33490	0
7 FeS	22800 *	3-5	8370	26790	0
*For reactions (3) and (4)		others	0	0	0

Table 3 : Comparison between experimental and calculated values of Cs.

N° Sample	$Cs.10^4$ exp.	$Cs.10^4$ th.mod	$Cs.10^4$ opt.bas.	Deviation % th.mod.	opt.bas.
1	4.92	3.10	7.28	37	- 48
2	1.80	1.04	2.59	42	- 44
3	1.58	1.10	2.10	30	- 33
4	1.85	1.84	2.73	0.5	- 48
5	1.90	1.91	2.78	0.5	- 46
6	2.05	0.95	2.05	55	0
7	1.70	0.79	1.67	54	2
8	1.70	0.97	1.82	43	- 7
9	1.70	1.43	1.75	16	- 3
10	4.00	0.70	0.95	83	76
11	1.06	0.75	1.22	29	15
12	0.64	0.78	1.34	- 22	-110
13	0.69	0.72	1.18	- 3	- 69
14	0.47	0.71	1.02	- 51	-117
15	0.36	0.63	0.72	- 75	-100
16	0.75	0.65	0.81	13	- 8
17	0.42	0.55	0.52	- 31	- 24
18	2.26	0.64	0.40	72	82
19	0.45	0.49	0.43	- 9	4
20	1.39	0.60	0.30	57	78
21	0.39	0.54	0.37	- 38	5
22	0.26	0.53	0.39	-100	- 50
23	0.40	0.50	0.31	- 25	22
24	0.50	0.43	0.27	14	46
25	0.26	0.35	0.19	- 35	27
26	0.26	0.30	0.25	- 15	4
27	0.16	0.27	0.19	- 69	- 19
28	0.26	0.25	0.16	4	38
mean value and standard deviation				3 ± 46	-12 ± 52

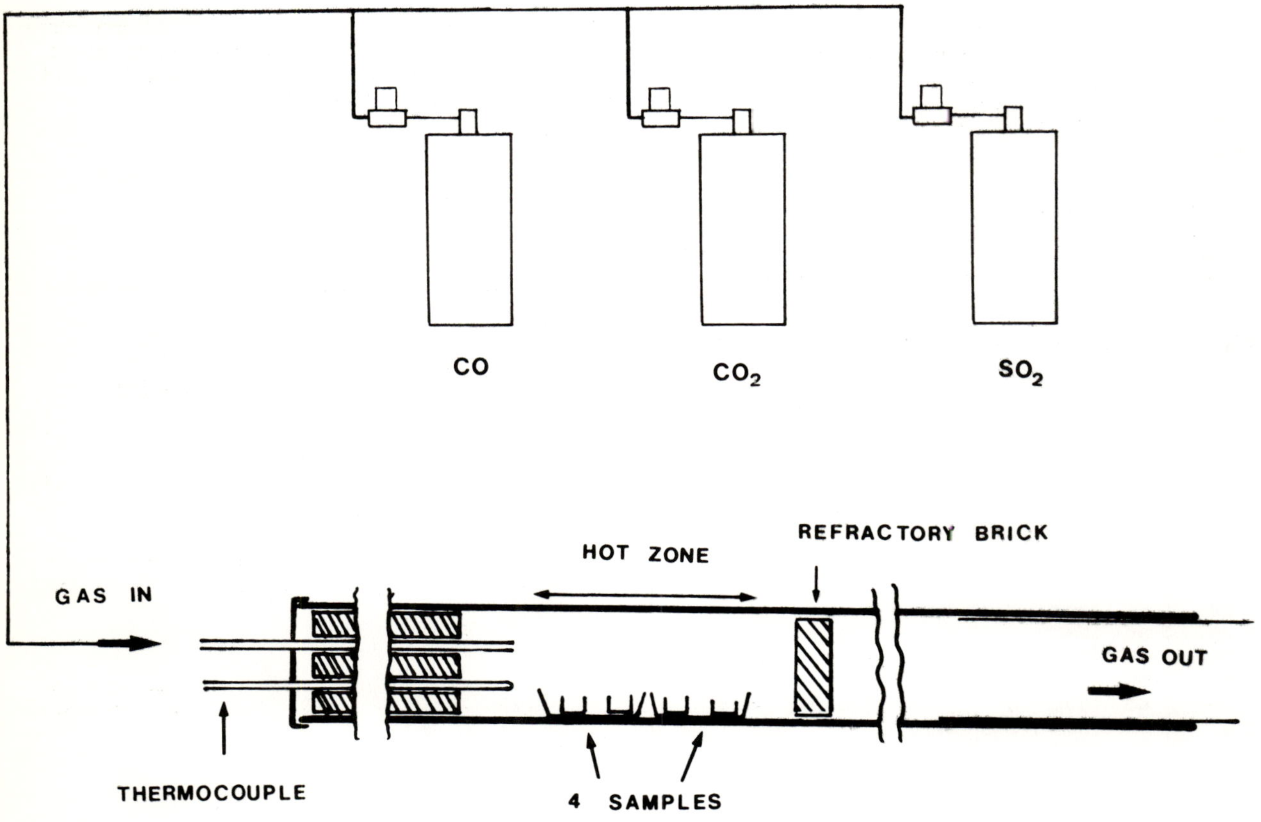

Fig.1 Schematic diagram of the experimental set-up.

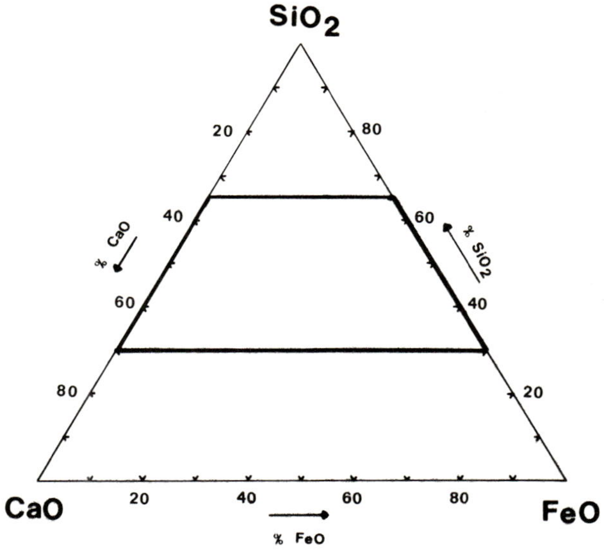

Fig.2 Studied composition domain.

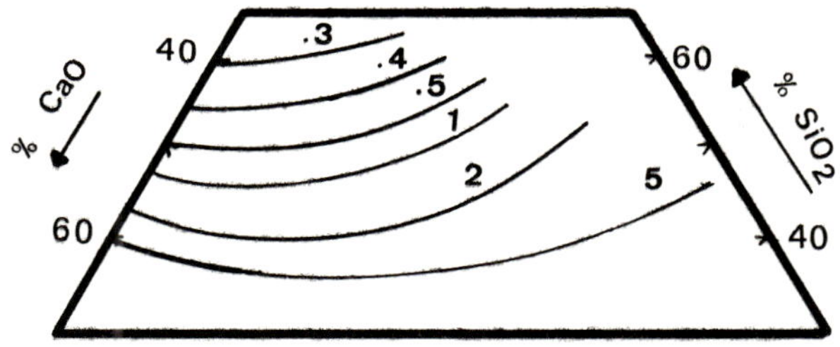

Fig.3 Experimental values of Cs x 10^4 obtained at 1480°C.

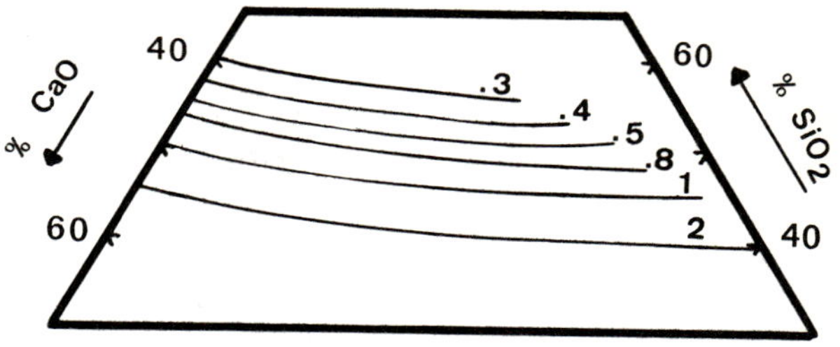

Fig.4 Iso-Cs lines deduced from SOMMERVILLE and SOSINSKY's optical basicity model at 1480°C.

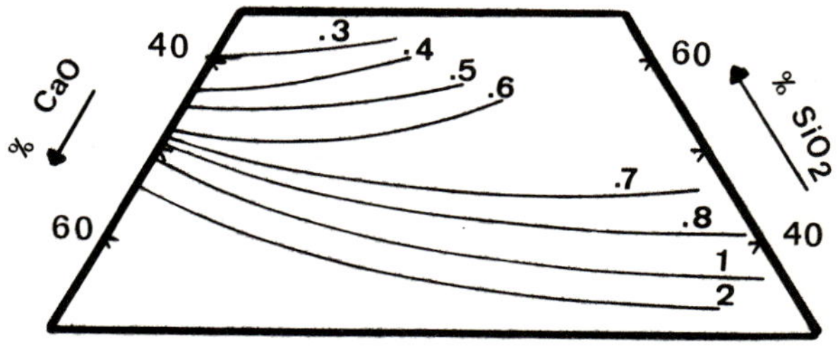

Fig.5 Representation of FINCHAM and RICHARDSON's data by the authors' thermodynamic model and SOMMERVILLE and SOSINSKY's relationship with optical basicity.

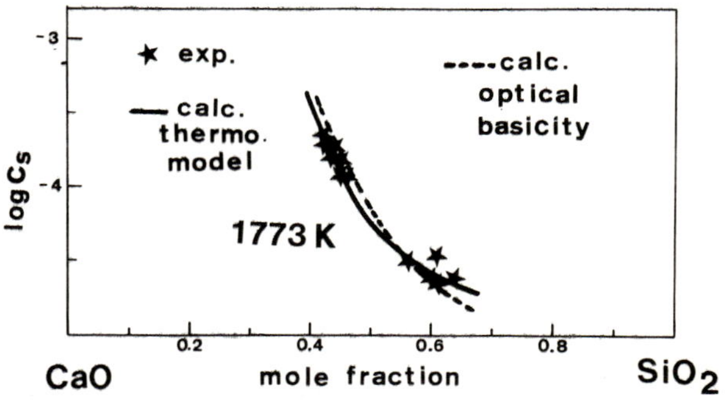

Fig.6 Iso-Cs lines calculated from the authors' thermodynamic model using the "monoanionic cells" concept.

Sulfide capacities of $CaO-SiO_2$ melts containing K_2O

ZHENQI HUANG and G R ST. PIERRE

G R St Pierre is Professor and Chairman, Department of Metallurgical Engineering, The Ohio State University, Columbus, Ohio 43210;
Z Huang, formerly Visiting Instructor, is now Instructor at Northeast University of Technology, Shenyang, Peoples Republic of China.

I. Introduction.

With very few exceptions, sulfur is considered undesirable in steel, and there is an increasing demand for steel with lower sulfur content. Sulfur may be removed from molten iron and steel by reaction with fluid, highly basic slags of low iron oxide content. Because of the importance of the $CaO-SiO_2$ system, it has been the subject of several studies, both theoretical and experimental, which have provided data on the thermodynamic properties of $CaO-SiO_2$ melts. (1-9)

The reactions by which a lime-containing slag absorbs sulfur from a gas containing SO_2 and S_2 may be conveniently expressed by the reactions:

$$(CaO) + 1/2\ S_2 = (CaS) + 1/2\ O_2 \qquad (1)$$

$$(CaO) + 1/2\ S_2 + 3/2\ O_2 + (CaSO_4) \qquad (2)$$

where parentheses denote reactants dissolved in the slag phase. It is clear that sulfide formation will be favoured by reducing conditions, whereas sulfate formation will occur more readily under the oxidizing condition. According to the results of Fincham and Richardson(5), when P_{O_2} is less than 10^{-5} atm, sulfur transfer takes place almost entirely by sulfide formation; whereas under more oxidizing conditions ($P_{O_2} > 10^{-3}$atm), sulfate formation predominates. For the sulfide capacities of the $CaO-SiO_2$ slag system, the effects of CaF_2, B_2O_3, FeO, MgO, Al_2O_3 and TiO_2 have been investigated by A. Bronson and G.R. St.Pierre (6,7) using an encapsulation method. They found that for melts of fixed CaO/SiO_2 ratio, the addition of CaF_2 increases sulfide capacity and the addition of B_2O_3 decreases it. The substitution of CaF_2 for CaO does not alter sulfide capacity significantly. The substitution of B_2O_3 for SiO_2 increases it slightly. When a basic oxide/acid oxide ratio of 1.21 is maintained, MgO additions decrease sulfide capacity, while FeO, TiO_2 and Al_2O_3 additions increase the capacity.

In silicate melts, the bonding of oxygen atoms depends on the relative proportions of silica and basic metal oxides. This may be illustrated by the Eq. (3)

$$2(Si\text{-}O\text{-}) = (Si\text{-}O\text{-}Si) + O^{2-} \qquad (3)$$

where (Si-O-) and (Si-O-Si) represent unshared and shared corners, respectively, of silicate tetrahedrons. The equilibrium constant for Eq. (3) is

$$K_3 = (Si\text{-}O\text{-}Si)\ (O^{2-})\ /\ (Si\text{-}O\text{-})^2 \qquad (4)$$

In acid slags, the concentration of (Si-O-Si) will be large; while in basic slags the concentration of free oxygen ions will be large. Therefore, basic slags, which have a higher free oxygen ion activity than acid slags, absorb more sulfur. On the other hand, from D.J. Sosinsky and I.D. Sommerville (9), the relationship between sulfide capacity and optical basicity at 1500°C is

$$\log C_s = 12.6\ B_{op} - 12.3 \qquad (5)$$

where B_{op} is the optical basicity. The B_{op} for K_2O and Na_2O are larger than the contribution from CaO. Therefore, it is expected that the additions of alkaline metal oxides to the $CaO\text{-}SiO_2$ system should increase sulfide capacity. The major objective of the present study is to measure the effects of alkaline metal oxides, Na_2O and K_2O, on the sulfide capacities of $CaO\text{-}SiO_2$ slags at 1500°C.

II. Background of the Method

Consider the exchange reaction given in Eq. 6

$$O^{2-}(slag) + 1/2\ S_2(gas) = S^{2-}(slag) + 1/2\ O_2(gas) \qquad (6)$$

The equilibrium constant is given by

$$K_6 = f_{S^{2-}}\ (S^{2-})\ (P_{O_2}{}^{1/2}) / a_{O^{2-}} P_{S_2}{}^{1/2} \qquad (7)$$

which may be rearranged as follows

$$(S^{2-})\ (P_{O_2}/P_{S_2})^{1/2} = K_6 / \alpha_{O^{2-}} / f_{S^{2-}} = C_s \qquad (8)$$

where C_s is well known as the sulfide capacity, (S^{2-}) and $f_{S^{2-}}$ represents the sulfide concentration and activity coefficient, respectively. All other terms have their usual meanings. It is clear that sulfide capacity, C_s is a property of slag melts dependent only upon slag composition and temperature. At a certain temperature, i.e. 1500°C, a reference slag, whose sulfide capacity $C_s{}^o$ is known, and other slags may be put into equilibrium in a closed capsule. In these experiments the capsule has been filled with Ar, SO_2 and O_2 gases. When the equilibrium state between all of the slags and the gas mixture is established, a given sulfurizing potential is also established throughout the capsule. That is,

$$(P_{O_2}/P_{S_2})^{1/2} = C_s{}^o\ /\ (S^{2-})_0 \qquad (9)$$

where $(S^{2-})_0$ is the final sulfur content in the reference slag. Therefore, the sulfide capacities of all other slags can be calculated from their sulfur contents, (pct S), and the given sulfurizing potential

$$C_s = C_s{}^o\ (pct\ S)\ /\ (pct\ S)^0 \qquad (10)$$

III. Experimental Procedure

The experiments were conducted in a furnace with tubular heating elements of silicon carbide at temperatures near 1500°C. The $CaO\text{-}SiO_2$ slags containing alkaline metal oxides and reference slags were contained in platinum cups and enclosed in a quartz capsule. Most of the slag samples in this study were contained in platinum cups of 3 mm radius and 4 mm height. The cups were hand-drawn from 0.127 mm platinum foil. A few slag samples were contained in platinum cups with 6 mm radius and 14 mm height.

The temperature was measured with a Pt-6%Rh-Pt-30%Rh thermocouple rising on the inside wall of the Al_2O_3 tube beside the quartz capsule. The master slags were prepared by fusing reagent grade SiO_2 and CaO powder in platinum cups in air. The sulfide capacity of the CaO-SiO_2 reference slags at 1500°C were taken from Abraham and Richardson(1), and are listed in Table I.

Table I. Sulfide Capacities of CaO-SiO_2 Reference Slags at 1500°C. Ref. (1)

%CaO/%SiO_2	XCaO/XSiO_2	-log C_S
1.06	1.14	3.94
1.28	1.37	3.65

The equilibration time was close to 5 hrs. In order to maintain the atmosphere within the capsule, two FeS-CaO-SiO_2 samples (FeS 10 wt. pct) and one FeO-CaO-SiO_2 sample (FeO 10 wt. pct) were charged in the capsule. The capsule was backfilled with a cleaned mixture of 5 pct SO_2-95 pct Ar gas prior to final sealing. Total pressure of the gas mixture in the capsule before heating was 0.2 atm. The experimental apparatus used in the present study is shown schematically in Fig. 1. All experiments were arranged by the method of regression-orthogonal experiments (10).

IV. Results and Discussion

The sulfur contents of the slag samples were determined by a LECO iodimetric titration method. The analysis of the errors is the same as in Ref. (6). The results of sulfur content in the slags are included in Table II. A second set of experiments for slags containing Na_2O failed because the capsule was not sealed properly during the equilibration.

Table II. Results of Sulfur Analyses

Sample No.	%CaO/ %SiO_2	%K_2O	W_s grams	Burrette Reading	%S_{app}	%S_{app}	%S_{true}	-logC_S
B 1.1	1.06	2	0.181	0.046	0.0204			
B 1.2	1.06	2	0.181	0.047	0.0209	0.0207	0.0396	-3.78
B 1.3	1.06	2	0.182	0.047	0.0208			
B 1.4	1.06	2	0.219	0.045	0.0164			
B 2.1	1.06	10	0.185	0.093	0.0454			
B 2.2	1.06	10	0.194	0.096	0.0448			
B 2.3	1.06	10	0.173	0.087	0.0451			
B 2.4	1.06	10	0.182	0.089	0.0439	0.0448	0.0856	-3.45
B 3.1	1.28	2	0.172	0.087	0.0453			
B 3.2	1.28	2	0.165	0.089	0.0485	0.0454	0.0867	-3.44
B 3.3	1.28	2	0.193	0.091	0.0425			
B 3.4	1.28	2	0.187	0.086	0.0412			
B 4.1	1.28	10	0.162					
B 4.2	1.28	10	0.179	0.034	0.1899			
B 4.3	1.28	10	0.183	0.035	0.1912	0.1879	0.2348	-3.04
B 4.4	1.28	10	0.186	0.034	0.1829			
B 5.1	1.06	0	0.173	0.032	0.0133			
B 5.2	1.06	0	0.179	0.033	0.0134	0.0137	0.0261	-3.94
B 5.3	1.06	0	0.191	0.037	0.0146			
B 5.4	1.06	0	0.184	0.031	0.0119			
B 6.1	1.28	0	0.185	0.065	0.0302			
B 6.2	1.28	0	0.179	0.064	0.0307	0.0301	0.0575	-3.65
B 6.3	1.28	0	0.163	0.057	0.0294			
B 6.4	1.28	0	0.192	0.063	0.0281			

In Table II, B5 and B6 are reference slags. %S_{app} = F_s/W_e (Reading - Blk) , where Fs = 0.1, represents the strength of the titer (0.0444 g/l of KIO_3 solution). Blk is the average blank value as shown in Table III. Samples B 4.1 to B 4.4 were determined using a standard sample in which the sulfur content was 0.19 %. For those samples, Fs = 1 (0.444 g/l of KIO_3 solution), Blk = 0.000, and F_{corr} = 1.215. The results for standard samples and blanks are listed in Tables III and IV. According to Eq. (9), the values of 1/2 lg (PO_2/PS_2) are -2.36 and -2.41 from B5 and B6, respectively. The average is -2.38. From Eq. (10), the sulfide capacities calculated using the determined sulfur contents are given in the last column of Table II.

Table III. Results of Blanks and Standards for B 1,2,3,5,6.

Sample No.	W_s grams	Burrette Reading	Blk	$\%S_{app}$	$\%S_{app}$	$\%S_{true}$	F_{corr}
Blk 1.1		0.007					
Blk 1.2		0.009					
Blk 1.3		0.008					
Blk 1.4		0.011	0.009				
Std 1.1	0.631	0.040		0.00491		0.0090	
Std 1.2	0.682	0.042		0.00484		0.0090	
Std 1.3	1.030	0.051		0.00429	0.00486	0.0090	
Std 1.4	1.024	0.053		0.00408		0.0090	1.91

Table IV. Results of Blanks and Standards for B4

Sample No.	W_s grams	Burrette Reading	Blk	$\%S_{app}$	$\%S_{app}$	$\%S_{true}$	F_{corr}
Blk 1-4		0.000	0.000				
Std 2.1	0.501	0.086		0.1716			
Std 2.2	0.559	0.087		0.1556			
Std 2.3	0.616	0.092		0.1494			
Std 2.4	0.583	0.087		0.1492	0.1564	0.19	1.215

V. Discussion

As the slags absorb sulfur from the gas phase, the following reactions adjust to establish a mutual equilibrium:

$$SO_2 = 1/2\ S_2 + O_2 \qquad (11)$$

$$O^{2-}(slag)+1/2\ S_2(gas)=S^{2-}(slag)+1/2\ O_2(gas) \qquad (6)$$

$$(FeS) + 1/2\ O_2 = (FeO) + 1/2\ S_2 \qquad (12)$$

The shift in P_{SO_2} is compensated by Eqs. (13) and (14),

$$P_{02} = K_{11}\ P_{SO_2} / P_{S_2}^{1/2}\ , \qquad (13)$$

$$P_{O_2} = \alpha^2_{FeO}\ P_{S_2}/\alpha^2_{FeS}\ K_{12}^{\ 2} \qquad (14)$$

The combinations of Eqs. (13) and (14) yields

$$P_{S_2}^{3/2} = K_{11}\ K^2_{12}\ P_{SO_2}\ a^2_{FeS} / \alpha^2_{FeO} \qquad (15)$$

or

$$3/2 lg\ PS_2 = lgK_{11} + lgPSO_2 + 2lg\ K_{12} + 2lg\alpha_{FeS} - 2lg\ FeO \quad (15a)$$

Where, lgK_{11} = -6.8536(10) and lgK_{12} = 2.9837 (11). Since the total pressure of the gas mixture (5 pct SO_2-Ar) in the capsule is 0.2 atm at room temperature, the P_{SO_2} at 1500°C is 0.065 atm. If the activity coefficient of FeO is assumed to be independent of temperature, the activity of FeO in the system of $CaO-SiO_2-FeO$ (CaO/SiO_2 = 1.06, FeO = 10%) is 0.2 (12). If the activity of FeS is assumed to be equal to its mole fraction in the system of $CaO-SiO_2-FeS$ (CaO/SiO_2 = 1.06, FeS = 10%), then α_{FeS} = 0.068. Substituting the above values into Eq. 15a, yields lg PS_2 = -2.0070, or P_{S_2} = $9.84X10^{-3}$ atm. Substituting this value into Eq. (14), one obtains

$$\lg PO_2 = -7.0372, \text{ or } PO_2 = 9.10x10^{-8} \text{ atm.}$$

Substituting these values for PS_2/PO_2 into Eq. (9),

$$(PO_2/PS_2)^{1/2} = 3.05x10^{-3} \quad \text{or}$$

$$1/2 \lg (PO_2/PS_2) = - 2.52 \qquad (16)$$

which is close to the measured value of lg $C_s{}^{o}$ (pctS)° = -2.38 in the present experiments. The measured value should be considered much more reliable than that calculated from the assumed FeS and FeO activities described in the preceding paragraphs. However, the calculation serves as a general consistency check.

The optical basicity of a slag can be calculated by use of the relationship[8]

$$B_{op} = X_A B_{opA} + X_B B_{opB} + \ldots \qquad (17)$$

where X_i is the equivalent i-cation fraction, based on the fraction of negative charge "neutralized" by the charge on the cation concerned. The values of B_{op} of slags, whose basicities range from 1.06 to 1.28, and whose alkaline metal oxide contents vary from 2 wt pct to 10 wt. pct, are listed in Table V. The values of sulfide capacity calculated from Eqs. 17 and 5, as well as the values of sulfide capacity measured in the present study are also given in Table V. The calculated and the measured values of the sulfide capacities are reasonably close.

From the results determined in the present study, a regression equation for the effects of basicity and K_2O content on the sulfide capacity can be obtained as follows:

$$\lg Cs=-3.420+0.194(R-1.17)/0.11+0.192(\%K_2O-6)/4 \qquad (18)$$

Where R, the ratio wt % CaO/wt % SiO_2, ranges from 1.06 to 1.28, while $\%K_2O$ varies from 2.0 to 10.0. It can be seen from Eq. (18) that both basicity and K_2O content increase the capacity of the slags, and their relative effects are almost the same.

Table V. Calculated and Measured Sulfide Capacities

			lgCs	
R	K_2O	B_{op}*	Calc.	Meas.
1.06	0	0.668	-3.879	-3.94**
1.06	2	0.674	-3.801	-3.78
1.06	10	0.700	-3.469	-3.45
1.28	0	0.692	-3.586	-3.65**
1.28	2	0.698	-3.509	-3.44
1.28	10	0.724	-3.178	-3.01

* B_{op} is the Optical Basicity.

**Reference Slags.

The relative effects of oxides on sulfide capacity are compared in Table VI. It is clear that K_2O has the strongest effect on sulfide capacity in all cases.

Table VI. Change of Sulfide Capacity with Composition ($\delta \lg C_S$ / δ %oxide)

Oxide	R=1.06	R=1.28
K_2O	0.097	0.105
CaF_2	0.020	0.000
B_2O_3	-0.050	-0.075

*Except for K_2O, values are from Refs. 6 and 7.

VI. Conclusion

1. K_2O strongly increases the sulfide capacities of CaO/SiO_2 slags.
2. The calculated PS_2/PO_2 potential based on estimated values of activities of FeO and FeS agrees reasonably well with the measured value.
3. The predictions of sulfide capacity based on calculated values of the optical basicity agree reasonably well with the measured values.

References

1. K.P. Abraham and F.D. Richardson: J. Iron and Steel Inst., 1960, Vol. 196, pp. 313-17.
2. R.J. Hawkins, S.G. Meherali, and M.W. Davies: J. Iron and Steel Inst., 1971, Vol. 209, pp. 646-57.
3. C. Wagner: Metall. Trans. B, 1975, Vol. 6B, pp. 405-09.
4. G.R. St.Pierre and J. Chipman: Trans. AIME, 1956, Vol. 206, pp. 1474-83.
5. C.J.B. Fincham and F.D. Richardson: Proc. Roy. Soc., Series A, 1954, Vol. 223, pp. 40-62.
6. A. Bronson and G.R. St.Pierre: Metall. Trans. B, 1979, Vol. 10B, pp. 375-80.
7. A. Bronson and G.R. St.Pierre: Metall. Trans B, 1981, Vol. 12B, pp. 729-31.
8. F.D. Richardson: Physical Chemistry of Melts in Metallurgy, Vol. 2, p. 293, Academic Press, New York, 1974.
9. D.J. Sosinsky and I.D. Sommerville: Metall. Trans. B, 1986, Vol. 17B, pp. 331-37.
10. W.Y. Zhu: Orthogonal and Regression Orthogonal Method of Experiments and Its Application (in Chinese), 1978.
11. JANAF Thermochemical Tables, 1982.
12. C.R. Taylor and John Chipman: Trans. AIME, 1943, Vol. 154, pp. 228-47.

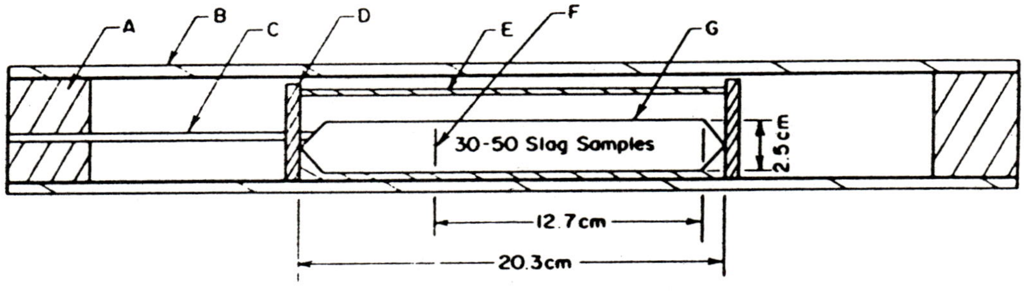

Fig. 1 Schematic disgram of the furnace system: A-alundum refractory; B-SiC furnace tube; C - Pt(6%Rh)/Pt(30%Rh) thermocouple; D - alundum radiation shield; E - alumina tube; F - platinum radiation shields; G - quartz capsule containing platinum cups.

Kinetics of gas/slag reactions

G R BELTON

The author is Director, Central Research Laboratories, BHP Co. Ltd., Shortland, NSW, Australia.

SYNOPSIS

Recent work on the interfacial rates of reaction of CO_2 and CO with iron oxide-containing melts is reviewed. Where equivalent systems have been studied, it is shown that rates measured by isotope exchange, oxidation, and reduction are in good accord. The dependence of the apparent first order rate constant for oxidation by CO_2 on the state of oxidation is found to be closely described by direct proportionality to $(Fe^{3+}/Fe^{2+})^{-2}$ for all high iron oxide content melts for which full information is available. Other reasonable dependences are found not to have wide applicability. Rates of reaction of H_2O with "FeO"-SiO_2 (sat) melts, obtained by measurement of the steady-state activity of oxygen in flowing H_2O-CO mixtures or by measurement of the rate of deuterium exchange between H_2O and H_2 at the surface, are found to be in good accord and to show a dependence of the apparent first order rate constant on the state of oxidation of the melt similar to that for reactions with CO_2. The effect of the surface active component P_2O_5 on the rate of reaction of CO_2 with liquid "FeO" is found to be consistent with a limiting model in which the depression of the rate is attributed to changes in the surface potential caused by the segregating phosphate ions.

INTRODUCTION

In the previous symposium in this series,[1] the author had the opportunity to review the available information on the rates and rate constants of the interfacial chemical steps in reactions of gases with liquid slags. Observed rate phenomena in the desulfurization of blast furnace type slags by oxygen[2,3] and water vapor[4] were taken as examples to show the complex phenomena which can occur in some overall reactions. The implicit conclusion was that an understanding of such rate phenomena, ie., some progress from empirical towards rational rate laws, may depend on a knowledge of:

(a) ionic constitution/chemical potential relationships

(b) surface segregation phenomena, and

(c) rates and rate laws of possible elementary steps.

The transfer of oxygen to or from a slag surface by the reaction of a gaseous (molecular) species was noted as a likely common step in many gas-slag reactions and, possibly, some metal-slag and carbon-slag reactions. In the case of reactions involving CO and CO_2, the author and his colleagues[5,6] had carried out isotope exchange measurements of the rate of dissociation of CO_2 on liquid iron oxide and some simple liquid slags. Values for the apparent first order rate constant, k_a, were found to be a function of the state of oxidation of the melt. For "FeO", "FeO"-SiO_2 (sat), "FeO"-CaO (sat) and MnO-SiO_2 (sat), k_a was found to be inversely proportional to the activity of oxygen in the melt, ie,

$$k_a = k_a^o a_o^{-1} \quad [1]$$

with rates given by:

$$v = k_a^o (pCO_2 a_o^{-1} - pCO) \quad [2]$$

where k_a^o is a system and temperature dependent constant and a_o is the oxygen activity of the melt, expressed as the equilibrium CO_2/CO ratio. Values of a_o were in the range 0.2 to 10 and temperatures were between about 1180 and 1550°C. Resulting values for k_a^o were almost two orders of magnitude higher for the "FeO"-CaO (sat) system than for the "FeO"-SiO_2 (sat) system, with intermediate values for iron oxide and unsaturated iron silicates (assumed to follow rate law [2]).

Measurements by Ban-ya, Iguchi, and Nagasaka[7] of the rates of reduction of liquid iron oxide to metallic iron in CO and CO-CO_2 mixtures at 1400°C were shown to be in close accord with the results of the isotope exchange studies in both the detailed form of the rate law and the value of the rate constant. Good agreement was also found between the studies with respect to rate constants for reaction with iron silicates and on the marked effect of calcium oxide on the rate. Thus, the available evidence was consistent with the dissociation and formation of CO_2 being the rate determining steps in the oxidation and reduction of these melts.

In the present paper, studies of the rates of reaction of CO_2 and CO with some ternary and quaternary iron oxide-containing slags are reviewed. Next, the dependences of the rates on the states of oxidation of the slags are examined. Measurements of the interfacial rates for the transfer of oxygen from H_2O to "FeO"-SiO_2 (sat) melts by both steady-state and deuterium exchange measurements are then reviewed. Finally, the influence of the surface active component P_2O_5 on interfacial rates is examined.

REACTIONS OF CO_2 AND CO WITH COMPLEX SLAGS

Nagasaka et al[8] have reported measurements of the rates of reaction at 1400°C of CO with FeO-CaO-SiO_2 slags contained in iron crucibles. As in previous studies[7], the weight-loss measurements were carried out at sufficiently low partial pressures of CO and at sufficiently high gas flowrates to avoid mass transfer limitations in either the liquid or gaseous phases. The first order rate law:

$$v = k\ pCO \qquad [3]$$

was observed. Values of k, in units of g-oxygen $cm^{-2}\ s^{-1}\ atm^{-1}$, are shown as a function of the iron oxide content in Fig.1.

The authors examined the dependence of the rate constant on melt composition and concluded that the best correlation was with the value of the Fe^{3+}/Fe^{2+} ratio at iron-saturation for each melt. A regression analysis of these results, together with those[5-8] for liquid iron oxide and binary melts with Al_2O_3, TiO_2, SiO_2 and CaO, gave:

$$k = 8.4\times10^{-4}\ (N^2_{Fe^{3+}}/N^3_{Fe^{2+}})^{1/3} \qquad [4]$$

in units of g-oxygen $cm^{-2}\ s^{-1}\ atm^{-1}$ at 1400°C. Inclusion of data for other temperatures gave an apparent activation energy of about 33 kcal $mole^{-1}$ (138 kJ mol^{-1}).

In very recent work, Nagasaka[9] has shown that new results for "FeO"-Na_2O, "FeO"-MnO-SiO_2, and "FeO"-Na_2O-SiO_2 melts are also in good accord with the empirical equation [4]. Results for "FeO"-P_2O_5 melts, however, were found to give lower values than expected. A combination of the appropriate plots from both papers is presented in Fig.2.

Fine, Meyer, Janke, and Engel[10] and, more recently, Graenzdoeffer, Kim, and Fine[11] have studied the rates of reduction of "FeO"-CaO-SiO_2-MgO melts at 1600°C under impinging flows of CO and CO-inert gas mixtures, each containing small amounts of CO_2. Rates were obtained from continuous measurement of the CO_2 content of the exiting gas mixture. The experiments and a full treatment of the experimental data are to be presented elsewhere in the present symposium.

Graenzdoeffer et al examined their data in terms of the rate equation:

$$v = k_1\ (a_{FeO}pCO - a_{Fe}pCO_2/K_1) \qquad [5]$$

where K_1 is the equilibrium constant for the reaction:

$$CO + FeO = CO_2 + Fe \qquad [6]$$

and where the standard states for the activities were taken to be stoichiometric liquid FeO and pure liquid Fe. Fig.3 reproduces the results of their analysis of data obtained under conditions where mass transfer limitations could be ignored. Three MgO-containing slags with values of the molar ratio (MgO+CaO)/SiO_2 from 1.28 to 1.77, and one MgO-free slag with CaO/SiO_2 = 1.07 were used. Initial states of oxidation varied from equilibrium with liquid iron to equilibrium with a CO_2/CO ratio of 40. Iron oxide contents were in the range 5 to 90 wt pct. Clearly, the suggested rate law [5] is a good representation of the measured rates for this range of conditions.

The consistent value of k_1, 3.9×10^{-5} mol $cm^{-2}s^{-1}atm^{-1}$, is in satisfactory agreement with the extrapolated value for liquid iron oxide from the work of Sasaki et al[5], as indicated by the dashed line in Fig.3. However, the composition dependence indicated by Eq. [5] cannot hold for slags high in CaO. Liquid calcium ferrites[5] and slags of high CaO/SiO_2 ratio (Fig.1) would give an apparent first order rate constant as an inverse function of the iron oxide activity at iron-saturation or at a given oxygen activity.

El-Rahaiby et al[6] have used the isotope exchange technique to measure the rate of dissociation of CO_2 on some "FeO"-CaO-SiO_2 melts. Values of the apparent first order rate constant for melts with a molar ratio CaO/SiO_2 = 1 are shown as a function of oxygen activity in Fig.4. The values differ only slightly from those found for "FeO" at the same temperatures but the simple inverse dependence on oxygen activity observed for "FeO" and SiO_2 and CaO saturated melts appears not to hold.

For the equimolar melt, the data are consistent with:

$$k_a = k^o_a\ a_o^{-0.75} \qquad [7]$$

at 1420°C. Accordingly, the apparent first order rate constant for the reduction of such a melt to metallic iron in a CO atmosphere should be given by:

$$k = k^o_a\ (a_o)_e^{0.25} \qquad [8]$$

where $(a_o)_e$ is the value of the CO_2/CO ratio which would be in equilibrium with the melt and metallic iron. Accepting a short extrapolation and taking available thermodynamic data[12], gives:

$$k \simeq 0.6\ k^o_a \qquad [9]$$

The decrease in the rate constant for the addition of equimolar amounts of CaO and SiO_2 to "FeO" would be reasonably consistent with the Graenzdoeffer et al suggestion, since $a_{FeO}\simeq0.45$ at 1450°C,[12] or the slight decrease observed by Nagasaka et al (Fig.1).

DEPENDENCE OF THE RATE CONSTANTS ON THE STATE OF OXIDATION

Sasaki et al[5] and El-Rahaiby et al[6] noted that for liquid iron oxide and its saturated melts with CaO and SiO_2, the available thermodynamic data[13-15] were closely consistent with the relationship:

$$(Fe^{3+}/Fe^{2+})^2 \propto a_o \qquad [10]$$

for each system. Accordingly, over the range of conditions of the isotope exchange measurements, the apparent first order rate constant could be written as

$$k_a = k' (Fe^{3+}/Fe^{2+})^{-2} \qquad [11]$$

where k' is a system and temperature dependent constant. For the equimolar "FeO"-CaO-SiO_2 melts, where relationships [1] and [10] are not followed, it was shown that expression [11] still gave a close description of the variation of k_a at 1420°C over the range of experimental conditions.

Sun and co-workers[16] have further examined the dependence on the state of oxidation by isotope exchange measurements of the rate of dissociation of CO_2 at 1300°C on a calcium ferrite melt with Ca/Fe=0.3. In addition, measurements were made of the interfacial rate of oxidation of thin films of melts with Ca/Fe≃0.33 at 1362°C in CO_2/CO mixtures, the melts having been pre-equilibrated at a series of CO_2/CO ratios. The equilibrium Fe^{3+}/Fe^{2+} ratios were determined for melts at 1300 and 1360°C as functions of the CO_2/CO ratio and found to be in close agreement with the data of Takeda, Nakazawa and Yazawa[17].

The kinetic data are shown logarithmically in Fig.5 as values of the apparent first order rate constant versus the value of the CO_2/CO ratio which would be in equilibrium with each melt. Despite the uncertainties in the oxidation measurements, the accord is sufficiently close to be able to conclude that the rate of oxidation is predominantly that for the rate of dissociation of CO_2 at the surface, ie, involves the same rate determining steps. The first order rate constant for oxidation at iron-saturation at 1400°C, derived[16] from the work of Nagasaka et al[8] on the reduction by CO, is also in satisfactory accord. This is also indicated in Fig.5.

The best straight line through the isotope exchange data is given by:

$$\log k_a = -4.54 - 0.79 \log (pCO_2/pCO) \qquad [12]$$

with an uncertainty in the slope (two standard deviations) of ±0.06. Combination with the equilibrium data of Takeda et al[17] leads to:

$$\log k_a = -5.20 - 2.10 \log(Fe^{3+}/Fe^{2+}) \qquad [13]$$

with an uncertainty in the value of the slope of ±0.14 (two standard deviations), if no uncertainty is assumed for the equilibrium data. The kinetic data are thus in experimental agreement with expression [11] and are in accord with the observations for a wide range of melts and conditions, as is summarized in Fig.6.

An alternative expression for the dependence of the apparent first order rate constant on the state of oxidation which would be consistent with equation [1] for "FeO" and "FeO"-SiO_2 (sat) melts is:

$$k_a \propto a_{Fe} \qquad [14]$$

This would also be consistent with the suggested rate law of Graenzdoeffer et al, Eq.[5], and could be rationalized by a model for the dissociation which includes virtual adsorption equilibrium for CO_2 at iron sites in the surface.

In Fig.7, the apparent first order rate constants for the dissociation of CO_2 on "FeO"-CaO (sat) melts at 1500 and 1550°C and the calcium ferrite melt at 1300°C are plotted logarithmically as a function of the activity of iron, calculated from the equilibrium CO_2/CO ratios and the appropriate thermodynamic data for the particular systems.[17,18] The dependence is clearly not in accord with the first order relationship [14], particularly for the calcium ferrite melts.

In an earlier discussion,[5] it was noted that the inverse first order dependence of k_a on the oxygen activity for liquid iron oxide and other melts could be explained by strong surface coverage by adsorbed oxygen. The surface concentration of vacant sites, Γ_v, remaining for reaction might then be presumed to be given by a limiting expression of the form:

$$\Gamma_v \propto (pCO_2/pCO)^{-1} \qquad [15]$$

In recent work, Kidd and Gaskell[19] have remeasured the surface tension of liquid iron oxide as a function of the ferric oxide content by the Padday's cone technique. Their results for 1460°C, re-expressed as a function of the logarithm of the CO_2/CO ratio which would be in equilibrium with each melt, are shown in Fig.8. The values were sensitive to the available information on the densities of the melts. Taking the lower curve and applying the Gibbs adsorption isotherm to the value of the slope at the steepest part of the curve yields an excess surface concentration of oxygen of about 1×10^{-10} mol cm^{-2} or an area occupied per oxygen atom of about 160(A)^2. Clearly, this is very strong evidence that a high coverage by adsorbed oxygen cannot be presumed in the analysis of the rate phenomena - at least for liquid iron oxide.

As previously discussed[1,5,6] the empirical expression [11] is consistent with a rational rate law based on a predominant rate determining step which involves the dissociation of a doubly charged (negative) adsorbed molecule or activated complex which includes one CO_2 molecule. The surface concentration of the transient charged species, formally represented as CO_2^{2-}, is given in terms of pCO_2 and the concentrations of electron donors and acceptors by a virtual electrochemical equilibrium of the form:

$$CO_2(g) + 2Fe^{2+} = CO_2^{2-}(ad) + 2Fe^{3+} \qquad [16]$$

as

$$\Gamma_{CO_2{}^{2-}} = K_a\ pCO_2\ (Fe^{2+}/Fe^{3+})^2 \quad [17]$$

for low surface coverage and where K_a is a temperature dependent constant for each system. The rate of dissociation is then proportional to $\Gamma_{CO_2{}^{2-}}$.

So far, all the systems which have been studied (Fig.6) and which follow the empirical expression [11] are known to exhibit significant semiconductivity[16]. It seems reasonable to expect that at very low transition metal oxide contents, the rate of charge transfer may be insufficient to maintain the virtual equilibrium [12]. Other rate determining mechanisms would then be expected.

RATES OF H_2O-H_2 REACTIONS WITH LIQUID SILICATES

It was noted in the previous review that, in principle, information on the relative rates of interfacial reaction of H_2O and CO_2 can be obtained from a knowledge of the steady-state value of the activity of oxygen in a melt when exposed to flowing H_2O-CO or CO_2-H_2 gas mixtures. Initial studies with lithium silicates[22] were reviewed.

Sasaki and Belton[23] have applied this technique in a study of reactions with silica-saturated iron silicates at 1250°C. H_2O/CO and H_2O-CO-CO_2 mixtures were jetted onto the surfaces of shallow melts held in platinum crucibles. The ferrous and ferric iron contents of the quenched melts were then determined. Values of $(Fe^{3+}/Fe^{2+})^2$ at flowrate independence are shown in Fig.9(a), for H_2O-CO mixtures, and in Fig.9(b), for H_2O-CO-CO_2 mixtures with $pCO_2 = 0.45 \pm 0.03$ atm. Combination of the slope of the reasonable straight line in Fig.9(a) with the equilibrium measurements of Michal and Schuhmann[14] gave the straight line relationship:

$$a_o = 20\ (pH_2O/pCO) \quad [18]$$

for the steady-state oxygen activity (standard state unit CO_2/CO ratio).

Sasaki and Belton assumed that oxygen transfer reactions in H_2O-H_2 atmospheres would be described by an apparent first order rate law of the form:

$$v = k_2\ pH_2O\ a_o^{-y} - k_3\ pH_2\ a_o^{1-y} \quad [19]$$

and those in CO_2-CO atmospheres by rate law [2]. For negligibly small pressures of H_2 and CO_2, ie., high flowrates of the H_2O-CO mixtures, it was shown that the steady-state oxygen activity should be given by:

$$a_o^y = \frac{k_2}{k_a^o}\ \frac{pH_2O}{pCO} \quad [20]$$

This would hold if the competing reactions are essentially separable, ie., the rate constants for the dissociation of H_2O and the reaction of CO are functions only of the oxygen activity.

For y=1, consistent with the straight line relationship [14], it was concluded that, at a given state of oxidation, the rate constant for the oxidation of the melts by water vapor (k_2) was thus about twenty times that for carbon dioxide (k_a^o) at 1250°C over the range of conditions of the experiments. The consistent rate equation for the reaction of the iron silicate in H_2O-H_2 mixtures at 1250°C was shown to be:

$$v = 4.5 \times 10^{-5}\ [pH_2O\ (a_o')^{-1} - pH_2] \quad [21]$$

mol cm^{-2} s^{-1}

where the standard state for the oxygen activity, a_o', is now unit H_2O/H_2 ratio.

For H_2O-CO-CO_2 mixtures, it was shown that steady-state values of the oxygen activity should be given by:

$$a_o = (pCO_2 + 20pH_2O)/pCO \quad [22]$$

or

$$(Fe^{3+}/Fe^{2+}) =$$

$$1.3 \times 10^{-3}(pCO_2 + 20pH_2O)/pCO \quad [23]$$

The dashed curve in Fig.9(b) shows the good accord with Eq.[23] for the data with $pCO_2 = 0.45$ atm.

In order to further examine the assumptions in the above analysis, Glaws and Belton[24] have carried out studies of the rate of deuterium exchange between water vapor and hydrogen on silica-saturated iron silicates at chemical equilibrium with the gas mixture. The rate of the overall exchange reaction:

$$HDO(g) + H_2(g) = H_2O(g) + HD(g) \quad [24]$$

was measured by passing H_2O-H_2 mixtures, enriched in HDO, over the silicate and determining the HD content of the exiting gas mixture by mass spectrometry. The method of containment of the melt in an inductively heated Pt-Rh susceptor was similar to that used in the $^{14}CO_2$ exchange studies[5,6] and details on the preparation of the enriched gas mixtures and mass spectrometric analysis procedure will be published elsewhere[24].

It can be shown[24] that the overall rate of exchange of hydrogen atoms, $\mathfrak{R}$, is given, independently of the mechanism, by the expression:

$$\mathfrak{R}\,\alpha = -\frac{2\dot{V}}{RT}\ \frac{pH_2O+pHDO}{Y}\ \ln\ [1 + Y\frac{pHD}{p^oHDO}] \quad [25]$$

where $Y = 1+B/K_i$, $B = (pH_2O+pHDO)/(pH_2+pHD)$, K_i is the isotope equilibrium constant (Eq.[23]), $\dot{V}$ is the volume flowrate, p^oHDO is the initial partial pressure of HDO, pHD is the resulting partial pressure of HD, and α is the value of any kinetic isotope effect. K_i is very close to unity at or above about 1250°C.[25]

In Fig.10, the derived rate of exchange of hydrogen atoms on the silicate at 1300°C is shown as a function of the pressure of water vapor in $Ar-H_2O-HDO-H_2$ mixtures at $(pH_2O+pHDO)/pH_2=0.9$. The results, which are demonstrated to be independent of flowrate, are consistent with the rate being first order with respect to pH_2O.

More extensive results, obtained at 1400°C, are presented logarithmically in Fig.11 as values of the apparent first order rate constant for the dissociation of H_2O, k_2' versus a_o (pH_2O/pH_2). The dissociation of H_2O is taken to require the exchange of two hydrogen atoms. The value of the slope is -0.99 ± 0.14 (two standard deviations). Accordingly, the data are consistent with the expression:

$$k_2' = k_2 \, (a_o')^{-1} \qquad [26]$$

where k_2 is a temperature dependent constant. The value of k_2 at 1400°C, 10^{-4} mol cm^{-2} s^{-1} atm^{-1}, should be corrected for the kinetic isotope effect. However, the theoretical upper limit for this effect at high temperatures[25] would introduce a constant factor (α) of only 1.4 and experimental evidence[26] suggests that the factor is likely to be significantly lower than this.

The deduced values of k_2' from the H_2O-CO steady-state measurements are also indicated in Fig.11. The magnitudes of the rate constants, as well as the dependences on a_o, are sufficiently close to conclude that the assumptions made in the interpretation of the steady-state measurements are justified. The principal finding of the studies, however, is that the apparent first order rate constant for the dissociation of H_2O (transfer of oxygen to the surface) has the same dependence on the state of oxidation of the melt as that for the dissociation of CO_2.

The only other study of H_2-H_2O reactions with a silicate melt, from which an interfacial rate constant has been deduced, is that of Pal, DebRoy, and Simkovich[27] who derived a value of 6×10^{-6} mol cm^{-2} s^{-1} atm^{-1} for the first order rate constant for the reduction of $PbO-SiO_2$ melts at 900°C in hydrogen. The PbO/SiO_2 ratio was 3 and the data were obtained from the initial rates of reduction at liquid lead saturation, corrected for mass transfer limitations in the gaseous phase.

THE INFLUENCE OF P_2O_5 ON THE INTERFACIAL RATES

Pal et al[27,28] found that additions of P_2O_5 to lead silicate melts caused a significant decrease in the rate of reduction by H_2. Values of the initial rate, uncorrected for mass transfer effects, for the reduction of a melt with $PbO/SiO_2=3$ in H_2 at 900°C are reproduced in Fig.12. The depression of the rate was attributed to surface active phosphate groups covering reaction sites at the surface. It was noted earlier in this paper (Fig.2) that Nagasaka[9] observed much lower rates of reduction of "FeO"-P_2O_5 melts in CO than would be expected from the empirical correlation of rate constants with the Fe^{3+}/Fe^{2+} ratio at equilibrium with metallic ion (Eq.[4]).

To investigate this further, Sun[29] has carried out $^{14}CO_2$ isotope exchange measurements of the rate of dissociation of CO_2 on "FeO"-P_2O_5 melts at 1400°C. The apparent first order rate constants for one set of experiments at a CO_2/CO ratio of 1 with a zirconia containment tube are presented in Fig.13(a). Experiments with an alumina containment tube gave higher results but showed a closely parallel depression of the rates with P_2O_5 content. Broadly, the rate constant decreased by a factor of about 3 for the addition of 3.5 mol pct P_2O_5. The magnitude of the effect appears to be in good accord with the observations of Nagasaka[9].

If it is assumed that reaction is negligibly slow on the fraction of the surface which is covered by 'P_2O_5' and that the adsorption of 'P_2O_5' is ideal, ie, Langmuirian, we may write:

$$k_a^p = k_a(1-\theta_p) \qquad [27]$$

where k_a^p and k_a are the apparent first order rate constants in the presence and absence of P_2O_5, respectively, and θ_p is the fractional coverage by "P_2O_5". θ_p is then given by:

$$\frac{\theta_p}{1-\theta_p} = K_p \, a_p \qquad [28]$$

where K_p and a_p are the adsorption coefficient and activity of P_2O_5, respectively. It follows that:

$$\frac{1}{k_a^p} = \frac{1}{k_a} + \frac{K_p a_p}{k_a} \qquad [29]$$

Elegant work by Kambayashi, Awaka, and Kato[30] has shown that the activity of P_2O_5 closely follows Henry's Law in "FeO"-P_2O_5 melts (at iron-saturation) up to about 8 mol pct P_2O_5. Accordingly, we may take a standard state of 1 mol pct P_2O_5. The kinetic data are shown plotted in accordance with this and Eq.[28] in Fig.13(b). From the values of the slope and intercept, the consistent value of K_p is 0.6.

It has been shown elsewhere[31] that ideal adsorption of a solute, at low concentration, leads to the following expression for the depression of the surface tension:

$$\sigma^o - \sigma = RT \, \Gamma_p^o \, \ln(1 - K_p a_p) \qquad [30]$$

where, in this case, σ^o and σ are the surface tensions of "FeO" and "FeO"-P_2O_5 melts, respectively, and Γ_p^o is the saturation coverage by P_2O_5. The available data for 1350-1400°C and up to 10 mol pct P_2O_5, as assessed by Mills and Keene[32], are plotted in the appropriate form for $K_p=0.6$ in Fig.14. The best slope of -125 mN m^{-1} yields a value for Γ_p^o of about 9×10^{-10}mol cm^{-2} or an area occupied per P_2O_5 "molecule" of about 19 $(A)^2$. This is of the order of what would be expected if the adsorbed species were PO_4^{3-} ions (2 per molecule) with a P - O bond length[33] of 1.55 A. Pal et al[28] have discussed their data for the reduction of $PbO-SiO_2-P_2O_5$ melts by H_2 at 900°C in terms of the same ideal model. The resulting value of K_p was about 0.2 per mol pct P_2O_5. Unfortunately, surface tension data are not available for these melts.

Although this surface coverage model describes the data, an alternative limiting model can be proposed which avoids the assumption of negligible rates on phosphate covered areas and which is consistent with the earlier suggestions that the rate determining step involves dissociation of a doubly charged adsorbed species - at least for reactions with CO_2.

Strictly, virtual electrochemical equilibria of the type:

$$CO_2(g) + 2Fe^{2+} = CO_2^{2-}(ad) + 2Fe^{3+} \qquad [16]$$

should be considered in terms of electrochemical potentials. If the CO_2^{2-} is assumed to be adsorbed on the surface of the condensed phase, we may express the equilibrium constant, K_a, as:

$$K_a = K_a^o \exp[-2F\,\eta/RF] \qquad [31]$$

where the term $-2F\eta$ in the exponential is the Gibbs free energy change accompanying the transfer of two moles of negative charges from the inner electrical potential of the melt, ϕ_b, to the adsorbed molecule assumed to be at the electrical potential immediately outside the melt, ϕ_s. F is the Faraday, K_a^o is the formal value of the equilibrium constant when $\eta = \phi_b - \phi_s = 0$, and η is variously described as the surface potential or space charge potential.

For small additions of P_2O_5 to an iron oxide melt, we assume K_a^o to be constant and attribute changes in K_a to changes in the surface potential, $\Delta\eta$, accompanying the segregation of phosphate ions.

We further assume that the steady-state concentration of the transient adsorbed charged species on the surface is sufficiently small for the contribution to the change in the smoothed surface potential by these dipoles to be ignored. Thus:

$$\ln \frac{K_a^p}{K_a} = \frac{-2F\Delta\eta}{RT} \qquad [32]$$

where K_a^p is the value of the virtual equilibrium constant for the P_2O_5 - containing melt, and K_a is that for the P_2O_5 - free melt. Direct proportionality of the rate of dissociation to the surface concentration of CO_2^{2-} leads to:

$$\ln \frac{k_a^p}{k_a} = -\frac{2F\Delta\eta}{RT} \qquad [33]$$

The surface potential may be expressed in terms of the total dipole moment per unit area, M_t, as:

$$\eta = M_t/\varepsilon_o \qquad [34]$$

where the value of ε_o is approximately $(36\pi \times 10^9)^{-1}$ Farad m^{-1} in SI units. Accordingly, we may express the change in surface potential as:

$$\Delta\eta = N\Theta_p\,\Gamma_p^o\,\Delta M_p/\varepsilon_o \qquad [35]$$

where ΔM_p is the change in dipole moment per segregated "molecule" of P_2O_5 (as ions), Γ_p^o is the saturation segregation (excess surface concentration), Θ_p is the fractional saturation, and N is Avagadro's number.

It was shown earlier that the surface tension data were consistent with a value for the adsorption coefficient, K_p, of about 0.6, and a saturation coverage, Γ_p^o of about 9×10^{-10} mol cm^{-2} (9×10^{-6} mol m^{-2}). In Fig.15, the kinetic data from Fig.13(a) are plotted in the appropriate form for Eq. [33] and [35] with values of Θ_p calculated for values of K_p of 0.5 and 1.0. The linear behavior is in accord with the model and would give, for K_p = 0.6, a slope of about -0.72 and a value of $\Delta\eta$ of 0.12V at saturation. The consistent value of ΔM_p is 2×10^{-31} C m or about 0.06 debye.

The order of magnitude of the change in dipole moment for the substitution of phosphate ions into the surface of the molten oxide is physically reasonable but, unfortunately, no independent data on slag systems, eg., work function measurements, are available for comparison.

CONCLUDING COMMENTS

P_2O_5 is the least basic of the common component oxides of slag systems, ie., it has the lowest optical basicity and highest electron density.[33] Despite this, the deduced effect of the strongly segregating phosphate ions on the magnitude of the change in surface potential is relatively small. Accordingly, it is to be expected that the small changes in surface constitution which might accompany changes in the state of oxidation of a given slag melt would have little effect on the surface potential and, hence, on the value of the rate constant. Changes in the surface concentrations of charged transient species would be expected to be dominated by changes in the electrochemical potential of electrons, given by electron donor/acceptor equilibria in the melt. This is consistent with the observations summarized in Fig.6, ie., for a given melt:

$$k_a = k'\,(Fe^{3+}/Fe^{2+})^{-2} \qquad [11]$$

for the dissociation of CO_2.

The limited information available on the rate of reaction with "FeO"-SiO_2 (sat) melts suggests that similar phenomena hold for reactions of H_2O with iron oxide-containing melts.

The basicity of an "FeO"-containing melt may be taken to indicate the ease with which charge can be transferred from the Fe^{2+} ions to anionic species in the melt and, presumably, on the surface. Broadly, the Fe^{3+}/Fe^{2+} ratio at fixed oxygen activity may be taken as a measure of basicity, and it has been shown elsewhere[6] that the apparent first order rate constant for the dissociation of CO_2 increases with this measure for melts in the system "FeO"-SiO_2-CaO. Similarly, the correlation obtained by Nagasaka et al[8,9] (Fig.2) indicates that with the exception of "FeO"-P_2O_5 melts there is a reasonably close dependence on basicity for the first order rate constant for the reduction of melts by H_2. In this case, the measure of basicity is the value of the function $(Fe^{3+})^2/(Fe^{2+})^3$ at iron saturation.

In view of the wide range of systems included in the above correlation, it appears that effects due to changes in surface constitution are relatively minor, except when there is a strongly segregating component which has a "basicity" which is significantly different from that of the melt. The alkali oxides are known[32] to be surface active in iron-silicate melts. It would be of interest to see if small additions of these basic oxides significantly increase rates of reaction of CO_2 and CO with these acidic melts.

ACKNOWLEDGEMENTS

The author is indebted to his co-worker Mr Sun Shouyi for stimulating discussions and his provision of unpublished data. Professor H. Alan Fine of the University of Kentucky is thanked for providing the results of his work before formal publication. The author also wishes to thank Mrs Pat Bourne for her patient preparation of the camera-ready manuscript.

REFERENCES

1. G.R. Belton : Proceedings, International Symposium on Metallurgical Slags and Fluxes, H.A. Fine and D.R. Gaskell, eds., Met. Soc. AIME, Warrendale, PA., 1984, pp. 63-85.

2. A.D. Pelton, J.B. See and J.F. Elliott : Metall. Trans., 1974, vol. 5, pp. 1163-71.

3. T. Mori, A. Moro-oka and H. Kokubu : Australia/Japan Extractive Metallurgy Symposium 1980, pp. 469-77, Australasian Institute of Mining and Metallurgy, Victoria.

4. B. Agrawal, G.J. Yurek and J.F. Elliott : Metall. Trans. B, 1983, vol. 14B, pp. 221-30.

5. Y. Sasaki, S. Hara, D.R. Gaskell and G.R. Belton : Metall. Trans. B, 1984, vol. 15B, pp. 563-71.

6. S.K. El-Rahaiby, Y. Sasaki, D.R. Gaskell and G.R. Belton : Metall. Trans. B, 1986, vol. 17B, pp.307-16.

7. S Ban-ya, Y. Iguchi and T. Nagasaka : Tetsu-to-Hagane, 1983, Vol. 69, p. S761 and 1984, vol. 70, pp. A21-24.

8. T. Nagasaka, Y. Iguchi and S. Ban-ya : Proceedings, 5th International Iron and Steel Congress, vol. 3, pp.669-78, Iron and Steel Soc., AIME, Warrendale, PA., 1986.

9. T. Nagasaka : Tetsu-to-Hagane, 1987, vol. 73, p. S773.

10. H.A. Fine, D. Meyer, D. Janke and H-J. Engel: Ironmaking Steelmaking, 1985, vol. 12, pp. 157-62.

11. S. Graenzdoeffer, W.M. Kim and H.A. Fine : Paper presented at the 7th Process Technology Conference, Toronto, April 17-20, 1988. Iron and Steel Society, AIME.

12. M. Timucin and A.E. Morris : Metall. Trans., 1970, vol. 1, pp. 3193-3201.

13. L.S. Darken and R.W. Gurry : J. Am. Chem. Soc., 1946, vol. 68, pp. 798-816.

14. E.J. Michal and R. Schuhmann : Trans. AIME, 1952, vol. 194, pp. 723-28.

15. H. Larsen and J. Chipman : Trans. AIME, 1953, vol. 197, pp. 1089-96.

16. S. Sun, Y. Sasaki and G.R. Belton : Metall. Trans. B, in press.

17. Y. Takeda, S. Nakazawa and A. Yazawa : Can. Met. Quart., 1980, Vol. 19, pp. 297-305.

18. J.F. Elliott, M. Gleiser and V. Ramakrishna : Thermochemistry for Steelmaking, vol. II, Addison-Wesley Publ. Co., 1963.

19. M. Kidd and D.R. Gaskell : Metall. Trans. B, 1986, vol. 17B, pp. 771-76.

20. K. Mori and K. Suzuki : Trans. Iron Steel Int. Jpn., 1969, vol. 8, pp. 382-85.

21. K. Irie, S. Hara, D.R. Gaskell and K. Ogino : quoted in ref. 19.

22. Y. Sasaki and G.R. Belton : Metall. Trans. B, 1983, vol. 1983, pp. 267-72.

23. Y. Sasaki and G.R. Belton : The Reinhardt Schuhmann International Symposium on Innovative Technology and Reactor Design in Extractive Metallurgy, pp. 583-95, Met. Soc. AIME, Warrendale, PA., 1986.

24. P.C. Glaws and G.R. Belton : in preparation for publication.

25. L. Melander : Isotope Effects on Reaction Rates, The Ronald Press Co., New York, NY, 1960, also, A.S. Freidman and L. Haar : J. Phys. Chem., 1954, vol. 22, pp. 2051-58.

26. W.G. Henderson, Jr. and R.B. Bernstein : J. Am. Chem. Soc., 1954, vol. 76, pp. 5344-46.

27. U.B. Pal, T. DebRoy and G. Simkovich : Trans. Inst. Min. Metall. C., 1984, vol. 93, pp. C112-17.

28. U.B. Pal, T. DebRoy and G. Simkovich : Metall. Trans. B, 1983, vol. 14B, pp. 693-99.

29. S. Sun : private communication.

30. S. Kambayashi, H. Awaka and E. Kato : Tetsu-to-Hagane, 1985, vol. 71, pp. 1911-18.

31. G.R. Belton : Metall. Trans. B, 1976, vol. 7B, pp. 35-42.

32. K.C. Mills and B.J. Keene : Intl. Mater. Rev., 1987, vol. 32, pp. 1-120.

33. T. Nakamura, Y. Ueda and J.M. Toguri : Nippon Kinzoku Gakkaishi 1986, vol. 50, pp. 456-461.

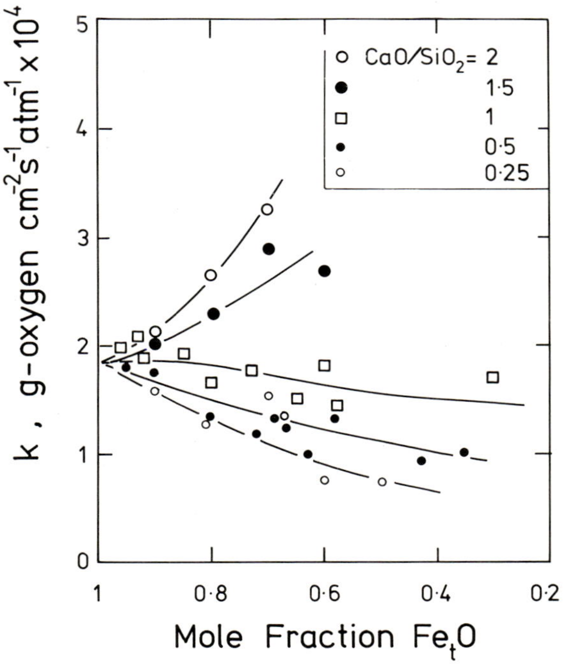

Fig. 1. First order rate constants for the reduction of "FeO"-CaO-SiO$_2$ melts by CO at 1400°C, after Nagasaka et al.[8]

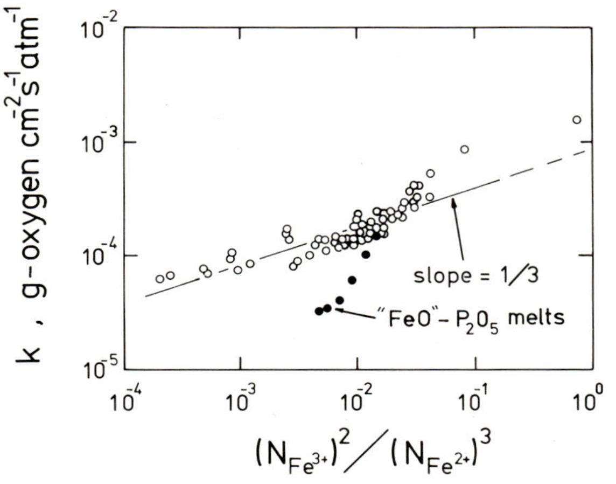

Fig. 2. Dependence on Fe^{3+} and Fe^{2+} contents of the first order rate constants for the reduction of "FeO"-containing melts by CO at 1400°C, after Nagasaka et al.[8,9]

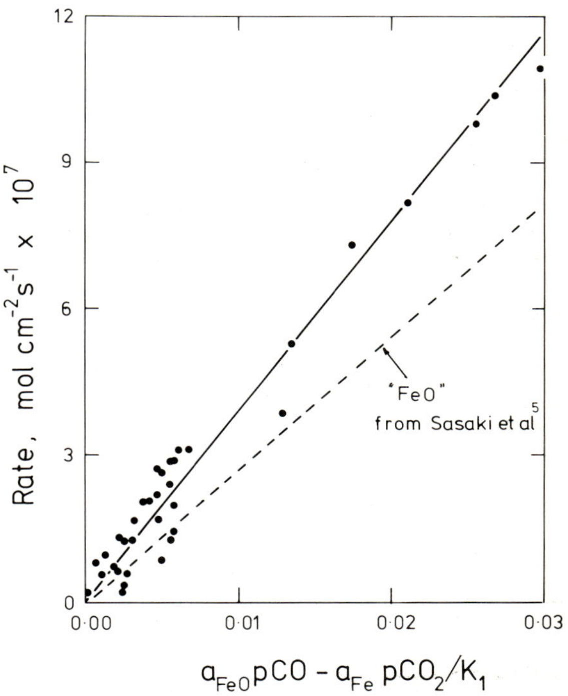

Fig. 3. Rates of reduction of "FeO"-CaO-SiO$_2$-MgO melts at 1600°C, after Graenzdoeffer et al.[10]

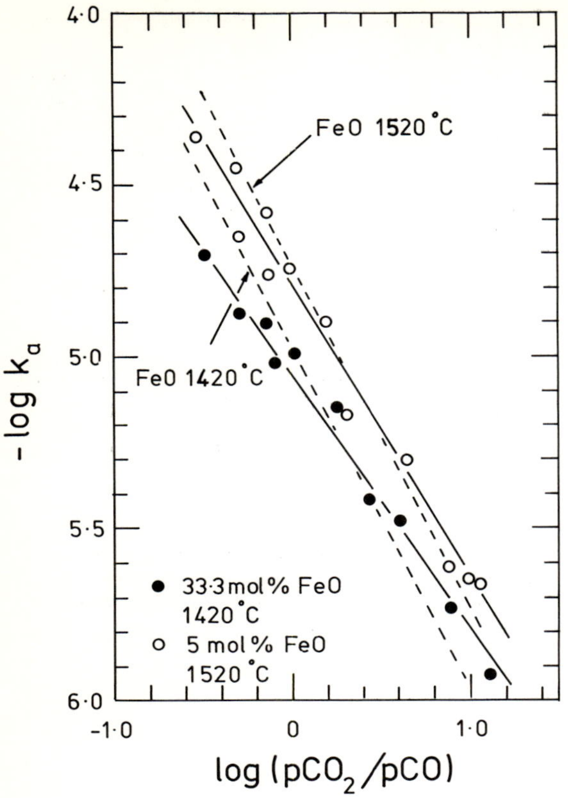

Fig. 4. Dependence of the apparent first order rate constant for the dissociation of CO_2 on "FeO"-CaO-SiO_2 melts (CaO/SiO_2=1) on the equilibrium CO_2/CO ratio, after El-Rahaiby et al.[6]

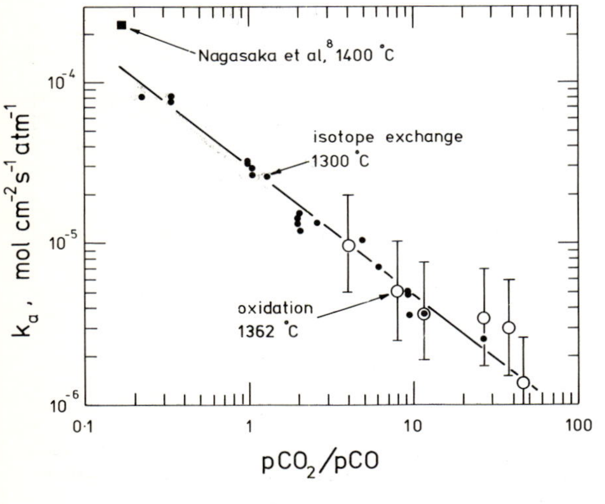

Fig. 5. Apparent first order rate constants as a function of oxygen activity for the reaction of CO_2 with a calcium ferrite melt (Ca/Fe=0.3) from the work of Sun et al.[16]

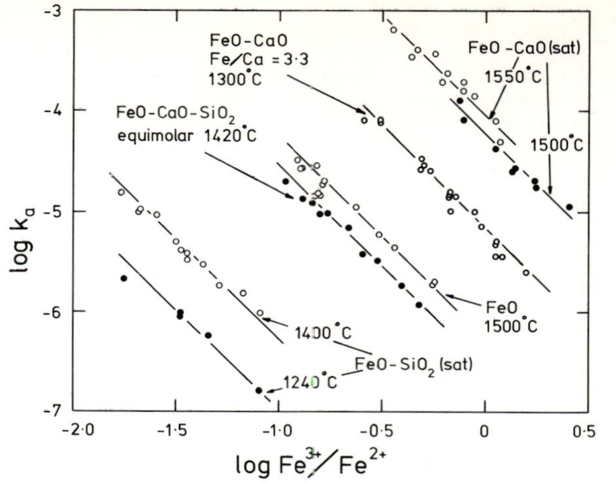

Fig. 6. Dependence of the apparent first order rate constant for the dissociation of CO_2 on the Fe^{3+}/Fe^{2+} ratio for the melts for which data are available. Lines are drawn with a slope of −2. Overlapping data for FeO-CaO (sat) at 1230°C and "FeO" at 1420°C are omitted for clarity but show an equally close fit with the slope of −2

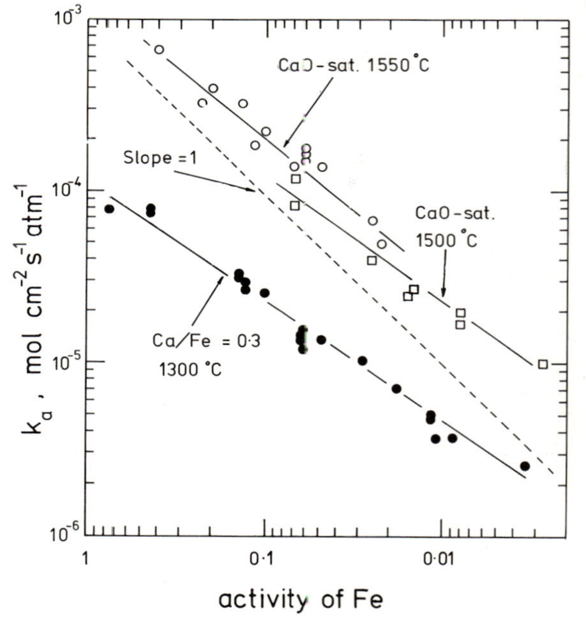

Fig. 7. Apparent first order rate constants for the reaction of CO_2 with "FeO"-CaO (sat) melts from Sasaki et al[5] and on unsaturated calcium ferrite melt from Sun et al[16] plotted as a function of the activity of iron in the melt.

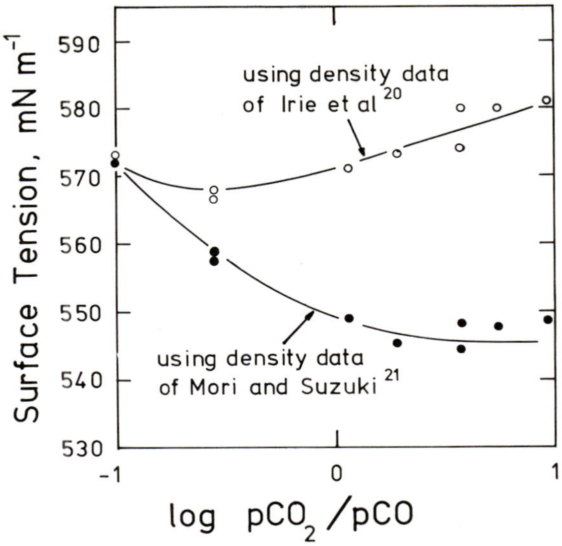

Fig. 8. Surface tension of liquid iron oxide at 1460°C from the work of Kidd and Gaskell[19] plotted as a function of the logarithm of the oxygen activity.

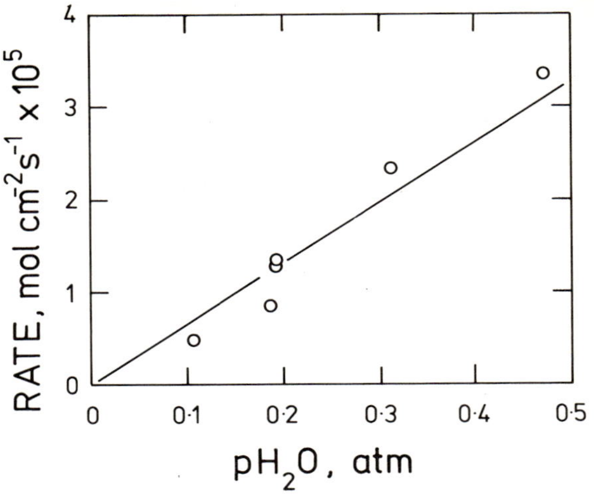

Fig. 10. Rate of exchange of hydrogen atoms at $(pH_2O+pHDO)pH_2=0.9$ on "FeO"-SiO_2 (sat) melts at 1400°C in H_2O-HDO-H_2-Ar mixtures, from Glaws and Belton.[24]

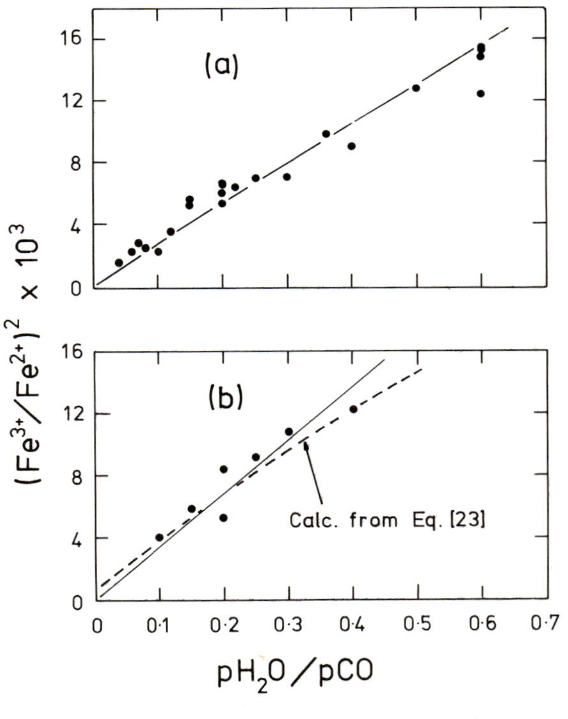

Fig. 9. Values of the Fe^{3+}/Fe^{2+} ratio at steady-state in "FeO"-SiO_2 (sat) melts at 1250°C exposed to flowing mixtures of H_2O and CO (a) and H_2O-CO-CO_2 mixtures with $pCO_2=0.45$ atm (b), after Sasaki and Belton.[23]

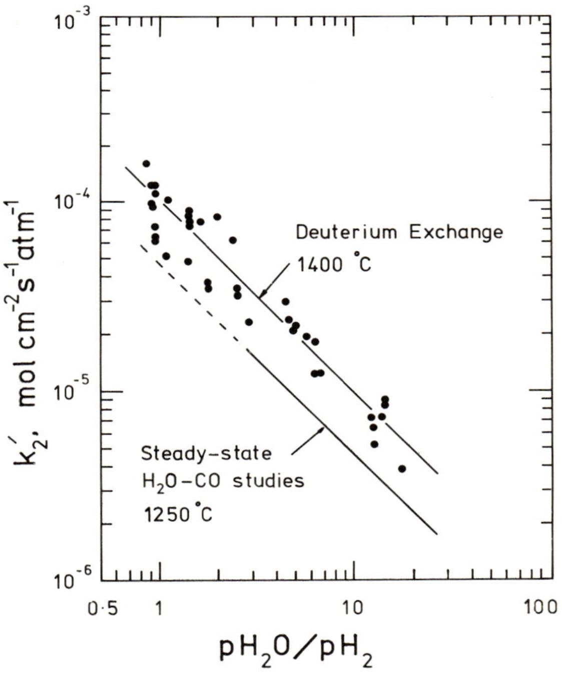

Fig. 11. Values for the apparent first order rate constant for the dissociation of H_2O on "FeO"-SiO_2 (sat) melts at 1400°C as a function of oxygen activity, from Glaws and Belton[24], in comparison with the deduced dependence at 1250°C from the steady-state measurements of Sasaki and Belton.[23]

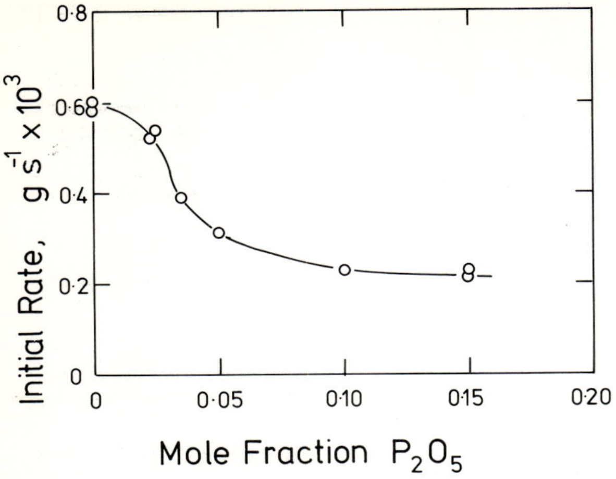

Fig. 12. The influence of P_2O_5 content on the initial rate of reduction of a lead silicate melt ($PbO/SiO_2=3$) in H_2 at 900°C from the work of Pal et al.[28]

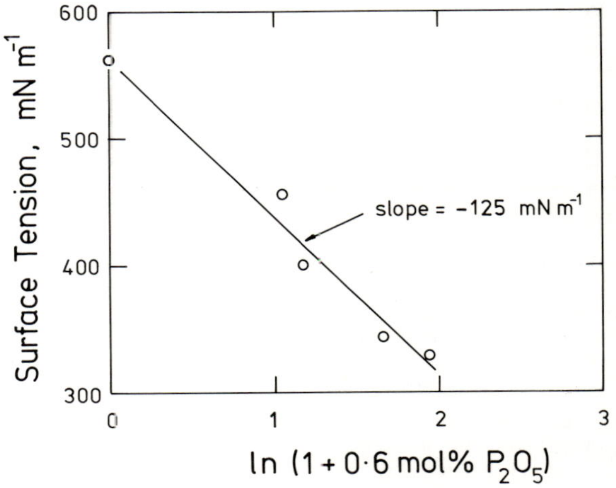

Fig. 14. Surface tension of "FeO"-P_2O_5 melts at 1350-1400°C from the assessment of Mills and Keene[32], plotted in accordance with Eq. [30] and K_p=0.6.

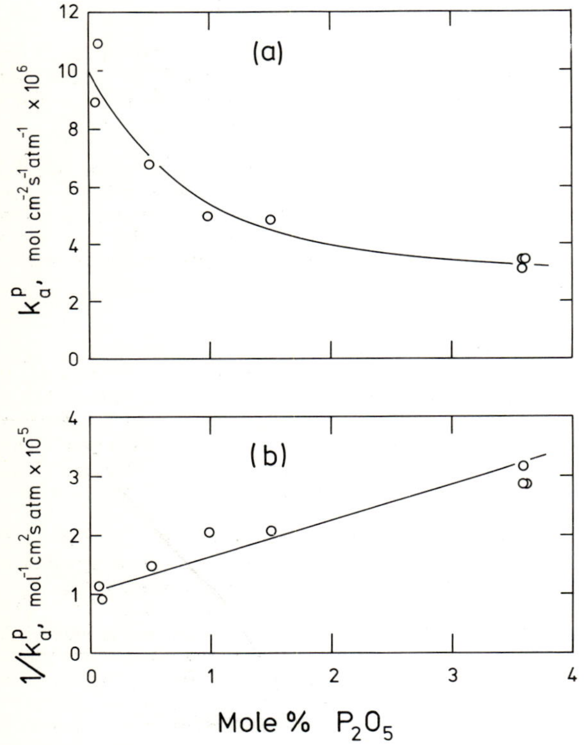

Fig. 13. (a), apparent first order rate constants for the dissociation of CO_2 on "FeO"-P_2O_5 melts at 1400°C from the work of Sun.[29]
(b), plot of the data in accordance with Eq. [29].

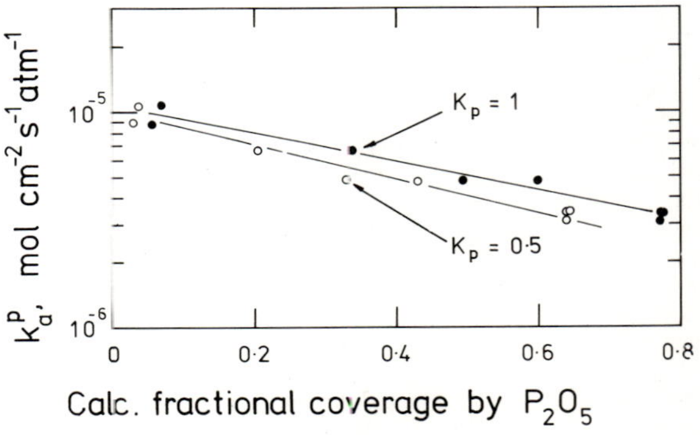

Fig. 15. Apparent first order rate constants from Fig. 13(a) plotted in accordance with Eqs. [33] and [35] for two assumed values of K_p.

Carbon dioxide dissolution in CaO based molten fluxes

M MAEDA and T IKEDA

The authors are with the Institute of Industrial Science, University of Tokyo, Tokyo, Japan.

SYNOPSIS

A thermogravimetric technique was used to measure carbon dioxide solubility in molten $CaO-CaCl_2-CaF2$ fluxes. The effects of temperature as well as flux composition on carbon dioxide dissolution were studied.

Experiments were conducted over a temperature range of 900-1500^oC. The solubility of carbon dioxide increased with increasing mole fraction of CaF_2 from 0.44 to 0.97 wt% at 1100^oC and decreased from 2.95 to 0.21 wt% with increasing temperature from 900 to 1300^oC at $XCaF_2$ = 0.6, for a flux with XCaO = 0.12 in $CaO-CaCl_2-CaF_2$ system.

Results obtained in this study were compared with those for other systems appearing in the literature. Carbonate capacities calculated from the solubility data were compared with sulphide, phosphate, phosphide, nitride and cyanide capacities in the $CaO-CaF_2$ system. Reasonable linear relationships were observed for carbonate capacities and other capacities.

1 Introduction

In the recent steelmaking processes, the refining of molten iron to extremely low impurity contents is required. Those impurities, especially phosphorus and sulphur, must be removed to levels as low as 10 ppm. For this purpose highly basic fluxed e.g. $CaO-CaCl_2$ or $CaO-CaF_2$ melts have been used for hot metal pretreatment. The basicity index, CaO/SiO_2, that has been used for a long time in conventional iron and steelmaking processes, is not an appropriate expression for these fluxes, because the amount of SiO_2 is so small yielding extremely high values for the index. Therefore, a new basicity index which can be applied to such fluxes is required.

The activity of O^{2-} in molten fluxes, which should be thought of in a similar manner to pH in an aqueous solution(1), must be the definition of basicity, even though it is theoretically impossible to measure.

Several basicity indices have been suggested such as:

1) Optical basicity,

2) Redox equilibrium for transition metal oxides,

3) Carbonate capacities, calculated from the amount of carbon dioxide absorbed in molten fluxes(2) assuming that carbon dioxide dissolves according to the reaction:

$$CO_2 + O^{2-} = CO_3^{2-}$$

Because a few reports demonstrated the close relationship between carbonate capacities and other capacities, carbonate capacity was chosen as a good measure of basicity in this study.

When these fluxes are actually used in the plant, transition metal oxides, such as FeO or MnO, will be present in the fluxes because they co-exist with molten metal under high oxidizing conditions. These metal oxides are not added to fluxes in the present study.

2 Experimental Aspects

A thermogravimetric technique was used to measure the dissolution of carbon dioxide in molten $CaO-CaF_2-CaCl_2$ fluxes by equilibration with $Ar-CO-CO_2$ gas mixture. In the experiments with nickel crucibles or boats, the gas mixture contained a small amount of CO gas (p_{CO} = 0.1) to prevent oxidation of the nickel.

Fluxes were prepared by melting in a nickel boat or platinum crucible after mixing reagent grade $CaCO_3$, CaF_2 and $CaCl_2$ which were dried at 130^oC in a vacuum drying oven. CaO was obtained by decomposition of $CaCO_3$ in an argon atmosphere at 1100^oC, a portion of the fluxes prepared in nickel boats was remelted in a nickel crucible (19mm inside diameter) and was used as the sample for an experiment. Fluxes prepared in platinum crucibles (25mm inside diameter) were used directly for experiments. Sample weights of about 8g were used in all experiments.

Weight changes were recorded with a personal computer through RS-232C interface and these values were used to compute carbonate capacities. The electronic balance was covered with a plastic case which was filled with argon gas at a constant temperature controlled at 29±1°C. Premelted samples in a metallic crucible were set in an alumina support crucible and suspended by the thermobalance with a Ni-Cr wire (0.2mm dia.) and an alumina tube (4mm I.D.). Pt crucibles were used for experiments at temperatures over 1350°C. A closed end reaction tube made of mullite was used with the reacting gas mixtures.

This furnace can be elevated to a position where an appropriate temperature gradient is obtained. The furnace temperature was controlled within ±2°C.

Water cooling was used to avoid an increase in temperature by radiation from the furnace. A shutter was placed to prevent incoming reaction gases and vapour from the sample leaking into the thermobalance.

The surface temperature of the sample was kept slightly higher than the bottom by about 10°C per one centimeter to avoid convection in the melt. Reaction gases were introduced to the surface of the sample through an alumina tube (4mm I.D.) placed at a distance of 20mm above the top of the crucible. The temperature was monitored continuously during an experiment by a 30% Rh-Pt/6% Rh-Pt thermocouple. After a sample was introduced, the tube was completely filled with argon or a mixture with carbon monoxide. The furnace was then raised to the position for an experiment. When a stable condition was observed absorption of carbon dioxide was started by changing gas to that containing carbon dioxide. Carbonate capacities are calculated from weight changes obtained after equilibrium was attained.

3 Results and Discussion

3.1 Preliminary experiment

Typical weight changes observed are illustrated in Figure 1. Effects of the buoyancy on weight change measurements were considered negligible from a comparison of weight changes caused by dissolution of carbon dioxide. As expected, the effect was not detected when Ar-CO gas mixtures were changed to Ar-CO-CO_2 gas mixture. The reaction of carbon dioxide dissolution reached saturation in about one hour. The weight decreased constantly after saturation as shown by (A)-line in Figure 1. This weight loss was attributed to the vaporization of $CaCl_2$. This was so small (about 10^{-7} g/cm^2/sec) that the composition of sample did not change within the experimental time, but was considered for the calculation of dissolved carbon dioxide. Solubility was calculated by the corrected weight changes as shown in (B)-line.

From the above experiments, the following results were obtained: (1) effects of the buoyancy on weight changes were neglible. (2) weight losses of sample were small enough to neglect the effect on the composition, however, it should be taken in account for the calculation of the carbonate solubility.

3.2 Effect of CaF_2 content on the solubility of carbon dioxide

Figure 2 shows the solubility of carbon dioxide in CaO-CaF_2-$CaCl_2$ melts as a function of CaF_2 content at a fixed XCaO of 0.12. Solubility of carbon dioxide increased with increasing molar fraction of CaF2 replacing $CaCl_2$ from 0.44 wt% to 0.97 wt% at 1100°C and decreased from 2.95 wt% to 0.21 wt% with increasing temperature from 900°C to 1300°C at $XCaF_2$ = 0.6, for the flux of X_{CaO} = 0.12 in CaO-CaF_2-$CaCl_2$ system.

3.3 Carbonate in the fluxes

It can be considered that the electrolytes, CaO, $CaCl_2$ and CaF_2 are fully dissociated to their ions in the flux used on the present study, because these fluxes do not contain strong network formers such as SiO_2 or P_2O_5. When carbon dioxide dissolves as in the following equation:

$$CO_2 + O^{2-} = CO_3^{2-} \qquad [1]$$

the anion fraction of each ions can be calculated from the initial composition and carbon dioxide content obtained by experiment(4).

The equilibrium constant $K_{(1)}$ is defined as:

$$K_{(1)} = a_{CO_3^{2-}}/(a_{O^{2-}} \cdot P_{CO_2})$$

$$= X_{CO_3^{2-}} \cdot f_{CO_3^{2-}}/(X_{O^{2-}} \cdot p_{CO_2} \cdot f_{O^{2-}})$$

$$= C_{carb} \cdot f_{CO_3^{2-}}/(f_{O^{2-}} \cdot X_{O^{2-}})$$

where, the molar carbonate capacity is defined as:

$$C_{carb} = X_{CO_3^{2-}}/p_{CO_2}$$

$$= K_{(1)} \cdot (f_{O^{2-}}/f_{CO_3^{2-}}) \cdot X_{O^{2-}} \qquad [2]$$

If temperature and $f_{O^{2-}}/f_{CO_3^{2-}}$ are constant, Ccarb should be proportional to $X_{O^{2-}}$ and a straight line passing through the origin should be obtained.

Figure 3 shows the relation between carbonate capacity and anion fraction of oxygen ion after dissolution of carbon dioxide in CaO-CaF_2-CaCl2 molten fluxes. Similar results were reported for the CaO-$CaCl_2$ system elsewhere(4). The linearity of these relations reflects the assumption that Ccarb should be proportional to $X_{O^{2-}}$. This figure also shows that the value of $f_{O^{2-}}/f_{CO_3^{2-}}$ is constant and independent of the amount of the basic oxide, when the flux contained only one kind of anion, however, the ratio (slope of the line in Figure 3) increased when CaF_2 was added to the CaO-$CaCl_2$ binary.

4 Relation between carbonate capacity and other capacities

Hypothetical exchange reaction between carbonate ion and sulphide ions is useful to understand the relations between capacities:

$$\tfrac{1}{2}S_2 + CO_3^{2-} = S^{2-} + CO_2 + \tfrac{1}{2}O_2 \qquad [3]$$

$$K_{(2)} = (a_{S^{2-}} \cdot P_{CO_2}) \cdot (P_{O_2})^{\frac{1}{2}}/[(P_{S_2})^{\frac{1}{2}} \cdot a_{CO_3^{2-}}]$$

$$= X_{S^{2-}} \cdot (P_{O_2}/P_{S_2})^{\frac{1}{2}} \cdot P_{CO_2}/X_{CO_3^{2-}} \cdot f_{S^{2-}}/f_{CO_3^{2-}}$$

$$C_{carb} = X_{S^{2-}} \cdot (P_{O_2}/P_{S_2})^{\frac{1}{2}} \cdot f_{S^{2-}}/f_{CO_3^{2-}}/K_{(3)}$$

where $C_{carb} = X_{CO_3^{2-}}/P_{CO_2}$

If the value of $f_{S^{2-}}/f_{CO_3^{2-}}$ does not depend on the composition, we may write

$$C_{carb} = \text{const} \cdot C_S^{2-}, \quad \text{or}$$

$$\log C_{S^{2-}} = \log C_{carb} + \text{const},$$

where

$$C_{S^{2-}} = X_{S^{2-}} \cdot (P_{O_2}/P_{S_2})^{\frac{1}{2}}$$

There was good linearity observed between the sulphide capacities by Hawkins et al(5) and Kor et al(6) and the carbonate capacities of the present work (Figure 4). It should be noted, however, that sulphide capacities in Figure 4 decreased with increasing sulphur content in the flux. This means sulphur itself influences the nature of the flux. In other words, we should examine fluxes which already contained sulphur for complete understanding of the relation between carbonate capacity and basicity. However, in practical use to develop better fluxes of higher basicity, carbonate capacity considering no their element such as S or P works well as an indication of the basicity.

We may also write the relationship between carbonate capacity and phosphate (7), phosphide(7), nitride and cyanide capacities(8) in the same manner as sulphide.

$$\tfrac{1}{2}P_2 + 3/2CO_3^{2-} + 5/4O_2 = P_{O_4}^{3-} + 3/2CO_2 \quad [4]$$

$$C_{carb} = (C_{P_{O_4}^{3-}})^{2/3} . f_{P_{O_4}^{3-}}/f_{CO_3^{2-}}/K_{(4)}$$

where

$$C_{P_{O_4}^{3-}} = (\text{wt \% } P_{O_4})/(P_{O_2})^{5/4}/(P_{P_2})^{\frac{1}{2}}$$

$$\tfrac{1}{2}P_2 + 3/2CO_3^{2-} = P^{3-} + 3/4O_2 + 3/2CO_2 \quad [5]$$

$$C_{carb} = (C_P{3-})^{2/3} . f_{P^{3-}}/f_{CO_3^{2-}}/K_{(5)}$$

where

$$C_{P^{3-}} = (\text{wt \% } P^{3-}) \cdot (P_{O_2})^{3/4}/(P_{P_2})^{\frac{1}{2}}$$

$$\tfrac{1}{2}N_2 + 3/2CO_3^{2-} = N^{3-} + 3/4O_2 + 3/2CO_2 \quad [6]$$

$$C_{carb} = (C_{N^{3-}})^{2/3} . f_{N^{3-}}/f_{CO_3^{2-}}/K_{(6)}$$

where

$$C_{CN^{3-}} = (\text{wt \% } N^3) \cdot (P_{O_2})^{3/4}/(P_{N_2})^{\frac{1}{2}}$$

$$\tfrac{1}{2}N_2 + \tfrac{1}{2}CO_3^{2-} + C = CN^- + 1/4O_2 + \tfrac{1}{2}CO_2 \quad [7]$$

$$C_{carb} = (C_{CN^-})^2 \cdot f_{CN^-}/f_{CO_3^{2-}}/K_{(7)}$$

where

$$C_{CN^-} = (\text{wt \% } CN^-) \cdot (P_{O_2})^{1/4}/(P_{N_2})^{\frac{1}{2}}$$

Figure 5 and Figure 6 show the relationship between these capacities. The relation between Ccarb and $C_{S^{2-}}$, $C_{PO_4^{3-}}$ and $C_{N^{3-}}$ were in good agreement with that expected theoretically. In comparison for C_{carb} with $C_{P^{3-}}$ and C_{CN^-}, good linearity was observed, however, slopes were greater than those expected theoretically.

It is necessary to clarify the effects of these ions of interest such as sulphide or phosphate, on carbonate capacities. The experimental data, however, suggest that introduction of carbonate capacities as a measure of qualitative basicity is reliable and effective.

5 Conclusions

1) The solubility of carbon dioxide in molten CaO-$CaCl_2$-CaF_2 flux was measured by thermogravimetric technique with thermobalance over the temperature range of 900°C to 1500°C.

2) The solubility of carbon dioxide increased from 0.44 to 0.97 wt% with increasing CaF_2 at 1100°C.

3) The solubility decreased from 2.95 to 0.21 wt% with increasing temperature from 900°C to 1300°C at fixed composition.

4) The relationships between carbonate capacity and other capacities were in good agreement with those expected theoretically.

References

1) N.Sano, Private communication

2) C. Wagner, "The Concept of the Basicity of Slags"., Metallurgical Transactions B., SEP. (1975), pp405-409

3) T. Kawahara, K. Yamagata and N. Sano, "The CO_2 solubilities of highly basic melts", Steel Research, 57, (1986), pp160-165

4) M. Maeda and A. McLean, "Dissolution of Carbon Dioxide in CaO-$CaCl_2$ Melts.", Transactions of the ISS., SEP. (1986), pp61-65

5) R.J. Hawkins, S.G. Meherali and M.W. Davies, "Activities, Sulphide Capacities and Phase Equilibria in CaF_2-Based Slags". Journal of The Iron and Steel Institute, AUG, (1971), pp646-656

6) G.J.W. Kor and F.D. Richardson, "Sulfide Capacities of Basic Slags Containing Calcium Fluoride". Transactions of the Metallurgical Society of AIME, 245 (1969), pp319-327

7) S. Tabuchi and N. Sano, "Thermodynamics of Phosphate and Phosphide in CaO-CaF_2 Melts", Metallurgical Transactions B, 15B, (1984), pp351-356

8) E. Martinez and N. Sano, Private communication

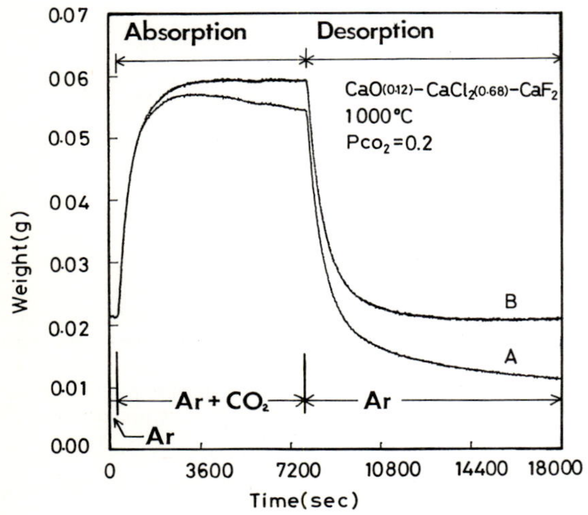

Figure 1

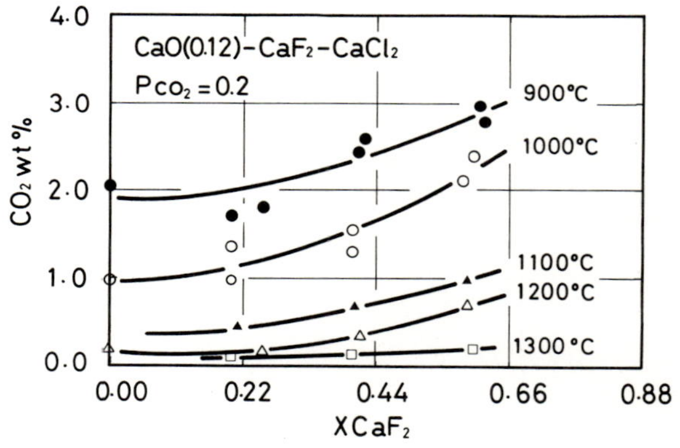

Figure 2

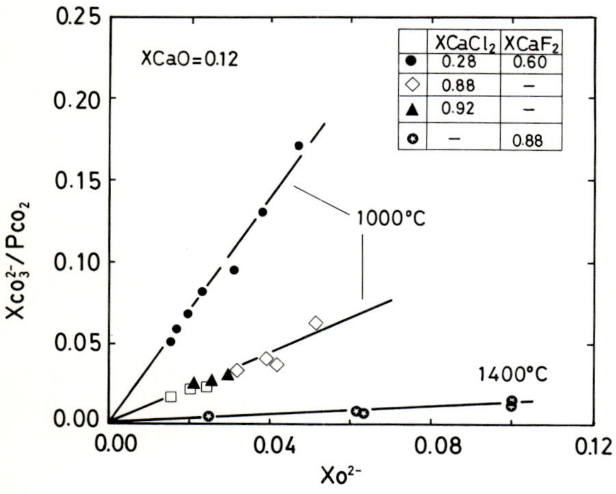

Figure 3

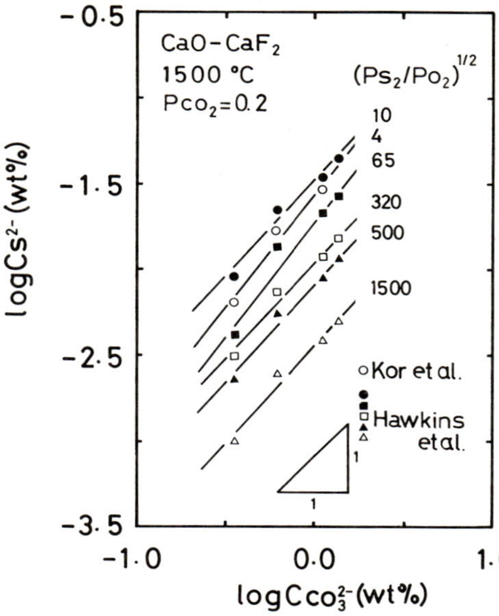

Figure 4

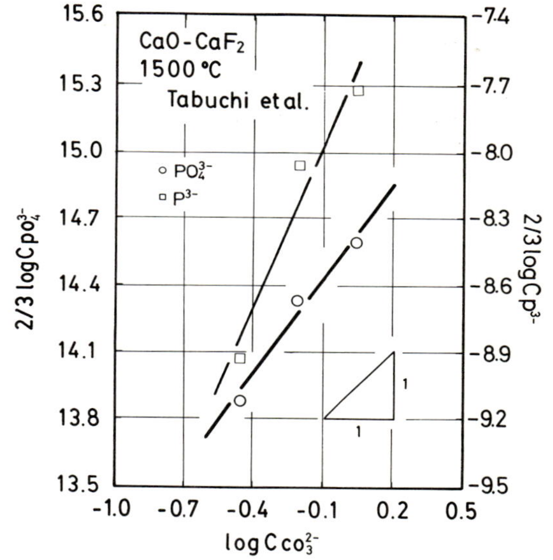

Figure 5

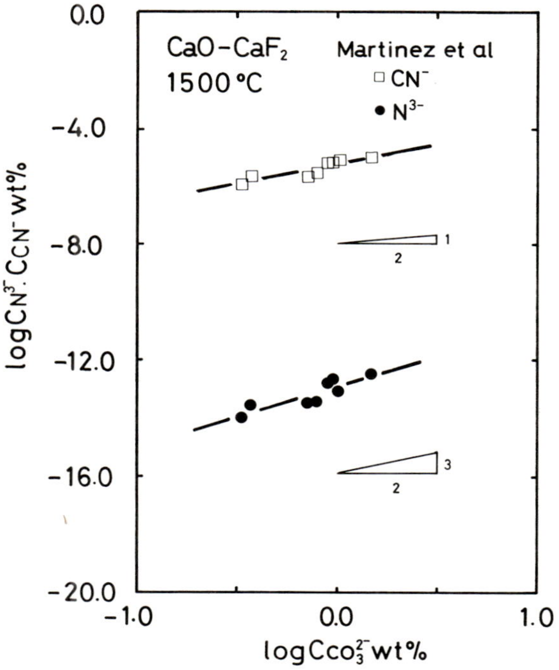

Figure 6

Fourier transform infra-red spectroscopy (high resolution range) of water containing $CaO-Al_2O_3$ and $CaF_2-CaO-Al_2O_3$ slags and applications

G LEEKES, N NOWACK and F SCHLEGELMILCH

Ms Leekes, formerly a Research Assistant at the Fachhochschule Niederrhein (FHN), is now with Spectro GmbH Kleve; Dr Nowack is Professor of Chemical Engineering (FB04) at the FHN, Krefeld; Dr Schlegelmilch is a Professor of Analytical Chemistry (FB01) at the FHN.

SYNOPSIS

In liquid $CaO-Al_2O_3$ and $CaF_2-CaO-Al_2O_3$ slags soluted water vapour (fluorine-hydrogen) was determined with Fourier-Transform-Infra-Red-Spectroscopy (FT-IR). Liquid slags were quenched in liquid deuterium oxide, (D_2O), to form a glassy phase. Interpretation of the vibration range yielded information about the chemical bonding. The solution enthalpy of water vapour in $CaO-Al_2O_3$ slags was estimated by variation of temperature. To determine the diffusion coefficient $D(OH^-)$ a capillary technique can be used and to measure the OH^- concentration gradient, the FT-IR technique requires only little sample substance. The FT-IR spectroscopy is quick and therefore suitable for control of metallurgical processes.

INTRODUCTION

At high temperatures the slag can take up molecules from the gas phase which dissociate in solution; this reaction type represents a Lewis acid base reaction. Gas solubility can also take place by redox reactions. The reactions products of gas solubility in liquid slags like OH^-, X-O-H (X can be P, B, Si...), CN^-, CO_3^{2-} and SO_4^{2-} are IR-active and can be measured by FT-IR spectroscopy.

IR-spectroscopy investigations of water vapour solubility in liquid salt melts have already been done (1-3) with discs of silicate glasses. The model slag system of borates and phosphates have also been investigated(4).

In basic slags the following reaction takes place(5):

$$H_2O_{(g)} + O^{2-}(slag) \rightleftarrows 2OH^-(slag) \qquad [1]$$

with the equilibrium constant K:

$$K.a(O^{2-}) = c^2(OH^-)/p(H_2O)$$

The OH^- ions absorb infra-red light while passing through a slag layer if its frequency corresponds to the vibration frequency. So the IR-spectrum is the plot of transmission or extinction (absorbance) as a function of the wave number $\tilde{v}$.

OH^- as a diatomic ion is considered as an isolated unit to have only one basic vibration wave number. The range of OH^- vibration covers 3700-3000 cm^{-1} depending on the existence of hydrogen bonding: $O^{2-}...H-O^-$. In liquid slags hydrogen bonding plays no important role because of the low bonding energy of up to 20 kJ/mol OH^-. An exception is hydrogen difluoride HF_2^- with a bonding energy $F-H...F^-$ of about 110 kJ/mol HF.

Soluted H_2O-molecules have three basic vibrations: asymmetric and symmetric stretching vibrations (3750-3300 cm^{-1}) and the deformation vibration of the crystal water (1600-1650 cm^{-1}).

The water content of the slags is quantitatively investigated by measurement of the integral extinction E_i of the OH^- band. This method is essentially more accurate than previously used basis line methods(1). The OH^- group in glassy solidified slags takes no definite lattice site whilst their vibrations are influenced by statistically distributed neighbour groups. In the FT-IR-spectra the OH^- band arises therefore as a broad band. Measurement of E_i pays regard to the distribution of all available OH^- groups. Indeed this method presumes high resolution of the IR-signal and a low statistical error of E_i (lower than +1%) using the FR-IR spectrometer. The integral extinction is defined by the equation [2]:

$$E_i = \int_{(band)} E(\tilde{v})\, d\tilde{v} \qquad [2]$$

The OH^- band is integrated from $\tilde{v}_1$ = 3700 to $\tilde{v}_2$ = 3000 cm^{-1}. The connection between E_i and the water content of the slag $c(OH^-)$ is shown in equation [3]:

$$E_i - E_i(0) \sim c(OH^-) \sim \sqrt{p(H_2O)} \qquad [3]$$

($E_i(0)$: integral extinction of water free slag sample)

EXPERIMENTAL PROCEDURE

<u>Preparation of the slag samples</u>

The water containing slags are ground to a medium grain size below 2µm. In the system $CaF-Al_2O_3$ 9.5 mg, in the system $CaF_2-CaO-Al_2O_3$ 5.0 mg of ground slag with over 300 mg water free potassium bromide (KBr: no

IR absorbance) as carrier material are prepared by the KBr-press technique. The produced discs have a thickness of 0.90 ± 0.01 mm. The sample is measured with FT-IR (Nicolet, 5DX-FT-IR) immediately after preparation. To avoid errors in the quantitative evaluation of the OH^- band by the small water content in the KBr, a pure KBr-disc is measured parallel to each slag disc. This disc is taken as a reference and is accurately substracted by means of a computer function from the sample spectrum.

Equilibrium adjustment

The $CaO-Al_2O_3$ slags are equilibrated at 1500 and 1600°C with definite water vapour partial pressures for which equilibrium is reached after more than eight hours. The liquid $CaF_2-CaO-Al_2O_3$ slags are equilibrated with definite H_2O-HF atmospheres at 1600°C. Additional to reaction [1] equilibrium of the components

$$CaO\ (slag) + 2HF_{(g)} \rightleftarrows CaF_2\ (slag) + H_2O^{(g)} \qquad [4]$$

has to be taken into consideration. The equilibrium constant of the reaction [4] at 1600°C is $K_p(1) = 61\ bar^{-1}$.

HF can dissolve in liquid slags by the following reaction:

$$HF_{(g)} + F^-(slag) \rightleftarrows HF_2^-\ (slag) \qquad [5]$$

Figure 1 shows the experimental apparatus for this equilibrium adjustment. The equilibrium is adjusted by a gas mixture of CF_4 and N_2. This gas mixture, with an additional fixed water vapour partial pressure, is passed through a platinum capillary. At high temperatures CF_4 and excess H_2O react completely:

$$CF_4(g) + 2H_2O(g) \rightleftarrows CO_2(g) + 4HF(g) \qquad [6]$$

The equilibrium constant of reaction [6] at 1600°C is $K_p(2) = 2.5 \times 10^{16}\ (bar)^2$. If $p(H_2O) > 2p(CF_4)$, reaction [6] reaches completion and a definite relation

$$\sqrt{p(H_2O)}/p(HF) \qquad [7]$$

can be adjusted.

After equilibrium is reached the liquid slags are removed from the furnace very quickly and are quenched in liquid deuterium oxide. During quenching the quenching speed of the middle of the sample is about 400°C/s. The slag solidifies as a glass and takes up no D_2O during fast quenching. Otherwise the OD-band has to be seen in the FT-IR spectrum. A doping technique is tried to calibrate the water content of $CaF_2-CaO-Al_2O_3$ slags in the following way. Pure CaF2 is premelted in a graphite crucible. A CaF_2^- sample produced in this way is weighed. The $CaO-Al_2O_3$ slag is equilibrated with definite water vapour partial pressure at 1600°C. The water content is known from (5). The water free CaF_2 sample is suddenly brought into the melt. The melt is stirred (c. 20s) and then quenched in D_2O. These experiments are based upon the previously made observation: the water reception respectively delivery of the slags is very slow.

RESULTS AND DISCUSSION

In Figure 2 the FT-IR spectra of the slag composition 44 wt% CaO-56 wt% Al_2O_3 is shown for three different water vapour partial pressures at 1600°C. With increasing $p(H_2O)$ the OH^- band gets broader and deeper.

A further proof is the missing deformation vibration band. Experiments, in which the slag are equilibrated with $D_2O(g)$ instead of $H_2O(g)$ under identical conditions were made to prove that the measured band from 3700-3000 cm^{-1} is really the OH^- band. The transmission minimum has shifted from 3450 cm^{-1} (Figure 2) for the OH^- band to 2618 cm^{-1} for the OD^- band. The form of the two peaks is analogous. The linear relationship between $E_i-E_i(0)$ and $p(H_2O)$, equation [3], investigated at 1500 and 1600°C is shown in Figure 3(a).

The dependence of the integral extinction on temperature allows the calculation of the solution enthalpy H_L. The increase of water vapour solubility with increasing temperature proves that reaction [1] is endothermic ΔH_L is calculated from the Van't Hoff relationship as:

$$\Delta H_L = -R \left[\frac{\ln \frac{E_i - E_{i(0)}}{\sqrt{p(H_2O)}}\Big|_{T_1} - \ln \frac{E_i - E_{i(0)}}{\sqrt{p(H_2O)}}\Big|_{T_2}}{1/T_1 - 1/T_2} \right] \qquad [8]$$

The values of the solution enthalpies are shown in Table 1 for the three slag compositions investigated. In the measured concentration range the solution enthalpies are positive and nearly constant.

In the fluoride slags vibrations of the OH^- group lie in the same range as in the system $CaO-Al_2O_3$ (transmission minimum 3482 cm^{-1}). The integral extinction and its dependence upon $p(H_2O)$ and p(HF) is shown in Figure 3(b). To calibrate the water content the described doping technique was used. The points $E_i = f$ (wt% H_2O) in Figure 3(b) are results of this calibration. In the system $CaF_2-CaO-Al_2O_3$ an additional weak deformation vibration band of the H_2O molecule at 1630 cm^{-1} arises. This bands indicates that H_2O molecule partially exist in the slags as crystal water. In the system $CaO-Al_2O_3$ this peak does not arise. The peak is only weak and an accurate quantitative interpretation is not possible. Indeed, an increase of the integral extinction proportional to the p(HF) is measurable from experiments. Thus, it is possible, that during quenching HF soluted in the slag, equation [5], partially reverses with the OH^- ions. At 1450 cm^{-1} a weak band arise in the spectrum of $CaF_2-CaO-Al_2O_3$ slags, which can be caused by overlapping of CO_3^{2-} and HF_2^- (residual concentrations).

Additional information about the structure of slags can be gained from an analyses of the vibrational range of Al-O and Al-O-F. In Figure 4 this vibrational range of the three glassy $CaO-Al_2O_3$ slags (curve 3-5) and for comparison a $CaF_2-CaO-Al_2O_3$ slag are represented. Typical absorbance maxima of substances with AlO_4 - tetrahedra are observed: 850-700 cm^{-1} for the asymmetric stretching and c. 400 cm^{-1} for the deformation vibration. The asymmetric stretching vibration can be used for identification of the slag. The position of the absorbance maximum of this band shifts with increasing CaO content continuously to higher wave numbers. Figure 5 shows the dependence of the position of this vibration from the Al_2O_3 content. The reason for the shift to higher wave numbers with increasing CaO and therefore increasing O^{2-} ion content is the loosening of the connection of AlO_4 tetrahedra upon a common oxygen atom. Thus the bonding grade Al-O and therefore the force constant increases.

The three $CaO-Al_2O_3$ slags investigated solidify as disordered glass. (Glassy quenched slags are investigated, for the glassy state is, in a thermodynamic point of view, is a frozen, supercooled melt). They are transformed by means of a heat treatment into their cyrsalline ordered equilibrium phases. Compacts of quenched 52 wt% CaO-48 wt% Al_2O_3 slag are heated at c. 800°C. The increasing grade of order ($C_{12}A_7$ and C3A) of the slag becomes evident very well in the FT-IR spectra: Figure 6. The crystallisation process begins at 800°C after five minutes spontaneously but at 780°C only after 15 minutes (activation energy 440 kJ/mol). The process is complete after one hour. Figure 6 shows this slag at different heat treatment times. At 1096 cm^{-1} a new strong band of C_3A arises. The structure of C_3A is well known from cement research (6). (6) describes the crystal structure of C_3A as consisting of $Ca_2Al_6O_{24}$ groups, based puckered Al_6O_{18} rings and two CaO_6 octahedra, respectively. The O^{2-}-co-ordination tetrahedra of AlO_4 in the Al_6O_{18} ring are considerably irregular, the shortest Al-O distance being 172 pm, the longest is 178 pm. The shorter bonding distance of 172 pm is not normal; which 178 pm. The strong wave number shift from 850 cm^{-1} to 1096 cm^{-1} is established for the decrease of distance. According to Badger's relationship(7) this wave number shift can be approximately calculated as:

$$\tilde{\nu}(AlO_4 \text{ in } Al_6O_{18})/\tilde{\nu}(AlO_4 \text{ regular}) = 1.56 \quad [8]$$

The measured one amounts to 1.73.

The complete work, including the diffusivity measurements by application of FT-IR spectroscopy and a capillary technique, will be published later (8).

ACKNOWLEDGEMENTS

This work was financed by the ministry of sicence and research of Nordrhein Westfalen (FRG), Nr. 9712/20300387. We thank the Ministry for this support. Our special gratitude belongs to Professor Dr. rer. nat. H.J. Engell for his help and interest in this work.

REFERENCES

1) Scholze, H.: Glastech. Ber. 32 (1959) 81/88

2) Scholze, H.: Glastech. Ber. 32 (1959) 142/152

3) Scholze, H.: Glastech. Ber. 32 (1959) 278/281

4) Kohnke, C., N. Nowack: Arch. Eisenhuttenwes. 52 (1981) 271/273

5) Schwerdtfeger, K., H.G. Schubert: Met. Trans. B. (1978) 9B, 143/44

6) Morikawa, H., S. Iwai, F. Marumo, H. Kim: Review 29th general meeting, Cement Association of Japan, Tokyo (1974)

7) Badger, R.M.: J. Chem. Phys. 2 (1934) 128

8) Leekes, G., N. Nowack, F. Schlegelmilch: Steel research (in press), III, quarter 88, (two parts)

Table 1 Solution Enthalpies in the system CaO-Al_2O_3 (1500-1600°C)

CaO (wt %)	Al_2O_3 (wt %)	H_L (kj/ mol OH^-)
52	48	69 + 6
47	53	66 + 5
44	56	72 + 10

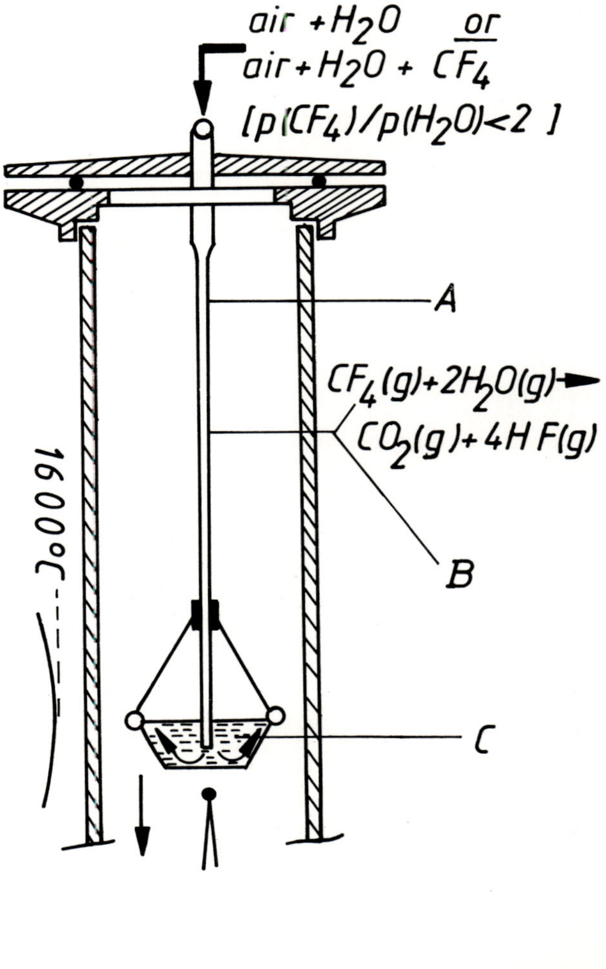

Figure 1 Experiment to adjust H_2O-HF equilibria in CaF_2-CaO-Al_2O_3 slags. A: platinum capillary, diameter 2 mm; B: reaction distance for transformation of the gas mixture CF_4-H_2O-air; C: platinum crucible containing slag

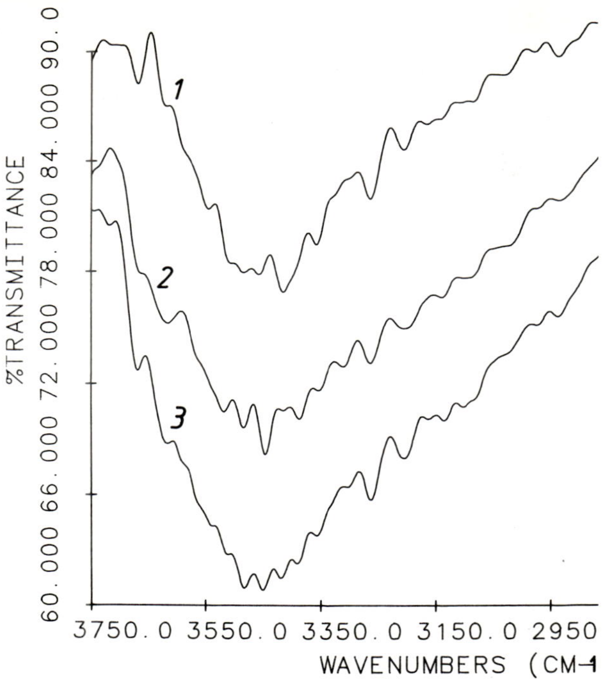

Figure 2 FT-IR spectra of the OH^- peaks of the slag 44 wt% CaO-56 wt% Al_2O_3 at three different water vapour partial pressures, 1600°C.
1. $\rho(H_2O)$=0.03 bar, 2. $p(H_2O)$=0.1 bar,
3. $p(H_2O)$-0.2 bar

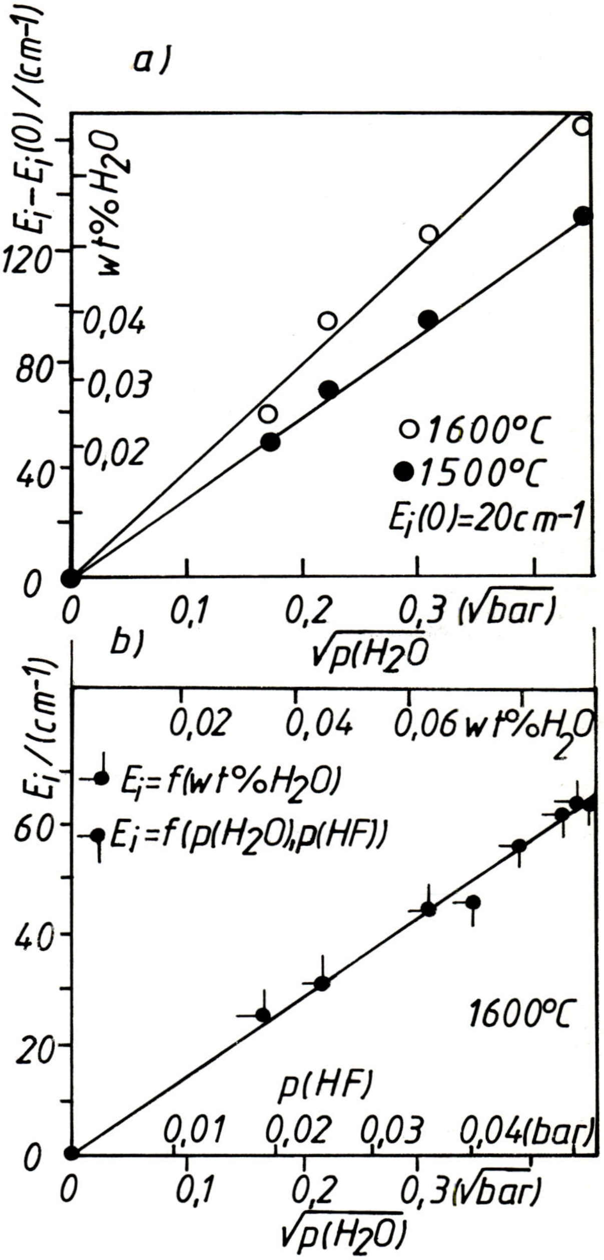

Figure 3 Dependence of the integral extinction E_i on the partial pressure $p(H_2O)$ and $p(HF)$;
a) 44 wt% CaO-56 wt% Al_2O_3
b) 13.5 wt% CaF_2-47 wt% CaO-39.5 wt% Al_2O_3 and from H_2O calibrations

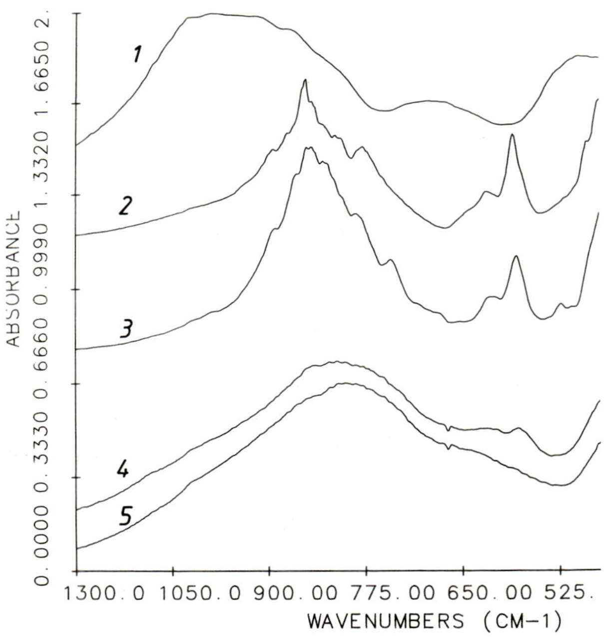

Figure 4 FT-IR spectra (in absorbance) vibrational range Al-O-F and Al-O-Si, $p(h_2O)$=0.2bar, 1600°C:
1. 40 SiO_2-40 CaO-20 Al_2O_3 (wt%)
2. 13.5 CaF_2-47 CaO-39.5 Al_2O_3 (wt%)
3. 52 CaO-48 Al_2O_3 (wt%)
4. 47 CaO-53 Al_2O_3 (wt%)
5. 44 CaO-56 Al2O3 (wt%)

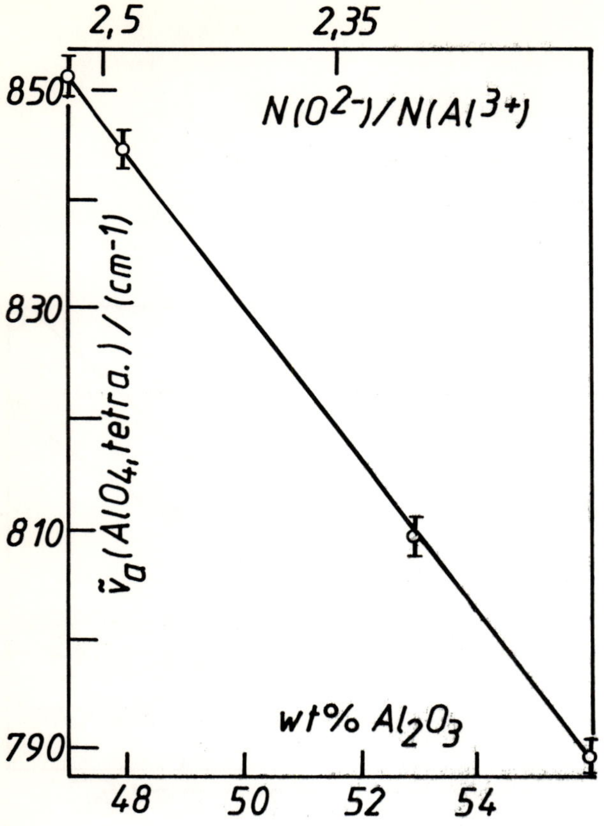

Figure 5 Position of the asymmetric stretching vibration of the AlO_4 tetrahedra dependent on the Al_2O_3 content of the CaO-Al_2O_3 slags investigated, quenched from 1500 and 1600°C on room temperature, 400°C s^{-1}

Figure 6 FT-IR spectra of the slag 52 wt% CaO-48 wt% Al_2O_3, synthesised at 1600°C; at different heating times:

1. t 3 min at 800°C or t 10 min at 780°C
2. t = 5 min at 800°C or t = 15 min at 780°C
3. t - 6 h at 800°C or 780°C

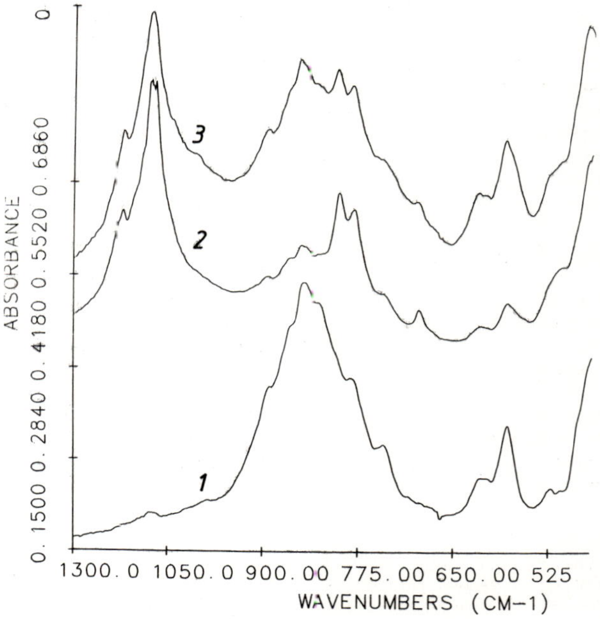

Determination of solubility and apparent diffusion coefficient of water vapour in synthetic slags

T T DO and K W LANGE

Dr-Ing Do is Scientific Co-worker; Dr-Ing Lange is Professor of Ferrous Metallurgy in the Department of Metallurgy, Technical University of Aachen, Aachen, FRG.

SYNOPSIS

The simultaneous determination of solubility and apparent diffusion coefficients of water vapour in molten slags in one single experiment carried out in a Sieverts' cell according to the volumetric method is described.

Both the solubility and the apparent diffusion coefficient increase with increasing contents of lime and decrease with increasing contents of silica and alumina. The effects of the molar ratio $B = N_{CaO}/(N_{SiO_2} + N_{Al_2O_3})$ and of the optical basicity Λ are presented.

I. INTRODUCTION

Hydrogen in steel can cause serious defects. It has been found that the knowledge about the solubility and the diffusion coefficient of water vapour in liquid slags is very helpful for the control of the hydrogen concentration in steel. Especially in ladle refining processes the slag composition is very important. In this work the solubility and the apparent diffusion coefficient of water vapour in liquid slags are determined.

II. EXPERIMENTAL APPARATUS AND PROCEDURE

For this purpose the volumetric method was chosen in using a Sieverts' cell. Figures 1 and 2 show the Sieverts' cell and the experimental apparatus, respectively.

The pressure change dp in the Sieverts' cell can be registered through a pressure gauge, which is joined to a 5 kHz-measuring amplifier. A calibrication resulted in

$$dU = \alpha_1 \cdot dp \qquad (1)$$

where dU is the voltage change, measured at the outlet of the amplifier. α_1 is about 0.263 V/mbar.

A determination of the hot volume of the Sieverts' cell with a gas burette yielded

$$dV = \alpha_2 \cdot dp \qquad (2)$$

with, for example, T = 1,550°C, α_2 = 0.0561 ml/mbar.

With the equation of state the equivalent number of gas moles dn belonging to dV can be computed according to

$$dn = \frac{1}{22.413 \cdot 10^3} \cdot \frac{P}{1,013} \cdot \frac{273}{T_{bu}} \, dV \qquad (3)$$

(T_{bu}: temperature of gas in burette in K, P: atmospheric pressure in mbar)

Inserting eq. (2) into eq. (3) gives

$$dn = \alpha_3 \cdot dp \qquad (4)$$

with

$$\alpha_3 = \frac{1}{22.413 \cdot 10^3} \cdot \frac{P}{1,013} \cdot \frac{273}{T_{bu}} \cdot \alpha_2, (mol/mbar^{-1}) \qquad (5)$$

This means that the change of the number of gas moles in the gas phase of the Sieverts' cell can be determined by the pressure change dp.

Experimental Procedure

A slag sample of about 7 g, contained in a Pt/Rh 70/30 crucible with an outer diameter of 30 mm, thickness 0.3 mm, is put into the Sieverts' cell (Fig. 1). The sample is held at the experimental temperature and dried with pure flowing argon (region I in Fig. 3). The magnetic valves MV 1 and MV 2 are shut off as soon as the experiment begins. Afterwards the pressure change in the Sieverts' cell can be observed (region II). When the pressure has remained constant for 10 minutes, it is assured that there is no longer an exchange between the dry argon gas and the liquid slag. The system stays in equilibium (at the start of this investigation this waiting period had been extended up to 60 minutes).

The measurement is started by pressing a certain key of the microprocessor (left hand side of region III in Fig. 3). First the magnetic valves MV 4 and MV 5 as well as MV 1 and MV 2 are opened and at the same time the magnetic valve MV 3 is shut off. Now the humid argon with a given water vapour pressure ($p_{H_2O,e}$ = 21.95 mbar) passes over the slag sample for 30 seconds. After this, the magnetic valves MV 1 and MV 2 are shut off again. The humid argon now passes through the flask into the open.

Then the dried slag sample tends to come into equilibrium with the humid argon within the Sieverts' cell. It absorbs water molecules and causes a decrease of pressure (starting from pressure level p^*, point A_1 in Fig. 3), which can be recorded by the pressure gauge.

Through this dissolution, the water vapour content in the Sieverts' cell and the driving force of the reaction decrease. When the pressure level reaches to p_R (Point B_1, Fig. 3), the two magnetic valves MV 1 and MV 2 are opened once again for 10 seconds. Thus the partial pressure of the water vapour at the interface gas/slag melt is kept quasi-constant (this is an important marginal condition for the later evaluation). Simultaneously, the first interval of the measurements comes to an end, as the pressure change can only be recorded with a closed Sieverts' cell. Therefore A_1B_1 indicates the pressure change in the Sieverts' cell for the period in which the two magnetic valves MV 1 and MV 2 are shut off. Since the slag melt continues to absorb water vapour with the magnetic valves open, $A_1^*A_1B_1$ is the real progress of the total pressure; it reveals the development of the water vapour partial pressure only between A_1 and B_1. In order to ascertain the absorption of the water vapour during the time-span in which the magnetic valves are opened, the stretch A_1B_1 is extrapolated beyond point A_1 to point A'_1. Here it is assumed that the change of the water vapour between A'_1 and B_1 on the other hand are analogous.

This assumption is justified for the following reason. If we compare the short periods in which the magnetic valves are opened (at first 30 seconds, then 10 seconds each) with the measurement intervals in which the valves are closed (at first 60 seconds, later up to 5 minutes), and if we consider the insignificant maximum difference of the reaction force between p^* and p_R (2.19 mbar at the most), we realise that the maximum water vapour partial pressure is always guaranteed when the magnetic valves are opened and that this partial pressure is a maximum of 2.19 mbar lower when the valves are closed. The deficiency due to the impossibility of recording the change of the partial pressure of the water vapour during the time of the open valves can thus be compensated with good approximation.

Thus, humid gas with p_{H_2O} = 21.95 mbar passes again and again over the slag melt, when the magnetic valves are opened repeatedly for 10 seconds as soon as the pressure level has reached p_R. Thus the "old" gas in the Sieverts' cell is repeatedly exchanged with fresh gas with the initial water vapour partial pressure and the total pressure rising again to the initial value p^*. The two magnetic valves are shut off again, after the span of opening (10 s) has elapsed. The next measurement interval begins at point A_2 and terminates at point B_2, when the pressure again reaches p_R. A'_2 is determined by the extrapolation described.

The measuring process continues like this. It is necessary to recognize that the time interval needed in order to reach the difference of the total pressure $p^* - p_R$ = 2.19 mbar becomes the greater, the longer the experiment lasts (see Curve I in Fig. 4). The measuring process is stopped as soon as the new equilibrium is obtained: the total pressure does not change any more in the closed reaction chamber. Apparently, all recorded pressure values vary only in the range between p^* and p_R.

The total change of the partial pressure (Curve II, Fig. 4), caused by the dissolution of water vapour, can be determined by continously adding the various curves showing the decrease of the pressure at all the intervals of measurement (through parallel shifting of, e.g., A_2B_2 downwards till A'_2 coincides with B_1). Only a few intervals are schematically shown in Figure 4; during the experiment approximately 14 to 40 intervals can be observed.

Thus, the pressure drops from p^* (initial equilibrium) with progressing experimental time and at decreasing velocity to p_e (new equilibrium); Δp_t (= $p_1^* - p_t$) expresses the resulting pressure drop (dependent on the time) which is registered from the beginning of the experiment till the time t. Δp_{tot} (= $p_1^* - p_e$) is the maximum of the pressure drop that is caused by the dissolution of the water vapour in the liquid slag.

As a consequence of eq. (1) a voltage axis as the right ordinate can be added to the upper diagram in Figure 3. Here the voltages U^*, U_1^*, U_2^*, and U_R correspond to the pressures p^*, p_1^*, p_2^*, and p_R.

Typical values for the pressure differences $p^* - p_R$ or $p_1^* - p_e$ are around 2.19 or in the range between 30 and 105 mbar. Correspondingly, the voltage difference $U^* - U_R$ reaches values of approximately 570 mV. Values between 8 and 27 V were registered for $U_1^* - U_e$.

Experiments with an empty crucible (blind experiments) show that the whole apparatus is inert towards the change from a dry argon to an argon-H_2O-gas mixture.

The advantages of this method are: there is no need to take a sample; the measurements are done automatically and quasi-continuously; the influence of the fluctuation of the room temperature and the adsorption of the water vapour at the internal walls of the Sieverts' cell are eliminated, because the Sieverts' cell, the pressure gauge, the magnetic valves, and the appropriate gas train are contained in heated water (45 ± 0.1 °C). The sensitivity is around ± 0.86 μg water vapour; the solubility and the diffusion coefficient of the water vapour are determined simultaneously in only one single experiment. The solubility is computed from the total pressure change. The apparent diffusion coefficient of the water vapour in the liquid slag is evaluated from the change of the pressure with time. Both will be explained in the following.

Evaluation of Data

Solubility

The liquid slag phase contains as many moles again of water vapour as the gas phase delivered to it, therefore one can write:

$$(dn_{H_2O,g} = - dn_{H_2O,l})_t \qquad (6)$$

g and l indicate the gas and liquid phase, respectively.

According to Curve II in Figure 4 above, the pressure decreases by $\Delta p_t = (p_1^* - p_t)$ at time t after starting the experiment owing to the absorbtion of the water vapour by liquid slag. Thus the decreasing number of moles of the water vapour in the gas phase can be computed (according to eq. (4)) as follows:

$$(dn_{H_2O,g} = \alpha_3 \cdot dp)_t \qquad (7)$$

(It is assumed that the pressure change takes place only through the exchange of the molecules of water vapour with the liquid slag.)

According to eqs. (6) and (7)

$$(dn_{H_2O,l} = -\alpha_3 \cdot dp)_t \qquad (8)$$

Therefore the changes of the concentration of the water vapour dc_{H_2O} (ppm) in the liquid slag can be computed as follows

$$(dc_{H_2O} = - \alpha_3 \cdot M_{H_2O} \cdot \frac{10^6}{G} \cdot dp)_t \qquad (9)$$

with G : weight of the liquid slag (g)
M_{H_2O} : molecular mass of the water vapour.

By setting

$$\alpha_4 = \alpha_3 \cdot M_{H_2O} \cdot 10^6 \quad , \quad (g/mbar) \qquad (10)$$

$$(dc_{H_2O} = \Delta c_{H_2O} = - \alpha_4 \cdot \frac{1}{G} \cdot dp)_t \qquad (11)$$

is obtained. According to eq. (11), the problem of determining the time dependent change of the concentration of water vapour in the liquid slag is simplified to the determination of the time dependent pressure change in the gas phase of the Sieverts' cell.

By means of the conversion rate α_4/G it is possible to derive the concentration-time curve (the lowest curve in Fig. 4) from the pressure-time curve (Curve II) in Figure 4 above: At the start of the experiment the pressure is p_1^*, and the liquid slag has the initial concentration c_1 of the water vapour. If the pressure drops from p_1^* to p_e, the concentration of the water vapour in the liquid slag rises simultaneously from c_1 to c_e. Integrating eq. (11) gives

$$c_e - c_1 = - \frac{\alpha_4}{G} \cdot (p_e - p_1^*) \qquad (12)$$

Inserting Sieverts' law

$$c_e = K_{H_2O} \cdot \sqrt{p_{H_2O,e}} \qquad (13)$$

$$c_1 = K_{H_2O} \cdot \sqrt{p_{H_2O,1}} \qquad (14)$$

where

K_{H_2O} is the mass action law constant in ppm/bar$^{-1/2}$ and

$p_{H_2O,1}$, $p_{H_2O,e}$ are the water vapour partial pressures of the dried and of the humid argon,

gives

$$K_{H_2O} = - \frac{\alpha_4}{G} \cdot \frac{p_e - p_1^*}{\sqrt{p_{H_2O,e}} - \sqrt{p_{H_2O,1}}} \qquad (15)$$

According to eq. (1) and Fig. 4

$$p_e - p_1^* = \frac{1}{\alpha_1} \cdot (U_e - U_1^*) \qquad (16)$$

Thus eq. (15) becomes

$$K_{H_2O} = - \frac{\alpha_4}{\alpha_1} \cdot \frac{1}{G} \cdot \frac{U_e - U_1^*}{\sqrt{p_{H_2O,e}} - \sqrt{p_{H_2O,1}}}$$

$$= -7.066 \cdot \frac{\alpha_4}{\alpha_1} \cdot \frac{U_e - U_1^*}{G} \qquad (17)$$

($p_{H_2O,e}$ = 21.95 mbar, $p_{H_2O,1}$ = 0.044 mbar)

Therefore, the determination of the solubility constant is simply bound to the voltage change of the pressure gauge.

Diffusion Coefficient

After the diffusing time t the average concentration $\bar{c}(t)$ of the water vapour in the slag with a height l is [1])

$$\frac{\bar{c}_t - c_e}{c_e - c_1} = - \frac{8}{\pi^2} \sum_{n=0}^{\infty} \frac{1}{(2n+1)^2} \cdot EXP_n \qquad (18)$$

with

$$EXP_n = \exp(-(2n+1)^2 \cdot \frac{D \pi^2}{4 \cdot l^2} \cdot t)$$

The eqs. (11), (12) and (18) result in

$$\frac{\bar{c}_t - c_e}{c_e - c_1} = \frac{p_t - p_e}{p_e - p_1^*} \qquad (19)$$

or (see also Fig. 4)

$$p_t - p_e = - \Delta p_{tot} \cdot \frac{8}{\pi^2} \sum_{n=0}^{\infty} \frac{1}{(2n+1)^2} \cdot EXP_n \qquad (20)$$

The distance between a measured point ($p_e - p_{t,msd}$, t) and the point ($p_e - p_t$, t), which is theoretically calculated by substituting some value of t into eq. (20), amounts to (see Fig. 5):

$$\Delta p_i = p_t - p_{t,msd} \qquad (21)$$

The yet unknown value of D must be determined in such a way that the sum

$$\Delta^2 = \sum_{i=1}^{N} (p_t - p_{t,msd})^2 \qquad (22)$$

of the squares of all the distances at N measured points becomes a minimum. With the abbreviation

$$\frac{\partial(\Delta^2)}{\partial D} = F(D) \qquad (23)$$

this condition (Δ^2 = minimum) results in

$$\frac{\partial(\Delta^2)}{\partial D} = F(D) = 0 \, . \qquad (24)$$

D is the zero of eq. (24). It is found by iteration and with the Newton method

$$D_{j+1} = D_j - \frac{F(D_j)}{F'(D_j)} \qquad (25)$$

at which

$$F(D) = \frac{\partial(\Delta^2)}{\partial D} = 2\sum_{i=1}^{N}(p_t - p_{t,msd}) \cdot \frac{\partial(p_t)}{\partial D} \qquad (26)$$

and

$$F'(D) = 2\sum_{i=1}^{N}\left[(p_t - p_{t,msd}) \cdot \frac{\partial^2(p_t)}{\partial D^2} + \left(\frac{\partial(p_t)}{\partial D}\right)^2\right] \qquad (27)$$

with

$$\frac{\partial(p_t)}{\partial D} = 2\,\frac{\Delta p_{tot}}{l^2} \cdot t \cdot \sum_{n=0}^{\infty} EXP_n \qquad (28)$$

and

$$\frac{\partial^2(p_t)}{\partial D^2} = \frac{-\Delta p_{tot} \cdot \pi^2}{2 \cdot l^4} \cdot t^2 \cdot \sum_{n=0}^{\infty} (2n+1)^2\, EXP_n \qquad (29)$$

The computation was done by using a suitable BASIC programm. To save computing time the initial value $D_{j=1}$ should be as close as possible to the final value: By taking only the leading term of the series in eq. (20)

$$\ln\left(\frac{p_e - p_t}{\Delta p_{tot}}\right) = -\frac{D \cdot \pi^2}{4 \cdot l^2} \cdot t + \ln\left(\frac{8}{\pi^2}\right) \qquad (30)$$

a reasonable initial value can be found readily from the slope of the straight line according to eq. (30).

III. RESULTS AND DISCUSSION

Slag samples containing lime, alumina, silica and magnesia with different compositions were investigated. In this paper only the results for the system $CaO - Al_2O_3$ will be presented. Results for other systems will be published elsewhere; also the arrangement for preventing the natural convection of the liquid slag.

Table 1 shows the compositions, the molar ratio $B = N_{CaO}/(N_{SiO_2} + N_{Al_2O_3})$ and the optical basicity of the investigated slags. The measurements were conducted between 1,500 and 1,600 °C and with a partial pressure of the water vapour of 21.95 mbar. The values of the solubility and apparent diffusion coefficients for each slag at the temperature 1,600 °C are then obtained by a regression analysis (last two columns in Table 1).

Because of the validity of the Sieverts' law, the solubility constant K_{H_2O} ($ppm \cdot bar^{-1/2}$) can be calculated.

As shown in Fig. 6 the solubility increases with increasing molar ratio B. The values agree very well with the investigation of K. Schwerdtfeger et al. [2] and Y. Iguchi et al. [3]. T. El Gammal et al. [4] found smaller solubility values.

The presentation in Fig. 6 does not reveal the effect of the silica content nor does the plotting against the optical basicity. Only in Fig. 7 are the decreasing effects of silica and alumina on the solubility evident.

It is very interesting to recognize within Fig. 8 and 9, that the behaviour of the diffusion in these slags is similar to that of the solubility. Also the apparent diffusion coefficients grow with increasing molar ratio B, except in the slag with the running number 1, and decrease with growing content of alumina and silica (Fig. 9).

The apparent diffusion coefficient is considerably greater (in the order of magnitude) than those for the cations Si^{4+}, Al^{3+}, Ca^{2+} or the anions O^{2-} which have been reported by others [5-7]. In a subsequent paper the high mobility and the transportation form of the water vapour in liquid slags is discussed with reference to our results, which are obtained by measuring also the diffusion coefficient of heavy water (D_2O).

IV. CONCLUSIONS

A description has been given of the simultaneous determination of the solubility and the apparent diffusion coefficient of water vapour in liquid slags, done in a single experiment in a Sieverts' cell according to the volumetric method. Further advantages of this method are: there is no need to take a sample; the measuring is done automatically and quasi-continuously; the sensitivity is as high as $\pm 0.86\ \mu g\ H_2O$; an influence of the changes of the room temperature is excluded.

The solubility and the apparent diffusion coefficients of water vapour in liquid slags of the system $CaO - Al_2O_3$, containing up to 15 mole-% SiO_2 are strongly increased by lime and decreased by silica and alumina. The solubility lies between 225 and 385 wt.-ppm, and the diffusion coefficient between 5.5 and $7.72 \cdot 10^{-5}\ cm^2/s$, respectively (for T = 1,600 °C and P_{H_2O} = 21.95 mbar).

V. ACKNOWLEDGEMENT

This study was supported by the **German Research Association (DFG)**, to which the authors are greatly obliged.

VI REFERENCES

1) Crank J.: The Mathematics of Diffusion, Oxford, 1956
2) Schwerdtfeger, K. and H. G. Schubert: Met. Trans. 9 B (1978), pp. 143/4
3) Iguchi Y., S. Ban-Ya and T. Fuwa: Trans. ISIJ, 9 (1969), pp. 189/194

4) El Gammal T. and A. López: Fachber. Hüttenpraxis Metallverarbeitung 24 (1986), Nr. 10, pp. 1/7
5) King T. B. and P. Koros: Trans. Metall. Soc. AIME 224 (1962), pp. 299/306
6) Henderson J., L. Yang and G. Derge: Trans. Metall. Soc. AIME 221 (1961), pp. 56/60
7) Towers H. and J. Chipman: Trans. AIME 209 (1957), pp. 769/73

Table 1 : Solubility and apparent diffusion coefficients of water vapour in slags of the system CaO - Al_2O_3 with up to 15 mol -% silica (T = 1,600 °C, P_{H_2O} = 21.95 mbar).

Nr	N_{CaO}	$N_{Al_2O_3}$	N_{SiO_2}	B	Λ	c	$D \cdot 10^5$
-		(mol - %)		-	-	ppm	cm^2/s
1	69	31	-	2.23	0.776	387	5.50
2	65	35	-	1.82	0.757	286	7.33
3	60	40	-	1.49	0.739	252	6.47
4	55	45	-	1.21	0.722	228	6.03
5	60	30	10	1.50	0.732	246	6.63
6	55	35	10	1.23	0.715	225	6.20
7	65	25	10	1.86	0.752	313	6.97
8	68	22	10	2.13	0.765	354	7.72
9	60	25	15	1.50	0.728	242	6.40

$B = N_{CaO} / (N_{SiO_2} + N_{Al_2O_3})$; Λ: optical basicity

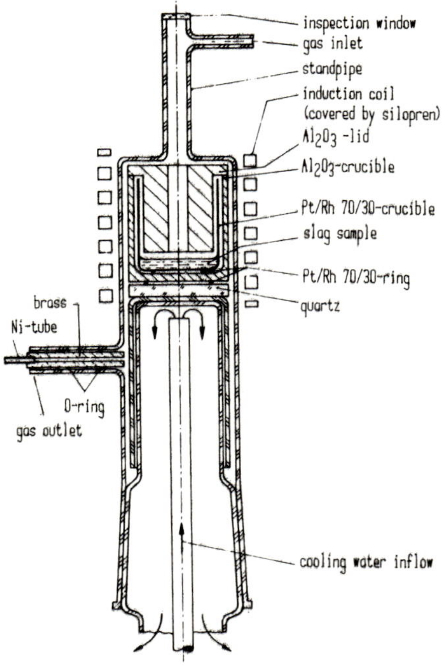

Figure 1 : Sieverts' cell for the measurement of the solubility and the diffusion coefficient of water vapour in liquid slags

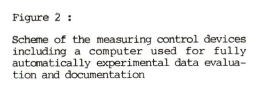

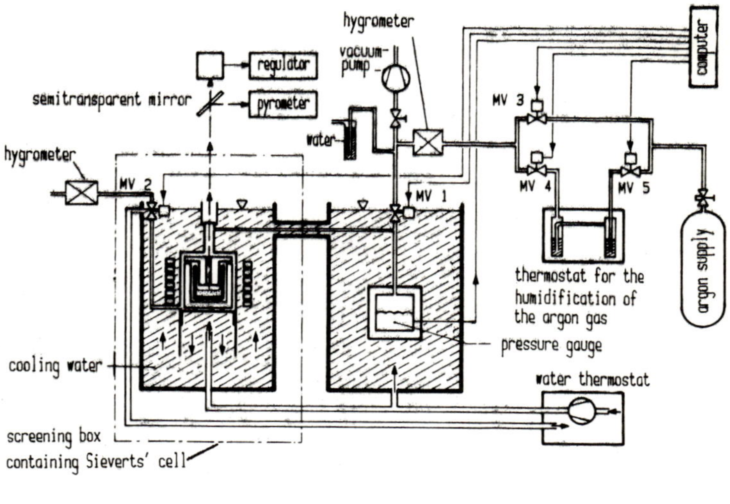

Figure 2 :

Scheme of the measuring control devices including a computer used for fully automatically experimental data evaluation and documentation

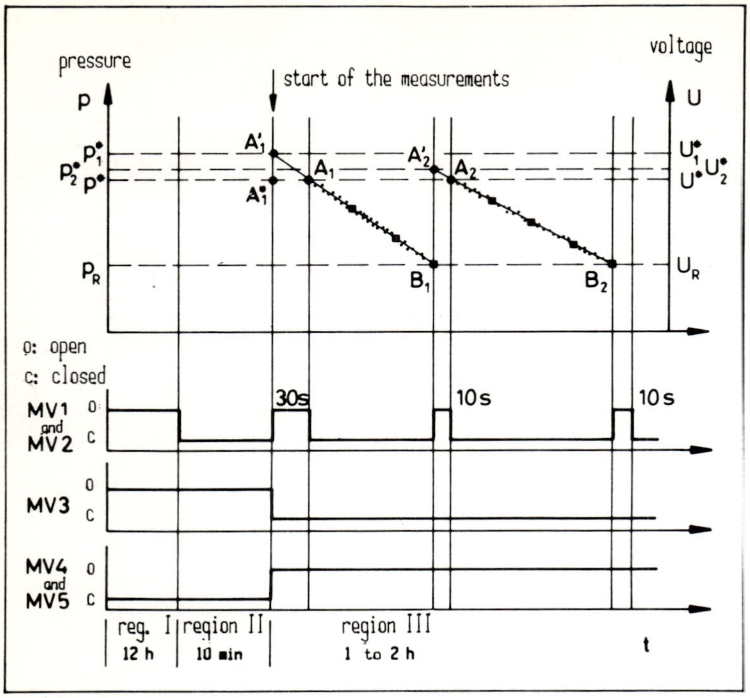

Figure 3 :

Schematic representation of the time dependence of pressure and voltage as well as the states of the magnetic valves

region I : Drying
region II : Testing the initial equilibrium
region III: Reading the new equilibrium

- mean value of 10000 measurements/s
- Determined from the straight line through the mean values

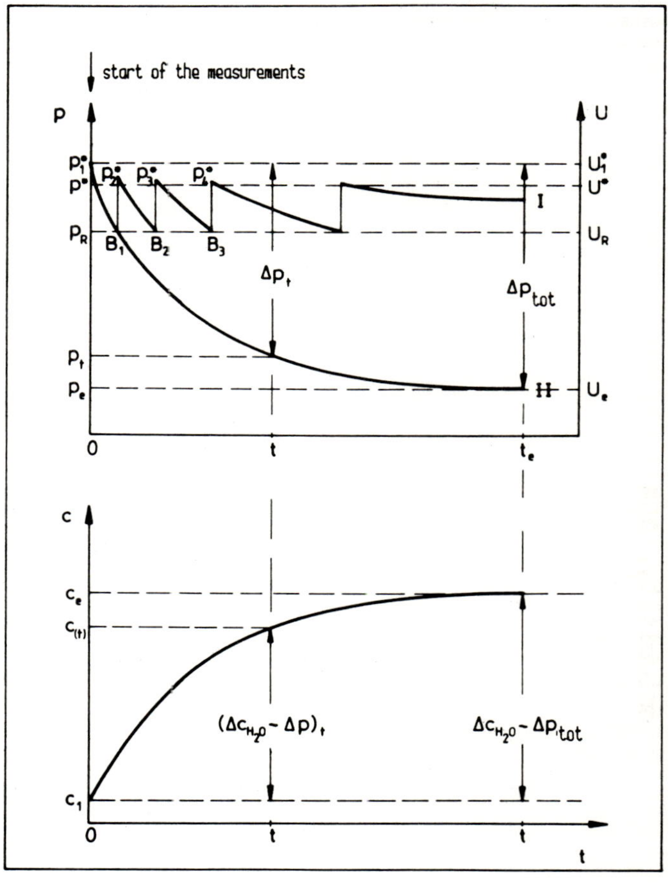

Figure 4 :

Schematic representation of the relationship between the time dependence of pressure and concentration

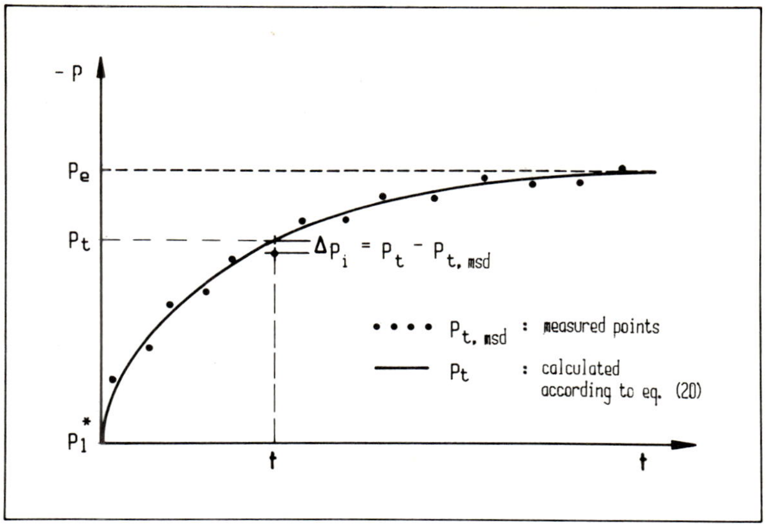

Figure 5 :

Distance Δp_i between the measured pressure $p_{t,msd}$ and the value p_t calculated according to eq. (20) at the time t

Figure 6 :

Influence of the molar ratio $B = N_{CaO} / (N_{SiO_2} + N_{Al_2O_3})$ on the solubility of water vapour in liquid slags of the system CaO-Al_2O_3 containing up to 15 mol - % SiO_2

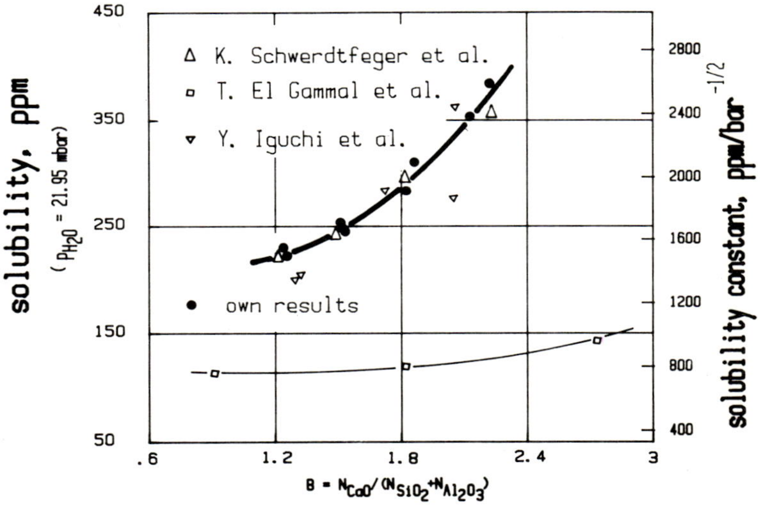

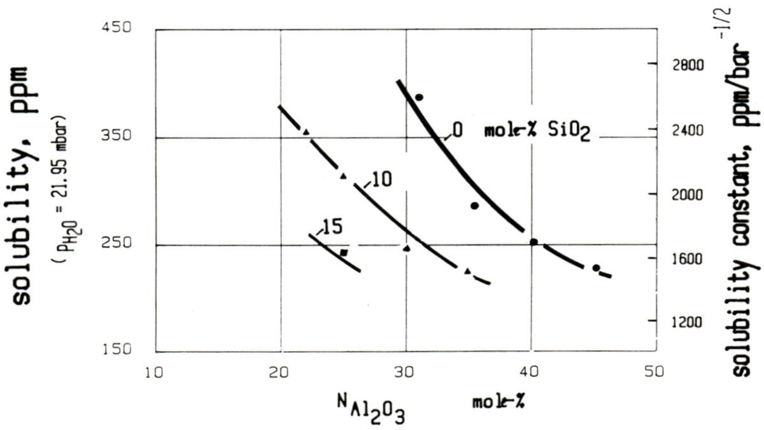

Figure 7 :

Influence of alumina and silica on the solubility of water vapour in liquid slags of the system CaO-Al_2O_3 containing up to 15 mol - % silica

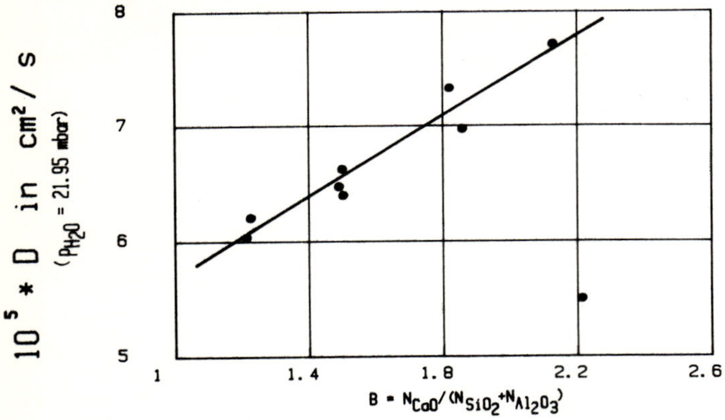

Figure 8 : Influence of the molar ratio $B = N_{CaO} / (N_{SiO_2} + N_{Al_2O_3})$ on the apparent diffusion coefficient of water vapour in liquid slags of the system $CaO-Al_2O_3$ containing up to 15 mol - % silica

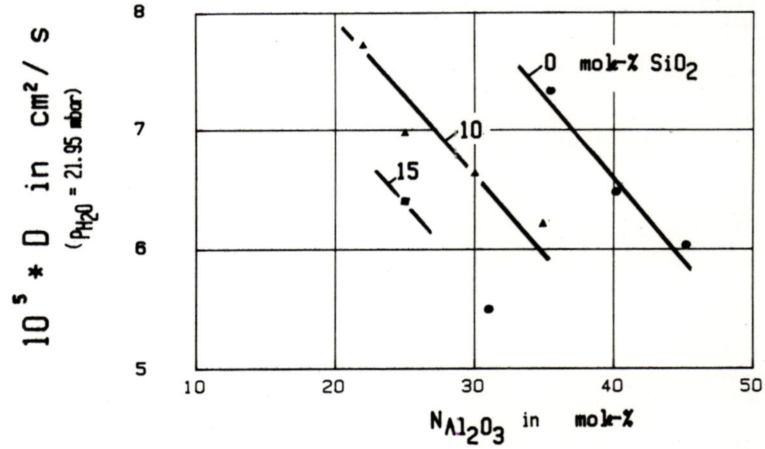

Figure 9 : Influence of alumina and silica on the apparent diffusion coefficient of water vapour in liquid slags of the system $CaO-Al_2O_3$ containing up to 15 mol -% silica

Behaviour of nitrogen in oxide melts

A HENDRY

The author is in the Division of Metallurgy and Engineering Materials, University of Strathclyde, Glasgow, UK.

SYNOPSIS

The behaviour of nitrogen in molten oxides has important consequences for two quite different types of new engineering materials. The implications of a knowledge of the role of nitrogen in slags for refining high-nitrogen stainless steels is obvious but similar nitrogen-containing oxide melts are of critical importance in the densification of silicon nitride engineering ceramics. In the latter case, an oxide liquid acts as a flux for dissolution of nitrides and reprecipitation of nitrogen-ceramic phases. The published data on the thermochemistry of nitrogen in oxide melts is reviewed in the present paper and the implications of this data discussed in terms of nitrogen steelmaking and fabrication of nitrogen ceramics. The behaviour of nitrogen in liquid alumino-silicates containing CaO or MgO is discussed.

INTRODUCTION

It is accepted that the solubility of nitrogen in liquid oxides in equilibrium with nitrogen gas is extremely small [1] except when reducing conditions prevail. In particular, considerable attention has been devoted [2-4] to the situation where the oxide melt is in equilibrium with carbon or graphite-saturated iron and a nitrogen gas containing carbon monoxide; that is under blast furnace conditions. This has been reviewed by Turkdogan [5] where much of the early work is also summarised. The conclusion of these studies is that nitrogen is accommodated in the slag under reducing conditions in the form of nitride anions (N^{3-}) or as cyanide anions (CN^-) and that there is a corresponding solubility of carbide. Expressions for nitride, cyanide and carbide capacities have been derived by Schwerdtfeger & Schubert [3] and the relative amounts of nitrogen accommodated in the liquid oxides as nitride or cyanide depend on the chemistry of the slag. Thus in $CaO-Al_2O_3$ of 40% CaO the majority of nitrogen is present as nitride anion [3] but on increasing to 50% CaO for the same total nitrogen, the cyanide anion accounts for all the nitrogen present (0.052%). In the case of $CaO-Al_2O_3-SiO_2$ slags [2] no cyanide ion was detected in a 52% CaO slag containing 0.421% nitrogen. In Na_2O-SiO_2 slag the nitrogen is held to dissolve exclusively as cyanide anions under reducing conditions [6].

In the present work however, reducing or high carbon activity conditions are not of interest as the investigations are aimed at the refining of high-nitrogen, low-carbon stainless steels by electroslag remelting (ESR) and by nitrogen gas remelting, or at the densification of nitrogen ceramics which is carried out in the absence of carbonaceous species. The following discussion considers the limited amount of published data on steelmaking and describes the behaviour of oxynitride fluxes and the glasses which are formed from them during densification of silicon nitride and related ceramics. Similar conclusions to those discussed above regarding the variation of nitrogen solubility with flux chemistry apply to very low carbon content systems but a clear distinction is drawn between the solubility of silicon nitride in oxides and that of gaseous nitrogen which has important conclusions for the behaviour of nitrogen in the liquid oxide structure.

Results and discussion

(i) **Solubility:** Although little information exists on solubility in slag systems some data are available from studies of glass-forming oxide melts. Mulfinger [7,8] studied the reaction kinetics of nitrogen absorption by reaction of silicon nitride with soda-glass melts and proposed a mechanism for incorporating nitrogen in the melt under reducing conditions. The upsurge of interest in silicon nitride technology during the 1970's led to increasing interest in nitrogen glasses and the melts from which they form, and several systems have been discussed by Drew, Hampshire & Jack [9]. These liquids were produced by melting Si_3N_4 and/or AlN with suitable oxide-alumino silicate mixtures in the absence of carbon and extremely high solubilities were obtained [9]. In the present case the behaviour of calcium and magnesium oxynitride melts is of particular importance and the work of Drew et al [9] has shown that up to 11 at% nitrogen can be accommodated in oxide compositions which will form glasses on quenching. Although it is possible that even higher solubilities of nitrogen may occur in compositions which do not form glasses on quenching, the limited liquid-phase forming regions of the M-Si-Al-O-N systems below 1700°C make it unlikely that significantly higher solubilities will be obtained. The maximum solubility in these systems [9] can then be expressed as given in Table 1. These compositions are equivalent to 6.7 wt% nitrogen in the case of CaO melts and 7.1 wt% nitrogen for MgO melts. Thus the solubility of nitrogen in these oxide melts, when produced by dissolution of silicon nitride, is considerably higher than that obtained by equilibration with gaseous nitrogen [1] even when reducing conditions are utilised [2]. For example, the maximum solubility obtained by Davies & Meherali [2] was 0.673 wt% (47% CaO, 37.9% SiO_2, 12.1% Al_2O_3) which rose to 0.706% when 7.3% MgO was added.

In the present work studies of the liquid systems formed when CaO and MgO are used as densification additives for preprepared "sialons" (phases in the Si-Al-O-N and related systems) have also provided information on nitrogen solubility. The data was obtained by similar techniques to those described by Drew et al [9] but also by direct measurement of concentrations in the transmission electron microscope and by analysis of the sintering behaviour using ternary oxide phase diagrams. The measurements made by these techniques are in agreement with those of Drew et al. and the measurements made in the electron microscope confirm that the concentrations determined are those at saturation with respect to the appropriate oxynitride solid. An example is shown in Figure 1 which shows a crystal of silicon oxynitride in a glassy matrix produced by quenching a CaO-containing sample from 1600°C. The nitrogen concentration is close to those given in Table 1 and is equivalent to approximately 7% Si_3N_4 (the phase which is precipitated and is shown in Figure 1 is however Si_2N_2O, silicon oxynitride). The oxide composition in this case is however very silica rich and analysis in the TEM gives the concentration shown in the energy dispersive X-ray spectrum of Figure 2. (It should be noted that potassium, magnesium and aluminium are present from impurities present in the silicon oxynitride used in the experiments). The CaO:MgO ratio is similar to those in the experiments of Davies & Meheralo [2] but the silica:alumina ratio is much greater.

The oxide composition in which the nitride phase dissolves during sintering of nitrogen ceramics is initially extremely high in silica as it is formed by reaction of the oxide sintering aid (CaO, MgO, Y_2O_3 etc) with the silica which is always present on the surface of silicon nitride, silicon oxynitride and sialons. This liquid composition range is very narrow at 1600-1700°C as can be seen, for example, from Figure 3 and therefore the high solubility of nitrogen demonstrated above is initially accommodated in this range of composition. In studies of silicon nitride and silicon oxynitride the liquid is restricted to this composition range throughout the high-temperature cycle but in the case of Si-Al-O-N (sialon) processing the composition gradually becomes more aluminium rich as the silicon aluminium oxynitride dissolves in the initial liquid phase. The maximum silicon:aluminium atomic ratio of β'-sialon, the phase used in the present experiments is Si:Al = 1:2 ($Si_2Al_4O_4N_4$) which in terms of silica:alumina ratio is SiO_2:Al_2O_3 = 1:1 (37 wt% SiO_2). Thus in Figure 3 the initial liquid phase composition on the MgO-SiO_2 binary moves across the upper part of the diagram (depending on sialon composition) as the sialon dissolves but the maximum nitrogen solubility does not appear to change significantly so long as the liquid remains in equilibrium with a β' solid phase. Still greater nitrogen solubilities are obtainable in the Si-Al-O-N system in the absence of additional metal oxides [10,11] but the melting points of these materials are too high to be of practical interest in the present context.

(ii) **Structure:** The dissolution of silicon nitride or related silicon compounds into metal oxide-silica systems lowers the melting temperature for that system [12] and similarly the addition of metal oxides to sialon liquids lowers their melting points as shown in the present work. It has also been qualitatively shown that the viscosity of the liquid is increased by nitrogen addition. This latter effect may be inferred for liquids from the measurements of glass viscosity and other properties on quenched sialon liquids reported by Drew et al [9]. The viscosity of nitrogen glasses increases by three orders of magnitude as the nitrogen content increases from zero to the maximum value (Table 1) and the glass transition temperature for the calcium- and magnesium-containing glasses increases by 70°C as the nitrogen content increases (see Figure 4).

In the present work attention has been devoted to the recrystallisation behaviour of quenched oxynitride liquids in order to identify the phases precipitated. The recrystallisation temperature for the calcium samples does not change with nitrogen content but the phases in equilibrium change with overall composition. This is shown in Figure 5. Samples containing liquid corresponding to composition 1 will produce an oxide phase, anorthite (An) or mullite (Mu), in equilibrium with O'-sialon when recrystallised from a glass, whereas samples of composition 2 under similar treatment will give gehlenite (Ge) and β'-sialon. The important point to consider in terms of structures of the glass and therefore of the liquid from which it formed is that a nitrogen-containing crystalline phase with structural units of $Si(O,N)_4$ is formed in all cases. It has also been reported that silicon oxynitride, Si_2N_2O, precipitates during recrystallisation of Mg-Si-O-N glasses and phases related to β'-sialon form from Mg-Si-Al-O-N glasses [13].

The observations reported above and in the literature point to the accommodation of nitrogen within the liquid oxide network as the reason for the high solubilities observed. Accommodation of N^{3-} anions in dilute solution has already been reported [1-8] and the physical properties of increased viscosity and glass transition temperature together with the chemical effects reflected in precipitation of crystalline oxynitride phases from quenched liquids support this proposition. In oxide liquids, Si and Al are network forming cations with four-fold coordination (MO_4) and whereas, in oxide melts, Mg and Ca are held to break down the polymerised network by preference for six-fold coordination, when nitrogen is present such cations are four coordinated (MN_4) and thus become network formers. These effects explain the observed physical and chemical phenomena. Mulfinger [7] has shown that substitution of oxygen by nitrogen even at low concentration does lead to higher average coordination and increased crosslinking

$$\equiv Si - O - Si \equiv \rightarrow \equiv Si - N \begin{matrix} \diagup Si \\ \diagdown Si \end{matrix}$$

A similar proposal is made by Davies & Meherali [2] where the need for reducing conditions is explained as being necessary to remove oxygen and replace by nitrogen from the gas. The high solubilities of nitrogen observed when nitrides are dissolved in the slag may then be explained as simply providing a source of nitride ions which do not require reducing conditions to effect the exchange reaction.

Conclusions

Extremely high solubilities of nitrogen in slags and fluxes are obtained by dissolution of nitrides. The solubilities observed are much higher than those obtained with gaseous nitrogen in non-reducing conditions because the nitrogen is accommodated in the liquid structure as N^{3-} anions within the oxide network with a corresponding increase in physical and chemical stability of the oxynitride network.

In high-nitrogen stainless steel manufacture addition of silicon nitride to the melting flux or slag results in a high nitrogen capacity in the slag and conditions suitable for rapid transfer to the metal. An important corollary which follows from the large difference in solubility between gaseous nitrogen and silicon nitride is that the reverse reaction of transfer of nitrogen from metal to slag will not occur as the accommodation of nitrogen from the metal into the slag would require reducing conditions.

In the sintering of nitrogen ceramics the high solubility of nitrogen in the liquid oxide is an essential requirement for rapid dissolution of the starting powder and precipitation of the final ceramic microstructure. The high flux of cationic and anionic species necessary for rapid material transport is then obtained although the variation of viscosity and diffusivity with different sintering oxides and nitrogen solubilities leads to important changes in rate controlling mechanisms of densification [12].

References

1. H-O Mulfinger & H. Meyer, Glasstech.Ber., 1963, 36, 481.
2. M.W. Davies & S.G. Meherali, Met.Trans., 1971, 2, 2729.
3. J. Schewerdtfeger & H. Schubert, Met.Trans. 1977, 8B, 535.
4. T. Shimoo, T. Iida, H. Kimura & M. Kawai, Nipon Kinzoku Gakkai-shi, 1972, 36, 723.
5. E.T. Turkdogan, "Physico-chemical properties of molten slags and glasses", 1983 (The Metals Soc,. London).
6. F. Tsukihashi & R.J. Fruehan, Met.Trans., 1986, 17B, 535.
7. J-O. Mulfinger, J.Am.Cer.Soc., 1966, 49, 462.
8. T. Kelen & H-O. Mulfinger, Glasstech.Ber., 1968, 41, 230.
9. R.A.L. Drew, S. Hampshire & K.H. Jack, in "Special Ceramics 7" Eds. D. Taylor & P.Popper, (Inst. of Ceramics, Stoke-on-Trent) 1981, 110.
10. K.H. Jack, in "Ceramics & Civilisation: High Technology Ceramics" (The American Ceramic Soc., New York) 1987, 259.
11. I.K. Naik, L.J. Gaukler & T.Y. Tien, J.Am.Cer.Soc., 1978, 61, 332.
12. S. Hampshire & K.H. Jack, in "Special Ceramics 7" Eds. D. Taylor & P. Popper (Inst. of Ceramics, Stoke-on-Trent) 1981, 37.
13. S. Wild, G. Leng-Ward & M.H. Lewis, J.Mat.Sci. Letters, 1984, 3, 83.

TABLE 1

Maximum solubility of nitrogen in Ca and Mg sialon liquids

Formula	Equiv. formula	MO	Al_2O_3	SiO_2	Si_3N_4
		mol% (wt%)			
$Ca_{18}Si_{18}Al_6O_{47}N_{11}$	$18CaO.9.75SiO_2.3Al_2O_3.2.75Si_3N_4$	53.8 (44.1)	8.9 (13.4)	29.1 (25.6)	8.2 (16.9)
$Mg_{22.9}Si_{16.7}Al_{3.7}O_{46.4}N_{10.3}$	$22.9MgO.8.98SiO_2.1.85Al_2O_3.2.58Si_3N_4$	63.3 (45.7)	5.1 (9.4)	24.7 (26.8)	7.1 (18.1)

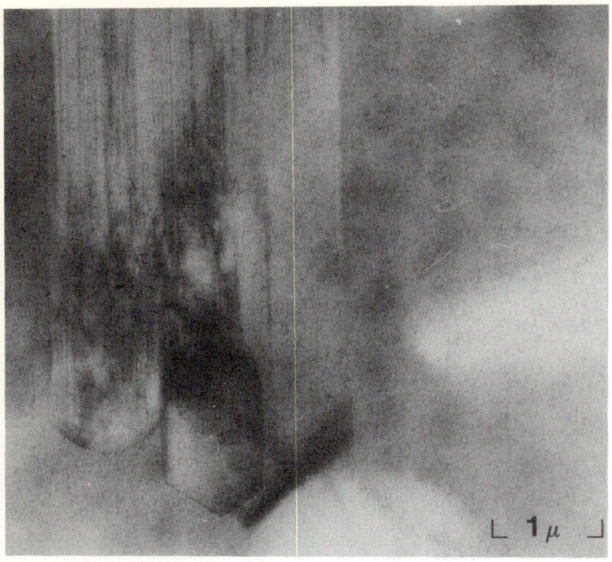

Fig. 1 Transmission electron micrograph of a crystal of Si_2N_2O in a Ca oxynitride glass.

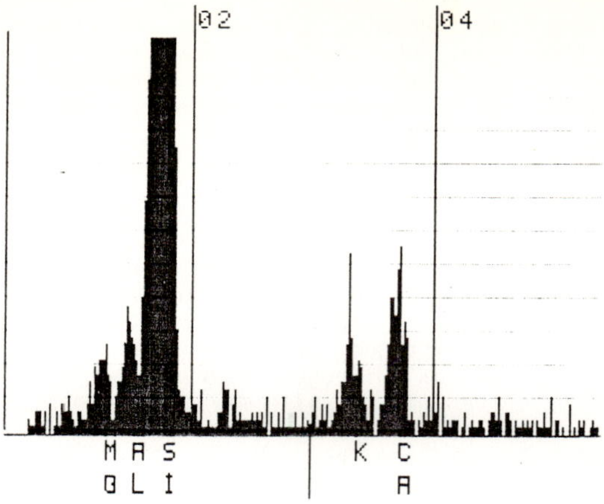

Fig. 2 Energy dispersive analysis spectrum of the glass region in Fig. 1.

Fig. 3 MgO-Al_2O_3-SiO_2 polythermal diagram.

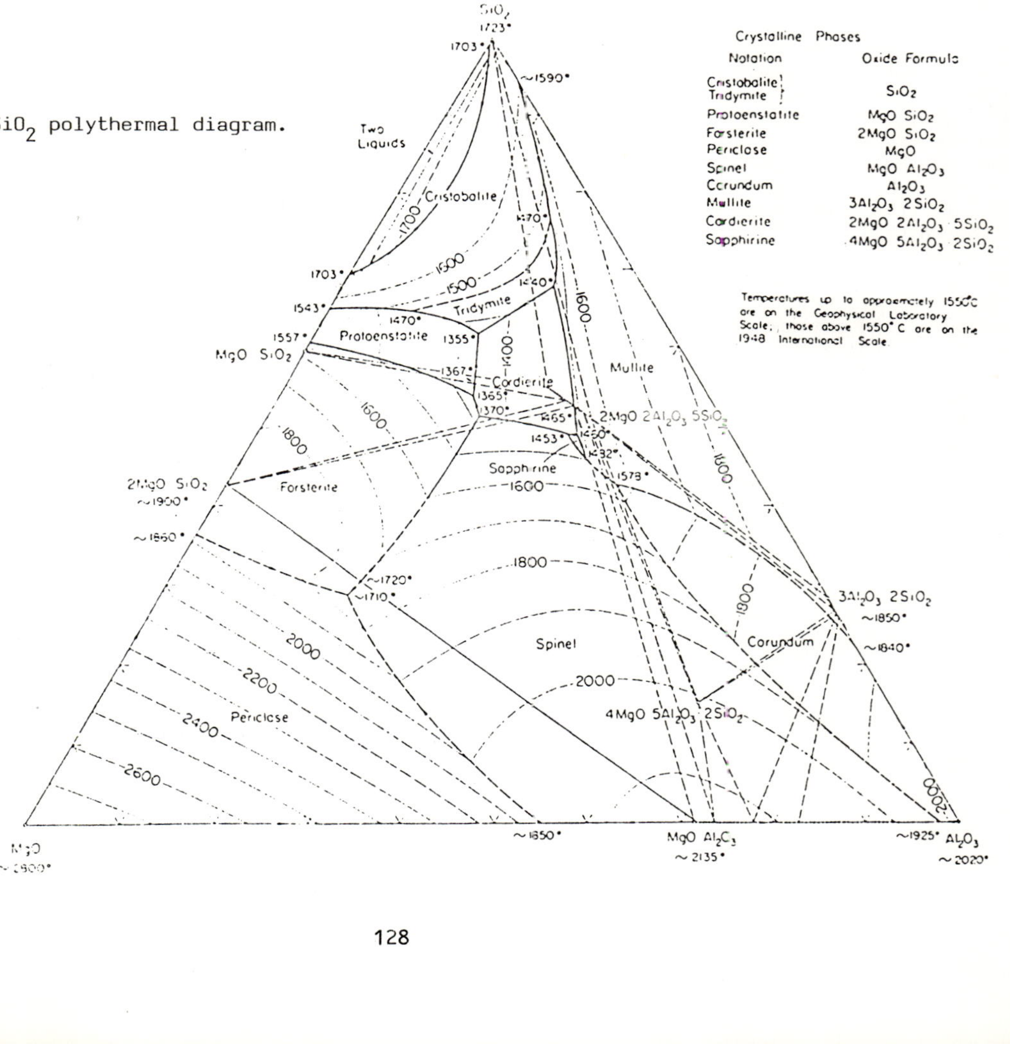

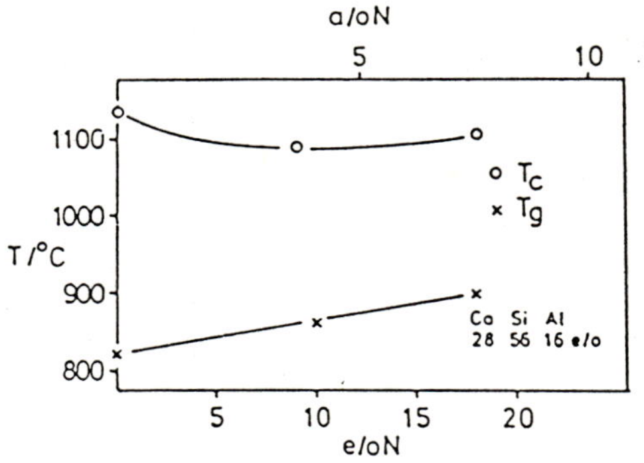

Fig.4 Variation of glass transition temperature (Tg) with equivalent nitrogen concentration (the maximum contents for Mg and Ca are those in Table 1). After Drew et al [9].

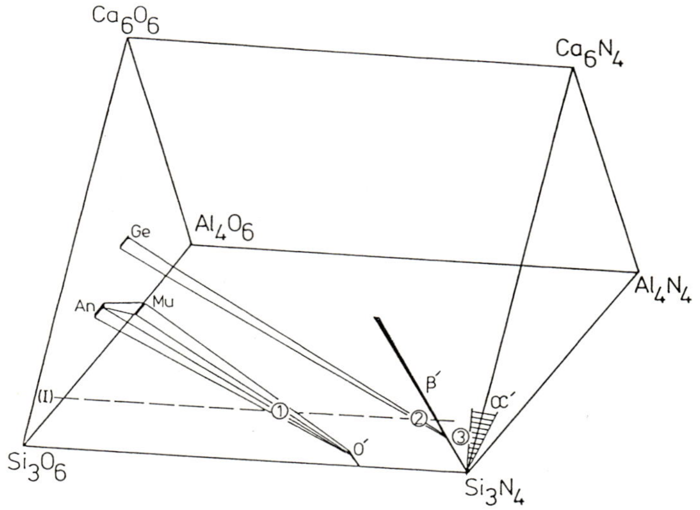

Fig.5 Ca-Si-Al-O-N behaviour diagram.

Kinetics of reduction of titania from blast furnace type slags by iron-carbon melt

H P SUN, K MORI, N SHINOZAKI and Y KAWAI

Mr Sun is in Graduate School, Kyushu University; Drs Mori and Shinozaki are in the Department of Iron and Steel Metallurgy, Kyushu University; Dr Kawai is with Nippon Steel Corporation.

SYNOPSIS

The kinetic behaviour of reactions between titania-containing silicate slags and Fe-C or Fe-C -Si melt has been examined. It was found that the reduction of titania and silica took place simultaneously. As a result of kinetic analysis based on the two-film theory, the reactions were concluded to be a chemical reaction-controlled one.

INTRODUCTION

Titania-bearing burdens are regarded to cause some trouble in blast furnace operations. For the purpose of clarifying the behaviour of titania reaction in a blast furnace, the distribution of titanium between silicate slags and carbon-saturated iron [1,2] and thermodynamic properties of TiO_2 in slag [3] and titanium in iron alloy have been studied. As a result, it has been shown that the reductions of titania and silica proceed interdependently. [1] However, any quantitative evaluation on the rate of titania reduction have not been done sufficiently.

In the present work, some laboratory experiments were carried out to clarify some kinetic features of titanium transfer from slag to iron.

EXPERIMENTAL METHOD

An electric resistance furnace was employed for heating and melting iron-carbon alloy and slag samples. About 200g of iron-carbon alloy was first melted in a graphite crucible of 30 mm I.D. and 100mm long in an argon stream. After the experimental temperature was attained, 20 g of synthetic slag was added to the iron melt. The progress of the reaction was followed by a chemical analysis of samples taken from the metal bath at certain time intervals.

Slags were prepared by premelting a mixture of required portion of each chemical reagent. The chemical compositions of initial slags and iron melts are listed in Table 1.

EXPERIMENTAL RESULT

The behaviour of reduction of titania and silica from slags of different slag basicities (CaO/SiO_2= c/s) at 1550°C is shown in Fig.1, which shows that titanium and silicon contents in iron increase approximately linearly with time. It is found that as the c/s ratio increases, the reduction rate of titania first decreases, reaches a minimum value around the c/s ratio of unity and slightly increases again at higher ratios. While the rate of silica reduction simply decreases with increasing slag basicity.

Delve et al.[1] showed that both reductions proceeded interdependently and a linear relationship between $[Ti]/(TiO_2)$ and $[Si]/(SiO_2)$ was observed. The similar plots are shown in Fig.2. A linear relationship holds for basic slags and the slope of the line is larger for more basic slag.

As seen in Figs.3 and 4, the rate of titanium transfer into iron is enhanced by the coexistence of silicon in iron. In some experiments, silicon is found to be oxidized, although the initial content of silicon is less than that in equilibrium with the slag used.

DISCUSSION

Such a silicon oxidation as seen in Fig.4 was found in the desulphurization of carbon-saturated iron by molten slags by Ramachandran et al[4]. This is explained electrochemically as follows; the anodic reaction ($Si=Si^{4+}+4e$) occurs together with the evolution of CO gas ($C+O^{2-}=CO+2e$) due to the slow rate of the latter reaction.

There are few data necessary for electrochemical treatment. So, the following reactions were assumed to proceed simultaneously in the present study.

$$TiO_2+2\underline{C}=\underline{Ti}+2CO(g) \quad (1)$$
$$SiO_2+2\underline{C}=\underline{Si}+2CO(g) \quad (2)$$
$$TiO_2+\underline{Si}=\underline{Ti}+SiO_2 \quad (3)$$

Then, the rate of reduction of titania at the slag-metal interface is expressed as follows;

$$\frac{W_m d[Ti]}{Adt}=k_1\{(Ti)_i-\frac{[Ti]_i}{K'_1}\}+k_3\{(Ti)_i[Si]_i-\frac{[Ti]_i(Si)_i}{K'_3}\} \quad (4)$$

Similarly, for silica reduction

$$\frac{W_m d[Si]}{Adt}=k_2\{(Si)_i-\frac{[Si]_i}{K'_2}\}-\frac{M_{Si}}{M_{Ti}}k_3\{(Ti)_i[Si]_i-\frac{[Ti]_i(Si)_i}{K'_3}\} \quad (5)$$

where $K'_1=[Ti]_e/(Ti)_e$, $K'_2=[Si]_e/(Si)_e$, $K'_3=K'_1/K'_2$

and subscript i denotes a quantity at the slag-metal interface.

When the two-film theory is applied, the rate of mass transfer of titanium and silicon in the slag and metal boundary layers can be given by Eqs. (6) and (7).

$$\frac{W_m d[Ti]}{A\ dt} = k_s \rho_s \{(Ti)-(Ti)_i\} = k_m \rho_m \{[Ti]_i - [Ti]\} \qquad (6)$$

$$\frac{W_m d[Si]}{Adt} = k_s \rho_s \{(Si)-(Si)_i\} = k_m \rho_m \{[Si]_i - [Si]\} \qquad (7)$$

The values of parameters, k_1, k_2, k_3, K'_1 and K'_3 determined by trial and error method are summarized in Table 1. The simulation results by this model are illustrated as solid lines in Figs. 1, 3 and 4. In the calculation, the following parameters were assumed as follows; $a_c=1$, $P_{CO}=1$ atm, $k_m \rho_m=0.130 g/cm^2 s$ and $k_s \rho_s=0.001 g/cm^2 s$.[5]

The overall rate of titania reduction can be expressed as follows;

$$\frac{W_m d[Ti]}{Adt} = k_t \Big\{ \{(Ti) - \frac{[Ti]}{K'_1}\} + \frac{k_3}{k_1} \{(Ti)[Si]_i - \frac{[Ti](Si)_i}{K'_3}\} \Big\} \qquad (8)$$

where $1/k_t=R_s+R_m+R_r$, $R_s=\{[Si]_i k_3/k_1+1\}/k_s \rho_s$, $R_m=\{(Si)_i k_3/k_1 K'_3+1/K'_1\}/k_m \rho_m$ and $R_r=1/k_1$.

The relative resistance to diffusion of TiO_2 in the slag layer ($R_s k_t$) was calculated from the values in Table 1 and plotted against the reaction time in Fig. 5, which shows that this resistance is only less than 10% for runs with a simple iron-carbon melt. The rates in such cases are considered to be controlled by chemical reaction rates.

When the silicon content is high (0.85%), the portion of titania reduced by silicon increases up to more than 90%. Under such a case, the resistance to transfer of titania in slag increases about 30% and then the reaction is regarded as a mixed control by chemical reaction and mass transfer.

The effect of slag composition on the rate of titania reduction may be explained by the increase in k_1 with increasing c/s ratio for basic slags. But for acid slags the contribution of reaction (3) may be the reason for the increase in the rate with decreasing c/s ratio.

As a result of Arrhenius plot of rate constant, apparent activation energies of reactions (1), (2) and (3) are calculated to be 260, 280 and 250kJ/mol, respectively. Such high values of activation energies also suggest that chemical reactions can be a rate-controlling step.

CONCLUSION

1) Reduction of titania and silica proceeded simultaneously.
2) The reduction of titania was enhanced by the addition of silicon.
3) Apparent activation energies of each reaction were 260, 280 and 250 kJ/mol, respectively.
4) The overall reaction was a chemical reaction-controlled. But the resistance to mass transfer in the slag phase cannot be neglected when the initial content of silicon in iron is high.

List of symbols

A	:interfacial area (cm^2)
k_j	:rate constant for reaction(j) ($g/cm^2 s$)
K'	:apparent equilibrium constant
k_s, k_m	:mass transfer coefficients in slag and in metal (cm/s)
W_m	:metal weight (g)
ρ_s, ρ_m	:densities of slag and metal (g/cm^3)

REFERENCES

1) F.D.Delve, H.W.Meter and H.N.Lander : Physical Chemistry of Process Metallurgy, (part 2), ed. G.R.St. Pierre, Interscience Pub. (1961) p.1111
2) E.Faust:Arch.Eisenhuttenwes.12(1939) p.1
3) K.Ito and N.Sano:Tetsu-to-Hagane,67(1981)p.2131
4) S.Ramachandran, T.B.King and N.J.Grant :Trans. AIME 206 (1956) p.1549
5) N.Shinozaki,K.Mori and Y.Kawai:Tetsu-to-Hagane, 67(1981)p.70

Table 1. Compositions of initial slags and metals, and parameters determined by the calculation.

Run No.	Temp. (°C)	c/s	Slag(mass%) CaO	SiO_2	Al_2O_3	TiO_2	Metal(mass%) Ti	Si	k_1 (kg m^{-2} h^{-1})	k_2	k_3	K'_3	K'_1 $\times 10^{-3}$
B-1	1550	0.60	27.6	46.2	19.7	4.55	0.000	0.014	0.213	0.850	12.7	3.25	2.58
B-2	1550	0.83	33.6	40.6	20.3	4.60	0.001	0.002	0.295	0.445	18.5	3.65	3.25
B-3	1550	1.01	36.5	36.0	18.3	4.68	0.002	0.008	0.373	0.335	20.5	4.35	3.89
B-4	1550	1.42	42.4	29.8	19.6	4.67	0.002	0.008	0.435	0.277	32.9	4.58	4.61
B-5	1550	1.64	46.1	28.1	17.3	5.00	0.001	0.007	0.570	0.223	35.8	6.48	5.19
S-1	1550	1.02	36.9	36.1	19.1	4.97	0.003	0.191	0.373	0.335	21.8	5.56	3.85
S-2	1550	0.96	37.1	38.5	18.5	4.91	0.001	0.385	0.373	0.335	23.0	6.64	3.71
S-3	1550	0.96	37.1	38.5	18.5	4.91	0.001	0.470	0.373	0.335	23.7	6.94	3.69
S-4	1550	1.07	38.9	36.5	20.2	5.00	0.001	0.854	0.373	0.335	24.5	6.98	3.68
S-5	1550	1.75	45.7	26.1	19.8	4.46	0.001	0.410	0.570	0.223	35.8	6.78	4.76
T-1	1350	0.83	33.6	40.6	20.3	4.60	0.000	0.015	0.035	0.062	2.50	5.44	0.005
T-2	1450	0.80	33.4	41.7	18.5	5.27	0.003	0.001	0.190	0.200	6.90	4.97	0.139
T-3	1600	0.83	32.4	39.2	20.4	5.17	0.002	0.015	0.488	1.120	29.7	3.03	10.69

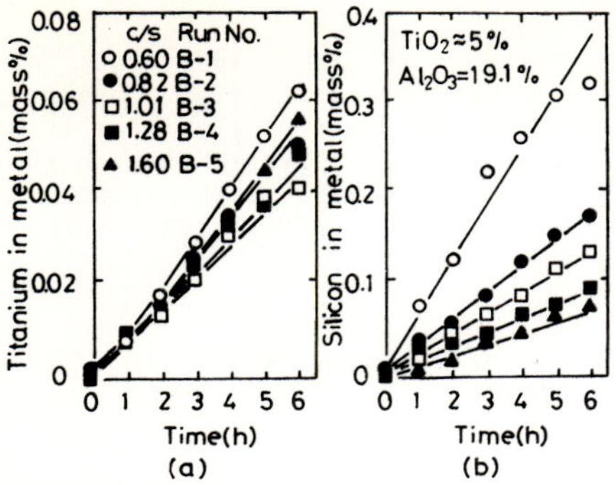

Fig. 1 Effect of slag basicity (c/s) on the rate of TiO_2 and SiO_2 reduction.

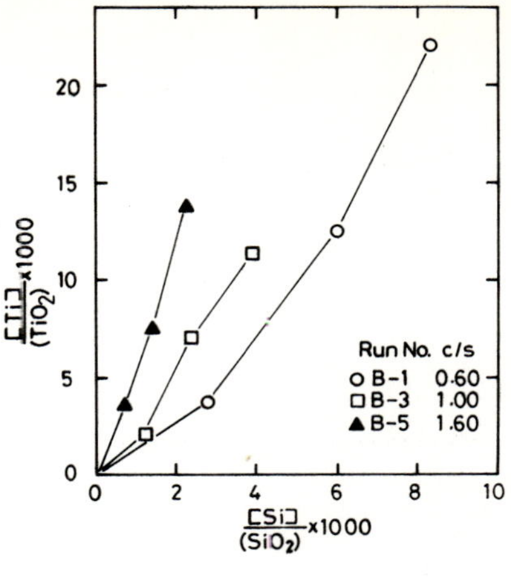

Fig. 2 Relation of [Ti]/(TiO_2) and [Si]/(SiO_2) during the reaction.

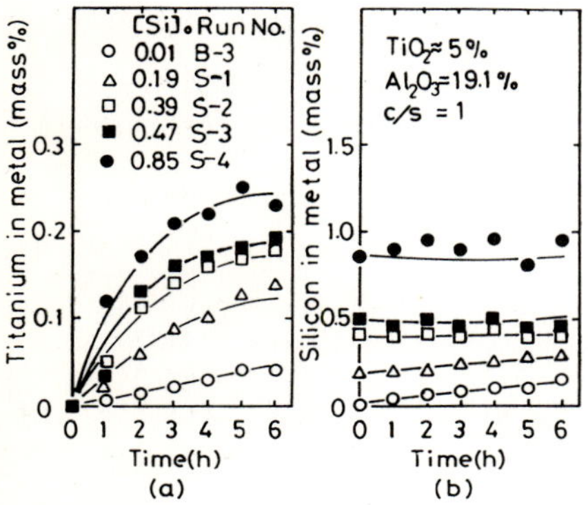

Fig. 3 Effect of silicon added to the iron on the reduction rates of TiO_2 and SiO_2 from slag with c/s=1 at 1550°C.

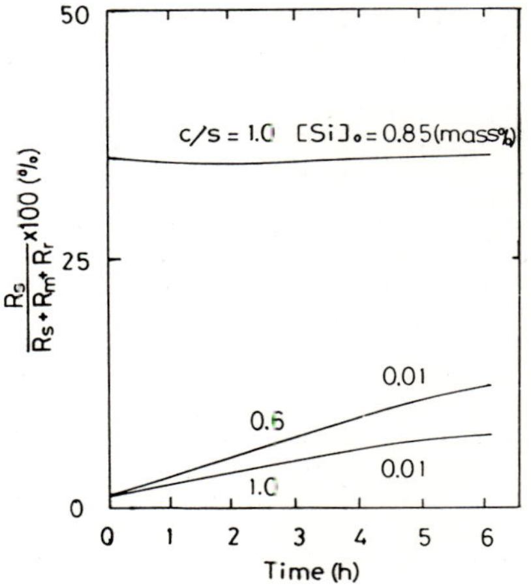

Fig. 5 The relative portion of the resistance to diffusion of TiO_2 in the slag layer.

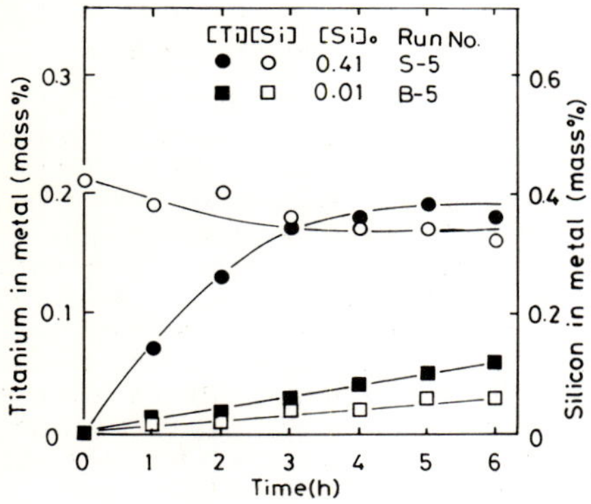

Fig. 4 Effect of silicon added to the iron on the reduction rate of TiO_2 and SiO_2 from basic slag at 1550°C.

Rate of phosphorus transfer from carbon saturated liquid iron to soda slags at 1400°C

J BORODE and H B BELL

The authors are in the Division of Metallurgy and Engineering Materials, University of Strathclyde, Glasgow, UK.

An experimental study has been made of the rate of transfer of phosphorus from liquid iron, contained in graphite crucibles, to soda sslags at 1400°C. The technique involved premelting Sorel iron with ferrophosphorus and once the melt temperature of 1400°C had stabilised a sample was taken. The slag constituents were then added, these consisted of sodium carbonate, silica and iron oxide. In some melts the iron oxide was added during the course of the experiments, normally after 6 minutes. Slag and metal samples were taken at two minute intervals during the experiments. The process of dephosphorisation was followed from the phosphorus content of the melts. The data from the slags is somewhat difficult to interpret since soda was lost by vaporisation during the process. This loss of soda is discussed later in the paper. It can be seen in Fig. 1 that phosphorus content of the melt initially fell rapidly to a minimum in about 10 m and thereafter began to increase slowly. This rephosphorisation was due mainly to two factors - the reduction in the sodium oxide concentration of the slag and the lowering of the oxygen potential of the slag. The iron oxide content of the slag arises from oxidation of iron by the CO_2 from decomposition of Na_2CO_3 and from any iron oxide added, reaction with carbon reduces the iron oxide. In the subsequent analysis of the initial rate process the minimum phosphorus content attained is used.

If mass transfer in the metal and slag phases is considered then in the metal phase

$$\frac{d[\%P]}{dt} = \frac{k_m A}{V_m}[\%P_i] - [\%P_b] \qquad (1)$$

and in the slag phase

$$\frac{d(\%P)}{dt} = \frac{k_s A}{V_s}(\%P_i) - (\%P_b) \qquad (2)$$

where suffix 'i' represents the initial and suffix 'b' the bulk concentrations.

From experiments in which the slag and metal depths were varied it appeared that mass transfer in the metal was rate controlling in the present work. Thus the process will be discussed on this basis.

Assuming equilibrium at the slag-metal interface then (1) leads to

$$\ln \frac{[\%P] - [\%P_e]}{[\%P_o] - [\%P_e]} = \frac{-k_m At[\%P_O]}{V_m\ [\%P_o] - [\%P_e]} \qquad (3)$$

where $[\%P_e]$ is the minimum phosphorus content reached in the experiment.

Effect of the initial phosphorus content of the melt.

The data for four melts in which the initial phosphorus content of the melt was varied is shown in Fig. 1. These metals were made in graphite crucibles 70 mm in diameter, with 1250 g metal and 150 g slag. The slag addition was 80% Na_2CO_3, 13.3% SiO_2 and 6.67% Fe_3O_4 was added after 6 minutes. The initial phosphorus content ranged from 0.148% to 0.704%. The data are plotted in Fig. 2 on the basis of equation 3. This shows that within experimental error the rates of these four experiments are the same and correspond to a km of 1.7 cm m^{-1}, i.e. the rate constant is independent of the initial phosphorus content of the melt. The percentage dephosphorisation at 10 minutes was almost identical in all four melts at 88%.

Effect of the initial silicon content of the melt.

Some melts were made in which the silicon content of the initial metal was varied. The three reported here were made using the same slag crucible and metal and slag weights as in the experiments reported above. The silicon contents ranged from 0.1% to 0.46%. The initial phosphorus concentration varied slightly but was in the range 0.27 to 0.3%. The changes in concentration of phosphorus with time are shown in Fig. 3. Also shown in Fig. 3 is the change in silicon content with time for the melt with an initial silicon content of 0.46%.

It is obvious from Fig. 3 that silicon has a considerable effect on the dephosphorisation process and that silicon is also removed in the process. Silicon and phosphorus are competing for oxygen i.e.

$$\underline{Si} + O_2 \rightarrow SiO2$$

$$2\underline{P} + 5/2\ O_2 \rightarrow P_2O_5$$

and the higher the initial silicon content the lower is the rate of dephosphorisation. This is shown in Fig. 4 where the data are plotted in accordance with equation 3. The rate constant k_m for 0.1% Si was 2.5 cm m^{-1} while for 0.46% Si it was 0.8 cm m^{-1}. The percentage dephosphorisation at the minimum value was 95% for the lowest silicon content to 72% for the highest. The silicon content fell to a lower value rapidly as is shown in Fig. 3; the percentage desiliconisation was 86% in the case of the highest silicon; in the case of the lowest silicon content it was below 0.01% after four minutes. The k_m for silicon expressed in a similar expression to equation 3 was 1.3 cm m^{-1} for the melt containing 0.46% Si initially.

It has been reported that dephosphorisation doesn't start till the silicon content is below 0.2-0.3%. Since the silicon content even in the melt with 0.46% Si was below 0.25% in two minutes this could not be confirmed in the present work, but it was obvious that the presence of silicon had a large effect on the dephosphorisation process. The final silica content of the slag was also raised by the presence of silicon in the initial melt and this also contributed to a lower degree of dephosphorisation.

Effect of the slag to metal ratio.

Data for three experiments made in 70 mm diameter graphite crucibles in which the ratio of slag addition to metal was varied are shown in Fig. 5. In each case the initial phosphorus content was between 0.28 and 0.29%, the silicon content was 0.15% and the same slag composition was used in each case. It is clear from the data that the extent of dephosphorisation is influenced by the slag-metal ratio. The percentage dephosphorisation ranges from 95% through 88% to 67% for the lowest slag-metal ratio.

The effect of sodium carbonate concentration.

The data for three metals made in 140 mm diameter graphite crucibles containing about 7 kg iron are shown in Fig. 6. The initial phosphorus content was in the range 0.29 to 0.3% and the slag to metal ratio was 0.07. The slags were added as sodium carbonate, silica and ferric oxide, the latter being about 7%. The considerable effect of the sodium carbonate concentration can be seen in Fig. 6 which shows dephosphorisation percentages of 31% through 43% to 73%. These melts were carried out in an induction furnace and were stirred by the induced currents.

Evaporation of Na_2O during the experiments.

Loss of soda from the slag was due to two effects - the evaporation of Na_2O from the slag and evaporation of sodium from the reaction

$$Na_2O + C = 2Na + CO$$

Fig. 7 shows the change in soda content of a slag with time. The loss depended on the initial soda content of the slag, for an initial Na_2O percentage of 75% the rate of loss was 3.6×10^{-2} g cm^{-2} m^{-1} in a graphite crucible at 1400°C.

This paper reports on the effect of several parameters on the rate and extent of dephosphorisation by soda based slags. Slags containing silica were used since in practice silicon would be oxidised in hot metal treatment. Other things being equal the rate of dephosphorisation was independent of the initial phosphorus content, and the phosphorus content reached a minimum in 10 m. As would be expected the presence of silicon in the melt lowered the rate of dephosphorisation and also the extent of dephosphorisation. This can be related to two main factors (a) the rate is lowered since initially silicon is oxidised in preference to phosphorus and (b) the final slag is less basic the higher the silicon content. This effect of basicity is shown in Fig. 6. As would be expected the extent of dephosphorisation was affected by the amount of slag used. Some data on the evaporation of Na_2O is given but no rigorous measurements were made to measure this and the data presented was based on slag analysis and a silica balance.

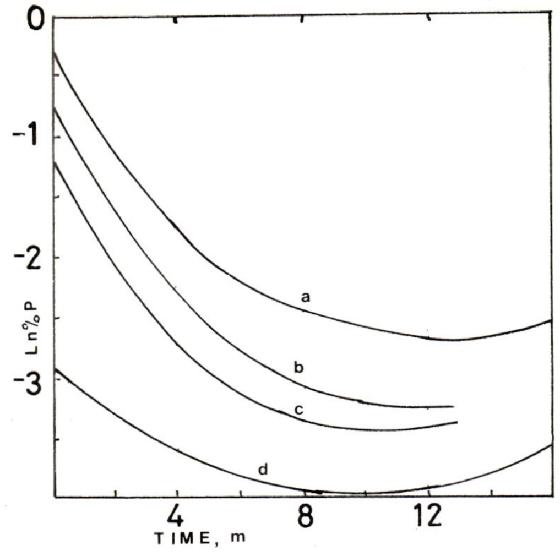

Fig. 1 Effect of initial phosphorus content on the rate of dephosphorisation, initial percentage phosphorus [(a) 0.7%, (b) 0.47%, (c) 0.3%, (d) 0.055%].

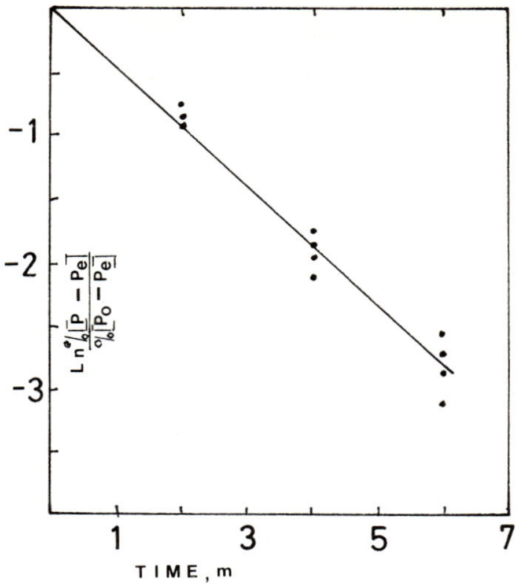

Fig. 2 The rate plot for data in Fig. 1.

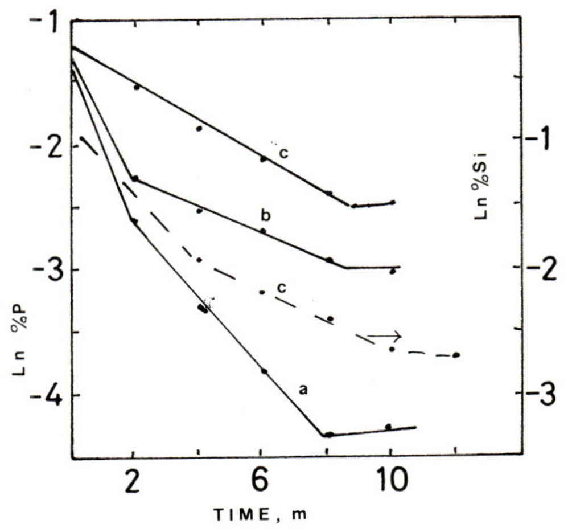

Fig. 3 Effect of initial silicon content on the rate of dephosphorisation and desiliconisation [(a) %Sc = 0.09, (b) %Sc = 0.40, (c) %Sc = 0.46].

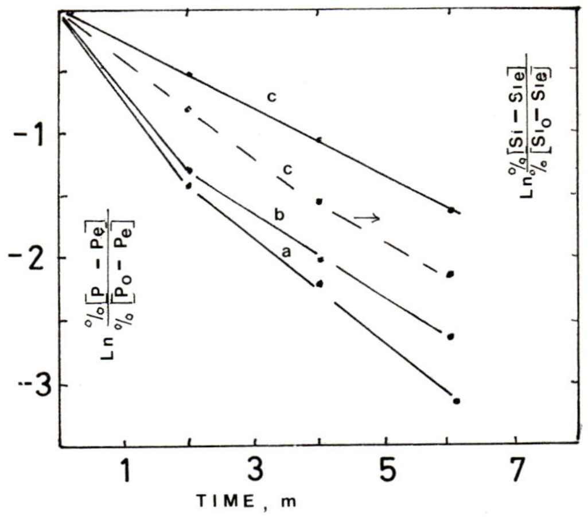

Fig. 4 Rate plots for data in Fig. 3. (key as in Fig. 3).

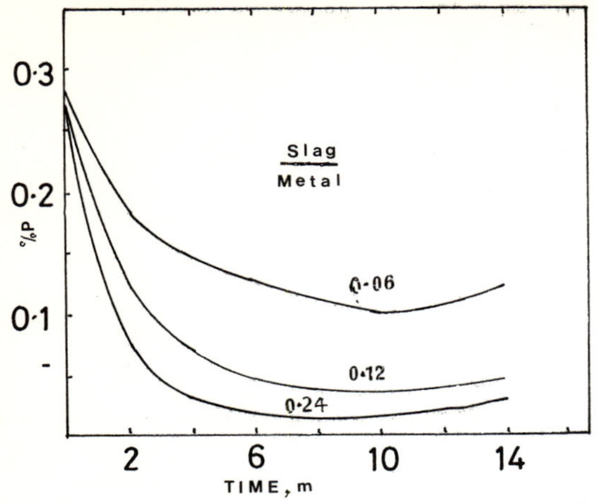

Fig. 5 Effect of slag to metal ratio on the rate of dephosphorisation.

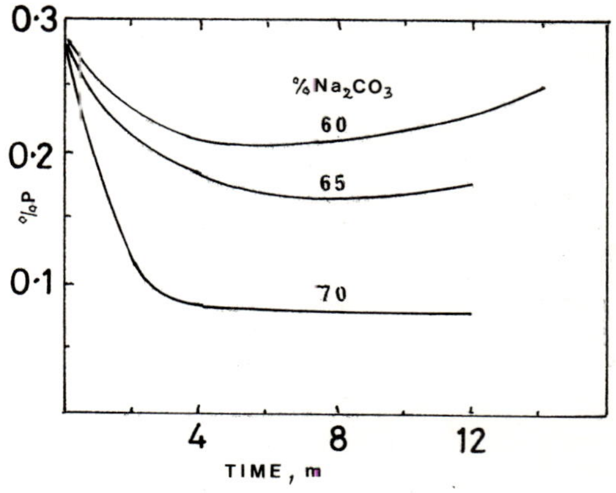

Fig. 6 Effect of the sodium carbonate concentration of the slag addition on the rate of dephosphorisation and desiliconisation.

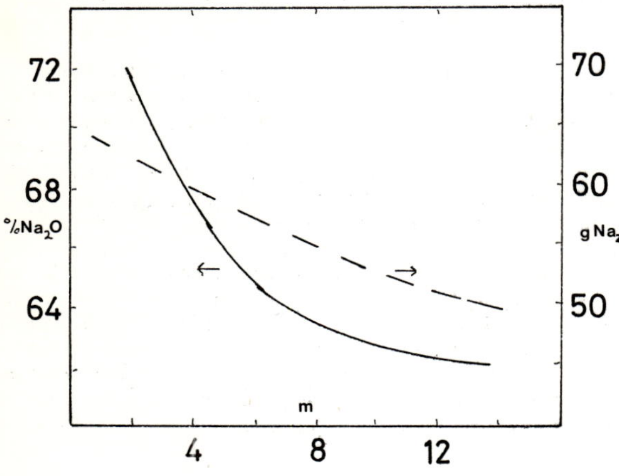

Fig. 7 The change in soda concentration with time for one melt.

Effect of sulphur on kinetics of reduction of iron oxide from molten slag by CO at 1873 K

W M KIM, G GRAENZDOERFFER and H A FINE

W M Kim is a Research Assistant in the Department of Metallurgical Engineering and Materials Science at the University of Kentucky, Lexington, KY; G Graenzdoerffer is a Research Engineer at the Max Planck Institute for Iron Research, Duesseldorf, West Germany (formerly a Research Assistant at the University of Kentucky); Dr Fine is an Associate Professor of Metallurgical Engineering at the University of Kentucky.

SYNOPSIS

The kinetics of reduction of iron oxide from molten synthetic slags containing sulphur were investigated. Results are presented for slags with $CaO/MgO/SiO_2$ ratios of 44.0/17.5/38.5 and with initial Fe_2O_3 concentrations of either 70 or 30 weight per cent. Partial pressures of CO in the reducing gas ranged from 0.0056 to 0.02 atm.

Sulphur at low concentrations reduced the rate of reduction of iron oxide from the slag. At less than approximately 0.1 weight per cent, the rate for a slag with sulphur, R, divided by the rate for the sulphur-free slag, $R_{0\%S}$, equalled

$$R/R_{0\%S} = 1 - 1.63\ \%S_{slag}$$

Higher sulphur concentrations resulted in an increase of almost 300 % in the rate of reduction of iron oxide from the slag.

Preliminary results of measurements of the sulphide capacity of the slag are also presented.

INTRODUCTION

Oxygen and sulphur, which are surface active in liquid metals, have been shown to retard the rate of interfacial reactions. (1,2) Surface active P_2O_5 reduced the rate of reduction of PbO from $PbO\text{-}SiO_2$ melts (3), and SiO_2, which is also surface active in slags, has been shown to slow the rate of reduction of FeO from slags in the $CaO\text{-}MgO\text{-}SiO_2\text{-}Fe_xO$ system (4).

Sulphur is known to be surface active in slags (5), and the rate of removal of sulphur from slags is directly related to the concentration of sulphur on the surface, i.e. the surface excess of sulphur, in the slag (6,7). The effect of sulphur on other reactions, such as the reduction of iron oxide from slag, is not known. As significant quantities of sulphur may be present in direct smelting processes for the production of iron, and since a large portion of the reduction of iron oxide occurs from liquid slag and/or ore in these processes, an investigation of the effect of sulphur on the rate of reduction of iron oxide was begun. Preliminary results of this study and measurements of the sulphide capacity of slag rich in iron oxide are presented in this paper.

EXPERIMENTAL

The apparatus and technique employed in this investigation for the determination of the effect of sulphur on the rate of reduction of iron oxide from slag were almost identical to those of earlier studies on sulphur-free slags in the $CaO\text{-}MgO\text{-}SiO_2\text{-}Fe_xO$ system. (4,8) The only minor change to the apparatus was the addition of a capillary flowmeter to the gas mixing system to allow known concentrations of SO_2 to be mixed with the entering gas.

In the previous work, the state of oxidation of the sample, i.e. the ferrous/ferric ratio, was established prior to the start of a reduction experiment by equilibrating the sample at a fixed CO/CO_2 ration for an hour. In the current study, the sample was equilibrated with a gas of known $CO/CO_2/SO_2$ ratios to fix the sulphur content of the slag, as well as the state of oxidation. Sample preparation and all other aspects of the experiment were identical to that described earlier. (4,8)

A limited number of measurements were also performed to establish the sulphur content and sulphide capacity of the slag being investigated. Samples for these measurements were prepared using the apparatus for the reduction experiments. Some samples were made exactly like the samples for the reduction experiment. The only difference was that the sample was solidified under a flowing stream of argon following equilibration with the $CO/CO_2/SO_2$ mixture. Other measurements were made on the reduction experiment samples which remained after short reduction experiments. All of the samples were broken out of the MgO crucibles and crushed. The sulphur content of the slag was then found using a Leco combustion sulphur analyser.

RESULTS

Experiments were performed on a prefused master slag with fixed $CaO/MgO/SiO_2$ ratios of

44.0/17.5/38.5 on a weight basis. Either 70 or 30 weight per cent Fe_2O_3 was added to the master slag to make samples for the reduction or sulphide capacity experiments. All of the slags were equilibrated with a gas having a CO/CO_2 ratio of 5. All experiments were done at 1873+2 K. The results of the reduction and sulphide capacity experiments are presented below.

Reduction Experiments - Experiments were done at partial pressures of CO in the reducing gas of either 0.0056, 0.01 or 0.02 atm. The balance of the reducing gas was ultra high purity (UHP) argon.

The effect of CO partial pressure in the reducing gas on the rate of reduction of iron oxide was established for samples which were equilibrated with gas containing either 2, 20 or 40 p.p.m SO_2. A linear dependence of the rate on the quantity ($a_{FeO}\ P_{CO} - a_{Fe}\ P_{CO_2} / K$) was found, as was the case for slags which did not contain sulphur. (4) The results of the experiments for slags initially containing 70 per cent Fe_2O_3 show that increasing the SO_2 content in the equilibrating gas reduced the rate of reduction of iron oxide in the ensuing reduction experiment.

A series of reduction experiments were also done at a CO partial pressure of 0.01 atm for slags initially containing 70 per cent Fe_2O_3 and which had been equilibrated at SO_2 levels between 100 and 2000 p.p.m. in the entering gas. As seen in Figure 2, the rate of reduction of iron oxide in the ensuing reduction experiments increased sharply as the SO_2 level increased from 100 to 500p.p.m. and approached a constant value between 500 and 2000 p.p.m. A similiar trend would be anticipated from the single result for the slag with an initial 30 per cent Fe_2O_3.

Sulphide Capacity - The weight percentage of sulphur in slags which initially contained 70 per cent Fe_2O_3 and which were equilibrated with gas having between 30 and 1940 p.p.m. SO_2 are given in Table 1. Increasing the SO_2 concentration in the equilibration gas from 30 to 40 p.p.m. produced a corresponding linear increase in the sulphur content of the slag, i.e. the content at 40 p.p.m.equalled the concentration at 30 p.p.m. times 40/30. Further increases in SO_2 level, resulted in an increase in the sulphur content of the slag, but at less than a linear rate.

DISCUSSION

The primary focus of the research program was the effect of sulphur on the rate of reduction of iron oxide from slag. As few and contradictory data were found on the sulphide capacity of slags of interest, a limited number of measurements were performed to establish the sulphur content of the slags being investigated. These results are discussed first, followed by those of the reduction experiments.

Sulphide Capacity - The sulphide capacity, C_s, is an intrinsic property of a slag and equals

$$C_s = (\%\ S)_{slag} * (P_{O_2}/ P_{S_2})^{0.5} \qquad (1)$$

where $(\%\ S)_{slag}$ is the weight per cent of sulphur in the slag, and P_i is the partial pressure of i in the gas phase in equilibrium with the slag. (9) In this investigation, the oxygen partial pressure was fixed by a CO/CO_2 ratio of 5 (~50%CO/10%CO_2) in the equilibration gas according to the reaction

$$CO + 1/2\ O_2 = CO_2 \qquad (2)$$

The partial pressure of sulphur was then established by the concentration of SO_2 in the equilibration gas according to the reaction

$$SO_2 + 2\ CO = 1/2\ S_2 + 2\ CO_2 \qquad (3)$$

Data for the free energies of formation of the various compounds were taken from the JANAF Thermochemical Tables (10).

The sulphide capacity should be independent of the gas composition for a fixed slag composition. It was therefore expected that at fixed oxygen potential the slag composition, i.e. ferrous/ferric ratio, would remain fixed and a linear relationship would result between $\%S_{slag}$ and the concentration of SO_2 in the equilibration gas. The slope of this line would then yield the sulphide capacity of the slag. This was not, however, found. As seen in Table 1, the sulphide capacity decreased as the SO_2 level increased above approximately 40 p.p.m.

In an attempt to explain the dependence of sulphide capacity on gas composition, the partial pressures of sulphur and oxygen were also calculated assuming that the gas reached equilibrium with COS, SO and SO_3. Using these values, the suphide capacity was found to be approximately a factor of 2 higher, but still dependent on gas composition.

The lower sulphide capacity at high SO_2 levels may have resulted from the removal of sulphur by CO during the reduction experiment, as these measurements were made on previously reduced samples. Alternatively, the samples may not have reached equilibrium. Additional work on this subject is needed.

Reduction Kinetics - The rate of reduction of iron oxide from sulphur-free slags was found in a previous investigation (4) to follow a first-order rate law for the reversible reduction of stoichiometric FeO by CO to form Fe and CO_2, i.e.

$$FeO + CO \longrightarrow Fe + CO_2 \qquad (4)$$

The rate equation for reaction (4) equalled

$$R/A_o = \exp\ (-32300/RT-1.37)\ (1.0-0.7\ a_{SiO_2}{}^{1/3})$$
$$(a_{FeO}\ P_{CO} - a_{Fe}\ P_{CO_2}/\ K)\ ,\ mol/cm^2\ s \qquad (5)$$

where K is the equilibrium constant for reaction (4). The activities of FeO and Fe for this analysis, a_{FeO} and a_{Fe}, were calcualted using Lumsden's regular solution model. (11) The activity of silica, a_{SiO_2}, was estimated from data in the literature. (12) Equation (5) fits the results for sulphur-free slags similar to those of this investigation, as well as previously published results (13-15) for molten slags at 1593 and 1673 K with CaO/SiO_2 ratios less than or equal to one.

As sulphur reduces the surface tension of slags (5) and is therefore surface-active, a similar approach to that employed in the prior investigation for surface-active silica was tried to explain the slower rate of reduction which occurred at low concentrations of sulphur. In this analysis, the fractional decrease in the rate of reduction of iron oxide in slag was assumed to be linearly dependent on the concentration of sulphur in the slag;

$$(R - R_{0\%S})/R_{0\%S} = c\ \%S_{slag} \qquad (6)$$

where $R_{0\%S}$ is the rate found from equation (5) and c is a constant. The concentration of sulphur in the slag, $\%S_{slag}$, employed in this analysis was either the measured value for slags equilibrated at 30 or 40 p.p.m. or a linearly interpolated value for slags equilibrated at lower SO_2 levels.

Equation (6) fits the data for slags with an initial concentration of 70 per cent Fe_2O_3. The slope predicted by a least square best fit analysis to the data was 1.63. In addition, the data for slags with an initial 30 per cent Fe_2O_3, fell within the 95 per cent confidence limits. The concentration of sulphur in these slags was assumed to be that of the slag with higher Fe_2O_3 concentration; data on this slag were not available at this time.

The results of the experiments at higher sulphur concentrations were very unexpected. It was anticipated that retardation of the rate of reduction would be enhanced at higher sulphur concentrations and the the dependence on concentration would be higher than first order. Clearly, sulphur has enhanced the rate of reaction at higher concentrations.

It was anticipated that COS would be formed during reduction of the slags with higher sulphur concentrations. The infrared absorption bands for COS are, however, at different wavelengths than those for CO_2, and COS should not be detected by the analyser. It is, therefore, not believed that the generation of COS can explain the higher measured rate. A second explanation might be the formation of sulphate ions in the slag as a surface active species. The presence of these ions on the surface would make more oxygen readily available for reaction, thereby increasing the measured rate. Using the data in the JANAF Thermochemical Tables (10) for the reaction

$$FeO + SO_2 + CO_2 = FeSO_4 + CO \qquad (7)$$

it was found that the activity of $FeSO_4$ would be on the order of 10^{-12} at equilibrium with a gas having a CO/CO_2 of 5 and 1500 p.p.m. SO_2 at 1873 K. Calculations for the activity of FeS, however, showed that the activity of FeS would be equal to unity at approximately 100 p.p.m. SO_2 at a CO/CO_2 of 5. The presence of a suphur-rich matte phase was not, however, found.

As noted in the discussion on the sulphide capacity measurements, the slag at higher sulphur concentrations may not have been in equilibrium. Thus, not only could the sulphur concentration be too low, but the state of oxidation may be too high. Further experiments are needed to clarify this situation.

CONCLUSIONS

Small additions of sulphur to slag reduce the rate of reduction of iron oxide from the slag. The rate of reaction for a slag containing less than approximately 0.1 weight per cent sulphur is given as a fraction of the rate for the sulphur-free slag by the relationship

$$R/R_{0\%S} = 1 - 1.63\ \%S_{slag}$$

At higher concentrations, the rate of reaction was enhanced by the presence of sulphur. Rates nearly a factor of 3 higher than those for identical, but sulphur-free, slag were found. The mechanism for this enhancement is unknown.

ACKNOWLEDGEMENT

This research has been supported by the Department of the Interior's Mineral Institute Program administered by the Bureau of Mines through the Generic Mineral Technology Center for Pyrometallurgy under grant number USDI-G-1123729-Kentucky. The authors wish to thank these organizations for their financial support of this work.

REFERENCES

1. F. D. Richardson, Can. Met. Quat. 21 (1982), 111-9.

2. G. R. Belton, Can. Met. Quat. 21 (1982), 137-43.

3. U. B. Pal, T. DebRoy and G. Simkovitch, Trans. Inst. Min. Metall. 93(1984), C112-7.

4. G. Graenzdoerffer, W. M. Kim and H. A. Fine, in Proceedings of the 7th Process Technology Division Conference, ISS/AIME, Warrendale, PA, 1988, in press.

5. R. Boni and G. Derge, Trans. AIME 206(1956), 59-64.

6. A. D. Pelton, J. B. See and J. F. Elliott, Met. Trans. 5(1974), 1163-71.

7. B. Agrawal, G. J. Yurek and J. F. Elliott, Met. Trans. 14B(1983), 221-30.

8. H. A. Fine, D. Meyer, D. Janke and H.-J. Engell, Ironmaking and Steelmaking 12(1985), 157-62.

9. C Wagner, Met Trans. B 6B(1975), 405-9.

10. JANAF Thermochemical Tables, Second Edition, U. S. Government Printing Office, Washington, DC, 1971.

11. J. Lumsden, The Physical Chemistry of Process Metallurgy, Part 1, TMS/AIME, Warrendale, PA, 1959, 165-244.

12. M. Timucin and A. E. Morris, Met. Trans. 1(1970), 3193-201.

13. T. Nagasaka, Y. Iguchi and S. Ban-ya, in Proceedings of the Fifth International Iron and Steel Congress, in press.

14. Y. Sasaki, S. Hara, D. R. Gaskell and G. R. Belton, Met. Trans. B 15B(1984), 563-71.

15. S. K. El-Rahaiby, Y. Sasaki, D. R. Gaskell and G. R. Belton, Met. Trans. B 17B(1986), 307-16.

Table 1 Sulphur Concentration Measurement Results

P.P.M. SO_2	$\%S_{slag}$	$C_s x10^3$
30	0.085	4.4
40	0.12	4.7
120	0.22	3.8
570	0.50	3.5
1940	0.77	2.5

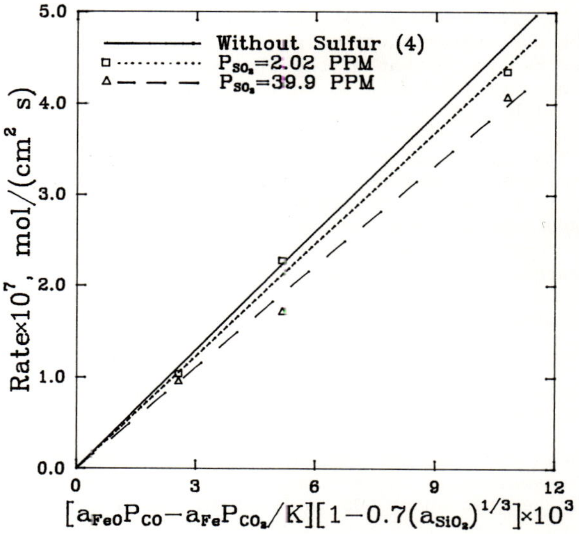

Figure 1

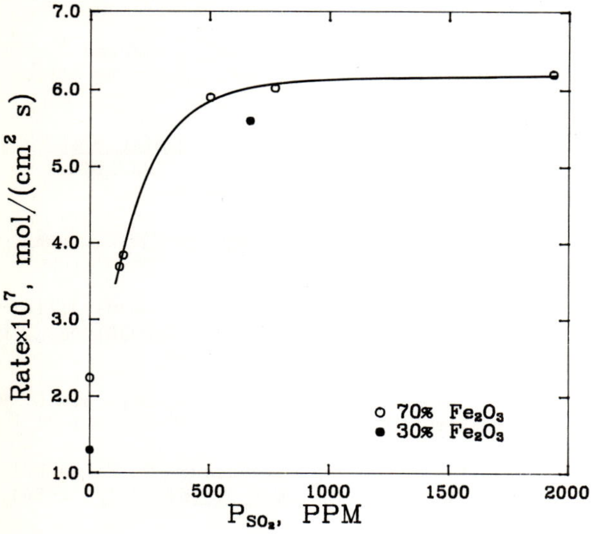

Figure 2

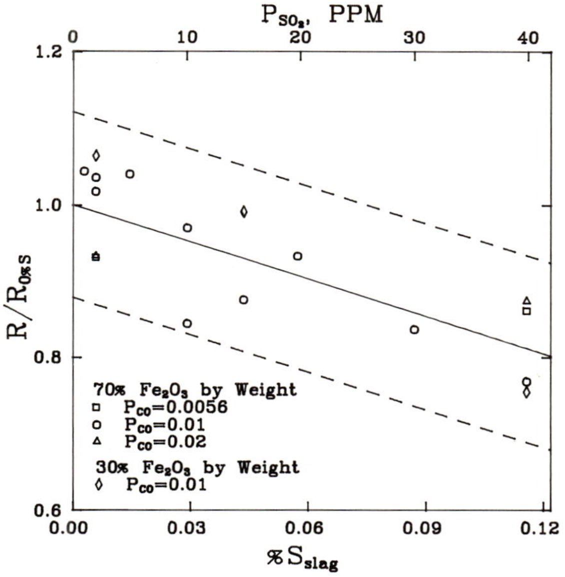

Figure 3

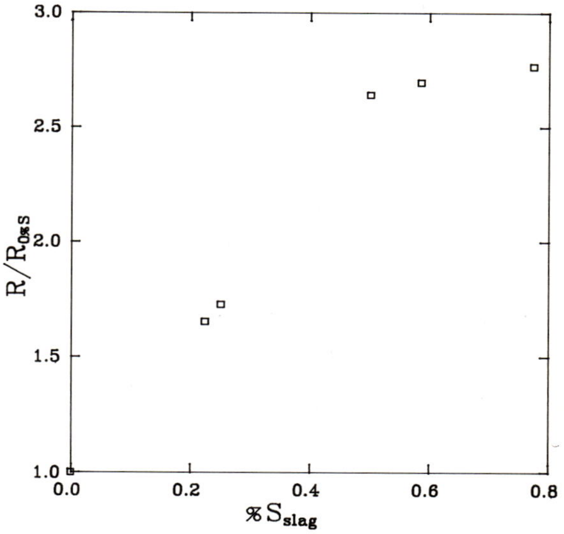

Figure 4

Measurement of respective interface in slag-metal-gas reaction and kinetic analysis of iron oxide reduction

K ISHII and Y KASHIWAYA

The authors are in the Department of Metallurgical Engineering, Hokkaido University.

Synopsis

FeO in silicate slag was reduced with carbon dissolved iron at 1450°C in the thin alumina crucible. The reaction was monitored by QMS gas analyser and the aspects of the interface were observed by X-ray fluoroscopy. The respective interfaces such as metal-slag, gas-slag and metal-gas could be measured by using a computer-aided image processor.

CO_2 was included in the evolved gas and oxygen potential of gas approached that of slag. If the gas reflected the oxygen potential of the slag-gas interface, it was supposed, from similarlity between gas evolution profile and variation of slag-gas interfacial area, that the slag-gas interface was mainly ruling the rate of reduction.

Introduction

In most smelting reactions, gas bubbles are inevitably formed and the interfacial area of slag-metal reaction deviates from the geometrical configuration of both phases. Therefore, the roles of gas-metal and gas-slag interfaces can never be neglected in thinking of what is the rate controlling step. In order to precisely investigate the mechanism of slag-metal reaction, the amounts and composition of involved gas should first be examined as well as the change of area of the respective interface between two from among three phases. In the present paper, reduction of FeO or MnO was pursued by gas analysis and the reaction interface was simultaneously observed by X-ray fluoroscopy, and these results were discussed to give the first step for understanding a slag-metal reaction.

Experimental

The experimental apparatus is illustrated in Fig. 1. The master silicate slags were containing CaO (18-25%), BaO (25-33%), Al_2O_3 (about 5%), and CaF_2 (about 5%) in mass%, while $(CaO+BaO)/SiO_2$ was kept at unity in mole ratio. BaO was added for increasing X-ray absorption. FeO in slag was varied from 2 to 20%. Slags reacted with carbon-saturated iron in most cases. The metals of 4g were settled in the thin crucible made of alumina so as to form two-dimensional interfaces and heated to 1450-1500°C in Ar. At the desired temperature, 2g of slag particle were dispensed. The evolved gas was introduced to a quadrupole mass spectrometer under an Ar carrier of 500ml/min. The amounts of CO and CO_2 were measured as intensities of spectra of m/e=8 and m/e=44, I_{28} and I_{44}, respectively. Then the flow-rates of CO and CO_2 were quantitatively determined according to the following relatinships[1].

$$I_{28}/I_{40} = a + bV_{CO}/V_{Ar} + cV_{CO2}/V_{Ar} \quad (1)$$

$$I_{44}/I_{40} = d + eV_{CO2}/V_{Ar} \quad (2)$$

Where, I_{40} is an intensity of Ar, V denotes the flow-rate and a-e are constants that should be decided prior to applying by standard gases.

The aspects of this reaction were observed in detail by the X-ray transmission technique and the courses were recorded in VTR or X-ray films. In order to separate three interfaces of slag-metal, gas-metal and gas-slag, or reveal the structural details of interfaces, the recorded pictures were treated with a computer-aided image processor. The picture was at first integrated up to the appropriate brightness and consequently the histogram of modified density was expanded to the full range of 8 bits (256 steps). The picture, moreover, was served with space-filtering treatment and edge treatment, applied to illustrate the interface structure. Because the interface was delicately different in brightness gradient, each phase was able to be bordered and separately stored into the memory as a coloured binary image. The respective area of interface was easily calculated from the sum of peripheral length of each phase such as slag(Ls), metal(Lm) and gas(Lg), according to equation (3)

$$\begin{aligned} As/m &= t*(Ls+Lm-Lg)/2 \\ As/g &= t*(Ls-Lm+Lg)/2 \\ Ag/m &= t\&(-Ls+Lm+Lg)/2 \end{aligned} \quad (3)$$

where, As/m, As/g, and Ag/m are areas of slag-metal, slag-gas and gas-metal, respectively, and t is a inner thickness of crucible.

Results and Discussions

Aspects of the interface, during slag-metal-gas reactions could be observed by applying computer-aided processing techniques to X-ray pictures. The computer program for processing is varied according to what information can be derived from original pictures. In the previous work for the reduction of MnO in slag, attention was focused on the microstructure of interfaces, and then the cellular structure was found in the slag-metal interface[2]. The maximum cell size appeared in Al-containing carbon-saturated iron; the rate of Mn transfer was quickest among a series of experiments in which the change of microstructure was investigated with the addition of a third element such as S, Si, Ti, Al or Cr. Gas bubbles formed in the vicinity of cell boundaries. The growth of bubbles closely relate to the motion of cells.

In the present study, the purpose is to measure interfacial area and, therefore, detail of interface is sacrificed for distinctly revealing each phase. In Fig. 2(a)-(d) illustrations of the process to make up a ternary image composed of slag, metal and gas are shown. Also Fig. 2(e)-(h) shows the aspects of reaction expressed in the ternary images, where slag containing 16.9%FeO has been reduced at 1450°C with carbon-saturated iron. As soon as the slag contacted metal, slag was emulsified by violent reduction of feO and immediately the metal surface became too flat (Fig. 2(e)). Secondly, gas in emulsion segregated and a huge gas phase appeared accompanying the irregular surface of metal (Fig. 2(f)). As the reaction proceeded, a number of small bubbles was produced at the node of the surface wave, which might also be the train of cells, and at last they were altered by large bubbles, while configuration of metal surface had changed from a flat shape to a round one (Fig. 2(g) and (h)).

The typical curves of CO and CO_2 evolved in FeO reduction are shown in Fig. 3. CO_2 was detected in the exhaust gas while the reduction of FeO continued and it had never been produced from decomposition of CO in the places outside the crucible. CO_2 had a tendency to be formed prior to CO. The prosperity and decay of evolution of gas could be described in terms of five-stage model by Gare et al[3], in which the decarburization curve was broken down into five periods. In this study, the first period of induction did not appear, but it seemed that the first peak of CO indicated in Fig. 3 corresponded to the second period of fast external decarburization. Consequently, the evolution of gas followed the nucleation period reaction sequence of lull period, external-internal nucleation period and slow internal period. The rates of oxygen removal from slag and decarburization were calculated on the basis of oxygen and carbon balances in the gas phase, respectively according to equations (4) and (5).

$$-d[O]/dt=(2V_{CO_2}+V_{CO})/22\ 414 \qquad (4)$$

$$-d[C]/dt=(\ V_{CO_2}+V_{CO})/22\ 414 \qquad (5)$$

Figure 4 shows typical decarburization curves. The amounts of carbon removal calculated by gas analysis agreed excellently with chemical analysis except for Si-containing metal. However, once Si was added to Fe-C metal, the reactions became more complicated and the total volume of evolved gas became inconsistent with the chemical analysis. It might be caused, in the case of 0.35%Si, by SiO_2 in slag reacting with Si and C to form SiO, and then SiO being reduced again by C, and gas evolution consequently being accelerated. On the other hand, in the case of up to 0.5 mass% Si, the oxidized compound of Si, SiO_2, tended to retard the decarburization by forming films on the slag-metal interface.

Since the CO_2/CO ratio of the evolved gas is also a barometer for oxygen pressure at the reaction interface, thermodynamic investigations were carried out in two extreme cases, the first in which evolved gas should be in equilibrium with FeO in slag, and the second with carbon in metal. The investigation of 16.9%FeO slag is shown in Fig. 5. It was clear from the figure that oxygen potential of the evolved gas was the same as that of slag except for the initial stage of reaction. In contrast, the carbon concentration in bulk metal was vastly different from the value estimated for CO_2/CO of gas. This may lead to the conclusion that there is a large boundary layer of carbon in metal if the evolved gas is in equilibrium with metal. Variation of respective interfacial area measured by the above-mentioned method is presented in Fig. 6. When FeO in slag was lower, each interfacial area was also smaller. However, even with this low FeO, the observed interfacial areas were at least a few times larger than those of geometrical configurations, and the characteristic features of the gas-slag area were very large in comparison with the two others, whose extent was a function of the reaction conditions. In Fig. 6, also the rate of gas evolution that was calculated as the sum of CO and CO_2, was plotted, too. It was suggested from the analogy of time-dependence that the slag-gas interface most closely related to the gas evolution and its area might be determining the overall reaction rate. However, this does not imply that the elementary reaction in the gas-slag interface, such as the reduction of FeO or the mass transfer of FeO, is controlling the overall reaction, but that something caused by the gas-slag interface area controls the reaction rate. It is possible to conclude from this that the sum of the interfacial area of gas-slag and gas-metal, that is, the total gas volume, indicates the most suitable correlation.

Concluding Remarks

The reduction of FeO in silicate slag was followed by gas analysis and X-ray fluoroscopy. The respective interface, such as slag-metal, metal-gas, and gas-slag, could be measured from X-ray pictures by using a computer-aided image processor.

(1) Both CO and CO_2 were included in evolved gas; the oxygen potential of gas approached that of slag.

(2) Of the three interfaces, the slag-gas interface was largest and greatly influenced gas evolution.

References

1. K. Ishii, et al. Proc. 5th IISC., Book 3, (1986), p.9
2. K. Ishii and T. Sinde: Trans. Iron and Steel Inst. of Japan; to be published.
3. T. Gare, et al. Ironmaking and Steelmaking, 8 (1981), 169.

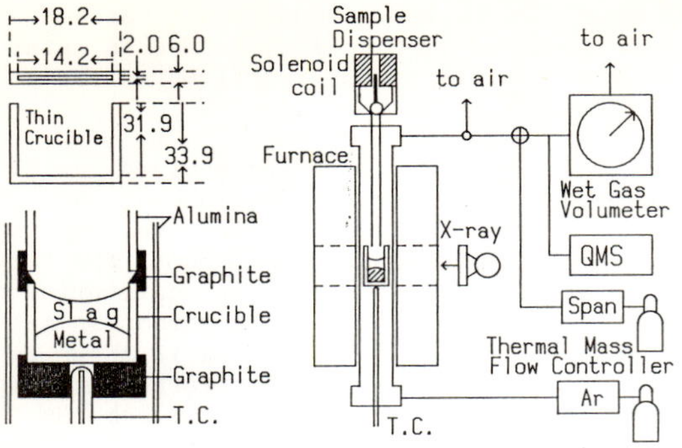

Fig. 1 Scheme of experimental apparatus.

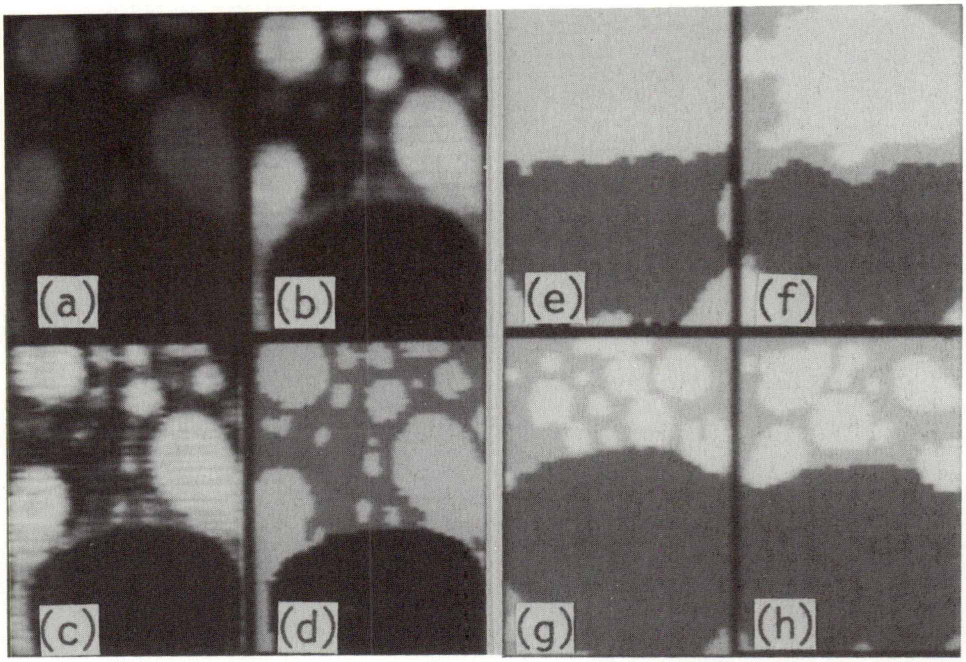

Fig. 2 Examples of image processing, (a)-(d), and ternary images during reduction of 16.9%FeO slag, (e)-(h). (a) Original digital picture, (b) band expansion (integration), (c) spatial filtering (sharpening), (d) ternary image.

(e) 30s, emulsified slag and flat interface,

(f) 250s, segregation and irregular interface,

(g) 390s, small bubbles and round interface, and

(h) 1520s, large bubbles and smooth interface.

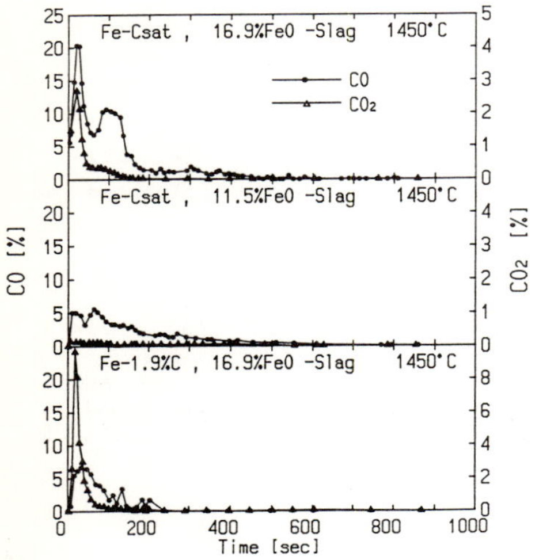

Fig. 3 Variation of gas evolution.

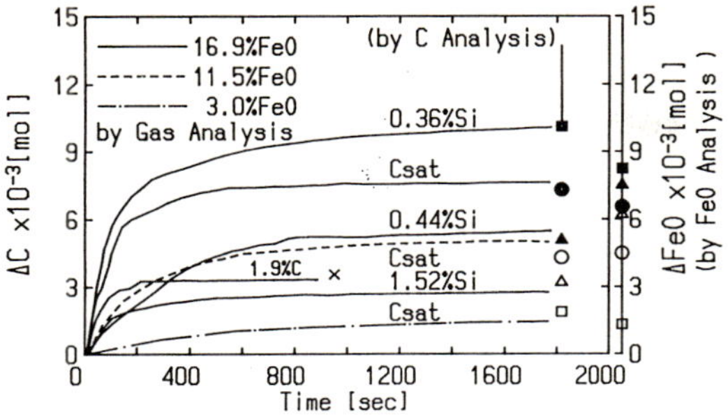

Fig. 4 the typical decarburization curves.

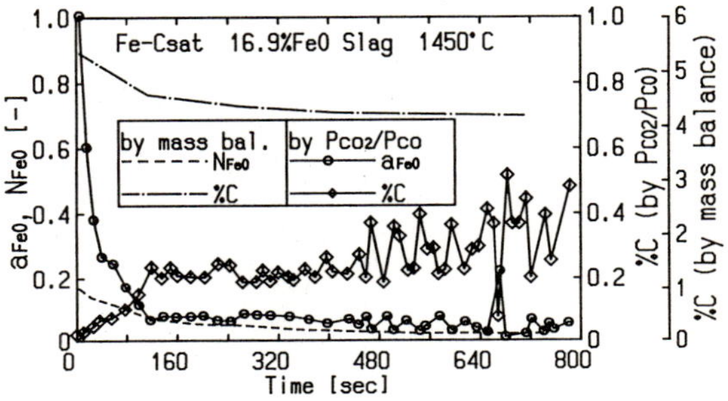

Fig. 5 Comparison of C and N_{FeO} obtained by mass balance with the activities estimated from the composition of evolved gas.

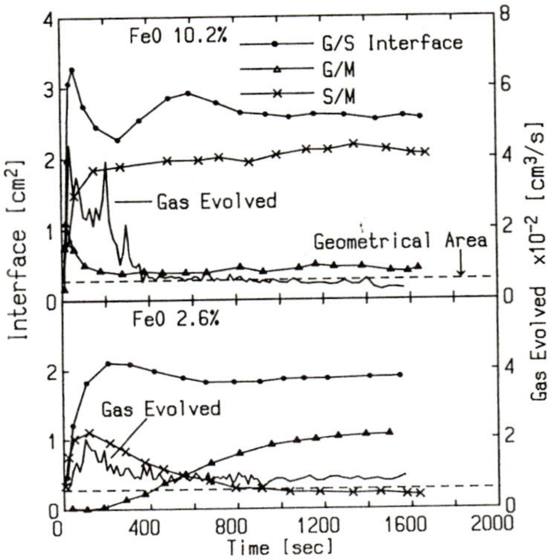

Fig. 6 Variation of respective interface and rate of gas evolution.

□

Critical review of optical basicity on metallurgical applications

T NAKAMURA, Y UEDA and J M TOGURI

Dr Nakamura and Prof Ueda are in the Department of Metallurgical Engineering at the Kyushu Institute of Technology; Prof Toguri is in the Department of Metallurgy and Materials Science at the University of Toronto.

SYNOPSIS

Duffy and Ingram have introduced the concept of the optical basicity scale based on the so-called theoretical optical basicity. This basicity concept has been applied to metallurgical slags. Attempts have also been made to correlate the theoretical optical basicity with various slag capacities such as sulphur. In general, these correlations are only valid over a limited slag composition range. In the present study, the formulation of the theoretical optical basicity scale is reviewed.

INTRODUCTION

Duffy and Ingram have developed a technique to determine basicity of glasses.(1) They also related the measured data to the values calculated from Pauling's electronegativity scale which they defined as the theoretical optical basicity. (2)

Following this development, they showed a correlation between the theoretical optical basicity values and the sulfide capacities in iron making slag systems.(3) Since then, the theoretical optical basicities have been correlated with other capacities such as phosphate capacity.(4)(5) Although successful correlations have been obtained in some slag systems, there has been very little discussion on this subject matter. Mori has made an attempt to give a thermodynamic meaning to optical basicity as applied to the phosphate capacity data.(6)

CONCEPT OF THE OPTICAL BASICITY

Duffy and Ingram established a basicity scale of the oxyacids based on the nephelauxetic effect of the 6s-6p transition of electrons in the probe ions such as Tl^{+}, Pb^{2+} and Bi^{3+}.(1) This transition is affected by the electron donor power of oxygen ions surrounding the probe ions and consequently peak shifts of the transition spectra can be observed in the UV wave length range. It was defined by the peaks of the probe ion in CaO and in the free state. For example, the optical basicity is calculated by equation (1) when Pb^{2+} is selected as a probe ion.

$$\Lambda = \frac{\nu_{free} - \nu_{sample}}{\nu_{free} - \nu_{CaO}} = \frac{60700 - \nu_{sample}}{31000} \qquad (1)$$

ν_{free}, ν_{CaO} and ν_{sample} are the peak wave numbers of Pb s-p spectra in the free state, in CaO and in the sample, respectively. The theoretical optical basicity values are calculated from the values of the Pauling type electronegativity. Duffy and Ingram also presented a calculation technique for determining the theoretical optical basicity in multi-component oxide systems.(2) Since it is difficult to measure the UV spectra of probe ions even in glassy samples, the theoretical optical basicity has been used for metallurgical applications. It should be emphasized that the theoretical optical basicity is different from the measured optical basicity and it reflects an average basicity of a slag.(7)

When the concept of theoretical optical basicity is applied to industrial slags and fluxes, the following problems have to be overcome.

(1) Treatment of transition metals, especially for metals which have several oxidation states.

(2) Treatment of a complex system containing both oxide and halide components.

Since the basicity moderating parameters are obtained from Pauling's electronegativity, there is no procedure to calculate the parameters of different valence states of transition metals. The present authors have proposed a new scale of theoretical optical basicity on the basis of the average electron density(8) instead of Pauling's electronegativity to solve this problem.(9) The selected values of the new theoretical optical basicity are shown in Table 1.

On the other hand, Sommerville et al. have empirically assigned the theoretical optical basicity values to the transition metal oxides based on the obtained correlation between the theoretical optical basicities and sulfide capacities in iron making slags.(10) Both values of the theoretical optical basicity of some transition metal oxides are compared in Table 2.

In general, large differences do not exist between Sommerville's and the present authors' values except for MnO.

APPLICATION OF OPTICAL BASICITY TO METALLURGICAL SLAGS AND FLUXES

In most cases involving the metallurgical application of optical basicity, there has been a tendency to relate basicity to the activity of oxygen ions which can never be measured. Many simple correlations between the theoretical optical basicity values and various capacities have been developed with partial success.(3)-(6)(11) These correlations are useful to predict the various types of capacities from an industrial prospective. However, since the optical basicity is a basicity and not the activity of oxygen ions in the slag, an unrestricted use of this term can be misleading. It is necessary that the essential meaning be considered in the apparent correlations between the optical basicity and various capacities. For example, sulfide capacity data(10) in the systems containing even transition metal oxides are well related to thoretical optical basicity values based on electron densities as obtained by the present authors as shown in Fig.1. Essentially two correlation lines were obtained. The lower line consisted of data from systems which contain CaO, MgO, Al_2O_3, TiO_2 and SiO_2. The upper line was obtained from data in $MnO-SiO_2$ and $CaO-FeO-Fe_2O_3-SiO_2$ slags. The correlations in Fig.1 become more consistent when using the theoretical optical basicity values used by Sommerville et al.(10) This results because the theoretical optical basicity values of FeO and MnO are larger than those predicted by the present authors, then consequently the data in $MnO-SiO_2$ and $CaO-FeO-Fe_2O_3-SiO_2$ slags move to larger values. However, there is no significance to a good correlation if there is no meanings of the plot. Two assumptions are required to obtain a successful correlation.(3) One is that the theoretical optical basicity reflects the activity of the free oxygen ions. The other is that activity coefficients of changes of S^{2-} are ignored. It is probably more likely that the cause for two lines in Fig.1 is a effect of the activity coefficient of S^{2-} in $MnO-SiO_2$ and $CaO-FeO-Fe_2O_3-SiO_2$ slags.

On the other hand, a poor correlation was found when the same theoretical optical basicity values were used for phosphate capacities. A plot of the phosphate capacity data in a $CaO-SiO_2-FeO-Fe_2O_3-MgO-CaF_2$ slag against the values of the theoretical optical basicity based on the average electron density is shown in Fig.2. No correlation exists. On the other hand, the phosphate capacity data in the systems containing even FeO_n have been well correlated with theoretical optical basicity based on Pauling's electronegativity as shown by Gaskell(4), Suito(5) and Mori(6). They used 0.51 as a value of theoretical optical basicity of FeO. In the present evaluation the results show that for sulfide capacities a value of 1.03 for FeO gives good agreement while a value of 0.51 gives a better fit for phosphate capacities.

It is suggested that this discrepancy is caused by changes in the activity coefficients of S^{2-} in sulfide capacity and PO_4^{3-} in phosphate capacity. Therefore, it is important to consider the significance of the correlations between various capacities and the theoretical optical basicity, especially to clarify the conditions when a good correlation is obtained, and also to ascertain the limit of the metallurgical application of optical basicity in terms of compositions and temperatures.

Although a few problems are pointed out, theoretical optical basicity is still a useful concept in slag chemistry. Sano et al. correlated data of CO_2 solubility in alkaline oxide binary slags with the theoretical optical basicity values of Duffy and Ingram.(12)

They obtained a good correlation except for the K_2O-SiO_2 and Na_2O-SO_3 systems. CO_2 solubilities in the same systems are replotted against the theoretical optical basicity values of the present authors in Fig.3. While most data have not changed, the data for the K_2O-SiO_2 system have shifted. Only data from the Na_2O-SO_3 system are out of the correlation in Fig.3. Since data on CO_2 solubility correspond to carbonate capacity which has been considered as a meaningful presentation of slag basicity (13), theoretical optical basicity works well as a basicity scale in these systems.

Another challenge that the application of the optical basicity concept to multi-component systems containing halide fluxes is disscussed in the conference.

CONCLUSION

It is important to consider the significance of the correlations between various capacities and the theoretical optical basicity. It is especially necessary to clarify the conditions and the limits of the application whenever the theoretical optical basicity is applied to metallurgical slags and fluxes.

REFERENCES

(1) J.A.Duffy and M.D.Ingram, J.Amer.Chem.Soc.93, (1971),6448.
(2) J.A.Duffy and M.D.Ingram, J.Inorg.Nucl.Chem., 37,(1975),1203.
(3) J.A.Duffy,M.D.Ingram and I.D.Sommerville, J. Chem.Soc.Farad.Trans.,74,(1978),1410.
(4) D.R.Gaskell, Trans.I.S.I.Japan,22,(1982),997.
(5) H.Suito and R.Inoue,Trans.I.S.I.Japan,25, (1985),118.
(6) T.Mori, Trans.Japan Inst.Metals,23,(1984),761.
(7) T.Yokokawa,T.Maekawa and N.Uchida, Trans.Japan Inst.Metals,25,(1986),3.
(8) R.T.Sanderson, Chemical Bonds and Bond Energy, 2nd Ed, Academic Press New York.(1976).
(9) T.Nakamura,Y.Ueda and J.M.Toguri, J.Japan Inst. Metals,50,(1986),456.
(10) D.J.Sosinsky and I.D.Sommerville, 2nd Internat.Symp. on Metallurgical Slags and Fluxes.,Eds. H.A.Fine and D.R.Gaskell, Met. Soc.AIME,(1984),1015.
(11) D.J.Sosinsky,I.D.Sommerville and A.Mclean, 5th I.I.S.Congress,Process Tech.Proc.,6, Washigton DC,ISS,(1986),697.
(12) T.Kawahara,S.Shibata and N.Sano, 5th.I.S. Congress, Process Tech.Proc.,6, Washigton DC,ISS,(1986),691.
(13) C.Wagner, Met.Trans.B.,6B,(1975),405.

Table 1 Selected values of the basicity moderating parameter and the theoretical optical basicity in oxide compounds.

Compound	γ	Λ	Compound	γ	Λ
Li_2O	0.95	1.05	CuO	1.12	0.89
Na_2O	0.91	1.10	B_2O_3	2.40	0.42
K_2O	0.87	1.15	Al_2O_3	1.47	0.68
Rb_2O	0.86	1.16	Fe_2O_3	1.44	0.69
Cs_2O	0.85	1.18	Cr_2O_3	1.44	0.69
MgO	1.09	0.92	As_2O_3	1.39	0.72
CaO	1.00	1.00	Sb_2O_3	1.25	0.80
SrO	0.96	1.04	Bi_2O_3	1.14	0.87
BaO	0.93	1.08	CO_2	2.49	0.40
MnO	1.05	0.95	SiO_2	2.11	0.47
FeO	1.07	0.93	GeO_2	1.64	0.61
CoO	1.07	0.93	TiO_2	1.56	0.64
NiO	1.14	0.87	P_2O_5	2.60	0.38
ZnO	1.13	0.89	SO_3	3.48	0.29

Table 2 Comparison of theoretical optical basicity values of transition metal oxides between those obtained by Sommerville et al.(10) and by the average electron density.(9)

Oxide	Theoretical optical basicity values	
	by Sommerville et al.	by average electron dencity
TiO_2	0.61	0.64
Fe_2O_3	0.70	0.69
FeO	1.03	0.93
MnO	1.21	0.95

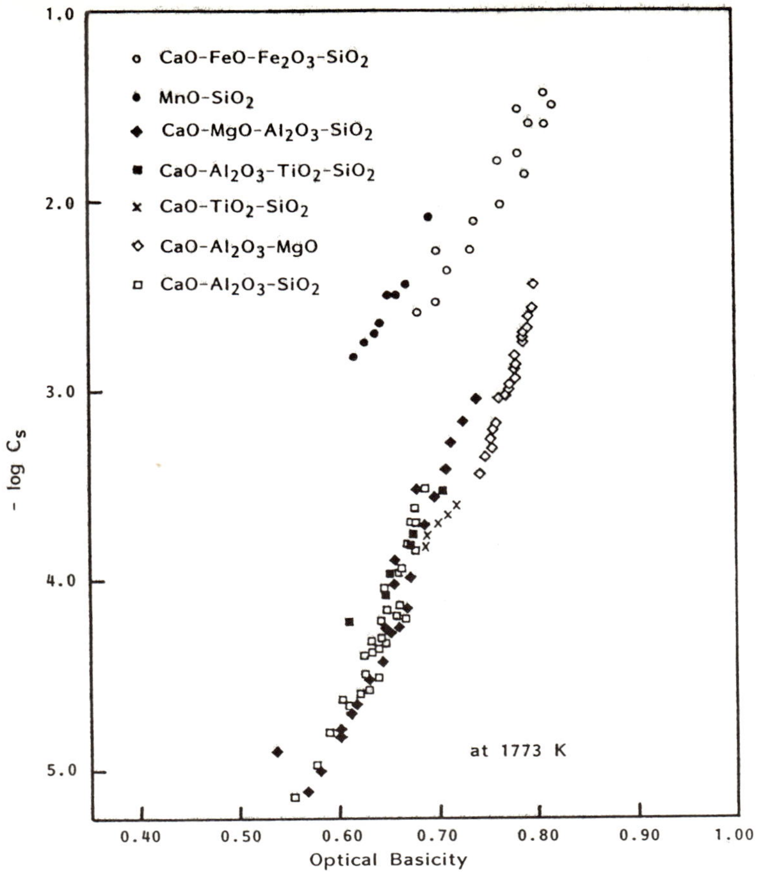

Fig.1 Plot of sulfide capacity data in the selected systems against theoretical optical basicity values at 1773 K.

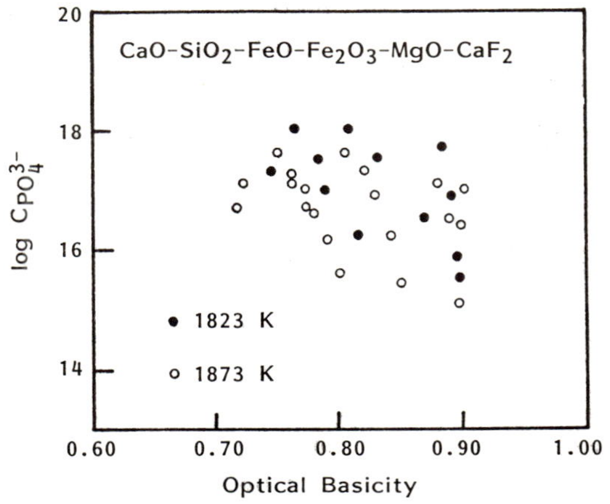

Fig.2 Correlation of phosphate capacity data in slag containing iron oxide against theoretical optical basicities by average electron density.

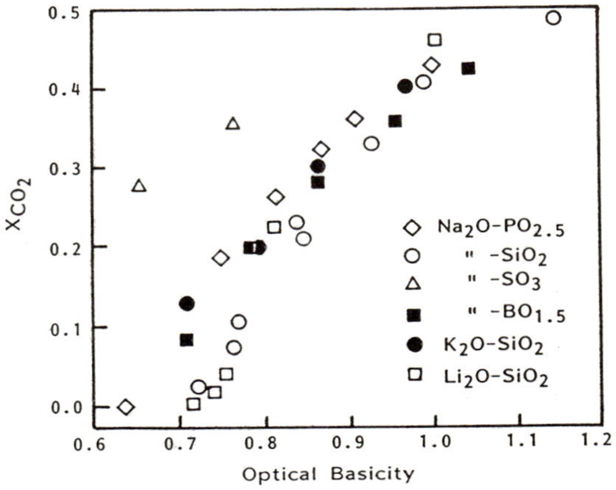

Fig.3 Relationship between CO_2 solubilities and values of theoretical optical basicity based on average electron density.

□

Use of optical basicity to calculate phosphorus and oxygen contents in molten iron

A BERGMAN and A GUSTAFSSON

Dr Bergman is at Swedish Steel, SSAB Strip Products Ltd., Domnarvet Works, Borlänge, Sweden; Mr Gustafsson, formerly at SSAB, Strip Products Ltd., is now with Smedjebacken - Boxholm Stål AB.

SYNOPSIS

The proportion of monomer and dimer phosphate ions in the slag, as related to the slag phosphorus content, is used accurately to relate the monomer phosphorus capacity to optical basicity in slags.

The oxygen concentration in molten iron is related to the slag composition, by using optical basicity, and used to calculate the molten iron phosphorus content from slag composition alone.

INTRODUCTION

In metallurgy, the concept of optical basicity has gained a continuously increasing interest ever since Duffy et al.[1] showed a straight line relation between optical basicity and sulphide capacity of blast furnace slags. In their work, the optical basicity of an oxide was empirically related to Pauling electronegativity of the cation with the exception of transition metal oxides. As a result, optical basicity has been applied not only in connexion with sulphide capacity,[1-10] but attempts have also been made to correlate data on: phosphorus distribution[11-18], manganese distribution[17, 19], carbonate capacities[20-21] and water capacities[22-23] with optical basicity. However, the success of these correlations has varied considerably.

Accordingly, some authors[9, 17, 18] have questioned the accuracy as well as the universality of optical basicity as a measure of chemical basicity. Nevertheless, in this work optical basicity is found to be an excellent measure when correlating metallurgical data.

SOME ASPECTS ON DEPHOSPHORIZATION

Phosphorus distribution has been investigated by many researchers with Mori[13], Turkdogan[24, 25] and Zhang et al. [26] recently listing most of the investigations in this field.

With phosphorus and oxygen dissolved in the metal and with the assumption that all phosphorus in the slag is present as PO_4^{3-} ions, the dephosphorization equilibria have been conveniently derived from the reaction:

$$[P] + \frac{5}{2}[O] + \frac{3}{2}(O^{2-}) \rightleftarrows (PO_4^{3-}) \quad (1)$$

The relation between the phosphate capacity and the equilibrium distribution ratio of phosphorus between slag and steel, L_P, is then given as

$$L_P = \frac{(\%P)}{[\%P]} = C_{PO_4^{3-}} \cdot [\%O]^{5/2} \cdot \frac{M_P}{M_{PO_4^{3-}}} \quad (2)$$

where

$$C_{PO_4^{3-}} = \frac{(\%PO_4^{3-})}{[\%P]\,[\%O]^{5/2}} \quad (3)$$

Although these relations are generally accepted, it could not be too strongly emphasized in view of the results of Selin[27], that the assumption of all phosphorus in the slag being present as monomer ions has a limited validity.

In his excellent piece of work Selin[27] approached the problem by considering the co-existence of monomer and dimer phosphate ions. The equilibrium between phosphorus in metal and dimer phosphate anions can be written

$$[P] + \frac{5}{2}(O^{2-}) \rightleftarrows \frac{1}{2}(P_2O_7^{4-}) \quad (4)$$

Seeing that, the equilibrium between monomer and dimer ions reads

$$(PO_4^{3-}) + (PO_4^{3-}) \rightleftarrows (P_2O_7^{4-}) + (O^{2-}) \quad (5)$$

From Eq 5 it is clear that the amount of the dimer ions will increase with increasing slag phosphorus content as well as with decreasing slag basicity.

In his mathematical interpretation of phosphorus distribution Selin[27] analysed his own and other people's data in view of Eq 1-5 and related the proportion of monomer phosphorus ($\%P_M$) in the slag to the total amount of phosphorus ($\%P$).

$$(\%P_M) = 1.6\sqrt{0.64 + (\%P)} - 1.28 \quad (6)$$

Accordingly, the phosphorus distribution is not constant, but a function of the slag phosphorus content. Here, it may be noted that Murkai et al.[28] also noticed an effect of phosphorus content on L_P. However, they interpreted it as a replacement of SiO_2 by P_2O_5.

The monomer/dimer relationship in Eq 6 can only be true in slags of high basicity, e.g. close to CaO saturation. The anionic constitution of calcium phosphate melts is complex (polymers 1 - 5) and this is a function of cation/anion ratio.

In this work we will assume the validity of Eq 6 and express the slag ability to absorb phosphorus as monomer phosphate ions as

$$C_{P_M} = \frac{(\%P_M)}{[\%P]\,[\%O]^{5/2}} = \frac{L_{P_M}}{[\%O]^{5/2}} \qquad (7)$$

The significance of this approach is clear from Fig. 1, in which the amount of monomer phosphorus is displayed as a function of the phosphorus content in the slag. In addition, a further piece of evidence that supports this approach is given in Fig 2. It can be clearly seen that the phosphorus distribution is a function of the phosphorus content itself.

OPTICAL BASICITY

Duffy et al.[1] derived values of "theoretical optical basicity" from Pauling electronegativity, transition metals excluded. Though, others[11-13] have since calculated values also for transition metal oxides by using the relation of Duffy et al.[1]

However, the situation is rather confusing as conflicting values have been reported. Sommerville and Sosinsky[3] derived their values by line-fitting to sulphide capacity data.

Nakamuka et al.[29], on the other hand, based their scale on the average electron density.

The three scales are compared in Table 1 and tested in Fig 3 by plotting calculated basicities of 130 different slags[27,30-32] against the logarithm of the monomer phosphorus capacity. Here, a clear linear grouping is found only for the first scale, i.e. that based on Pauling electronegativity.

RELATIONSHIP BETWEEN MONOMER PHOSPHORUS CAPACITY AND OPTICAL BASICITY

From the foregoing it is clear that a linear relation exists between phosphorus capacity and optical basicity. Another interesting comparison is provided by applying optical basicity to very complex steelmaking-type slags.

At 1600 °C experimental results are from the $CaO-SiO_2-MgO_{SAT}-FeO-Fe_2O_3-Al_2O_3-TiO_2-VO_2-P_2O_5$ system examined by Selin[27].

In making these calculations we used the values ΛTiO_2 = 0,55 and ΛVO_2 = 0,57 derived elsewhere[33]. As shown in Fig 4 an excellent linear relation is given by Eq 8 (R^2 = 0, 94 σ= 0,057)

$$\text{Log } C_{P_M} = 17.58\Lambda - 7.86 \qquad (8)$$

Regarding the temperature dependence, the data of Suito et al.[30-32] at 1550 °C, 1600 °C and 1650 °C of complex slags, including slags containing appreciable amounts of Al_2O_3, MnO, TiO_2, BaO and Na_2O, are used to correlate the monomer phosphorus capacity with optical basicity and temperature, Eq 9.

$$\text{Log } C_{P_M} = 21.55\Lambda + \frac{32912}{T} - 27.90 \qquad (9)$$

OXYGEN PARTITION BETWEEN SLAG AND METAL

Recently, interest has been focused on estimation of the oxygen contents of molten iron in equilibria with slag[6, 24, 26, 27, 34, 35]. Usually, the oxygen dissolved in the iron is related to the activity of FeO in the slag which, in turn, is calculated by using more or less complicated models.

In this work, the oxygen content is derived from the optical basicity and the iron content of the slag.

Regression analysis applied to the complex slags[27] yields

$$[\%O] = -0.6867\Lambda + =.00296\%\,FeO^* + 0.5106 \qquad (10)$$

where FeO^* is 1,2865 Fe_{tot} i.e. all iron calculated as Fe^{2+}. As shown in Fig 5, observed values of [%O] correspond well with the values calculated from Eq 10.

DISCUSSION

It has been shown that the monomer phosphorus capacity gives an adequate description of the slag ability to absorb phosphorus. It was also made clear that the oxygen content in molten iron could be accurately related to the slag composition.

Consequently, by combining Eq 8 and 10 it is possible to calculate the equilibrium phosphorus content of the metal from slag composition alone. The comparison in Fig 6 shows that calculated and experimental data match amazingly well.

REFERENCES

1. J.A. Duffy, et al. J. Chem. Soc. Faraday Trans. I 74 (1978), 1410.
2. S.D. Brown, et al. Ironmaking and steelmaking 9(1982),163.
3. I.D. Sommerville, D.J. Sosinsky: Proc. 2:nd Int. symp on Metallurgical slags and fluxes, Lake Tahoe Nevada USA., 11-14 Nov. (1984), 1015.
4. J.D. Young, I.D. Sommerville: Proc. Japan-Canada seminar on secondary steelmaking, Tokyo, 1985.
5. D.J. Sosinsky, I.D. Sommerville: Met. Trans.17B(1986),331.
6. T. Tsao, H.G. Katayama: Trans. ISIJ. 26 (1986), 717.
7. A.H. Chan, R.J. Fruehan: Met. Trans.17B (1986), 491.
8. H.G. Katayama, Tetsu-to-Hagane,73 (1987), 336.
9. R.G. Reddy, M. Blander: Met. Trans. 18B (1987), 591.
10. M. Merakib: Steel research 59(1988),16.
11. D.R. Gaskell: Trans. ISIJ 22 (1982), 997.
12. H. Suito, R. Inoue: Trans. ISIJ 24 (1984), 47.
13. T. Mori: Trans. Jpn. Inst. Met. 25 (1984), 761.
14. Y. Fukami, Y. et al. Nisshin Steel Tech. Rep.53,(1985),1.
15. T. Ting, et al. Tetsu-to-Hagane, 72 (1986), 225.
16. K. Kunisada, H. Iwai: Trans. ISIJ., 27 (1987),263.
17. K. Kunisada, H. Iwai: Trans. ISIJ., (1987), 332.
18. R. Inoue, H. Suito: Trans. ISIJ., 24 (1984) 816.
20. T. Kawahara, et al. Steel research, 57 (1986),160.
21. T. Kawahara, et al Proc. Sixth Process Technology Conf, Apr 6-9 1986, Washington.
22. D.J. Sosinsky, et al ibid.
23. I.D. Sommerville: Proc. conf. Scaninject IV, 1986.
24. E.T. Turkdogan: "Physiochemical properties of molten slags and glasses", the Metals Society, London, 1983.
25. E.T. Turkdogan: Trans. ISIJ. 24 (1984), 591.
26. X.F. Zang, et al ISS Trans, 6 (1985)29.
27. R. Selin, Dr. Diss, RIT, Stockholm, Sweden, 1987.
28. M. Muraki, et al Trans. ISIJ, 25 (1985), 1025.
29. T. Nakamura, et al J. Jpn Inst. Metals 50 (1986), 456.
30. H. Suito, et al Trans. ISIJ., (1981) 250.
31. H. Suito, R. Inoue: ibid 24 (1984), 257.
32. H. Suito, R. Inoue: ibid 24 (1984), 47.
33. Å. Bergman: Unpublished results.
34. J.D Shim, S. Ban-ya. Tetsu-to-Hagane, 67 (1981), 1745.
35. S. Ban-ya, M. Hino: Tetsu-to-Hagane 73 (1987), 74.

Table 1 A comparison of different scales of optical basicity.

OXIDES	THEORETICAL FROM PAULING ELECTRO-NEGATIVITY[1,9,13]	SOMMERVILLE et al[3] FROM SULFIDE CAPACITY	NAKAMURA et al[29] FROM ELECTRON-DENSITY
K_2O	1,40		1,16
Na_2O	1,15		1,11
BaO	1,15		1,08
Li_2O	1,0		1,06
CaO	1,0		1,0
MgO	0,78		0,92
TiO_2	(0,61*)	0,61	0,65
Al_2O_3	0,605		0,66
MnO	0,59	1,21	0,95
FeO	0,51	1,03	0,94
Fe_2O_3	0,48	0,70	0,72
SiO_2	0,48		0,47
B_2O_3	0,42		0,42
P_2O_5	0,40		0,38
SO_3	0,33		0,29

* Doubtful value from ref (23).

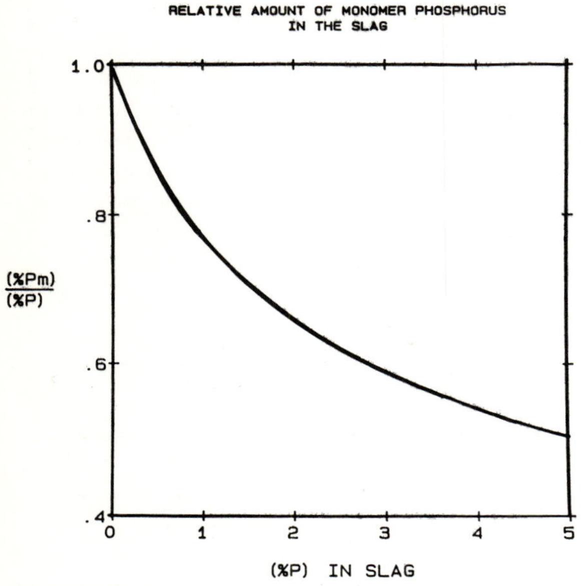

Fig. 1. The proportion of monomer phosphorus in relation to the amount of phosphorus in the slag.

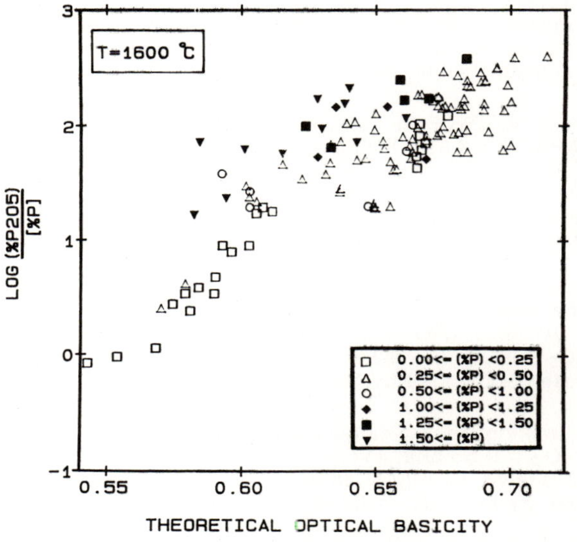

Fig. 2. Effect of the phosphorus content and optical basicity of the slag on the logarithm of the phosphorus distribution. Data from Selin[27] and Suito et al.[30-32].

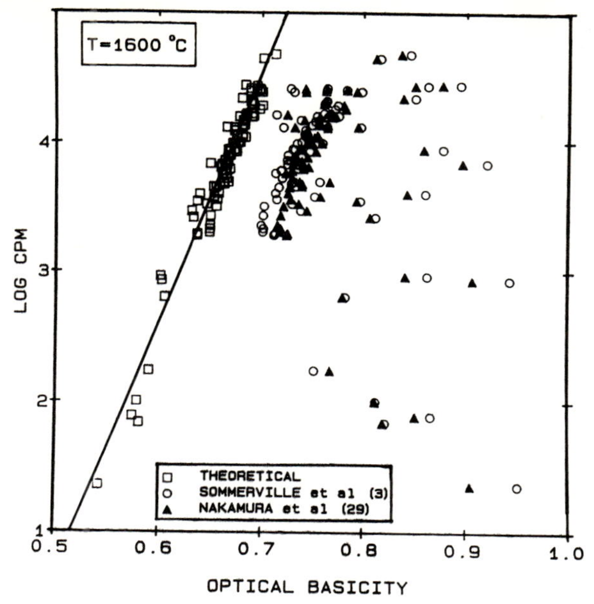

Fig. 3. The relation between the logarithm of the monomer phosphorus capacity and the three scales of optical basicity.

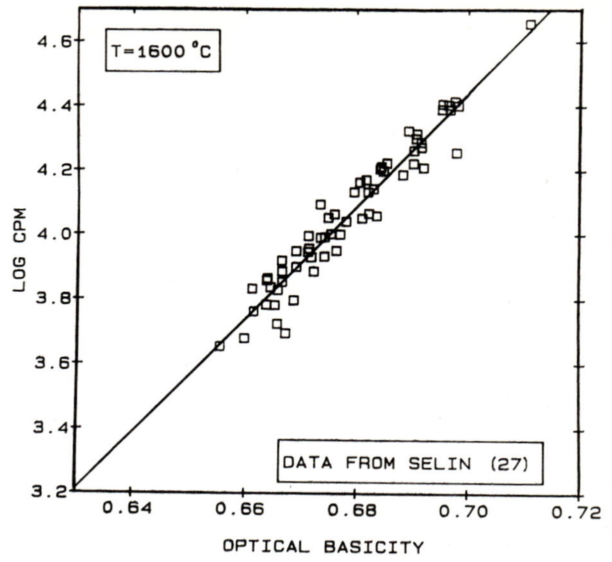

Fig. 4. The relation between monomer phosphorus capacity and optical basicity of 62 complex steelmaking-type slags. Data from Selin[27].

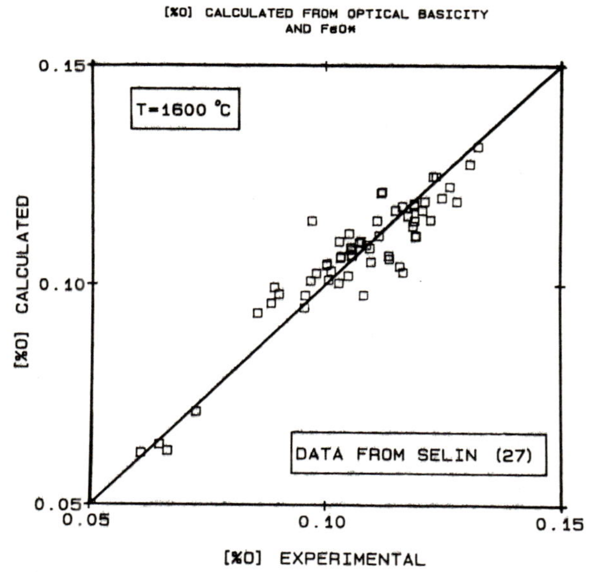

Fig. 5. Comparison of calculated and observed oxygen content in metal.

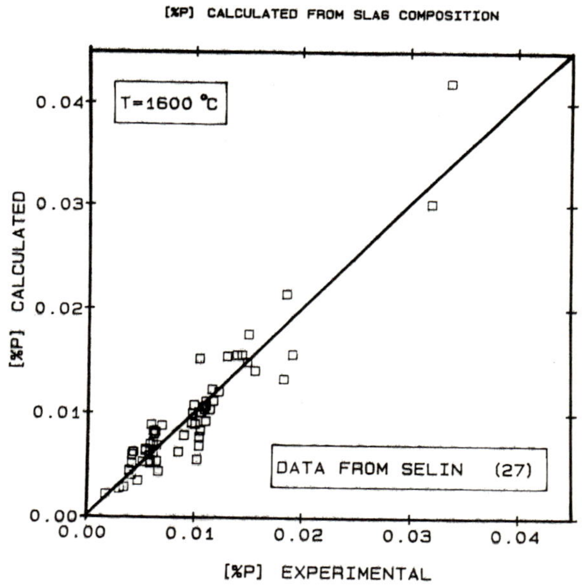

Fig. 6. Comparison of calculated and observed phosphorus content in metal.

Alternative methods of measuring optical basicity

J A DUFFY

The author is in the Department of Chemistry, University of Aberdeen, Scotland.

SYNOPSIS

Calculation of the optical basicity, Λ, of an oxidic slag relies on spectroscopic data of probe ions, e.g. lead(II), in the UV region. Media containing transition metal ions have very poor UV transparency, preventing these spectroscopic measurements, and Λ cannot be assigned to slags containing transition metal oxides. An alternative method of obtaining Λ is discussed, using the electronic polarisability of oxide(-II), $\alpha_{O^{2-}}$. For a series of silicate glasses there is a linear relationship between $1/\alpha_{O^{2-}}$ and Λ. It is suggested that glasses containing transition metal oxides are interpolated into this relationship to obtain Λ.

SLAG BASICITY

Oxidic slags are composed of oxide(-II) species, the average electron donor power of which determines the slag basicity. The electron donor power can be measured by probe ions, such as thallium(I) or lead(II), whose s-p absorption bands undergo a red shift in proportion to the electron density they receive from the oxide(-II)s. Ionic oxides are highly electron donating (and accordingly are very basic); and calcium oxide has been chosen as a reference material for expressing the electron donor power of materials composed of acidic and basic oxides such as slags or glasses, and for these media, the ratio

$$\frac{\text{electron donor power of medium}}{\text{electron donor power of CaO}}$$

has been called the "optical basicity".[1] Using thallium(I), etc. probe ions, the optical basicity, Λ, is given by

$$\Lambda = \frac{\nu_f - \nu_{medium}}{\nu_f - \nu_{CaO}} \qquad 1$$

where ν_f is the s-p ($^1S_0 \longrightarrow {}^3P_1$) frequency of the probe ion in the free state, ν_{medium} is the frequency observed in the medium under investigation, and ν_{CaO} is the frequency of the probe ion in CaO. The spectroscopic measurements are usually made at ambient temperature with samples that have been cooled to a glass.[2]

The basicity of a material depends on the nature of the cations (e.g. calcium(II) or silicon(IV)) present, and accumulated data have shown that the optical basicity depends on the proportion of oxide(-II) negative charge the cations neutralise, and the ability of the cations to use the oxide(-II) charge clouds for covalency. This ability is expressed in terms of the "basicity moderating parameter", γ, values of which are given for various cations in Table 1. Thus, the optical basicity of a glass or slag composed of oxides $AO_{a/2}$, $BO_{b/2}$,... depends on

(i) the equivalent fractions X_A, X_B,..., and

(ii) the basicity moderating parameters, γ_A, γ_B,..

such that:[2,3]

$$\Lambda = \frac{X_A}{\gamma_A} + \frac{X_B}{\gamma_B} + \ldots \qquad 2$$

Optical basicity calculated from this formula has been termed "theoretical" and given the symbol, Λ_{th}. Values of Λ_{th} for blast furnace slags have been shown to correlate quite well with their properties, for example, the distributions of sulphur and phosphorus at the slag/metal interface.[3,4]

The γ values in Table 1 are from UV spectroscopic measurements of probe ions and therefore require the material to have good UV transparency. This creates a difficulty in determining γ for transition metal ions since the presence of most transition metal oxides drastically impairs the UV transparency. For this reason, basicity moderating parameters are not known for transition metal ions. (Attempts to calculate them from electronegativity are not viable, since the relationship between γ and electronegativity, x,[5] ($x = 0.25 + 0.75\gamma$) has not been shown to extend to transition metal ions.) An alternative method for determining optical basicity is therefore desirable.

Several years ago, we showed the connection between optical basicity and oxide refractivity, $R_{O^{2-}}$.[2] This was more thoroughly investigated by Iwamoto et al.[6,7] whose results suggested that refractivity measurements might provide a method for optical basicity determination in certain cases.

For several reasons this work is more conveniently developed using electronic polarisability, α, rather than refractivity.

ELECTRONIC POLARISABILITY OF OXIDES

Molar polarisability, α_m, (in A^3) is related to molar refractivity, R_m, (in cm^3) by

$$\alpha_m = 3\,R_m / 4\pi N \qquad 3$$

(N is the Avogadro number) and is obtained from refractive index, n, and specific gravity, d, by writing the Clausius-Mossotti-Maxwell relationship in the form

$$\alpha_m = \frac{0.396\,M}{d}\left(\frac{n^2 - 1}{n^2 + 2}\right) \qquad 4$$

For example, CaO (formula weight, M: 56.1) has n = 1.84 and d = 3.32 and α_m is therefore 2.97 A^3; subtracting $\alpha_{Ca^{2+}}$ (0.46 A^3), $\alpha_{O^{2-}}$ is 2.51 A^3.

The basicity moderating parameter, γ, represents the tendency for a cation to use the oxide(-II) electron charge clouds for covalency or, in other words, to polarise the oxide(-II). Thus, it is anticipated that $\alpha_{O^{2-}}$ decreases with increasing γ_M. Table 1 indicates this, and when $1/\alpha_{O^{2-}}$ is plotted against the optical basicity for the oxides in Table 1, that is, against $1/\gamma_M$, the points lie reasonably close to a straight line (Fig. 1) which can be expressed by

$$\alpha_{O^{2-}} = \frac{\gamma_M}{\gamma_M - 0.6} \qquad 5$$

In terms of optical basicity this becomes

$$1/\alpha_{O^{2-}} = 1.0 - 0.6\,\Lambda_{th} \qquad 6$$

VITREOUS SILICATES

The applicability of this rudimentary relationship to silicate systems is now investigated by considering data for various silicate glasses (Table 2). As shown in Fig. 2, the plot of $1/\alpha_{O^{2-}}$ versus Λ_{th} gives a series of points which lie very close to the line, eq. 6. Indeed, if a line were drawn through the points, it would have the same slope as eq. 6 and would require only a small change in the intercept constant (from 1.0 to 0.975).

For present purposes, the importance of Fig. 2 is that it would provide an estimate of the optical basicity for any silicate glass from its oxide polarisability. Silicate glasses based solely on transition metal oxides generally have limited glass-forming properties, but alkali and alkaline earth metal silicate (and other) glasses will incorporate substantial quantities of transition metal oxides. Measurement of their refractive index and specific gravity would provide the molar polarisability and hence $\alpha_{O^{2-}}$. (Correcting for the total cation polarisability requires an "estimated" value for most transition metal ions, but errors introduced here are small.) Interpolation of $1/\alpha_{O^{2-}}$ into the trend in Fig. 2 would yield a value for Λ_{th}. Substitution into eq. 2 would then provide the basicity moderating parameter for the transition metal ion and the optical basicity of the transition metal oxide. In principle, such work can be extended to other glass forming systems, e.g. phosphates and aluminosilicates. Work of this type is presently being undertaken in our laboratory.

REFERENCES

1. J.A. Duffy and M.D. Ingram, J. Amer. Chem. Soc. 93 (1971) 6448.
2. J.A. Duffy and M.D. Ingram, J. Non-Cryst. Solids 21 (1976) 373.
3. J.A. Duffy, M.D. Ingram and I.D. Sommerville, J. Chem. Soc. Faraday Trans. 1, (1978) 1410.
4. D.R. Gaskell, Trans. Iron Steel Inst. Japan 22 (1982) 997.
5. J.A. Duffy and M.D. Ingram, J. Chem. Soc. Chem. Commun. (1973) 635.
6. N. Iwamoto, Y. Makino and S. Kasahara, J. Non-Cryst. Solids 68 (1984) 379.
7. Ibid., 68 (1984) 389.
8. J.R. Tessman, A.H. Kahn and W. Shockley, Phys. Rev. 92 (1953) 890.

Table 1

Values of $\alpha_M n+$, $\alpha_O 2-$ and γ_M for oxides

Oxide	$\alpha_M n+$ (Å³)	$\alpha_O 2-$ (Å³)	γ_M
BaO	1.59	3.58	0.87
CaO	0.46	2.51	1.00
Li_2O	0.03	2.08	1.00
MgO	0.04	1.79	1.28
Al_2O_3	0.05	1.46	1.65
SiO_2	0.03	1.42	2.09
B_2O_3	0.003	1.39	2.36
P_2O_5	-	1.35	2.50

Values of $\alpha_M n+$ (Pauling values) cited in ref. 7. Refractive indices and specific gravities for calculating polarsisabilities (eq. 4) are from ref. 8. Values of γ_M are from ref. 2.

Table 2

Polarisability and optical basicity data for silicate glasses

Glass system	% metal oxide	$\alpha_O 2-$	$1/\alpha_O 2-$	Λ_{th}
SiO_2	0.00	1.46	0.686	0.478
Li_2O-SiC_2	5.0	1.47	0.680	0.492
	25.0	1.54	0.650	0.553
	30.0	1.55	0.645	0.571
	35.0	1.58	0.634	0.589
Na_2O-SiC_2	10.0	1.51	0.664	0.514
	15.0	1.53	0.654	0.533
	20.5	1.56	0.642	0.555
	25.6	1.59	0.631	0.577
	30.0	1.63	0.612	0.597
	34.0	1.64	0.608	0.616
	40.0	1.72	0.581	0.646
	50.0	1.78	0.563	0.702
K_2O-SiO_2	10.1	1.55	0.645	0.526
	17.5	1.57	0.637	0.564
	21.5	1.61	0.623	0.586
	25.6	1.63	0.614	0.609
	29.0	1.68	0.594	0.630
	34.3	1.71	0.584	0.663
$CaO-SiO_2$	42.9	1.68	0.594	0.621
	50.0	1.74	0.573	0.652
	52.5	1.78	0.563	0.664
	55.0	1.80	0.555	0.676

Polarisabilities are calculated from eq. 4 using data from ref. 6.

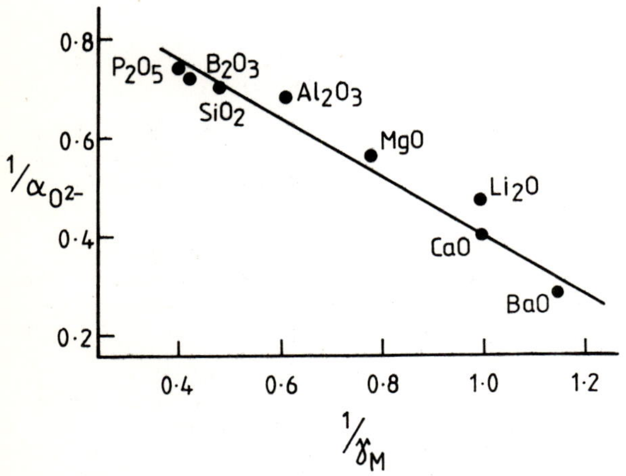

Fig. 1 Plot of $1/\alpha_O 2-$ versus $1/\gamma_M$ for oxides (Table 1)

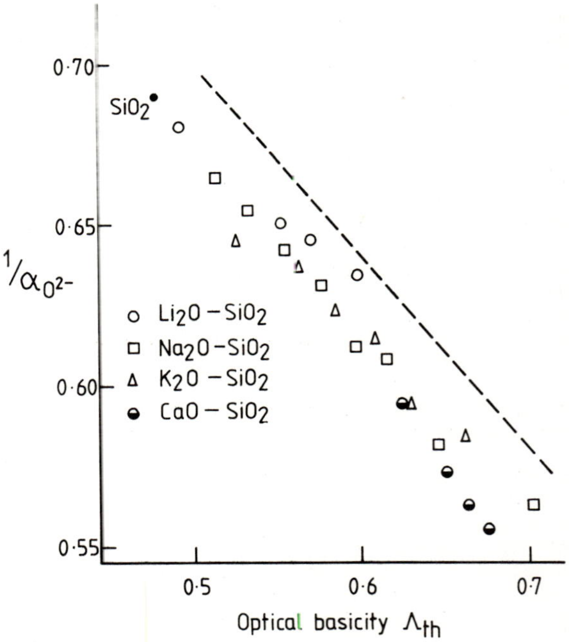

Fig. 2 Plot of $1/\alpha_O 2-$ versus Λ_{th} for silicate glasses (Table 2). The broken line is eq. 6.

Optical basicity: a flexible basis for flux control in steelmaking

C L CAREY, R J SERJE, K GREGORY and He QINGLIN

The authors are with BHP Steel International Group, Slab and Plate Products Division, Port Kembla, Australia.

1.0 INTRODUCTION

The BOF steelmaking operation at BHP International Steel Group, Port Kembla works is fully computer controlled[1]. An integral part of the system is the prediction of flux additions to the furnace.

The majority of conventional flux models used for the calculation of lime and dolomite additions to the Basic oxygen steelmaking process are based on empirical relationships or on the use of V-ratio (% CaO/%SiO_2) as a measure of slag basicity. The latter has been in use at Port Kembla since plant start-up. V-ratio was set manually in the model such that the product of slag volume and V-ratio was constant. With the increasing occurrence of low silicon metal and the continuing drive towards lower phosphorus levels in steel this type of flux model has become unsuitable to the point that it has been abandoned by several steel plants.

Recent research work[2,3] has shown very good correlation between sulphide and phosphate capacity and a factor known as the optical basicity. This concept has been found to give a very accurate measure of the behaviour of steelmaking slags[4]. The present work illustrates how optical basicity can replace V-ratio as the basis of a flux calculation model.

The individual and combined effect of the oxides normally present in steelmaking slags generated at the BHP Port Kembla plant on optical basicity has been quantified. The results obtained with the optical basicity based model are assessed using the current flux model as a standard.

2.0 OPTICAL BASICITY CONCEPT[5,6]

2.1 Background

The optical basicity concept was developed in the field of glass chemistry[5,6], and it expresses the basicity in terms of the electron donor power of the oxide ions in a glass. This is measured experimentally by using p-block metal ions, Tl^+, Pb^{2+} and Bi^{+3} which have an electronic configuration $d^{10}s^2$. When small quantities of these probe ions are dissolved in glass, the electron donation by the oxygen to the probe ion brings about a reduction in the 6s-6p energy gap and this in turn produces a shift in frequency in the ultraviolet band compared with that of the free probe ion. This effect is measured spectroscopically in transparent slags and consequently the technique is difficult to apply to steelmaking slags since the presence of transition metal oxides such as ferrous, manganese and titanium render the glass opaque.

From a large number of measurements using Pb^{2+} as the probe ion, it has been established[7] that the optical basicity Λ of an oxide is related to the Pauling electronegativity, X, of the cation by

$$\frac{1}{\Lambda} = 1.36\ (X - 0.26) \qquad (1)$$

It is possible, therefore, to calculate the value of Λ for any non-transition metal oxide using equation (1). The values obtained and used in the present work are given in Table 1.

The values for the transition metal oxides FeO, MnO, and TiO_2 were those derived from sulphide capacities containing these oxides[8] which are 1.03, 1.21 and 0.61 respectively.

The calculation of the theoretical optical basicity for slags containing multiple elements can be conducted by using the equation:

$$\Lambda = \Sigma X_i\, \Lambda_i \qquad (2)$$

where

$$X_i = \frac{\text{Mole fraction (i) x No.Oxygen atoms in(i)}}{\Sigma \text{Mole fraction (i) x No.Oxygen atoms in (i)}} \qquad (3)$$

$$\text{Mole fraction (i), } N(i) = \frac{Wt(i)/M(i)}{\Sigma Wt(i)/M(i)} \qquad (4)$$

In these equations Λ (i) is the optical basicity, Wt (i) is the mass, M(i) the molecular weight and X(i) the equivalent cation fraction of the oxide (i) in the slag.

2.2 Application to BOF Steelmaking Slags

The first approach taken to consider the viability of the optical basicity concept as a base for a flux calculation model in the steelmaking process was to:

(1) Analyse the influence on optical basicity of the major components of BOF slags,
(2) Assess the capacity of the optical basicity concept to predict final slag composition, and
(3) Compare the ability of V-ratio and Λ to predict turndown conditions, especially phosphorus and manganese.

The first step was accomplished by calculating Λ for the average slag composition generated at the

BHP Port Kembla plant. Equation (2) was applied to obtain:

$$\Lambda = 1/B\,(NCaO * \Lambda_{CaO} + NMgO * \Lambda_{MgO} + 2 * NSiO_2 * \Lambda_{SiO_2} + 5 * NP_2O_5 * \Lambda_{P_2O_5} + NMnO * \Lambda_{MnO} + NFeO * \Lambda_{FeO} + 3 * NAl_2O_3 * \Lambda_{Al_2O_3} + 2 * NTiO_2 * \Lambda_{TiO_2} + 3 * NSO_3 * \Lambda_{SO_3}) \quad (5)$$

Where $B = NCaO + NMgO + 2 * NSiO_2 + 5 * NP_2O_5 + NMnO + NFeO + 3 * NAl_2O_3 + 2 * NTiO_2 + 3 * NSO_3 \quad (6)$

The results obtained are illustrated in Fig.1. The basicity lines were calculated by increasing the mole fraction of the subject oxide and proportionately decreasing the remainder. From this figure it can be observed that Λ increases with any component whose partial basicity is greater than 0.61. The combined effects are shown on the ternary iso-optical basicity diagram, Fig.2 (based on the average BOS slag analysis) where it can be seen that Λ is most sensitive to changes in (FeO+MnO). This is due to the "effective" optical basicity values used for these transition metal oxides.

With respect to the second step, a comparison of predictability between optical basicity and V-ratio was made by examining the difference between the aim and achieved basicity values. The V-ratio and actual data were available in the BOF computer history file, aim Λ was calculated using the new flux model, from the fluxes actually added. The actual minus aim values were represented as an error ratio to counter the number difference between the two basicities. The results are shown in Fig.3. The mean and standard deviations for the Λ error are much smaller than for the V-ratio error, -0.337 (0.27), and -0.021 (0.019) respectively.

For the third step a comparison was made between V-ratio and Λ in their ability to predict turndown phosphorus and manganese. Λ gave a better correlation on both cases as shown in Table 2.

3.0 CALCULATIONS OF FLUX ADDITIONS FOR STEELMAKING

3.1 Explanation of Existing Flux Model

The current model calculates lime and dolomite requirements to achieve aim slag V-ratio and MgO levels.

The steel grades produced are divided into eight basic flux groups defined by tapping conditions such as steel temperature, phosphorus and carbon level. Each group is then sub-divided into hot metal silicon ranges each with an aim slag V-ratio and MgO. These values are set empirically to achieve the desired steel and slag chemistry.

The required flux additions are then calculated from a slag oxide mass balance and slag weight estimate.

3.2 Fundamentals of the Optical Basicity Based Model

The new model uses the optical basicity concept and a mass balance across the steelmaking process to calculate the lime and dolomite required to achieve an aim slag optical basicity and slag MgO.

The aim Λ is estimated by a regression equation obtained from the analysis of data for all steelgrades produced at Port Kembla in bottom stirred vessels, with a hot metal silicon level of up to 1%.

The aim Λ is calculated by the expression:

$$\Lambda = (692.2694 - .075 * HMSI + .328 * HMP + .0452 * HMRAT - 2.2 * PST - .0944 * M_1BC + .102 * TAPT + .0103 * AMNORE)/1000 \quad R^2 = 0.49 \quad (7)$$

In equation (7) HMSI, HMP and HMRAT are the hot metal silicon, hot metal phosphorus and hot metal ratio respectively. PST, M_1BC and TAPT are the aim turndown phosphorus, carbon and temperature respectively. AMNORE is the manganese ore addition computed for a heat of steel. The assumptions made in the new model were as follows:

1) All fluxes charged to the BOS vessel dissolve completely in the slag.
2) Silicon content in steel at the end of blow is zero.
3) The elements Ca, Mg, Si, P, Mn, Fe, Al, Ti and S exist in the slag in the form of CaO, MgO, SiO_2, P_2O_5, MnO, FeO, Al_2O_3, TiO_2 and SO_3 respectively.

The amount of fluxes to be charged in the vessel is calculated by solving the set of equations given in the appendix. This is achieved by using the multivariables constrained method (complex algorithm)[9].

The aim slag FeO and MgO levels are a function of steel grade and the expected MgO saturation level respectively. These values are calculated by equations E and F in the appendix.

3.3 Assessment of the Proposed Model

The accuracy of the new flux model was assessed using the existing V-ratio model as a base for comparison. In section 2.2 it was established that Λ was more accurate than V-ratio in predicting turndown phosphorus (Table 2). This result is confirmed when equation (7) is used in the Λ based model. Fig. 4 illustrates how successful the current model is in achieving the required phosphorus for all steel grades. The analysis of 2708 heats shows that 9% of these heats did not meet the specified phosphorus (quadrant I). However, the quantity of lime used was smaller than that calculated by the Λ based model as indicated by the negative value of the lime balance (actual lime used minus lime calculated by Λ model). The heats located in quadrant III (58%) met the required phosphorus level. In this case the model suggests that the same result might have been achieved by using less lime as indicated by the positive value of the lime balance.

The heats which fall in quadrant IV (24%) would have utilised more lime to achieve the same result had the proposed model been used. Of concern are the heats in quadrant II where the phosphorus level has not been met and the Λ model is calculating a smaller lime addition. This area is currently being examined by running the two models in parallel.

Overall it is considered that the proposed model would have given a better performance in 67% of the heats (quadrants I and III) and in 24% of the heats would have demanded extra lime.

A comparison of the sensitivity of the two models with hot metal silicon and hot metal phosphorus was made and is illustrated in Figs. 5 and 6. The correlations of lime addition with hot metal silicon are very similar (R^2 opt = 0.367 vs R^2 V-ratio = 0.400). However, the response of the two models to a change in hot metal phosphorus is completely different as indicated by the difference in regression coefficients (R^2opt = 0.147, R^2V-ratio = 0.0042). This result is explained by the presence of the hot metal phosphorus analysis in equation (7).

The lime calculated by the proposed model is also a function of tapping temperature (R^2 = 0.3). This is explained by the fact that the phosphorus partition between slag and metal is reduced with increasing temperature. Therefore, to meet phosphorus requirements the lime addition is increased proportionally with tap temperature.

The effect of the proposed model on the economics of steelmaking was also evaluated. An overall saving of 1.67kg of lime/tonne of steel would be achieved with the Λ based model. It is interesting to note that the saving obtained when low metal silicon is processed (0.09% - 0.3% Si) is more substantial (2.48 Kg/ton). With respect to the dolomite usage, the difference in this case is much smaller 0.38 Kg/ton of steel in favour of the proposed model.

4.0 CONCLUSIONS

An optical basicity based flux model that calculates the lime and dolomite additions required for the BOF Steelmaking process has been developed. When compared to the existing flux program the new model has proved to be more responsive to changes in hot metal phosphorus and tap temperature and as sensitive to variations in hot metal silicon.

An assessment made on the performance of the flux models in achieving the aim steel phosphorus showed that in 67% of the heats analysed the new model gave better results. However, there was a small proportion of heats (9%) that did not meet the specified phosphorus level. The conditions under which this occurs are currently under investigation. It has been established that the utilisation of the optical basicity based model will result in a saving of 1.67 Kg/t and 0.38 Kg/t of lime and dolomite respectively.

REFERENCES

1. K. Gregory , "Optimisation of BOS Furnace Control Practices at AIS Port Kembla", Proc. 68th AIME Steelmaking Conf., 1985.
2. I.D. Sommerville, "The Measurement, Prediction and use of Capacities of Metallurgical Slags", Scaninject IV, Part 1, 1986, pp 8:1-8:21
3. T. Mori, "On the Phosphorus Distribution between Slag and Metal", Transaction of the Japan Inst. of Metals, Vol. 25, No 11 (1984), pp 761-771.
4. R. Serje, K. Gregory and H. Qinglin, BHP Steelmaking Technology and Development - Technical Note No 87/5, 1987.
5. J.A. Duffy and M.D. Ingram, Journal Amer Chem. Soc., Vol. 93, 1971, pp 6448-6454.
6. J.A. Duffy and M.D. Ingram, J. Non-crystalline solids., Vol.21, 1976, pp 373-410.
7. J.A. Duffy and M.D. Ingram, J. Inor. Nucl. Chem., Vol. 37, 1975, pp 1203-1206.
8. D.J. Sosinsky and I.D. Sommerville, "The Composition and Temperature Dependence of the Sulfide Capacity of Metallurgical Slags", Met. Trans. Vol. 17B, 1986, pp 331-337.
9. J.L. Kuester and J.H. Mize, "Optimisation Techniques with Fortran", 1973.

APPENDIX

OPTICAL BASICITY FLUX MODEL

LIST OF SYMBOLS

Λ	Overall slag optical basicity
Λi	Partial optical basicity of slag component i
Wi	Weight of slag component i (Kg)
Mi	Molecular weight of slag component i
LIME	Lime to be charged (Kg)
CAOLM	CaO content of lime (%)
DOL	Dolomite to be charged (Kg)
CAODOL	CaO content of dolomite (%)
STONE	Limestone weight (Kg)
CAOST	CaO content of limestone (%)
BLIME	Burnt lime weight (Kg)
CAOBL	CaO content of burnt lime (%)
MGOLM	MgO content of lime (%)
MGODOL	MgO content of dolomite (%)
DSTONE	Dolomite stone weight (Kg)
MGODS	MgO content of dolomite stone (%)
HM	Hot, metal weight (100 Kgs)
IX	Slag rake flag having a value of 1 or 0
CAOBFS	CaO content of blast furnace slag (%)
MGOBFS	MgO content of blast furnace slag (%)
WSLAG	Slag weight (Kg)
HMSi	Hot metal silicon (%)
HMP	Hot metal phosphorus (%)
HMRAT	Hot metal ratio in vessel charge
PST	Aim steel phosphorus (%)
M_1BC	Aim turndown carbon (%)
TAPT	Tap temperature (C)
AMNORE	Manganese ore addition (Kg)
SMGO	MgO saturation level (0.01%)

The following is the set of equations used by the optical basicity based model:

$$\Lambda = (\frac{WCaO}{MCaO} * \Lambda CaO + \frac{WMgO}{MMgO} * \Lambda MgO$$

$$2 + \frac{WSiO_2}{MSiO_2} * \Lambda SiO_2 + \frac{5 * WP_2O_5}{MP_2O_5}$$

$$* \Lambda P_2O_5 + \frac{WMnO}{MMnO} * \Lambda MnO + \frac{WFeO}{MFeO} *$$

$$\Lambda FeO + \frac{3 * WAl_2O_3}{MAl_2O_3} * \Lambda Al_2O_3$$

$$+ 2 * \frac{WTiO_2}{MTiO_2} * \Lambda TiO_2 + \frac{3 * WSO_3}{MSO_3} *$$

$$\Lambda SO_3) /B/S \qquad (A)$$

where

$$B = (\frac{WCaO}{MCaO} + \frac{WMgO}{MMgO} + \frac{2WSiO_2}{MSiO_2} + \frac{5WP_2O_5}{MP_2O_5}$$

$$+ \frac{WMnO}{MMnO} + \frac{WFeO}{MFeO} + \frac{3WAl_2O_3}{MAl_2O_3} + \frac{2WTiO_2}{MTiO_2}$$

$$+ \frac{3WSO_3}{MSO_3}) /S$$

$$S = \frac{WCaO}{MCaO} + \frac{WMgO}{MMgO} + \frac{WSiO_2}{MSiO_2} + \frac{WP_2O_5}{MP_2O_5} + \frac{WMnO}{MMnO}$$

$$+ \frac{WFeO}{MFeO} + \frac{WAl_2O_3}{MAl_2O_3} + \frac{WTiO_2}{MTiO_2} + \frac{WSO_3}{MSO_3}$$

$$\Lambda = (692.2694 - .075 * HMSi + .328 * HMP + .0452 * HMRAT - 2.2 * PST - .0944 * M1BC + .102 * TAPT + .0103 * AMNORE) / 1000 \quad (B)$$

$$WCaO = (LIME * CAOLM + DOL * CAODOL + STONE * CAOST + BLIME * CAOBL) * 0.01 + HM (0.022 - 0.02IX) * CAOBFS \quad (C)$$

$$WMgO = (LIME * MGOLM + DOL * MGODOL + DSTONE * MGODS) * 0.01 + HM (0.022 - 0.02IX) * MGOBFS \quad (D)$$

$$WFeO = F * Wslag \quad (E)$$

Where F is a function of steel grade and Wslag is given by the following expression

$$Wslag = WCaO + WMgO + WSiO_2 + WP_2O_5 + WMnO + WAl_2O_3 + WFeO + WTiO_2 + WSO_3$$

$$WMgO = SMGO \times Wslag \quad (F)$$

Table 1: Optical Basicity values

Oxide	CaO	MgO	Al_2O_3	SiO_2	P_2O_5	SO_3
Electronegativity of cation	1.0	1.2	1.5	1.8	2.1	2.5
Optical basicity	1.0	0.78	0.60	0.48	0.40	0.33

Table 2: Correlation between steel analysis, V-Ratio and Λ

	R^2	
	V-Ratio	Λ
First bath Mn	-0.031	-0.49
P	-.28	-0.43

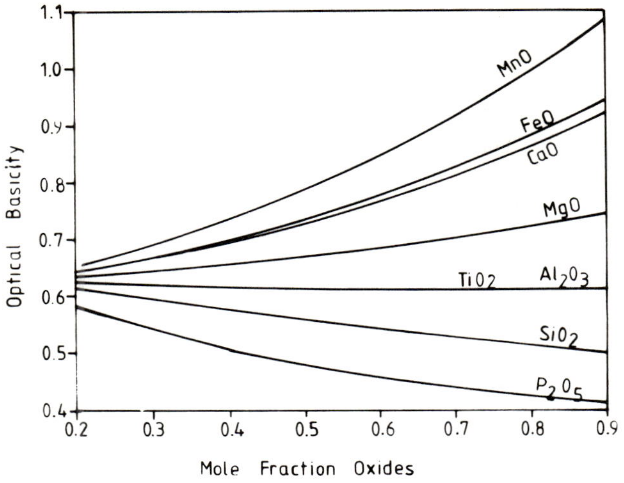

Figure 1: Effect of oxides on optical basicity in steelmaking slags

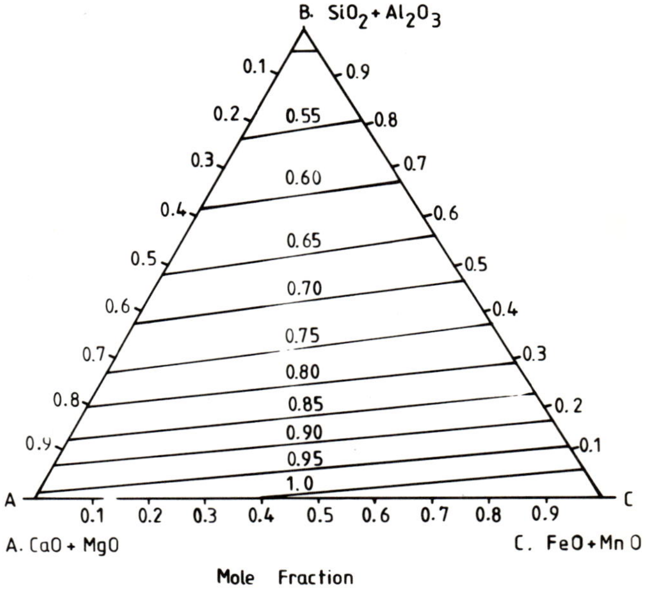

Figure 2: Iso-optical basicity diagram for BOS slag

A = 81.5% CaO, 18.5% MgO
B = 95.9% SiO_2, 4.1% Al_2O_3
C = 81.3% FeO, 18.7% MnO

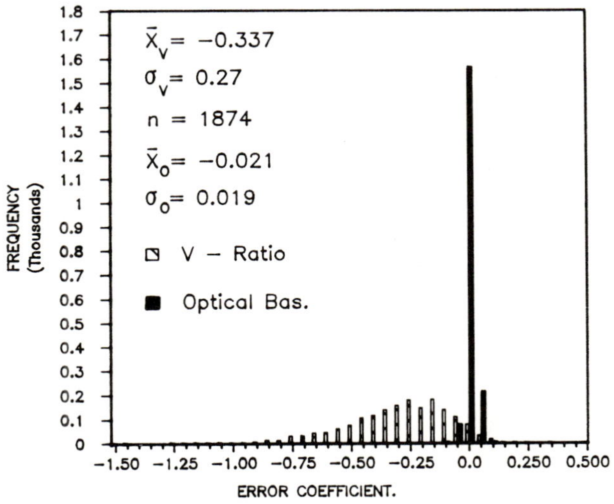

Figure 3: Error distribution for optical basicity and V-Ratio

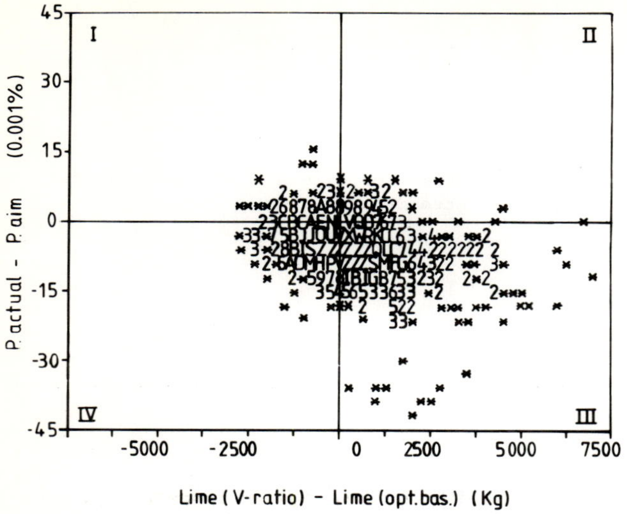

Figure 4: Comparison of phosphorus control performance between V-Ratio and optical basicity based models

Figure 5: Response of flux models to changes in hot metal silicon

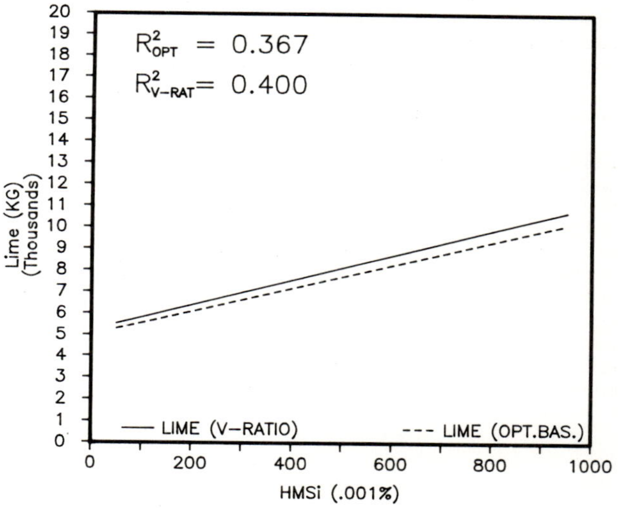

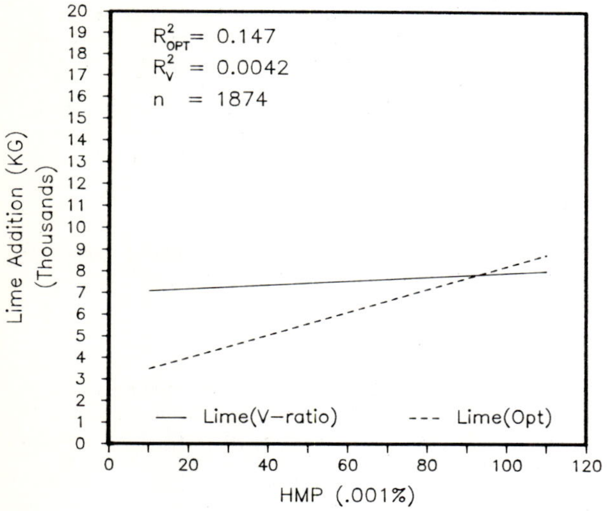

Figure 6: Response of flux models to changes in hot metal phosphorus

Thermodynamic properties of sodium borates and silicates

T YOKOKAWA, K KAWAMURA and T MAEKAWA

The authors are in the Department of Chemistry, Faculty of Science at the Hokkaido University, Sapporo, Japan.

SYNOPSIS

Electrochemical study applied to the acid-base and redox properties of slags is described. Their combination allows discussion on the constitution of slag to some extent. The function of alumina in sodium silicate is briefly discussed.

INTRODUCTION

Redox equilibria in metallurgical slags seem to be well established. Most metallurgical processes are, however quite complicated chemically by the variety of components and their compositions as well as kinetically by departure from the chemical equilibria. The recent process of producing Cr containing steel with oxygen injection is one example of unexpected results. Thus, practice is more important than equilibrium study.

Even so, the redox equilibria of multivalent ions in oxide melts are important for understanding these solutions. Since there is no supporting electrolyte, the thermodynamic activities of any charge bearing species cannot be defined. Consequently, direct correlation of redox with acid-base properties is not so productive as in aqueous electrolyte solutions. Unfortunately we cannot identify any ionic complex species from a thermodynamic experiment.

On the other hand, a wide field, from the glass industry and nuclear waste immobilization to magma, in addition to metallurgy and high temperature corrosion, await quantitative information on redox potential (or oxygen pressure) as functions of basicity of the solvent melts.

BASICITY

Various methods of determining thermodynamic activities are available. Among them the emf method of a concentration cell seems to be the best since the wide activity range from 1 to 10^{-15} or less can be covered easily without sacrificing precision even in the low value range. This low range is essential for measuring basicity since basicity is often defined in terms of activity. Construction of a cell without a liquid junction or of a cell with well defined liquid junction potential is a necessary condition.

In the last 20 years, this method has become quite popular, with the aid of various solid electrolytes. The results seem consistent with other equilibrium measurements whenever such comparison is possible. A limited number of ternary systems were also successfully covered by this method. These data on basicities of solvent melts motivated redox equilibrium studies in view of the fruitful application of Pourbaix diagrams in aqueous solutions.

REDOX POTENTIAL

The redox equilibrium of a multvalent ion pari is established at each temperature and oxygen pressure. The classical way of determining the equilibrium constant is by differentiate chemical analysis of the oxidized and reduced species in a sample after the melt is quenched to the glassy state. This procedure encounters the difficulties of quenching the melt without equilibrium shift and of quantitative analysis of often minor quantities in large amounts of the same element of different valency. As a consequence, it would be very helpful to be able to measure the equilibrium directly in high temperature melts. Linear sweep voltammetry satisfies this demand.

In our laboratory, several redox pairs have been studied [1-4]. The PbO/Pb pair has advantages that the electrode process is well understood and is applicable to a wide range of solvent systems. Three platinum electrodes are used in the voltammetry. A folded wire just in contact with the melt surface is used as the reference, where the following reaction takes place reversibly,

$$(1/2)\ O_2 + 2\ e = O^{2-}. \qquad (1)$$

Another coiled wire immersed in the melt functions as the working electrode. Here the potential responds reversibly to the reaction,

$$Pb^{2+} + 2\ e = Pb. \qquad (2)$$

The platinum crucible functions as the third counter electrode. When the working electrode is brought to a certain negative potential (relative to the reference electrode), the current rises abruptly. Here, Pb^{2+} starts to be reduced to metallic Pb, which reacts simultaneously with the platinum electrode to give a Pb-Pt liquid alloy of the liquidus composition on the electrode surface.

At this potential reactions (1) and (2) are electrochemically in equilibrium and the potential E is given by equation (3):

$$-2EF = \Delta G^o + RT \ln[a(Pb)\ P(O_2)1/2\ /\ a(PbO)], \quad (3)$$

where ΔG^o is the standard free energy of the reaction,

$$PbO \text{ (in melt)} = Pb\text{(in Pt)} + (1/2)\ O_2 \quad (4)$$

and a(i) is the activity of the ith component.

The literature value of ΔG^o and the activity of Pb in Pb - Pt melts (Schwerdtfeger [5]) allow us to determine a(PbO) from E values when $P(O_2) = 1$ as the experimental condition. At the same time this E value corresponds to the equilibrium $P(O_2)$ under the PbO/Pb-Pt system in the open circuit condition:

$$\ln P(O_2) = 4EF/RT. \quad (5)$$

RESULTS AND DISCUSSION

Figure 1 shows the results of potentials measured in our laboratory. Here the vertical axis is the polarographic half wave potential or similar quantity or equilibrium oxygen pressure in the open circuit condition. The horizontal axis is the basicity of the solvent melts defined by the euqation (6):

$$pO = -\log a(Na_2O) \quad (6)$$

Thus pO of $Na_2O\ 2B_2O_3$, $Na_2O\ 2SiO_2$ and $3.5Na_2O\ 1.0Al_2O_3\ 5.5SiO_2$, for example, are 10, 9 and 8.5 respectively (see ref. 6 for more detail). The horizontal line denoted as O_2/O^{2-} is the reference and the straight line denoted as Na^+/Na is the boundary where Na_2O of given activity is stable against decomposition. Similarly, SiO_2 or B_2O_3 have their stable region in their mixture with Na_2O.

The PbO/Pb pair shows the following features. Figure 1 shows the oxygen pressure at which 1 mol per cent PbO is in equilibrium with a Pb-Pt alloy as a function of the basicity. When the solvent is exchanged from Na_2O-B_2O_3 to Na_2O-SiO_2, Na_2O-Al_2O_3-SiO_2, log (PO_2) vs pO is not quite the same in detail. However, the rough coincidence indicates that the basicity of the melts plays the most decisive role in governing the ions' stability and the species of acidic oxide shows a minor effect. Another thing to be noted is that the slope is a little less than that of Na^+/Na. This means PbO is a weaker basic oxide than Na_2O.

As stated in the previous paragraph, it is not possible to specify the ionic formula of the Pb-O_n complex. However, it can be seen qualitatively that Pb^{2+} behaves as shown by:

$$PbO = Pb^{2+} + O^{2-} \quad (7)$$

rather than

$$PbO + (n-1)O^{2-} = PbO_n{}^{2(n-1)-} \quad (8)$$

If it is considered that the activity of O^{2-} is unknown and only that of Na_2O is known, reaction (7) and equation (8) should be replaced by:

$$PbO + 2Na^+ = Pb^{2+} + Na_2O \quad (9)$$

and

$$PbO + (n-1)Na_2O = PbO_n{}^{2(n-1)-} + 2(n-1)Na^+ \quad (10)$$

Again, reaction (9) is better substantiated rather than equation (10); thus the activity of PbO increases with increase of Na_2O. any MO/M system, if the slope is positive, satisfies a relation similar to equation (10) and it means that MO is an acid. Therefore there are three cases differentiated in principle by the slope in Figure 1:

(i) Positive slope;
The MO behaves as an acid in the melts (equation 10).

(ii) Intermediate slope between 0 and that of Na^+/Na;
The MO is basic though less so than Na_2O (reaction 9).

(iii) Steeper slope than Na^+/Na line;
The MO is a stronger base than Na_2O. However, the experimental slope will coincide with that of Na^+/Na, because the MO in minor quantity reacts with Na_2O without any substantial effect.

EFFECT OF ALUMINA

The diffusivity of Pb^{2+} ion can be evaluated from the peak current of the voltammogram. Addition of alumina diminished the diffusivity. In mineralogy it is well known that SiO_2 in the network can be replaced by $NaAlO_2$. If this is true also in melts, the part of Na_2O equivalent to Al_2O_3 content does not contribute to modification of the network. Rough estimation of diffusivity from the voltammogram indicates agreement with this notion. This means the reaction:

$$Na_2O + Al_2O_3 = 2NaAlO_2 \quad (11)$$

proceeds quantitatively. This suggests that Al_2O_3 is stronger or equally strong acid compared with SiO_2. This does not agree with the basicity of the Na_2O-SiO_2-Al_2O_3 melts, where Al_2O_3 does not reduce the activity of Na_2O so much as SiO_2[6].

CONCLUSION

Thermodynamic properties of borate and silicate melts were reviewed in terms both of acid-base character of solvents and of redox equilibria of the solutes. Extension of this method to various species is desired to elucidate the chemical properties of oxide slags and magmas.

REFERENCES

[1] A. Sasahira and T. Yokokawa: Electrochim. Acta 29, 533 (1984).
[2] A. Sasahira and T. Yokokawa: Electrochim. Acta 30, 441 (1985).
[3] K. Kawamura and T. Yokokawa: J. Electrochem. Soc. 135, 1447 (1988).

[4] A. Sasahira, K. Kawamura, M. Shimizu, N. Takada, M. Hongo and T. Yokokawa: J. Electrochem. Soc., to be published.
[5] K. Schwerdtfeger: Trans. Metall. Soc. AIME 236, 23 (1966).
[6] H. Itoh and T. Yokokawa: Trans. JIM 25, 879 (1984).

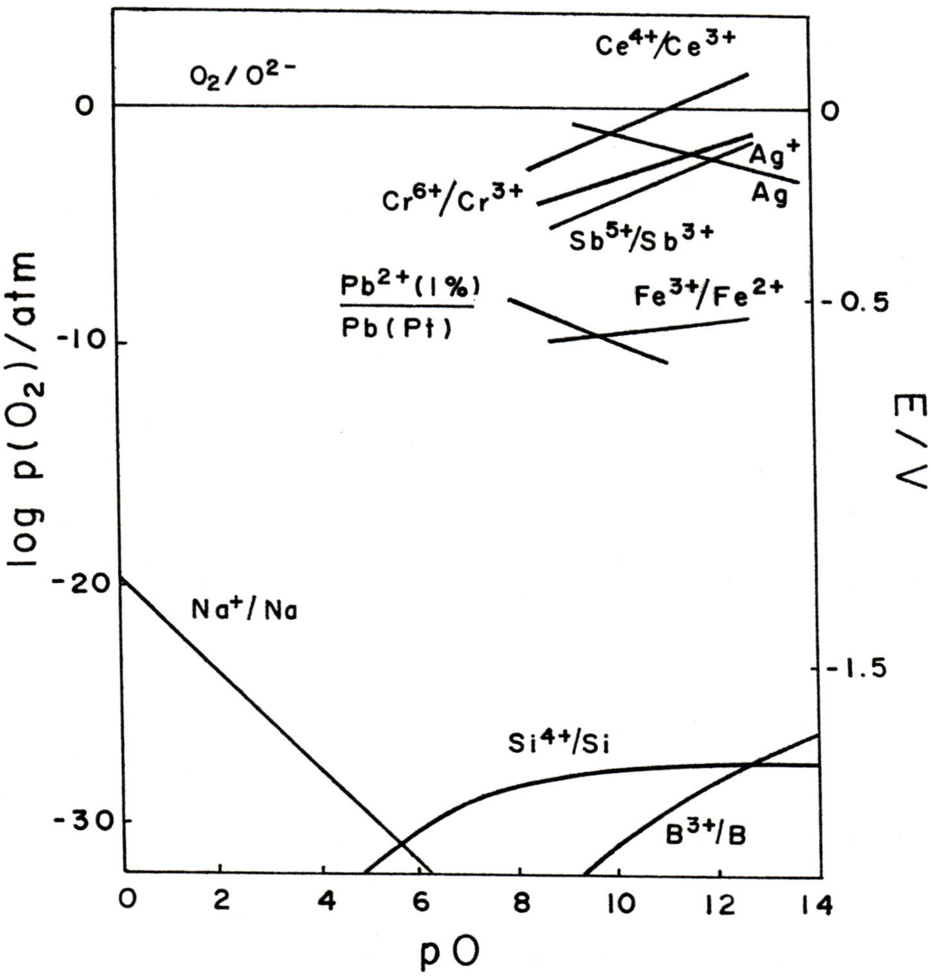

Figure 1. Redox potential or redox equilibrium oxygen pressure as functions of the solvent basicity defined by eq.(6) (ref.[4]).

Transition metal ions and thermochemical aspects of slag basicity

M D INGRAM

The author is in the Department of Chemistry, University of Aberdeen, Scotland.

SYNOPSIS

An outstanding problem in the application of the optical basicity concept to metallurgical slags concerns the assignment of basicity values to transition metal oxides. If Pauling electronegativities are used to generate the corresponding basicity moderating parameters, γ, then the calculated basicities (λ is typically around 0.5) are too low to account for the heats evolved in 'model' acid-base or slag reactions. However, we can account for the 'high' basicities of MnO, NiO, etc. (typically close to unity) by noting the contribution of metallic bonding (i.e. cation-cation interactions) to the stability of these compounds. As a consequence we must also recognise that the effective basicity of transition metal oxides may vary depending on the nature of the chemical reaction under consideration.

INTRODUCTION

The optical basicities of metallurgical slags are usually calculated from the equation[1]

$$\lambda = X_1\lambda_1 + X_2\lambda_2 \qquad [1]$$

where X_1, X_2, etc. are *equivalent* fractions of the component oxides and λ_1, λ_2, etc. are corresponding oxide basicities, where by definition the basicity of CaO is unity. Equation [1] has been used to correlate the chemical properties (basicities) of a wide range of multicomponent slags.[1-5]

It has been known for some years[1,6] that the basicities of individual oxides, which ultimately derive from spectroscopic shifts measured in glass, can be related *via* basicity moderating parameters ($\gamma = \lambda^{-1}$) to the Pauling electronegativities, x, of the corresponding elements by the equation

$$\lambda^{-1} = \gamma = 1.35\ (x - 0.26) \qquad [2]$$

Since many textbooks of inorganic chemistry contain tables of electronegativity, see for example ref. (7), this might seem to provide a straightforward way of evaluating λ-values, and thus a convenient 'short cut' to unravelling the complexities of slag chemistry. The problem which was recognised at the outset[1] was that while λ-values can be evaluated in this way for oxides of the main group elements, there are difficulties in the case of the transition metal oxides. We can see this by considering some simple correlations between optical basicity and the heats of chemical reactions.

OPTICAL BASICITY AND THERMOCHEMISTRY

Consider first the electronegativities and basicity values included in Table (1). Some values are taken directly from ref. (1) which the remainder (i.e. those of the transition metal oxides) are derived from tabulated electronegativity values and equation [2]. If these λ-values really do reflect the *basicity* of the oxides, then we could expect some correlation with the heats evolved in simple acid-base reactions. Fig. (1) shows a plot of ΔH (298 K)[8] versus λ (oxide) for reactions like

$$MC + SO_3 = MSO_4 \qquad [3]$$

calculated per mole of SO_3 reacting. The idea is that SO_3 is a 'standard acid' which will react in similar manner with all the metallic oxides. The smooth curve shows the 'expected behaviour' for the main group oxides, the reaction becoming increasingly exothermic with increasing oxide basicity. On the other hand, the transition metal oxides lie off the line and seem to be more basic than equation [2] implies. Judging from the ΔH values one could estimate a 'typical' basicity (e.g. for NiO) of about 0.74. This is a little lower than that of MgO, but is significantly higher than the value of 0.5 which is indicated by the electronegativity calculation (Table (1)).

The discrepancy between calculated and 'thermochemical' basicities is even greater if we consider reactions of metallurgical interest like

$$S + MO = \tfrac{1}{2}O_2 + MS \qquad [4]$$

where again values of ΔH are standard values at 289K, and refer to the reaction of one 'mole' of S. For the oxides of main group elements this reaction becomes progressively *less* endothermic as the basicity of the oxide increases, and the points lie quite close to a straight line. However, if we adjust the basicity of the transition metal oxides to bring them into line, then we would need to assign λ-values in the range 0.95 - 1.10, which would imply, for example, that NiO is as basic as CaO. In the chemical context this result is entirely unexpected. However, it is noteworthy that similar high values for the basicity of FeO have been used to account for the sulphide capacities of certain steel making slags,[2,5] and are also indicated by an evaluation of optical refractivity data[9] in glasses and crystals which is presently in progress in our laboratory.

CONCLUSIONS

Probably the most important conclusion to be drawn from this discussion is that single, i.e. *invariant* λ-values cannot be assigned to the transition metal oxides. This statement implies that the properties of these oxides cannot be fully expressed in terms of their 'lime characters' - see ref. (1) - and that the chemistry of the corresponding slags is not solely a function of the 'state of bonding' of the constituent O^{2-} ions. This would appear to be the case for the oxides of main group elements, which is the underlying principle of optical basicity.

Recently, Duffy[10] has indicated the importance of *metallic* bonding in transition metal oxides[11] where stability is conferred by some kind of interaction between neighbouring cations. On this basis, we could speculate that the basicity of transition metal oxides remains high only in reactions in which this cation-cation interaction is *not* disturbed. Transition metal oxides will become less reactive (i.e. their basicity will decrease) if the cations need to become separated as a result of the chemical reaction. It seems that reaction [4], which broadly speaking corresponds to sulphur uptake in the slag, comes in the former category, while acid-base neutralisations like reaction [3], and also the related depolymerisation processes which occur in silicate slags[11] and which are discussed elsewhere in this conference,[12] are to be included in the latter category. Thus *for the same oxide* different basicities may be required depending on the process under consideration.

This paper has been concerned with laying down some ground rules which should be relevant to the discussion of basicity in steel making slags. The problem is almost certainly more complicated than has been previously supposed, in so far as the idea of 'basicity' has to be modified to take into account the presence of the transition metal ions. The nature of slag reactions will have to be considered in some detail even if new experimental methods are developed so as to provide a direct determination of optical basicities in these technologically important systems.[13] In a sense this is a justification for the largely empirical approach pioneered by Sommerville and coworkers.[2,12]

REFERENCES

1. J.A. Duffy, M.D. Ingram and I.D. Sommerville, J. Chem. Soc. Farad. Trans. 1 (1978) 1410.
2. D.J. Sosinsky and J.D. Sommerville, Met. Trans. B, 17 (1986) 331.
3. D.R. Gaskell, Trans. Iron & Steel Inst. Japan, 22 (1982) 997.
4. H. Suito and R. Inoue, Trans. Iron & Steel Inst. Japan, 24 (1984) 47.
5. T. Nakamura, Y. Ueda and J.M. Toguri, J. Japan. Inst. Metals, 50 (1986) 456.
6. J.A. Duffy and M.D. Ingram, J. Non-Cryst. Solids, 21 (1976) 373.
7. W.W. Porterfield, "Inorganic Chemistry, a Unified Approach", Addison-Wesley, Reading, Massachusetts (1984).
8. F.D. Rossini, D.D. Wagman, S. Levine and I. Jaffe, "Selected Values of Chemical Thermodynamic Properties", Circular 500 of the National Bureau of Standards, U.S. Govt. Printing Office, Washington, D.C. (1952).
9. J.A. Duffy and M.D. Ingram (unpublished work).
10. J.A. Duffy, Solid State Chemistry, 62 (1986) 145.
11. M.R. Masson, I.B. Smith and S.G. Whiteway, Can. Journ. Chem., 48 (1970) 1456.
12. D.J. Sosinsky, I.D. Sommerville and C.R. Masson (these proceedings).
13. J.A. Duffy (these proceedings).

Table 1

Calculated basicities of some metal oxides

Main group elements		Transition elements[‡]	
Oxide	Λ-value*	Oxide	Λ-value*
K_2O	1.4	MnO	0.59
Na_2O	1.15	NiO	0.475
Li_2O	1.00	CoO	0.475
BaO	1.15	CdO	0.51
CaO	1.00	ZnO	0.55
MgO	0.78	CuO	0.43
Al_2O_3	0.605	Tl_2O	0.48
BeO	0.605	PbO	0.48

‡ Some post-transition elements are included

* Calculated from equation [2], see text.

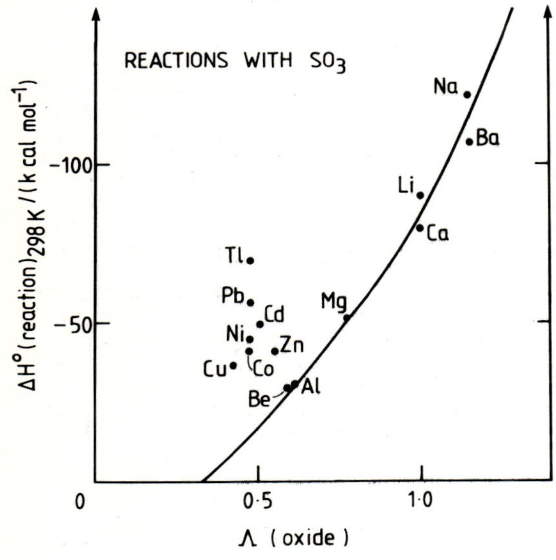

Fig. 1 Variation in the standard enthalpy change, ΔH^o, for the reaction $MO + SO_3 = MSO_4$, or its equivalent, with the basicity of the metal oxide, λ, as calculated using the simple electronegativity formula (see text).

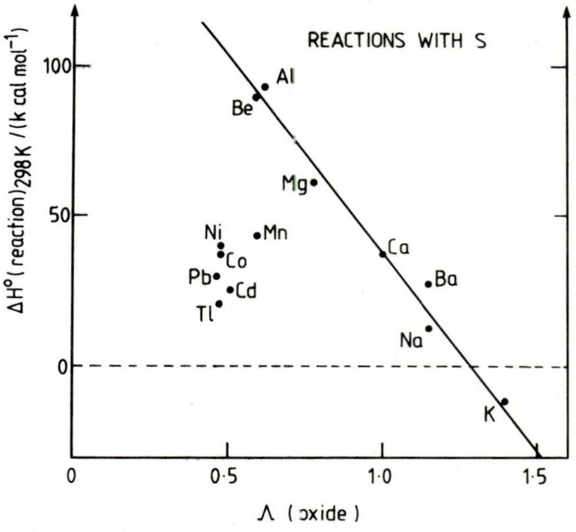

Fig. 2 Variation in the standard enthalpy change, ΔH^o, for the reaction $S + MO = MS + \frac{1}{2}O_2$, or its equivalent, with the basicity of the metal oxide as calculated in Fig. 1.

Raman spectroscopic study on structure of binary silicates and ternary oxide melts with and without fluoride

Y IGUCHI, K YONEZAWA, Y FUNAOKA, S BAN-YA and Y NISHINA

Profs Ban-Ya and Iguchi are in the Department of Metallurgy, Tohoku University; Messrs Yonezawa and Funaoka, formerly Graduate Students at Tohoku University, are now with Nippon Steel Corp. and Kobe Steel Ltd., respectively; Prof Nishina is in the Institute for Materials Research, Tohoku University.

SYNOPSIS

A Raman spectral study has been carried out to learn the structure of glassy binary alkali or alkaline metal silicates, ternary silicates with Al_2O_3 or TiO_2, and CaO-SiO_2 or CaO-Al_2O_3 system with CaF_2. Raman spectra were deconvoluted with Gaussian function, and the fraction of each silicate anion and the shift of wave number of Raman band were investigated relating to the number of NBO/Si. From these results, the behavior of cations and CaF_2 in various melts were discussed.

INTRODUCTION

Slag formation and slag-metal reactions in pyrometallurgy are closely related to the structure of complex silicate anions in molten slags. Many structural analysis of liquid slags and glasses have been carried out by spectroscopic and diffraction methods.[1~15] Our previous papers have reported the results of the Raman spectroscopic study on the structure of binary and ternary silicates consisting of alkali metal, alkaline earth metal, iron and manganese oxides in both the glassy and molten states.[16,17]

In this study, Raman spectra are deconvoluted, and the fraction of each silicate anion is evaluated for glassy binary silicates (Li_2O,Na_2O, K_2O,Rb_2O,Cs_2O,MgO,CaO,SrO,BaO-SiO_2), ternary CaO-SiO_2-Al_2O_3 or -TiO_2, and CaO-SiO_2 or CaO-Al_2O_3 containing CaF_2. The effects of cations on the silicate structures, bonding behavior of Al^{3+} or Ti^{4+} ion in CaO-SiO_2 system, and the role of CaF_2 in CaO-SiO_2 or CaO-Al_2O_3 system are discussed.

EXPERIMENTAL

Glassy samples for Raman spectroscopy were prepared by melting and quenching the mixture from silica powder, and alkali or alkaline earth metal carbonate, oxide(Al_2O_3,TiO_2) or fluoride(CaF_2) in our laboratory. Samples containing CaF_2 were melted in a closed platinum crucible to prevent fluoride gas forming and composition change.

Raman spectral measurements were carried out with a double monochromator at right angles to the incident beam of an argon ion laser of 4880 Å for the colorless and transparent samples, while the back scattering method was applied for the colored samples. Deconvolution of the observed spectra was done by using Gaussian function with reference to the previous results, including ours. Not only the intensity and half width of Raman band but also mass balance and ionic neutrality were considered for estimation of the fraction of silicate anions with the respective number of non-bridging oxygens per each silicon atom (NBO/Si).

At first, deconvolution of Raman spectra was applied to the simple binary alkali and alkaline earth silicates as shown in Fig. 1. As a result, it was confirmed that this method was valid for fairly quantitative discussion of the silicate structure in the range of 800~ 1200 cm^{-1} assigned to the stretching vibration of the Si-O bonding.

EXPERIMENTAL RESULTS AND DISCUSSIONS

Alkali metal and alkaline metal silicates

The relationships between the concentration of basic oxides and intensity or wavenumber shift of Raman bands with each number of NBO/Si were obtained by deconvolution of the Raman spectra. Figure 2 shows the results for CaO-BaO system as an example. From this figure, it was clear that basic oxides broke preferentially three dimensional silica network until the composition of disilicate. At highly basic composition than disilicate, the basic oxide started to modify the oxygen bridges in the silicate anions with NBO/Si=1. In disilicates, The Raman band due to NBO/Si=1 shifted to lower wavenumber with increasing ion-oxygen attraction of added basic oxides, while the Raman band with NBO/Si=2 shifted to lower wavenumber with increasing cation radia as shown in Fig. 3. From these figures, it could be considered that cations with higher coulombic force weakened more the Si-O bonding with NBO/Si=1 and bigger cations made weak the Si-O bonding with NBO/Si=2.

CaO-SiO_2-Al_2O_3 system

In SiO_2-Al_2O_3 and CaO-Al_2O_3 binary systems in the glassy or crystallized state, the Raman band at about 500 cm^{-1}, or 540 and 760 cm^{-1} was assigned to Al^{3+} in sixfold coordination with oxygen, or in fourfold coordination, respectively.

By the addition of Al_2O_3 to the acid CaO-SiO_2 system, Raman bands in the range between 800 and 1200 cm^{-1} shifted toward lower wave numbers. And the intensity of Raman band due to silicate with more NBO/Si increases with increasing Al_2O_3 contents. However, the addition of Al_2O_3 to the basic $CaO-SiO_2$ system showed the opposite effect to the silicates. The Raman band at about 500 cm^{-1} was observed in the acid $CaO-SiO_2-Al_2O_3$ but not in the basic system. From these results, the following consideration was given. When Al_2O_3 was added to the $CaO-SiO_2$ system with the basicity(N_{CaO}/N_{SiO_2}) below unity, the oxygen bridges in the silicate anions were broken. In this composition range, Al^{3+} was mostly in sixfold coordination with oxygen. On the other hand, in the higher basic slag with the basicity > 1, Al^{3+} became coordinated fourfold as a network former.

$CaO-SiO_2-TiO_2$ system

Raman spectra of $CaO-TiO_2$ and SiO_2-TiO_2 binary systems were investigated and Raman bands at 700 850 cm^{-1} and at 650 cm^{-1} were attributed to TiO_4 tetrahedral and TiO_6 octahedral structures, respectively.

In the ternary system, Raman bands in the range of 800 1200 cm^{-1} did not shift independently of basicity, so it was considered that Ti ion did not replace Si ion in silicates. This behavior was different from that of aluminum ion. The intensity of Raman bands due to silicate with more number of NBO/Si increased at higher acid composition but its effect was little. The behavior of Ti^{4+} in $CaO-SiO_2-TiO_2$ system might be concluded as follows. Titanium cations(Ti^{4+}), which were added to the $CaO-SiO_2$ system, were both in fourfold and sixfold coordination. Sixfold ratio increased with decreasing basicity, that is, induced oxygen ion(O^{2-}) modified complex silicate anions. In the basic composition with the basicity > 1, TiO_2 behaved as a diluent and did not affect the silicate anion structure.

$CaO-SiO_2-CaF_2$ and $CaO-Al_2O_3-CaF_2$ systems

By the addition of CaF_2 to the basic $CaO-SiO_2$ system, little shift and little change of relative intensity of Raman bands assigned to each number of NBO/Si were observed. So, CaF_2 in the basic $CaO-SiO_2$ slags behaved as a diluent and did not have any effect on the structure of silicates, in which the basicity was higher than unity. On the other hand, the addition of CaF_2 to the acid $CaO-SiO_2$ system caused increase in the intensity of Raman band due to silicates with higher number of NBO/Si. That is, a part of CaF_2 broke the silicate network in the acid slag. The fraction of CaF_2 which acted as a modifier increased with decreasing the basicity, and three dimensional network was broken to chains. But the degree of modification was much lower compared with typical basic oxides. A Raman band at about 700 cm^{-1} was observed, the intensity of which increased with increasing CaF_2 content and decreasing basicity. This Raman band might be assigned to Si-F bonding. Previous investigators were not in agreement on the existance of Si-F bonding in silicate melts.[2,18,19] However, in this study Si-F bonding was confirmed for the $CaO-SiO_2-CaF_2$ system with the basicity < 1, which was melted in the closed crucible. Addition of CaF_2 to the binary $CaO-Al_2O_3$ slags had no effects on the aluminate structures. CaF_2 behaved only as a diluent to the $CaO-Al_2O_3$ system.

CONCLUSION

1) In alkali metal and alkaline metal silicates, cations with higher coulombic force weakened the Si-O bonding with NBO/Si=1 and bigger cation made weak the Si-O bonding with NBO/Si=2.

2) Al_2O_3 in the $CaO-SiO_2$ system with N_{CaO}/N_{SiO_2} <1 broke the oxygen bridges in the silicate anions. In this composition range, Al^{3+} was mostly in sixfold coordination with oxygen, while in the higher basic slag with the basicity > 1, Al^{3+} went into fourfold as a network former.

3) Ti^{4+} in the $CaO-SiO_2$ system, was both in fourfold and sixfold coordination. Sixfold ratio increased with decreasing the basicity. In the basic composition with the basicity > 1, TiO_2 behaved as a diluent.

4) The existence of Si-F bonding was confirmed for the $CaO-SiO_2-CaF_2$ system with N_{CaO}/N_{SiO_2} < 1, which was melted in the closed crucible.

ACKNOWLEDGMENTS

The authors are grateful to Drs. N. Kuroda and Y. Sasaki(Institute for Materials Research, Tohoku University), and Messrs. Y. Sasaki, S. Nakahara, and H.Nagashima. A part of this work was supported by the Ministry of Education, Science and Culture of Japan under a Grant-in-Aid for Scientific Research No.60550467(1985~1986).

REFERENCES

1) D.Kumar,R.G.Ward and D.J.Williams:Trans. Faraday Soc.,61(1966),p.1850
2) T.Ito,T.Yanagase,Y.Suginohara and N.Miyazaki: J.Japan Inst.Metals,31(1967),p.290
3) P.Tarte:Spectrochem.Acta,23A(1967),p.2127
4) J.Etchepare:Spectrochem.Acta,26A(1970),p.2147
5) M.Hass:J.Phys.Chem.Solids,31(1970),p.415
6) I.B.Smith and C.R.Masson:Can.J.Chem.,49(1971), p.683
7) S.A.Brawer and W.B.White:J.Chem.Phys., 63(1975),p.2421
8) N.Iwamoto,Y.Tsunawaki,M.Fuji and T.Hattori: J.Non-Cryst.Solids,18(1975),p.2421
9) W.L.Konijendijk and J.M.Stevels:J.Non-Cryst. Solids,21(1976),p.447
10)Y.Waseda,H.Suito and Y.Shiraishi:J.Japan Inst. Metals,41(1977),p.1068
11)H.Nakamura,K.Morinaga and T.Yanagase:J.Japan Inst.Metals,41(1977),p.1300
12)S.K.Sharma,D.Virgo and B.O.Mysen:Canegie Inst. Year Book,77(1977/1978),p.649
13)Y.Waseda and J.M.Toguri:Metall.Trans.,9B(1978), p.595
14)B.O.Mysen,F.J.Ryerson and D.Virgo:Amer. Mineralogist,65(1980),p.1150
15)K.Kusabiraki and Y.Shiraishi:J.Japan Inst. Metals,45(1981),p.250,p.259,p.888
16)S.Kashio,Y.Iguchi,T.Goto,Y.Nishina and T.Fuwa: Trans.ISIJ,20(1980),p.251
17)Y.Iguchi,M.Wako,S.Ban-ya,Y.Nishina and T.Fuwa Metallurgical Slags and Fluxes,TMS-AIME,1984, p.975
18)T.Baak and A.Olander:Acta Chem.Scan.,9(1955)8, p.1350
19)P.Dumas,J.Corset,W.Carvalho,Y.Levy and Y.Neuman:J.Non-Cryst.Solids,47(1982),p.239

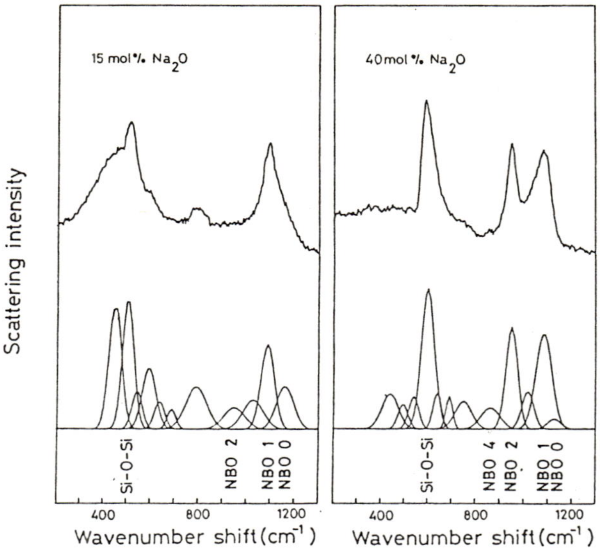

Fig. 1 Deconvolution of Raman spectra in Na_2O-SiO_2 system.
The upper line is the Raman spectrum treated with Bose factor to eliminate the Rayleigh effect.
The lower lines are deconvoluted Raman bands.

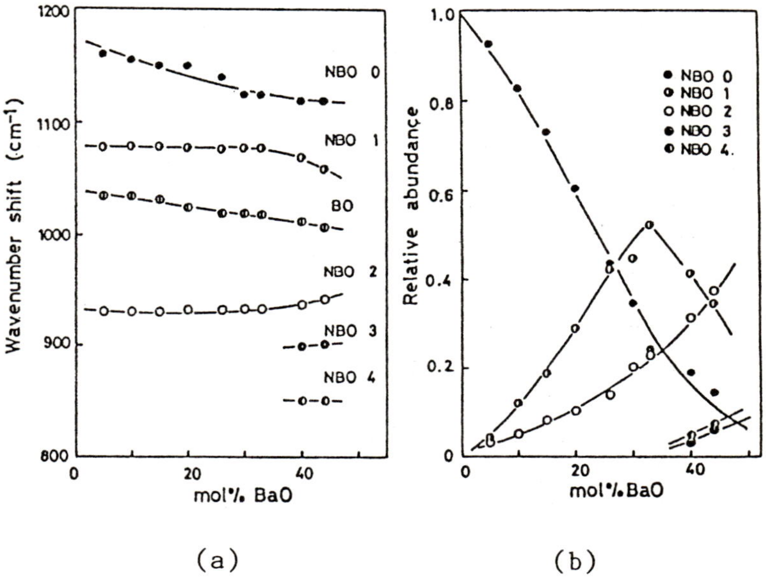

Fig. 2 Wavenumber shift (a) and relative abundance (b) of Raman bands assigned to silicate anions with each number of NBO/Si in the BaO-SiO_2 system.

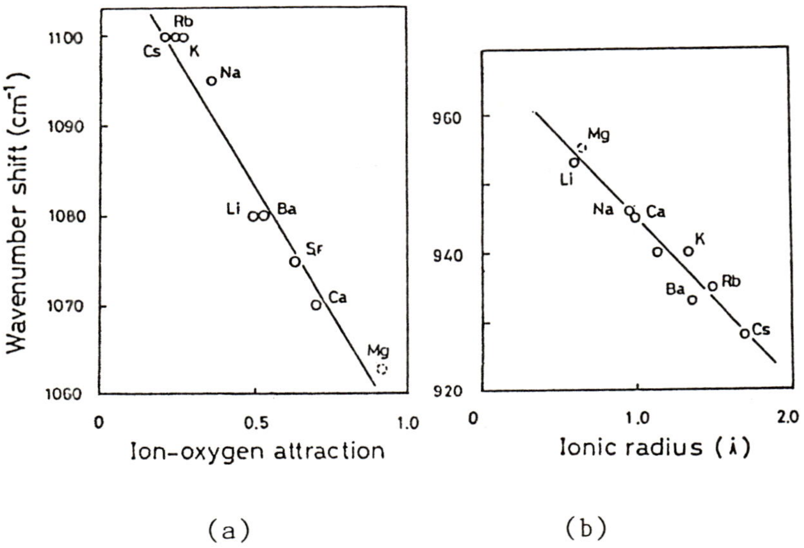

Fig. 3 Effect of cations on the silicate structure in disilicate.
(a) Wavenumber shift of Raman band due to the silicate anion with NBO/Si = 1 as a function of ion-oxygen attraction.
(b) Wavenumber shift of Raman band due to the silicate anion with NBO/Si = 2 as a function of inonic radius.

□

Structural analysis of ferrite slags by X-ray diffraction

K SUGIYAMA, IN-KOOK SUH and Y WASEDA

The authors are in the Research Institute of Mineral Dressing and Metallurgy (SENKEN), Tohoku University, Japan.

SYNOPSIS

The structure of molten $CaO-Fe_2O_3$ slag has been determined by a high temperature X-ray diffraction method. Iron atoms occupy a tetrahedral site of oxygens rather than an octahedral site in this melt with the higher CaO concentration region. The present structural information is consistent with the viscosity data of molten ferrite slags.

INTRODUCTION

Ferrite slags such as the $Na_2O-Fe_2O_3$ or $CaO-Fe_2O_3$ systems, are recently receiving much attention mainly due to the potential capabilities to dissolve larger amount of Fe_2O_3 or FeO compared to silicate slags and to remove inherent inconvenient elements, for example arsenic and phosphorus, effectively [1]. Also,the viscosity of these ferrite slags show much lower values. Because of these properties, ferrite slags can eliminate the troubles caused by solid magnetite precipitation and are favorable to the oxidation smelting process of copper or nickel sulfide ores. Numerous measurements of density, viscosity or electrical resistivity of these molten slags have been reported. The structural features, however, have not yet been revealed. The main purpose of this work is to present the results on the direct structural analysis for ferrite slags by means of a high temperature X-ray diffraction.

EXPERIMENTAL PROCEDURE

The CaO content was varied in the range of 33-60mol% and the intensity data were measured at temperatures about 100°C above the liquidus temperature under both argon and air atmosphere using Mo Kα radiation coupled with the diffracted beam monochromator. Detailed experimental arrangements and procedure of data analysis were almost identical to those employed previously for molten slag systems [2,3]. For convenience of discussion, the essential features of data analysis are given below.

A reduced interference function i(Q) is expressed in the following form using the coherent X-ray scattering intensity $I_{eu}(Q)$ obtained directly from the experimental data by the generalized Krogh-Moe-Norman method.

$$i(Q) = [I_{eu}(Q) - \sum_{uc} f_j^2]/f_e^2 \qquad (1)$$

where f_i is the atomic form factor of species i, f_e is the average scattering factor per electron, $Q=4\pi\sin\theta/\lambda$, 2θ is the scattering angle and λ is the wavelength. The interference function i(Q) is structurally sensitive part of the total scattering intensity, and related to the radial density function $\rho_j(r)$, which is the electron distribution around j atom at a distance of r, as follows:

$$Qi(Q) = 4\pi\int_0^\infty \sum_{uc} K_j r[\rho_j(r)-\rho_e]\sin Qr\, dr \qquad (2)$$

$$f_j = K_j f_e \qquad (3)$$

where, ρ_e is the electron density per cubic Ångstrom and K_j is the effective electron number of species j defined by equation (3), which is approximately equal to the atomic number Z_j. Using equations (1) and (2) , the following well-known relation is given:

$$\sum_{uc} K_j 4\pi r^2 \rho_j(r) = 4\pi r^2 \rho_e \sum_{uc} K_j + \frac{2r}{\pi}\int_0^\infty Qi(Q)\sin Qr\, dQ \qquad (4)$$

The interatomic distance r_{ij} and its coordination number N_{ij} of i-j pairs can be estimated by comparing a set of pair functions $P_{ij}(r)$ in equations (5) and (6) using least-squares analysis. The right hand side of the equation (5) corresponds to the experimental pair function, while the theoretical pair function is provided by equation (6).

$$\sum_{uc}\sum_{i}\frac{N_{ij}}{r_{ij}}P_{ij}(r) = 2\pi^2 r\rho_e \sum_{uc} Z_j + \int_0^{Qmax} Qi(Q)\sin Qr\, dQ \quad (5)$$

$$P_{ij}(r) = \int_0^{Qmax} \frac{f_i f_j}{f_e f_e} e^{-\alpha^2 Q^2} \times \sin Qr_{ij} \sin Qr\, dQ \quad (6)$$

where α^2 is a convergence coefficient.

RESULTS AND DISCUSSION

The correlation function (Qi(Q)) and the radial distribution function by applying the Fourier transformation described in the previous section, are plotted in Figures 1 and 2, respectively. Interatomic distances and coordination numbers estimated by the pair function analysis for the respective peak in the radial distribution function are summarized in Table 1.
Several well resolved peaks are evident in the correlation function of Fig.1. It may be worth mentioning that the second peak in the region between $Q=30nm^{-1}$ and $Q=55nm^{-1}$ mainly corresponds to the first neighbouring Fe-O pairs and this second peak becomes broader as the CaO content increases. A small peak around $Q=80nm^{-1}$ is also well correlated with increasing CaO content.
In the radial distribution function of Fig.2, the shift of the first peak corresponding to Fe-O pairs toward low r region is clearly detected and its coordination number gradually decreases with increase in the CaO content. They contrast to the unchanged behaviour of the Ca-O pairs in this system. The variation of the Fe-O pairs is attributed to the preference of a tetrahedral site of oxygens rather than an octahedral site and then the formation of the local ordering such as FeO_4^{5-} or $Fe_2O_5^{4-}$ is suggested in the melt structure with the higher CaO concentration region. Present structural information is consistent with viscosity data of molten ferrite slags including the CaO-Fe_2O_3 system [4]. For example the viscosity increase caused by the addition of CaO is possibly interpreted by the formation of the local ordering such as ferric anion complexes. However, the quantitative information about complexes is not finally identified yet from the present diffraction experiments. It should be noted that the interatomic distance and its coordination number of Fe-O pairs indicate that part of iron atom is still surrounded by the six oxygen atoms in the CaO rich region of the CaO-Fe_2O_3 system.

CONCLUSION

Present X-ray structural information provides the following fundamental features about the environment around an iron atom in molten CaO-Fe_2O_3. The variation of the interatomic distance of Fe-O pairs indicates a change in the position of iron atom from an octahedral site to a tetrahedral site of oxygens, and thus the formation of local ordering unit such as FeO_4^{5-} or $Fe_2O_5^{4-}$ is feasible in the higher CaO concentration region of this melt. This characteristic structural behaviour is consistent with the viscosity data where the viscosity increase is detected in the higher CaO concentration region. However, further systematic investigation is required to identify the local ordering unit in molten ferrite slags.

ACKNOWLEDGEMENTS

The financial support from Nippon Steel Corporation, is greatly appreciated.

REFERENCES

1. A.Yazawa, Erzmetall **30** 511 (1977).
2. Y.Waseda, Can. Met. Quart. **20** 57 (1981).
3. B.E.Warren, Addson-Wesley, Reading (1969)
4. S.Sumita, T.Mimori, K.Morinaga and T.Yanagase, J. Jpn. Inst. Metals **44** 94 (1980).

Table 1 : Interatomic distance r and coordination number N of Fe-O and Ca-O pairs in molten CaO-Fe_2O_3 slags.

CaO	(mol%)	33	40	50	60
Fe-O	r(nm)	0.203	0.202	0.197	0.196
	N	4.3	3.9	3.6	3.5
Ca-O	r(nm)	0.237	0.238	0.238	0.237
	N	6.4	5.4	6.4	5.5

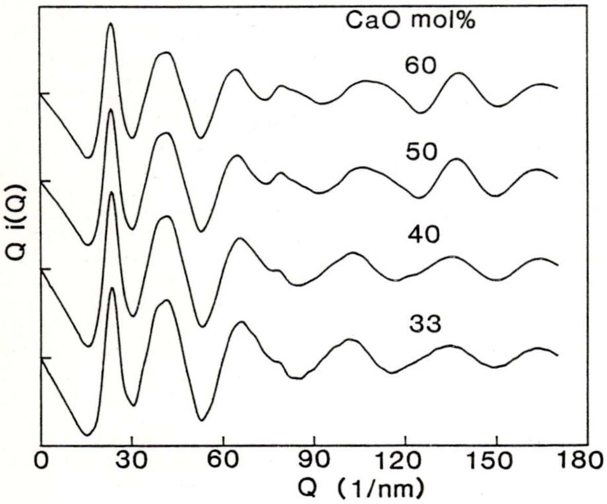

Figure 1 Correlation functions of molten CaO-Fe_2O_3.

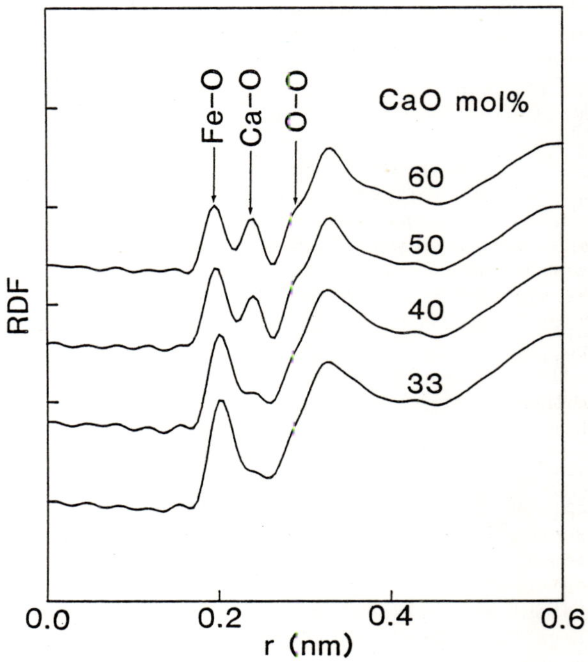

Figure 2 Radial distribution functions of molten CaO-Fe_2O_3.

Spectroscopic investigation of Cr (II), Cr (III) and Cr (VI) in silicate slags

H F MARSTON and B B ARGENT

Mr Marston is at the Swinden Laboratories of British Steel plc, Moorgate, Rotherham; Professor Argent is in the Division of Metals of the School of Materials, University of Sheffield, where the work was carried out.

SYNOPSIS

Transition metal ions in slags respond to the changing oxidation conditions (for example in steelmaking) by changing their valency and structural rôle. The behaviour of Cr in $CaO-SiO_2$ slags has been studied through chemical and spectroscopic analysis. In slags equilibrated at $p_{O_2}=1{\cdot}6.10^{-9}$bar, 71% of the Cr was present as Cr(II), reducing to an average figure of 14% at $p_{O_2}=1{\cdot}1.10^{-5}$bar. At $p_{O_2}=0{\cdot}1$ bar, 14% was Cr(VI). In each case, the balance was Cr(III). Spectroscopy provided evidence of Cr(III) in all slags. Varying quantities of Cr(II) and Cr(VI) were detected, but not simultaneously in the same sample. Within the compositional range studied, no consistent effect of the chromium level, the slag basicity or the presence of MgO or Al_2O_3 on either the chromium distribution or the molar absorptivity could be discerned.

INTRODUCTION

During the manufacture of stainless steel, the slag is initially subjected to moderately oxidising conditions in the electric arc furnace. Conditions are more strongly oxidising in a secondary steelmaking unit during carbon removal, but they become strongly reducing in the final reduction stage.

Transition metal ions can respond to such changes in the oxidation conditions by changing their valency. They may also change their structural rôle, for example by acting as a network modifier under reducing conditions, but as a network former when the slag is oxidised. Spectroscopic analysis can provide data about the electronic and physical configuration, and hence the structural rôle, of transition metal ions in slags. Such studies can also provide information about the slag matrix. In the present work, spectroscopic and chemical analyses have been used to investigate the behaviour of chromium in four vitreous lime-silica slags produced under reducing and oxidising conditions.

EXPERIMENTAL PROCEDURE

Lime-silica slags with 10 mol % magnesia and/or alumina were produced with chromium levels of 0·05 to 2%. After equilibration at 1850K in a Pt-Rh crucible, a portion of the molten slag was cast onto a copper plate. The beads of slag were stress-relieved by furnace cooling from 1050K.

The slag matrix compositions and designations used in the text are given in Table 1. This table also gives typical compositions of some of the slags encountered in stainless steelmaking, ignoring other species. The experimental compositions were selected to demonstrate the effects of changing basicity (acid CAS v. basic CAS), adding magnesia (Basic CAS v. CMAS) and removing alumina (CMAS v. CMS). However, there was also the practical requirement of the melt to be sufficiently fluid to be capable of being cast to form a vitreous sample. The acid CAS slag was found to be the most viscous (most acidic composition) that could be prepared. The compositions of the remaining slags were adjusted to have the highest CaO/SiO_2 ratio that would nevertheless allow the cast slag sample to cool without devitrification. The highest chromium contents that were used experimentally were comparable with those given for the arc furnace and final AOD slags in Table 1. These two slags were only marginally more basic than the three most basic experimental compositions.

The atmospheres used for equilibration and the designations used in the text are given in Table 2. Air was avoided to preclude interaction between chromium and nitrogen.

Spectroscopic samples 0·3 to 3 mm thick were prepared by grinding the slag beads. Transmission spectra were determined in the range 200-2500 nm using a Beckmann spectrophotometer. The intense absorption of Cr(VI) prevented transmission measurements at some ultra-violet peaks, even with the 0·3 mm samples. By using the spectrophotometer in reflection mode with powdered slag samples, the wavelengths of these peaks could be determined, although the absorption could not be quantified. For comparison, the spectra of aqueous solutions containing Cr(II), Cr(III) and Cr(VI) were also determined.

Portions of the cast samples were analysed for total chromium. For simplicity, the chromium level has been expressed in terms of elemental Cr rather than as an arbitrary oxide. Where the chromium level was above 0·3%, the Redox species (Cr(II) or Cr(VI), as appropriate) was determined by dissolving the slag under controlled conditions prior to potentiometric titration[1]. This procedure was insensitive at lower chromium levels.

RESULTS AND DISCUSSION

Spectra of Aqueous Solutions: The lowest line in Figure 1 shows the spectrum of a solution containing Cr(II), prepared by reducing acidic chromic chloride with amalgamated zinc. The peak of the single broad band is at about 710 nm. The band is asymmetric, with a long wavelength tail. Spectra obtained during progressive atmospheric oxidation of the Cr(II) in this solution to Cr(III) are also shown in Figure 1 (for clarity, the successive spectra in Figures 1 to 5 are displaced vertically). The single broad Cr(II) band is progressively replaced by two narrower but more intense Cr(III) bands with peaks at about 410 and 585 nm.

The spectrum of an acidic potassium dichromate solution (Figure 2) comprises a very intense band with peak at 350 nm, and a shoulder at 255 nm, on the ultra-violet cut-off edge. The absorption is probably[3,4] due to $HCrO_4^-$ rather than $Cr_2O_7^{2-}$. In alkaline solution, chromate ions[3,4] (CrO_4^{2-}) produce two distinct bands with one peak at 355 nm and another peak (rather than a shoulder) at 260 nm. The wavelengths of the absorption peaks and their intensities are given in Table 3.

Cr(VI) produces intense absorption bands because the electronic transitions involved are "permitted". These "charge-transfer" bands correspond to transfer of an electron from an orbital predominantly located on the metal ion to a predominantly ligand orbital[2]. In contrast, the observed Cr(II) and Cr(III) bands correspond to "forbidden" transitions for electron transfer from one metal ion orbital to another. The band is observed because of distortion of the (electronic) environment. Consequently, the absorption of a band gives an indication of the degree of distortion around the ion. The peak wavelength is determined by the change of energy of the electron as it undertakes the transition.

Spectra of Slags: Examples of the observed spectra of slags are shown in Figures 3 to 5. The absorptivity has been calculated relative to the total chromium content, as the absorbing species varies with wavelength. In Table 3, which summarises the spectroscopic results, the molar absorptivity is quoted relative to the (assumed) absorbing species.

The slags held at $p_{O_2}=1\cdot1.10^{-5}$ bar show two clearly resolved absorption bands with peaks at 440 and 635 nm (Figure 3). These bands, corresponding to Cr(III), have a longer wavelength and higher absorptivity than those observed in the aqueous solution. However, the shorter wavelength (higher energy) of the second peak compared with a sodium silicate glass[3] (Table 3) indicates that the presence of Ca^{2+} provides a stronger Cr(III)-O ligand field than does the Na^+ ion. Some slags produced at $p_{O_2}=1\cdot1.10^{-5}$ gave an indication of either Cr(II) or Cr(VI); none of the slags showed clear evidence of both Cr(II) and Cr(VI).

The slags equilibrated at $p_{O_2}=1\cdot6.10^{-9}$ bar characteristically show a broad band with a peak in the range 600 to 630 nm (Figure 4). The presence of some Cr(III) is indicated by a subsidiary peak or shoulder at about 440 nm. The 635 nm band observed at $p_{O_2}=1\cdot1.10^{-5}$ is masked by a broad Cr(II) band with a peak at about 605 nm, a shorter wavelength than in aqueous solution. The data for the Cr(II) band in Table 3 were determined by subtracting the calculated contribution of Cr(III) (based on observations at $p_{O_2}=1\cdot1.10^{-5}$) from the spectrum observed at $p_{O_2}=1\cdot6.10^{-9}$.

Both Cr(II) and Cr(III) are in octahedral co-ordination, in which the metal ion is at the centre of an octahedron with an oxygen ion at each of the six vertices. The oxygen ions are a combination of free oxygen ions (O^{2-}), non-bridging oxygen ions (at the corners of silica tetrahedra, ≡Si-O⁻) or bridging oxygen ions, ≡Si-O-Si≡. In the case of Cr(II), the octahedron is distorted by the Jahn-Teller effect[2]; the consequential asymmetry of the absorption band can be detected in the spectra of the slags as well as the aqueous solution.

An intense absorption band with a peak at 370 nm appeared under oxidising conditions ($p_{O_2}=0\cdot1$ bar, Figure 5, upper lines). Reflection spectroscopy detected a second Cr(VI) band as a shoulder at about 260 nm (Figure 5, lowest line). The much weaker bands corresponding to Cr(III) can be discerned as a shoulder at 440 nm and a minor peak at 635 nm, although 86% of the Cr was present as Cr(III).

In $3Na_2O.7SiO_2$ glasses, the Cr(VI) bands (peaks at 260 and 370 nm) have been ascribed to chromate (CrO_4^{2-}) ions[3,4], because of the similarity of shape of the bands, notably the relative heights of the peaks at 260 and 370 nm compared with those observed in aqueous solution (Table 3). In the present work, the intensity of the 260 nm band could not be determined, as required for unambiguous identification of the nature of the Cr(VI) complex.

The Cr(VI) complexes cited above are tetrahedral anionic complexes, except that $Cr_2O_7^{2-}$ contains a double tetrahedron. As in silica, the central ion has four neighbouring oxygen ions at the vertices of a regular tetrahedron. These tetrahedral complexes are anionic, and hence in these complexes Cr(VI) probably acts as a network forming oxide. It could for example build into the silicate network by changing a non-bridging oxygen ion ≡Si-O⁻ to a bridging oxygen ion, ≡Si-O-Cr≡.

From Figures 6 to 8, it can be seen that, for each absorption band, the absorption is proportional to the total chromium content, with no consistent deviation for different slag matrices. These observations indicate that the redox equilibria were independent of the chromium level and the composition of the slag matrix, within the range studied. This conclusion is of limited validity with respect to Cr(VI) in the oxidised slags because the 370 nm peak could only be resolved in three cases. However, the Cr(III) absorption at 635 nm in this series of slags was independent of the Cr level and the slag matrix. The linear regression lines through the origin (labelled L.R.) were calculated from all of the data points on the graph.

Chemical Analysis of Slags: Figure 9 shows that, within the experimental error of the determination of Redox species, there was a linear correlation between Cr(II) content and the total Cr level for the slags equilibrated at $p_{O_2}=1\cdot6.10^{-9}$ bar. The fraction of chromium present as Cr(II) was independent of the Cr level and also the slag matrix composition. The results were less clear at $p_{O_2}=1\cdot1.10^{-5}$, and for Cr(VI) at $p_{O_2}=0\cdot1$ (Figure 10). This is partly attributable to the wider scatter in chemical analysis at low levels of Cr(II) and Cr(VI). However, the upper curve in Figure 3 shows an unusual instance in which a slag equilibrated at $p_{O_2}=1\cdot1.10^{-5}$ contained Cr(VI) rather than Cr(II).

Effect of Equilibration Atmosphere: The effect of changing the oxidising conditions is controlled by the Redox equilibrium which is determined by reactions such as

$$(Cr^{2+}) - e = (Cr^{3+}) \qquad ...1$$

$$\{O_2\} + 4e = 2(O^{2-}) \qquad ...2$$

In Reaction 1, there is no change in the co-ordination of the chromium ion. The overall equilibrium under reducing conditions can therefore be expressed as the combination of Reactions 1 and 2:

$$4(Cr^{2+})+\{O_2\} = 4(Cr^{3+})+2(O^{2-}) \qquad ...3$$

$$K_3 = \frac{(a_{O^{2-}})^2\ (a_{Cr^{3+}})^4}{p_{O_2}\ (a_{Cr^{2+}})^4} \qquad ...4$$

For a given slag, the oxygen ion activity ($a_{O^{2-}}$) is constant as it is related to the slag basicity. Equation 4 can therefore be used to estimate the Cr(II) content in the slags produced at $p_{O_2}=1\cdot1.10^{-5}$ from the values determined for $p_{O_2}=1\cdot6.10^{-9}$. From the fraction Cr(II) of 0·71 (Figure 9, solid line), the fraction of Cr(II) at $p_{O_2}=1\cdot1.10^{-5}$ is predicted to be 0·21 compared with the least-squares value from the dashed line in Figure 9 of 0·14 ± 0·06. Similarly, the measured absorptivity of 1·20 m²/mol Cr_{total} at 440 nm for $p_{O_2}=1\cdot6.10^{-9}$ is close to the value of 1·29 predicted through Equation 4 from the measured absorptivity of 3·52 for $p_{O_2}=1\cdot1.10^{-5}$. Note that the values in Table 3 have been corrected for the fraction Cr(III) as determined by chemical analysis.

From this assessment, the fact that the Cr(II):Cr(III) ratio was apparently independent of the matrix composition implies that the basicity as indicated by the oxygen ion activity ($a_{O^{2-}}$) did not change significantly over the present range of slag matrix compositions. This cannot be expected to apply in substantially more basic slags, which could not be studied by the present technique.

In assessing the behaviour of Cr(VI), an equation like Equation 2 is inadequate because of the change of co-ordination. Accurate representation of the equilibrium requires knowledge of the complex that forms. Assuming Cr(VI) is present as tetrahedral $[Cr(VI)O_4]^{2-}$,

$$2\begin{bmatrix} 3\ \equiv Si\text{-}O^- \\ Cr^{3+} \\ 3\ \equiv Si\text{-}O\text{-}Si\equiv \end{bmatrix} + \tfrac{3}{2}O_2 + 2O^{2-} = 2\ [Cr(VI)O_4]^{2-} + 9\ \equiv Si\text{-}O\text{-}Si\equiv \qquad ...5$$

Octahedral Cr(III) Tetrahedral Cr(VI) Bridging Oxygen

Equation 5 suggests that the formation of the Cr(VI) complex will be favoured by oxidising conditions and a high basicity (i.e. high $a_{O^{2-}}$, with few bridging oxygen ions). This conclusion would not be reached if the change in the co-ordination of the Cr ion was ignored and an equation similar to Equation 3 had been derived. In reactions that can be postulated in which Cr(VI) builds into the silicate network, formation of the Cr(VI) complex is still favoured by oxidising conditions and a high basicity, as measured by either $a_{O^{2-}}$ or the number of bridging oxygen ions.

Because alumina is also an amphoteric oxide, an effect of alumina content on the chromium co-ordination could be anticipated, especially under oxidising conditions when they could be competing to build into the silicate network. As previously indicated, no change in the behaviour of Cr in alumina-free slag could be discerned.

Chemical analysis of a steelmaking slag to determine Cr(II) or Cr(VI) is complicated by the presence of other redox species, such as Fe(II), Mn(II) or Mn(IV). Spectroscopy provides a procedure for identifying the precise species that are present in a slag, provided that a vitreous sample can be produced. Vitreous samples of some slags, for example the slag produced by an ingot or continuous casting mould powder, can be recovered from plant. Because a species like Cr(II) requires more strongly reducing conditions than Fe(II), knowledge of the precise redox species that are present in a slag could be used to provide an indication of its oxidation state. In stainless steelmaking, knowledge of the oxidation state of the slag could subsequently be used to control blowing in the A.O.D. to minimise the oxidation of chromium.

CONCLUSIONS

Chemical and spectroscopic analyses have been used as complementary techniques for assessing the oxidation state of chromium in vitreous lime-silica slags. The distribution of chromium between Cr(II), Cr(III) and Cr(VI) in these slags was primarily dependent on the oxidation state. No consistent effect of chromium level, slag basicity or the presence of MgO or Al_2O_3 could be discerned within the slag composition range studied, although the absorption characteristics differed from those of aqueous solutions and sodium silicate glasses.

ACKNOWLEDGEMENTS

The authors are grateful to Dr A. Paul for helpful discussions and practical assistance, and the Science and Engineering Research Council for provision of a Research Studentship to HFM.

REFERENCES

1. H.F. Marston and D. Knight, Analyst, 1977, **102**, 745
2. L.E. Orgel, "An Introduction to Transition-Metal Chemistry: Ligand Field Theory". Methuen, London, 1966
3. P. Nath, A. Paul and R.W. Douglas, Physics Chem Glasses, 1965, **6**, 203
4. A. Paul and R.W. Douglas, Physics Chem Glasses, 1967, **8**, 151

Table 1 Compositions of Experimental and AOD Steelmaking Slags

Identity	CaO wt %	MgO wt %	Al_2O_3 wt %	SiO_2 wt %	Cr_2O_3 wt %	liqidus K
Acid CAS	30	0	16	54	0.05	1550
Basic CAS	41	0	16	43	to	1580
CMS	33	7	0	60	3.0	1600
CMAS	37	7	16	40		1580
Arc Furnace	40	5	5	35	5	
AOD (Blown)	30	5	2	15	30	
AOD (Final)	50	5	2	35	2	

Table 2 Equilibration Atmospheres

Identity	O_2	Argon	CO	CO_2	p_{O_2}, bar
$p_{O_2}=0{\cdot}1$	10	90	–	–	0·1
$p_{O_2}=1{\cdot}1.10^{-5}$	–	–	10	90	$1{\cdot}1.10^{-5}$
$p_{O_2}=1{\cdot}6.10^{-9}$	–	–	90	10	$1{\cdot}6.10^{-9}$

Table 3 Spectroscopic Results

Medium	Cr(II)		Cr(III)		Cr(VI)	
	Peak nm	ε m^2/mol	Peak nm	ε m^2/mol	Peak nm	ε m^2/mol
Aqueous (acidic)	710	0·5	410 585	1·7 1·7	255 355	Shr 100
Aqueous (alkaline)					270 370	260 370
$CaO-SiO_2$ slags	605	2·3	440 635	4·1 3·0	260 360	Ref 160
$3Na_2O.7SiO_2$ glass [3]			650	1·85	270 370	420

Key: ε = molar absorptivity, A/cl
A = Absorbance, $\log_{10}(100 / \%\ \text{transmission})$
l = path length, m
c = concentration, mol/m^3
Shr = shoulder
Ref = reflection (qualitative)

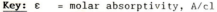

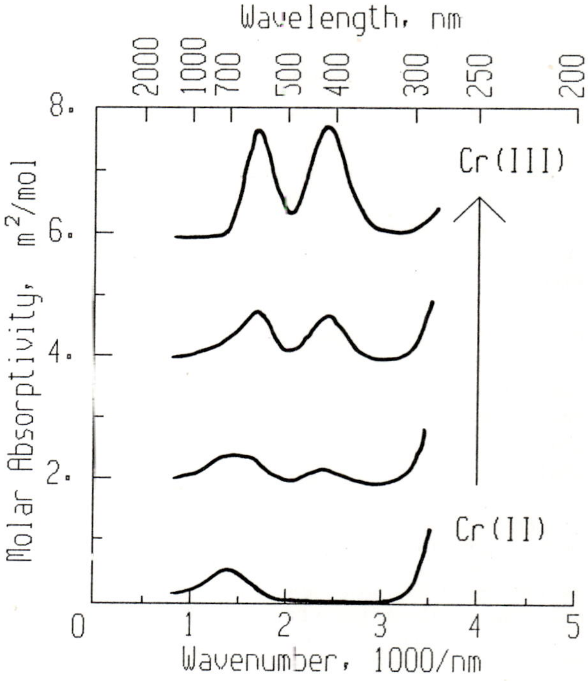

Figure 1 Progressive Oxidation of Cr(II) to Cr(III) in Aqueous Solution

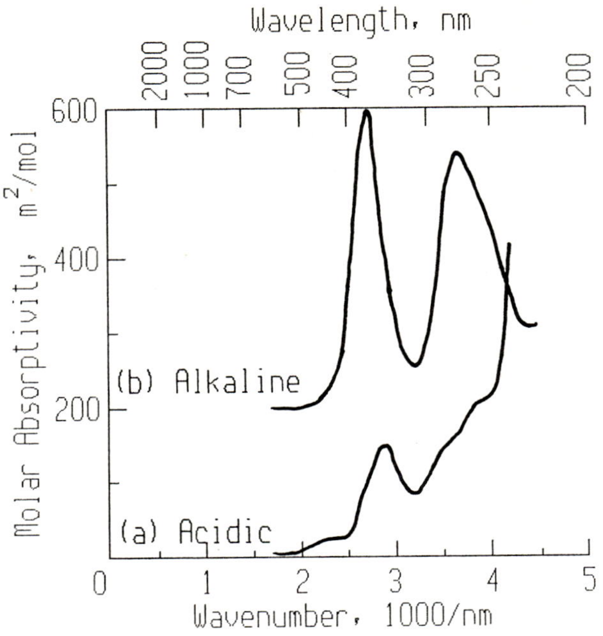

Figure 2 Cr(VI) in (a) Acidic and (b) Alkaline Aqueous Solution

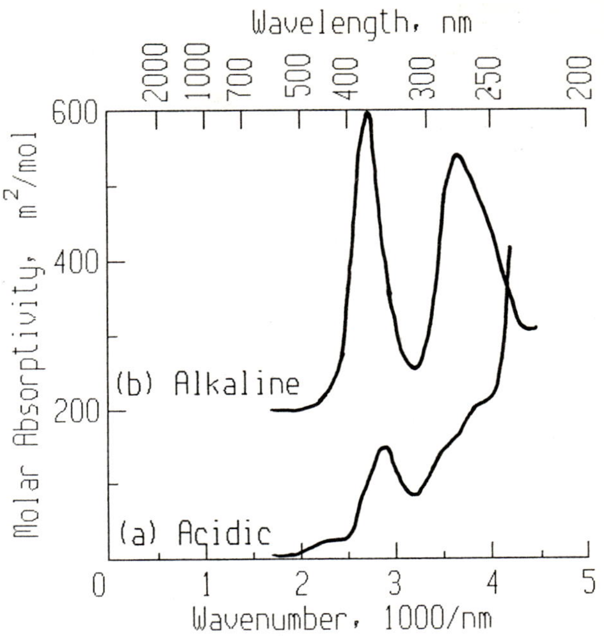

Figure 3 Predominantly Cr(III) Absorption in Lime-Silica Slag; p_{O_2}=1·1.10^{-5}bar

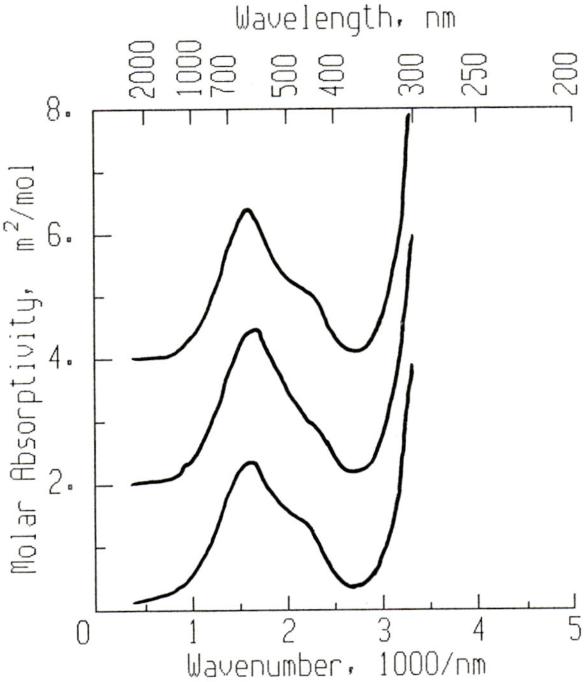

Figure 4 Cr(II) and Cr(III) Absorption in Lime-Silica Slag; p_{O_2}=1·6.10^{-9}bar

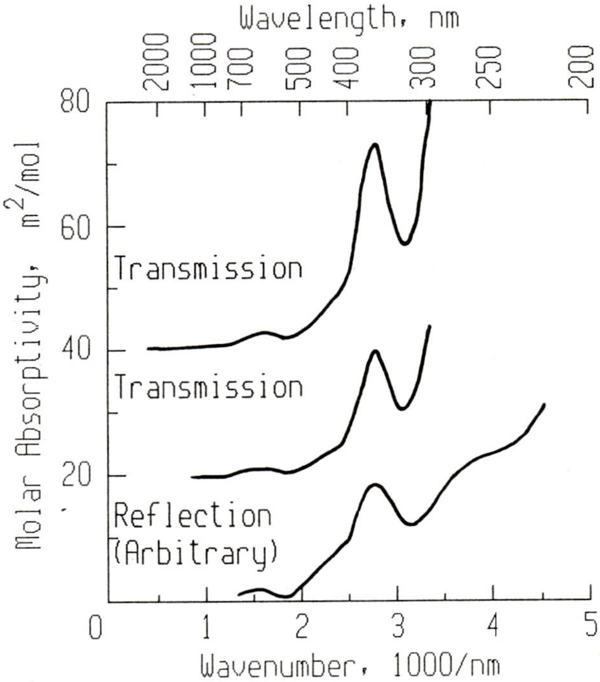

Figure 5 Cr(VI) and Cr(III) Absorption in Lime-Silica Slag; p_{O_2}=0·1 bar

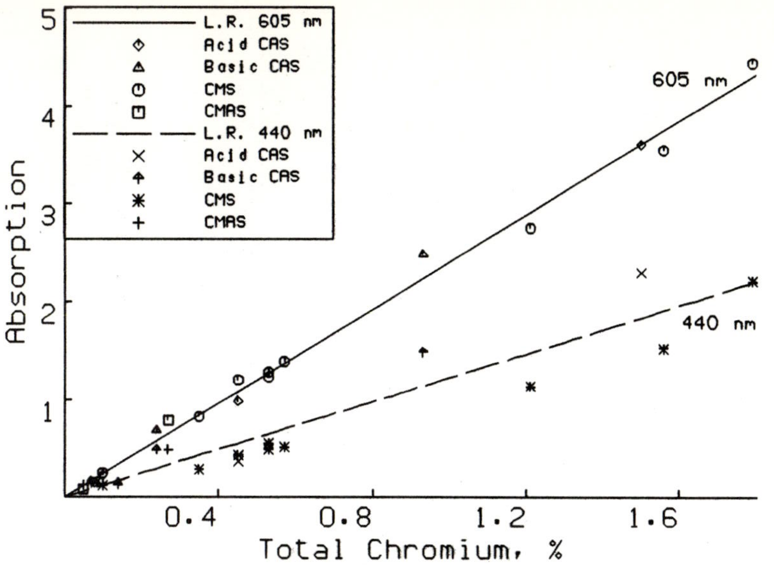

Figure 6 Dependence of Absorption on Cr content; Equilibrated at $p_{O_2}=1{\cdot}6.10^{-9}$bar

Absorbing Species: Cr(II) and Cr(III) at 605 nm
Predominantly Cr(III) at 440 nm

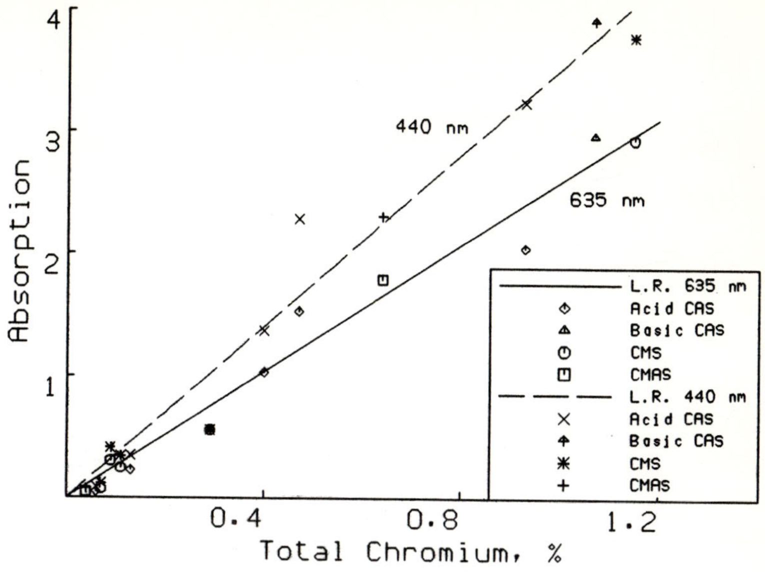

Figure 7 Dependence of Absorption on Cr content; Equilibrated at $p_{O_2}=1{\cdot}1.10^{-5}$bar

Absorbing Species: Cr(III) at 440 nm and 635 nm

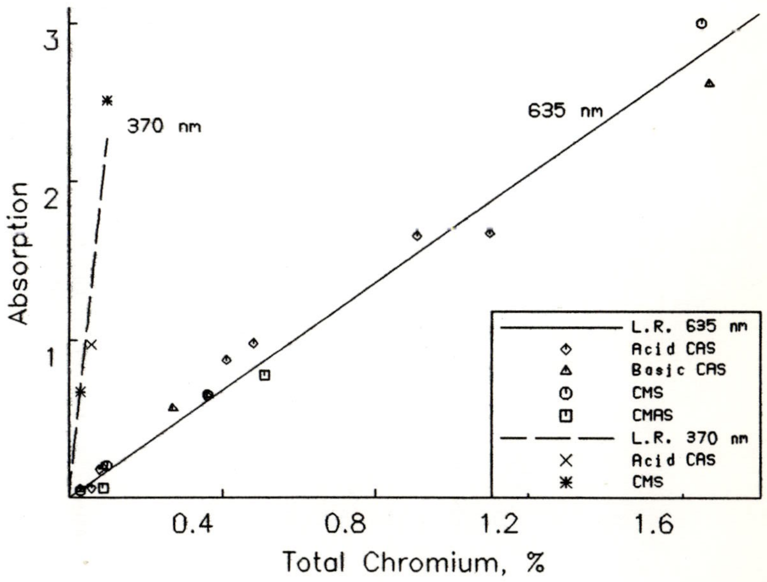

Figure 8 Dependence of Absorption on Cr content; Equilibrated at $p_{O_2}=0{\cdot}1$ bar

Absorption: Cr(VI) at 370 nm; Cr(III) at 635 nm

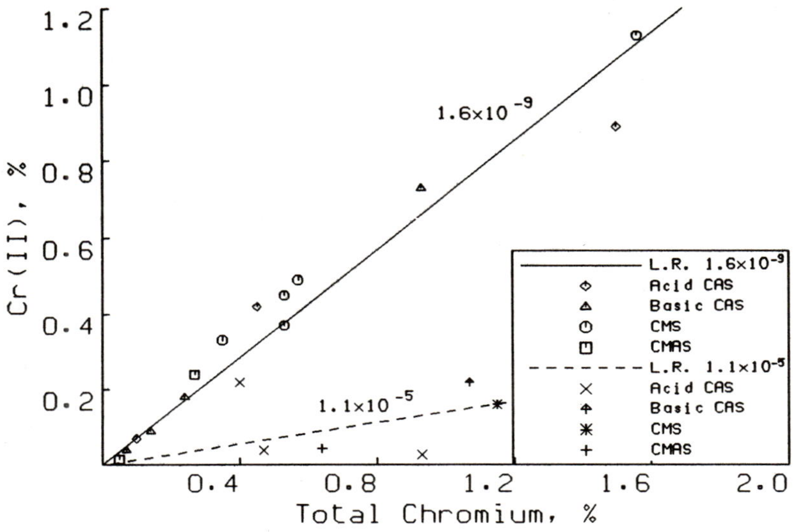

Figure 9 **Comparison of Fraction Cr(II) at p_{O_2}=1·6.10^{-9} and p_{O_2}=1·1.10^{-5}bar**

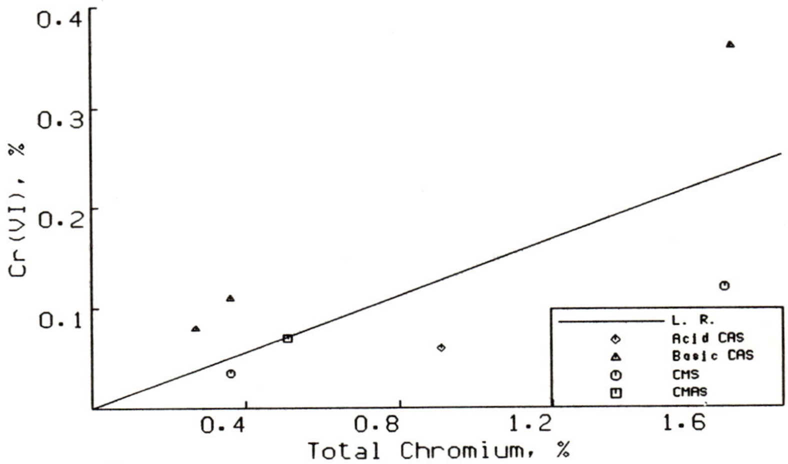

Figure 10 **Determination of Fraction Cr(VI) at p_{O_2}=0·1 bar**

Constitution of continuous casting fluxes

N C MACHINGAWUTA and P GRIEVESON

The authors are with the Department of Materials, Imperial College, London.

SYNOPSIS

The mineralogical phases formed during the simulation of continuous casting and after subsequent recrystallisation at 800 , 1000 and 1100°C are characterised by X-ray diffraction techniques. At the lower temperature, the formation of metastable phases is observed. Heat treatment at the higher temperature produces a number of changes in the crystalline phases formed. A constitution diagram is proposed which predicts the compositional dependence for the formation of such crystalline phases using results from about 40 commercial fluxes with a $CaO:SiO_2$ = 1 on a weight % basis.

Synthetic fluxes based on CaF_2-CaO-SiO_2 with lime/silica ratios of 0.8, 1.0 and 1.2 with soda and alumina additions are also investigated. Synthetic fluxes in some ways exhibit more complex crystallisation behaviour than the commercial fluxes. The results seem to indicate structural associations which may be related to the liquid structure of the fluxes.

INTRODUCTION

During experiments to simulate continuous casting with a water cooled copper mould, it was found that the heat flux through the slag infiltrated between the mould and the strand was dependent upon the composition of the flux used. The continuous casting fluxes in general use by the steel industry are based on the CaO-SiO_2-Na_2O-Al_2O_3 with fluoride additions and minor proportions of other oxides such as K_2O, MgO, MnO and iron oxide. Of the fluxes studied by simulation experiments, the vast majority, ~40, have a CaO/SiO_2 ratio very close to one. Samples of slag infiltrated between the copper mould and solidifying iron have been examined by X-ray diffraction using a Guinier focussing camera. In addition, slag samples were heat treated at 800 , 1000 and 1100°C to cause recrystallisation of the glassy phase, and re-examined. This communication describes our attempts to correlate the results of these experiments.

EXPERIMENTAL RESULTS FOR COMMERCIAL FLUXES WITH $CaO:SiO_2$ RATIO OF ONE

Examination of infiltrated slags using optical microscopy revealed a region of partially recrystallised material in a glassy matrix. Upon X-ray diffraction analysis of these samples, the only phase identified was cuspidine, $3CaO.2SiO_2.CaF_2$. After heat treatment at 800°C for 2 hours the intensity of the cuspidine diffraction pattern increased in all cases. The relative amounts of cuspidine present as determined by the relative intensity of the diffraction pattern varied directly with the fluoride content of the flux. The weakest pattern was produced with a flux containing 1.4% fluoride while the strongest pattern was given by fluxes with 6.8% fluoride. In addition to the increase in intensity of the cuspidine phase after heat treatment at 800°C, other diffraction patterns are observed depending on the soda and alumina concentrations. The conditions for the appearance of these phases is summarised.

Pseudowollastonite, $CaO.SiO_2$, is formed in all fluxes with a soda content less than ~5.5 wt% and Al_2O_3 < ~ 9wt%. Gehlenite, $2CaO.Al_2O_3.SiO_2$, is formed in fluxes in which the alumina content is > 6 wt% and the soda content is less than ~5.5 wt%.

With soda contents > 5.5 wt% a complex phase known as combeite was observed in all fluxes in this range. Combeite has a composition $Na_4(Ca,Al)_3Si_6O_{16}(OH,F)_2$ and has been identified in fluxes with alumina contents as low as 1.5 wt%. It seems certain that the presence of water in the molten flux is necessary for the formation of combeite. In the presence of reasonable quantities of soda and alumina the phase nepheline $Na_2O.Al_2O_3.2SiO_2$ is observed. A careful analysis of the conditions necessary for the appearance of nepheline indicates that the product of the soda and alumina contents in weight % must be greater than 27.5. With high soda contents the presence of disodium aluminosilicate $2Na_2O.Al_2O_3.2SiO_2$ is observed under the conditions that wt% Na_2O x wt % Al_2O_3 > 700.

From these observations a constitution diagram (see Figure 1) is proposed which predicts the compositional dependence for the formation of crystalline phases when heat treated at 800°C using ~40 different commercial fluxes with CaO/SiO_2~1. It should be noted that the additional phase, cuspidine, is observed in all cases but has been excluded from Figure 1. It is also interesting to report that several samples of infiltrated slag obtained from British Steel Corporation plants gave diffraction patterns which agree with these results when heat treated. It is also worth reporting that we never observed the phase pectolite, $Na_2O.4CaO.6SiO_2.H_2O$, although this has been reported by other investigations (1,2,3).

Following the heat treatment of infiltrated slags at 800°C for 2 hours to cause recrystallisation of the glassy phase, further experiments were carried out at 1000°C for 2 hours. Initially, these experiments were carried out to determine if all the glassy phase had recrystallised. The heat treatment was followed by X-ray diffraction analysis and some of our results are presented in Table 1.

The results of these X-ray diffraction analyses showed remarkable changes from those observed at 800°C. The general observations are that upon heat treatment at 1000°C the intensity of the cuspidine reflexions are very much reduced. The high temperature heat treatment also produces less sodium aluminosilicates whilst the presence of gehlenite appears or its intensity increases and sodium calcium silicate appears with some fluxes. These results were surprising but indicate the formation of metastable crystalline phases at low temperatures. At high temperatures, the nucleation and growth of thermodynamically stable phases occur.

EXPERIMENTAL RESULTS FOR SYNTHETIC FLUXES

The X-ray diffraction results from laboratory prepared (synthetic) slags of $CaO-SiO_2$ ratios of 0.8, 1.0 and 1.2 with varying quantities of soda and alumina are presented in Tables 2, 3 and 4. The observations made with synthetic slags differ in a number of ways from the commercial powder results. An attempt is made to summarise the general features observed with synthetic fluxes. In the silica rich $CaF_2-CaO-SiO_2$ slag (CaO/SiO_2 = 0.8) cuspidine is absent and the formation of pseudowollastonite and very weak CaF_2 is observed. Upon heat treatment, cuspidine is observed to form together with the presence of pseudowollastonite and wollastonite.

With the synthetic flux (CaO/SiO_2 = 1.0), the presence of cuspidine and wollastonite is observed and the intensity of the diffraction pattern increases upon heat treatment. With a CaO/SiO_2 = 1.2 slag, cuspidine is again noted with weak calcium silicate. Upon heat treatment the intensity of both phases increases.

As soda is added to synthetic slags with CaO/SiO_2 = 0.8, the presence of CaF_2 is still noted but cuspidine is also observed. In the heat treated specimens, the intensity of pseudowollastonite is usually stronger than that of wollastonite. However, with CaO/SiO_2 ratios greater than 0.8, wollastonite is always present to a greater extent than pseudowollastonite in heat treated specimens. For a slag with CaO/SiO_2 = 1.0 and 6 wt % Na_2O combeite is observed on heat treatment in agreement with observations with commercial fluxes. With CaO/SiO_2 = 1.2, a stronger combeite diffraction pattern is observed with 6 wt % Na_2O whilst this phase is absent in the equivalent CaO/SiO_2 = 0.8 flux. Thus an increase in CaO/SiO2 ratio requires less soda to stabilise the combeite phase upon heat treatment.

Additions of alumina to the synthetic fluxes resulted in the formation of more glassy infiltrated slag. Whilst both pseudo- and β-wollastonite were identified at low alumina concentrations (6 wt %), heat treatment of slags with higher alumina concentrations at 800°C produced a diffraction pattern for cuspidine only. Furthermore the intensity of the cuspidine pattern becomes stronger as the alumina content increases. Heat treatment temperatures were raised to 1000 and 1100°C to identify any crystalline phases formed. At temperatures of 1000 and 1100°C the phases gehlenite, $2CaO.Al_2O_3.SiO_2$, and anorthite, $CaO.Al_2O_3.2SiO_2$, were identified in the recrystallised material.

Careful inspection of the diffraction patterns of the infiltrated slag and after heat treatment revealed variations in the unit-cell dimensions of cuspidine and wollastonite as the alumina varied. Some of these results are shown in Figures 2 and 3. In contrast, the unit cell dimensions of pseudowollastonite showed no variation with temperature or alumina content, Figure 4. The results indicate that an addition of up to 4 wt% Al_2O_3 decreases the unit-cell dimensions of cuspidine by ~6%. Further alumina additions cause no additional change. These results support the view that cuspidine can accommodate some alumina in solid solution. When most of the samples were heat treated at 1000°C, the unit cell dimensions revert to those associated with alumina free cuspidine indicating the rejection of alumina as gehlenite and anorthite are formed. The presence of both gehlenite and anorthite in the heat treated specimens was surprising since in commercial fluxes the phase anorthite was never observed.

DISCUSSION

The formation of metastable crystalline phases in the first stages of crystallisation at low temperatures followed by nucleation and growth of stable phases is not unique as similar observations have been made on the subliquidus metastable immiscibility in glasses. Demixing of a glass, at temperatures between the liquidus and the glass transition temperature, is a special case of phase separation of a supercooled liquid which has failed to crystallise and is metastable with respect to a crystalline phase. The phenomenon of subliquidus immiscibility in the sodium borosilicate system has been recognised for some years(4). Upon suitable heat treatment at subliquidus temperatures, the initially homogeneous borosilicate glass separates into 2 glassy phases, one silica rich and the other approaching the sodium tetraborate composition. Thus the observations made in the present study seem to be similar. However, the more complex nature of the recrystallisation reactions seem to point to specific composition association related to the liquid phase. If it is accepted that the first phases to recrystallise at low temperatures are related to the liquid composition, the nucleation and growth of a metastable crystalline phase is caused by the lack of the long range diffusion necessary to produce equilibrium phases. Following this argument, the formation of cuspidine is due to the close association of fluoride ions with silica and lime. In the $CaF_2-CaO.SiO_2$ system Baak and Olander (5) interpreted the freezing point depression data as evidence of cyclic $Si_3O_9^{6-}$ ions in the melt. Kojima and Masson(6) using similar techniques concluded that with higher CaF_2 concentrations the structure of the melt was made up of chains based on the monomer SiO_3F3- thus:

$$SiO_3F^{3-} + Si_xO_{3x}F^{(2x+1)-} = Si_{(x+1)}O_{3(x+1)}F^{(2x+3)-} + F^-$$

Suito and Gaskell(7) concluded that for the ratio $CaO/SiO_2 > 1.5$, linear chain silicate anions and F^- are formed, whilst for CaO/SiO_2 between 0.4 and 1.5, polyions based on the monomer SiO_3F^{3-} exist.

The ease of formation of cuspidine in these slags certainly supports the concept of the presence of anions based on SiO_3F^{3-} and also suggests that there exists some association of these ions with Ca^{2+} in the liquid state. The observations made with synthetic slags with $CaO/SiO_2 = 0.8$ where the presence of CaF_2 is observed seems to support the hypothesis of Kojima and Masson rather than Suito and Gaskell as discrete fluoride ions must exist to form CaF_2. The formation of cuspidine in these slags upon heat treatment at 800°C supports the additional presence of oxyfluoride ions.

Another observation which was very surprising, was the presence of both wollastonite and pseudo-wollastonite after heat treatment in synthetic slags. At the heat treatment temperature, wollastonite in the stable phases and pseudowollastonite must be substantially metastable at 800°C. Some insight can be gained by considering the crystal structures of the two phases. At temperatures above 1125°C pseudowollastonite with a pseudo hexagonal structure is the stable phases whilst below this temperature wollastonite with a triclinic structure is stable. The structure of pseudowollastonite consists of discrete ring type anions of formula $Si_3O_9^{6-}$. The wollastonite structure comprises long chains. The experimental observations suggest very strongly that the liquid phases must contain both ring and chain anions which lead to the relatively easy formation of both phases. These observations tend to support the hypothesis of Baak and Olander. In the synthetic slags, the addition of soda seems to indicate very little effect on the fluoride-oxy anion distribution but supports the view that ring type anions are converted into chains.

The addition of alumina to the synthetic fluxes suggests that aluminium becomes incorporated into the silicon oxyfluoride anions as witnessed by the effects on the unit cell dimension of cuspidine. The variation in the unit cell dimensions of pseudowollastonite with alumina additions indicates the incorporation of aluminium in the silicate chains. The lack of change of unit cell dimensions of wollastonite suggests that aluminium is not incorporated into the silicate ring anions. In addition, the presence of both gehlenite and anorthite in the heat treated specimens was also surprising since in commercial fluxes the presence of anorthite was never observed. These observations indicate the presence of chains with variable Al/Si ratios. In commercial fluxes soda is always present and is a strong indication with the formation of nepheline that sodium is preferentially associated with silica rich aluminosilicate chains in the liquid. a perusal of the $CaO-Al_2O_3-SiO_2$ phase diagram indicates that the compositions are significantly removed from the equilibrium wollastonite-anorthite-gehlenite three phase region indicating the metastable nature of anorthite in these samples.

These results seem to be significant and perhaps throw some light on the structure of these complex liquid mixtures. It seems that further studies would be quite rewarding in leading to a full picture of melt structure and recrystallisation behaviour.

ACKNOWLEDGEMENTS
The authors are grateful to the National Physical Laboratory for the financial support for this work and wish to express their gratitude to Dr. S. Bagha and Dr. K.C. Mills for fruitful discussions of the results.

REFERENCES

1) T. Hiromoto, T. Shima and R. Sato. Steelmaking Proceeding ISS-AIME, 62, 1979, p40

2) R. Sato. Steelmaking Proceeding ISS-AIME, 62, 1979, p48

3) A. Olusanya. Ph.D. Thesis, University of London, 1983.

4) W.E.S. Turner, F.J. Winks. J. Soc. Glass Technology, 10, 1926, p102

5) T. Baak and A. Olander. Acta Chemica Scand., 9, 1955, p1350

6) H. Kojima and C.R. Masson. Can. J. Chem. 47, 1969, p4221

7) H. Suito and D.R. Gaskell. Metall. Trans. 7B, 1976, p567.

TABLE 1

Phases identified by X-ray diffraction analysis after further heat treatment at 1000°C for 2 hours

Slag	Cuspidine	Wollastonite	Gehlenite	Nepheline	$Na_4Al_2Si_2O_9$	Sodium Calcium Silicate
C	W (MS)	S (VW)	VS (WM)	- (-)	- (-)	- (-)
D	M (S)	- (-)	MS (-)	VW (VW)	VW (WM)	M (-)
V	VS (S)	- (-)	- (-)	W (W)	M (M)	- (-)
Z	VW (VS)	- (-)	S (-)	VW (W)	VW (M)	M (-)
B-A	VW (MS)	VW (VW)	VS (M)	- (W)	- (-)	- (-)

Identifications and intensities in brackets are results from heat-treatment at 800°C for 2 hours

TABLE 2

Phases identified in infiltrated slag for synthetic powders of varying soda content

CaO/SiO_2	wt% Na_2O	t_{slag} mm	Q MWm^{-2}	Heat Treatment °C	Cuspidine	Pseudo Wollastonite α	Wollastonite β	CaF_2	Combeite
0.8	---	0.65	1.30	---	---	S	---	VVW	---
"	---	"	"	800	S	MS	MS	---	---
"	2.0	0.60	1.37	---	M	---	MS	VW	---
"	"	"	"	800	MS	---	S	---	---
"	4.0	0.67	1.18	---	M	S	---	W	---
"	"	"	"	800	MS	MS	M	---	---
"	6.0	0.77	0.90	---	---	MS	VW	VW	---
"	"	"	"	800	S	MS	M	---	---
1.0	---	0.52	1.28	---	S	---	S	---	---
"	---	"	"	800	S	---	VS	---	---
"	2.0	0.56	1.28	---	MS	---	MS	---	---
"	"	"	"	800	MS	---	MS	---	---
"	4.0	0.48	1.38	---	M	MW	MW	---	---
"	"	"	"	800	S	M	S	---	---
"	6.0	0.43	1.30	---	M	MW	---	---	---
"	"	"	"	800	S	W	M	---	MW
1.2	---	0.52	1.40	---	M	W	W	---	---
"	---	"	"	800	S	W	M	---	---
"	2.0	0.62	1.34	---	S	W	S	---	---
"	"	"	"	800	MS	---	S	---	---
"	4.0	0.47	1.24	---	S	MS	VW	---	---
"	"	"	"	800	MS	M	MS	---	---
"	6.0	0.44	1.34	---	W	VVW	---	---	---
"	"	"	"	800	VS	W	S	---	M

VS = very strong   S = strong   MS = medium-strong   M = medium
MW = medium-weak   W = weak   VVW = very-very-weak

TABLE 3

Phases identified in infiltrated slag for synthetic powders of varying alumina content

CaO/SiO_2	wt% Al_2O_3	t_{slag} mm	Q MWm^{-2}	Heat Treatment °C	Cuspidine	Pseudo Wollastonite α	Wollastonite β	CaF_2
0.8	---	0.57	1.30	---	---	S	---	VVW
"	---	"	"	800	M	M	M	---
"	4.0	0.46	1.19	---	VW	---	---	---
"	"	"	"	800	S	M	M	---
"	6.0	0.63	0.92	---	VVW	W	---	---
"	"	"	"	800	M	---	---	---
"	8.0	0.89	1.13	---	---	---	---	---
"	"	"	"	800	VS	---	---	---
"	10.0	0.56	1.14	---	---	---	---	---
"	"	"	"	800	VS	---	---	---
1.0	---	0.52	1.28	---	S	---	S	---
"	---	"	"	800	S	---	VS	---
"	4.0	0.39	1.14	---	W	VW	---	---
"	"	"	"	800	W	W	---	---
"	6.0	0.47	1.18	---	W	VW	W	---
"	"	"	"	800	MW	W	VW	---
"	8.0	0.61	1.12	---	W	---	---	---
"	"	"	"	800	S	---	---	---
"	10.0	0.60	1.08	---	VW	---	---	---
1.2	---	0.52	1.40	---	M	W	W	---
"	---	"	"	800	S	W	M	---
"	4.0	0.56	1.06	---	MS	VW	---	---
"	"	"	"	800	VS	---	MS	---
"	6.0	0.56	1.17	---	VW	---	---	---
"	"	"	"	800	S	---	---	---
"	8.0	0.52	1.21	---	M	---	---	---
"	"	"	"	800	VS	---	---	---
"	10.0	0.40	1.10	---	W	---	---	---
"	"	"	"	800	VS	---	---	---

TABLE 4

Phases identified an infiltrated slag for synthetic powders of varying alumina content

CaO/SiO_2	wt% Al_2O_3	t_{slag} mm	Q MWm^{-2}	Heat Treat °C	Cuspidine	Pseudo Wollast α	Wollastonite β	CaF_2	Anorthite	Gehlenite
0.8	---	0.65	1.30	1000	W	M	VS	MS	---	---
"	---	"	"	1100	VW	MW	VS	---	---	---
"	4.0	0.46	1.19	1000	M	MW	VS	MS	M	W
"	"	"	"	1100	W	W	VS	W	W	---
"	6.0	0.63	0.92	1000	W	VW	VS	M	M	---
"	"	"	"	1100	MW	M	S	MW	M	W
"	8.0	0.89	1.13	1000	W	W	VS	M	M	VW
"	"	"	"	1100	M	M	MS	MW	M	VVW
"	10.0	0.56	1.14	1000	M	S	MS	MS	S	VW
"	"	"	"	1100	MS	M	MS	W	MS	VW
1.0	---	0.52	1.28	1000	MW	W	MS	---	---	---
"	---	"	"	1100	W	VW	VS	---	---	---
"	4.0	0.39	1.14	1000	M	MW	S	---	MW	VW
"	"	"	"	1100	S	W	VS	---	M	W
"	6.0	0.47	1.18	1000	S	M	S	---	VW	VVW
"	"	"	"	1100	M	M	S	---	VVW	MW
"	8.0	0.61	1.12	1000	MS	---	MS	---	M	VVW
"	"	"	"	1100	MS	W	MS	---	M	VW
"	10.0	0.60	1.08	1000	S	W	MS	---	MS	VW
"	"	"	"	1100	MS	MW	MS	---	M	W
1.2	---	0.52	1.40	1000	S	MW	S	---	---	---
"	---	"	"	1100	MS	M	MS	---	---	---
"	4.0	0.56	1.06	1000	MS	VW	MS	---	W	W
"	"	"	"	1100	S	W	S	---	MW	M
"	6.0	0.56	1.17	1000	S	MS	MW	---	M	M
"	"	"	"	1100	MS	S	VW	---	MW	MS
"	8.0	0.52	1.21	1000	S	MS	MW	---	M	M
"	"	"	"	1100	S	M	M	---	MW	MS
"	10.0	0.40	1.10	1000	M	W	M	---	M	M
"	"	"	"	1100	MS	W	M	---	MW	M

wollast = wollastonite

VS = very strong S = strong MS = medium-strong M = medium

MW = medium-weak W = weak VVW = very-very-weak

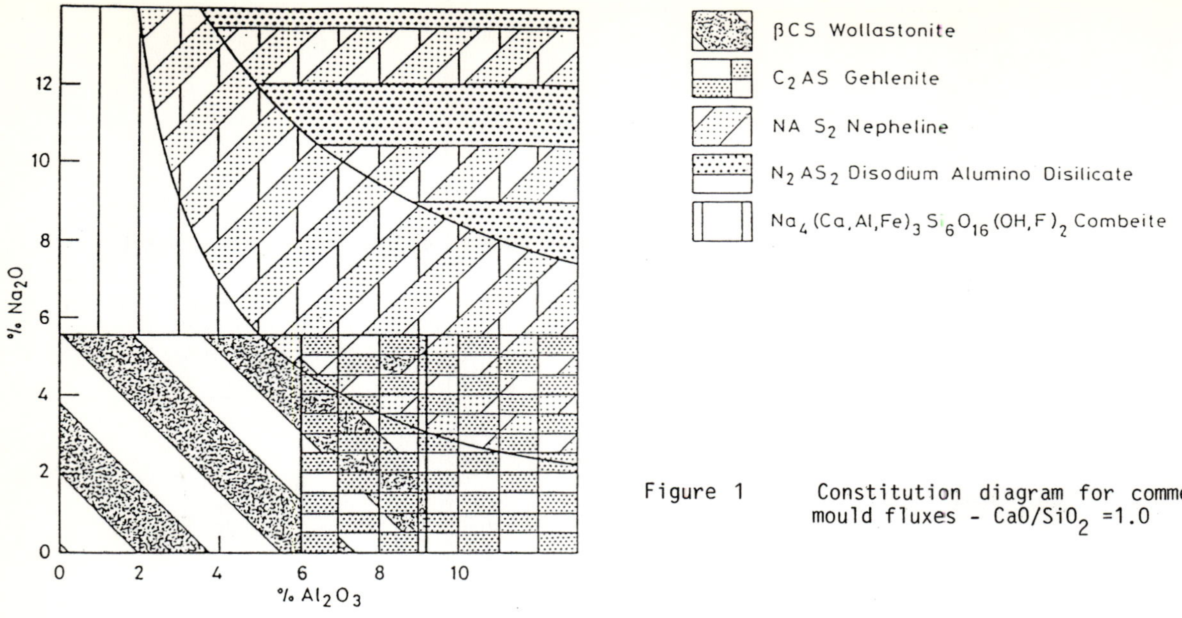

Figure 1 Constitution diagram for commercial mould fluxes - CaO/SiO_2 =1.0

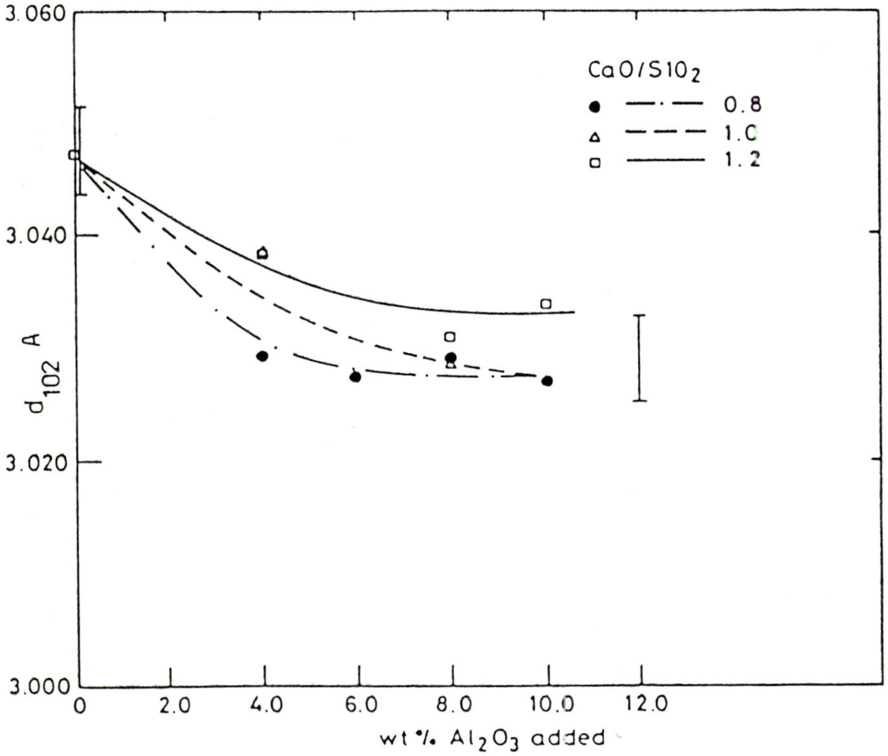

Figure 2 Variation of d spacing of cuspidine with alumina content - H.T. at 800°C

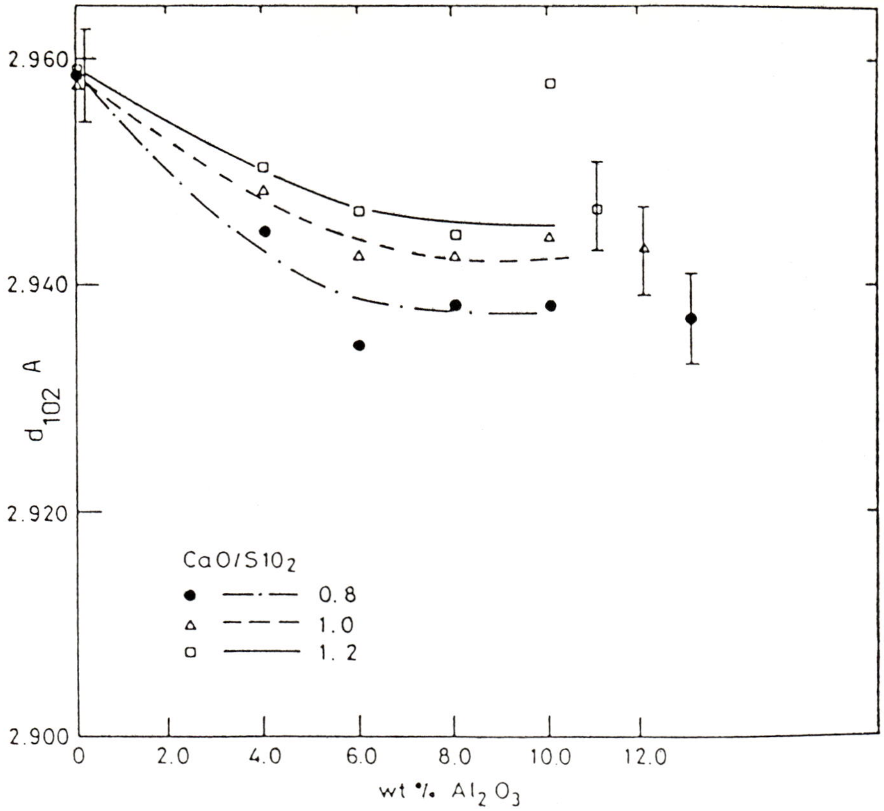

Figure 3 Variation of d spacing of wollastonite with alumina content - H.T. at 1000°C

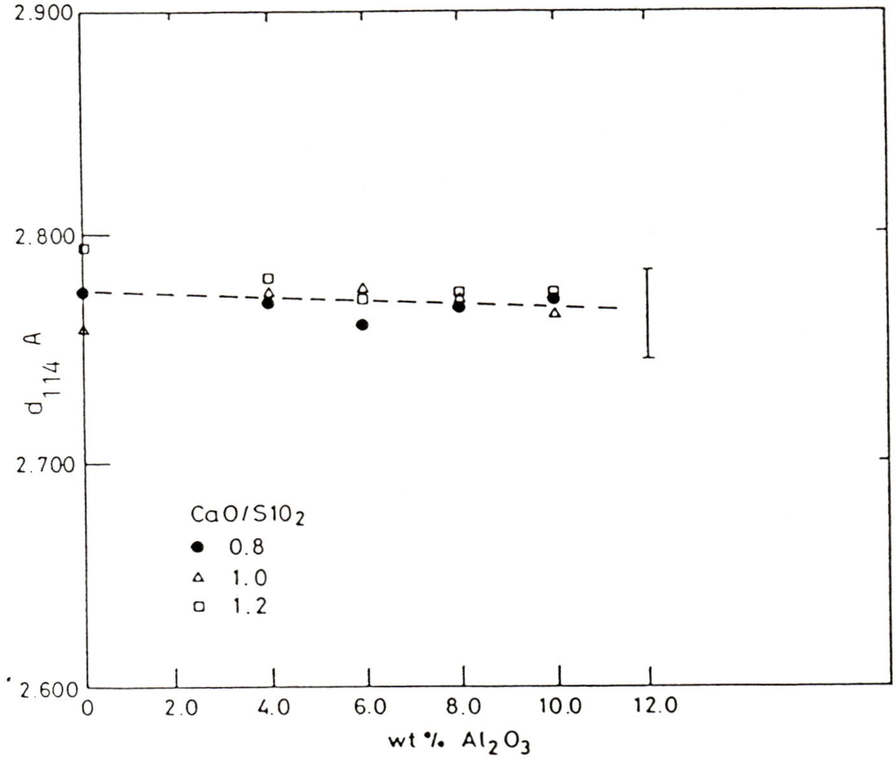

Figure 4 Variation of d spacing of pseudowollastonite with alumina content - H.T. at 1000°C

Viscosity of Na_2O-B_2O_3-SiO_2 system in glassy and molten states

Y SHIRAISHI and H OGAWA

The authors are in the Research Institute of Mineral Dressing and Metallurgy, (SENKEN), Tohoku University, Sendai, Japan.

SYNOPSIS

In order to study the network forming ability of boron trioxide in a silicate network and to compare the flow characteristics in the glassy and molten states, the viscosities of sodium boro-silicates have been measured in both the glassy and molten states by a parallel plate viscometer and a rotating cylinder viscometer over the viscosity range 10^{10} to 10^1 poise. The measurements were carried out on the two series of the samples, i.e. $Na_2O \cdot 2SiO_2$- B_2O_3(0 - 50 mol %) and $Na_2O \cdot SiO_2$- B_2O_3(10 - 70 mol %) systems. The results obtained were as follows:

1) The temperature dependence of the viscosity of sodium boro-silicate system does not obey the Arrhenius relationship, but the Fulcher relationship represents the data well in both the glassy and molten states.

2) Apparent activation energies for viscous flow exhibit a remarkable change from 100 - 200 kcal/mol in the glassy state to 20 - 30 kcal/mol in the molten state.

3) In the iso-viscosity curves of meta-silicate-borate system, a maximum can be observed clearly around the composition $Na_2O/B_2O_3 \cong 1$ in the glassy state but the corresponding change is slight in the molten state.

4) In the silica-rich composition of di-silicate-borate system, the effects of B_2O_3 on the viscosities are not the same in the two states. B_2O_3 acts as a network former in the glassy state but behaves as a network modifier in the molten state.

INTRODUCTION

One well - accepted view is that the structure of glass should be similar to that of the high temperature melt from which the glass is formed. This idea has been supported by data for silicate melts obtained using the *in situ measurements* by X-ray diffraction[1]), infrared emission spectrometry[2,3]), and Raman spectrometry[4,5]). The general concept for the structure of a glass may be stated as that of a frozen liquid structure which is characterized by a lack of the long range order and by no fluctuation of density with time. This concept leads to the importance of the viscosity which provides the most fumdamental differences between the states. On the other hand, viscosity measurements provide the knowledge of the behaviour of a network former in the melt and the glass. Consequently, it would be of interest to know the behaviour of the viscosity of a system which includes two or more network formers. In the present study, the viscosities of boro-silicate systems are measured to elucidate the effect of boron trioxide on the network structure of silicates. Measurements have been carried out in both the glassy and molten states.

EXPERIMENTAL

1) Procedure and Apparatus

The viscosities of glassy samples were measured by a parallel plate viscometer. This method is based on the following equation[6]);

$$\eta = 2\pi MgH^5/3V(-dH/dt)(2\pi H^3 + V) \quad (1)$$

where, η is the viscosity (poise), M is the load (g), by which the sample is deformed, g is the accelaration due to gravity (cm/sec^2), H is the height (cm), V is the volume (cm^3) and dH/dt is the deformation rate of the sample (cm/sec). This equation is derived using the assumption that the sample is incompressible and always keeps a cylindrical shape during deformation. To determine the time-height relationship, the combination of linear variable differential transformer (LVDT) and a micro-computer was adopted, because LVDT provides a very high sensitivity (1 μm/mV) for the height measurement and a microcomputer provides a high speed read out of LVDT output (80 msec). The essential part of the experimental apparatus is illustrated in Fig. 1(A). This viscometer operates well over the viscosity range from 10^{10} to 10^6 poise when the temperature is increased at a suitable rate.

The viscosities of the molten samples were measured using a "Rotovisco", rotating cylinder viscometer, with the rotor and measuring crucible being fabricated from Pt-6%Rh. The geometrical arrangement of the rotor and the crucible is shown schematically in Fig. 1(B). The following problems arose inevitably during the actual measurement; i) failure of alignment, ii) eccentric rotation, iii) offset of immersion depth due to the thermal expansion, and so on. These effects on the accuracy of the measurements were evaluated using a standard liquid to calibrate the viscometer at room tempeature. Errors due to these geometrical factors by the accumulation of the individual errors were estimated to be less than ±10 %.

2) Samples
The composition of the samples were selected as shown in Fig. 2 to clarify the effect of borate on the viscosities of silicate systems. The samples were prepared from guaranteed grade B_2O_3 and $Na_2O \cdot SiO_2 \cdot 9H_2O$ and reagent grade silica sand. Two series of samples were prepared in Pt crucibles. These were based on sodium meta- and di-silicates mixed with the selected amount of premelted boron trioxide. The glassy specimens were prepared by casting the molten sample into a copper mould and by machining into cylindrical shape, (8 mm diam. x 5 - 8 mm height), with the aid of a diamond wheel. The samples for the molten state measurements were obtained by the progressive melting of the sample in the crucible until the correct depth of sample was obtained.

Details of the experimental procedure and the apparatus are described elswhere[7,8].

RESULTS AND DISCUSSION

The viscosity of $Na_2O \cdot 2SiO_2$- 50 mol % B_2O_3 in the glassy and molten states is shown in Fig. 3 as a function of 1/T, where T is the absolute temperature. It is evident that the Arrhenius relationship does not hold over the entire temperature range of the measurements but the Fulcher relationship, equation (2), seems to represent the data well for both the glassy and molten states simultaneously.

$$\log \eta = A + B/(T - T_0) \qquad (2)$$

where, A , B and T_0 are constants. This equation can be compared with the equation based on the free volume theory. When the free volume is proportional to the temperature, both equations are identical. Thus, the meaning of T_0 may be considered as the temperature at which the free volume tends to zero. All the data, together with the Fulcher constants, are summarized in Table 1. As can be seen from Fig. 3, the activation energy for viscous flow can not be determined uniquely. The apparent activation energies obtained from the tangents on the curve strongly depend on temperatures, and decrease steeply from 100 - 200 kcal/mol in glassy state to 20 - 30 kcal/mol in molten state. This enormous decrease in activation energy implies the degradation of the network structure of the sample.

Iso-viscosity curves in the pseudo-binary $Na_2O \cdot SiO_2$-B_2O_3 system are shown in Fig. 4. In this meta-silicate system, the effects of borates are almost the same for both the glassy and molten states, at least over the mid-range of B_2O_3 content in the system. As shown in Fig. 4, a maximum in the iso-viscosity curve can be seen around 50 mol % B_2O_3 for the glassy state although the corresponding change in molten state is not so clear. This behaviour can be considered as a reflection of a change of the coordination number of oxygen around a boron atom. By the comparison with the investigation of boro-silicate glasses, it may be concluded that the increase of the viscosity corresponds to the formation of BO_4 structure and the decrease corresponds to the formation of BO_3 structure. The ratio of BO_4/BO_3 was estimated as nearly unity at the composition $Na_2O/B_2O_3 \cong 1$ by the NMR measurement.

However, iso-viscosity curves in the pseudo-binary system of $Na_2O \cdot 2SiO_2$-B_2O_3 are somewhat different to those in Fig. 4. As shown in Fig. 5, it can be seen that the iso-viscosity curves behave in quite a different manner for the glassy and molten states. In the glassy state, the viscosities increase with increase in borate content but they decrease in the molten state. These opposite results indicate the amphoteric nature of B_2O_3 in the two states, namely, B_2O_3 behaves as a network former in the glassy state but as a modifier in the molten state.

Similar results were reported by Fulcher[9] who has measured the viscosity of the $Na_2O \cdot 4.2SiO_2$-$Na_2O \cdot 3.6B_2O_3$ (0 - 50 mol %) system and by Mazurin and Shvaiko-Shvaikovskaya[10] who made measurements in silica-rich regions (SiO_2>60, B_2O_3<25, $Na_2O \cong 20-30$ mol %).

It is not easy to understand why such amphoteric behaviour should depend upon temperature. However, it might be useful to refer to observations on the thermal expansivity of the boro-silicate glasses. According to Samoe[11], the linear thermal expansivities of this system exhibit great differences between the values obtained above and below the glass transition temperature, Tg. The expansivities above Tg are much larger than those below Tg and increase vigorously with increase of B_2O_3 content although the latter is almost constant. These results suggest that the strength of the bonds in this system become weak with increasing temperature and borate content. If these tendencies are still valid for the condition studied here, the amphoteric behaviour should be stated in terms of the change of bond nature but a satisfactory explanation is not yet known.

Finally, the results obtained in this study can be summarized as Fig. 6, in which iso-viscosity curves are represented schematically both in glassy and molten states. The figure shows that borate behaves as a network former in the glassy state but as an amphoteric in the molten state even in the composition range $Na_2O/B_2O_3 > 1$, However, outside of this range, borate acts as a modifier in both states, which may be due to the formation of BO_3 group.

CONCLUSION

In order to determine the effect of B_2O_3 on the network structure of silicate, the viscosity of sodium boro-silicate system has been measured in both the glassy and molten states. The results obtained are as follows:

1) The temperature dependence of the viscosity can be represented well by the Fulcher relationship over the temperature range corresponding to the viscosity range of 10^{10} to 10^1 poise.

2) The apparent activation energies for viscous flow decrease remarkably from 100 - 200 kcal/mol in the glassy state to 20 -30 kcal/mol in the molten state.

3) Boron trioxide behaves as a network former both in the glassy and molten states within the

composition range, Na_2O / B_2O_3 >1 of the metasilicate.

4) However, in silica-rich compositions, borate behaves as a network modifier in molten state even in the composition, Na_2O / B_2O_3 > 1.

5) Outside of this composition range, i.e. Na_2O/ B_2O_3 < 1, B_2O_3 behaves as network modifier due to the formation of BO_3 structure.

REFERENCES

1) Y.Waseda and H.Suito: Tetsu to Hagane', 62 (1976), 1493.
2) J.R.Sweet and W.B.White: Phy. Chem. Glasses, 10(1969), 246.
3) K.Kusabiraki and Y.Shiraishi: Nippon Kinzoku Gakkaishi, 45(1981), 250.
4) a) S.Kashio, Y.Iguchi, Y.Nishina, T.Goto and T.Fuwa: Trans. ISIJ, 20(1980), 250.
b) N.Iwamoto, N,Umesaku and K.Doi: Nippon Kinzcku Gakkaishi, 47(1983), 251.
5) B.O.Mysen, D.Virgo and I.Kushiro: Amer. Mineralogist, 66(1981), 678.
6) E.H.Fontana: J. Amer. Ceram. Soc., 49(1970), 594.
7) Y.Shiraishi, L.Gra'na'sy, Y.Waseda and E.Matsubara: J. Non-Cryst. Sol., 95&96 (1987), 1031.
8) Y.Shiraishi and H.Ogawa: Tohokudaigaku Senken Iho, 44(1988), 8.
9) G.S.Fulcher: J. Amer. Ceram. Soc., 8(1925), 339.
10) O.V.Mazurin and T.P.Shvaiko-Shvaikovskaya: "Issledovanie stekloobraznykh sistem i sintez novykh stekla na ikh osnove", (1971), Moskva, 151.
11) M.O.Samoe: Ann. Pys., 9(1928), 35.

Table 1 Viscosity and Fulcher constants of $Na_2O - B_2O_3 - SiO_2$ system.

mol % oxide			Temp. (K) at η (P)				log η (P) at temp. (K)				Fitting	log η = A + B/(T - T_0)		
B_2O_3	Na_2O	SiO_2	10^{10}	10^9	10^8	10^7	1000	1100	1200	1300	temp.range	-A	B	T_0
10	45	45	(710)	729	746	765								
20	40	40	(738)	762	776	793								
30	35	35	(771)	797	814	830								
40	30	30						2.061	1.371	0.881				
50	25	25	(823)	844	863	881	(3.832)	2.425	1.623	1.047	840 - 1325	1.984	1899	400.3
60	30	30	(785)	807	827	846	3.571	2.425	1.666	1.127	803 - 1325	2.063	2206	335.2
70	15	15						2.415	1.700	1.267				
0	33	67	(783)	822	855	895	(5.545)	(4.616)	(3.961)	3.474	805 - 1438	-0.162	2583	247.0
10	30	60	(822)	850	874	902								
20	27	53	(833)	865	887	910	(5.194)	(3.801)	2.901	2.270	857 - 1422	1.299	2378	260.6
30	23	47	(843)	871	891	910	(5.163)	3.647	2.666	1.978	865 - 1478	1.982	2603	359.4
40	20	40	(846)	872	893	913	(5.077)	3.535	2.546	1.857	887 - 1416	1.982	2525	369.0
50	17	33	(794)	830	854	879	(4.550)	3.258	2.372	1.727	823 - 1401	2.376	3021	290.7

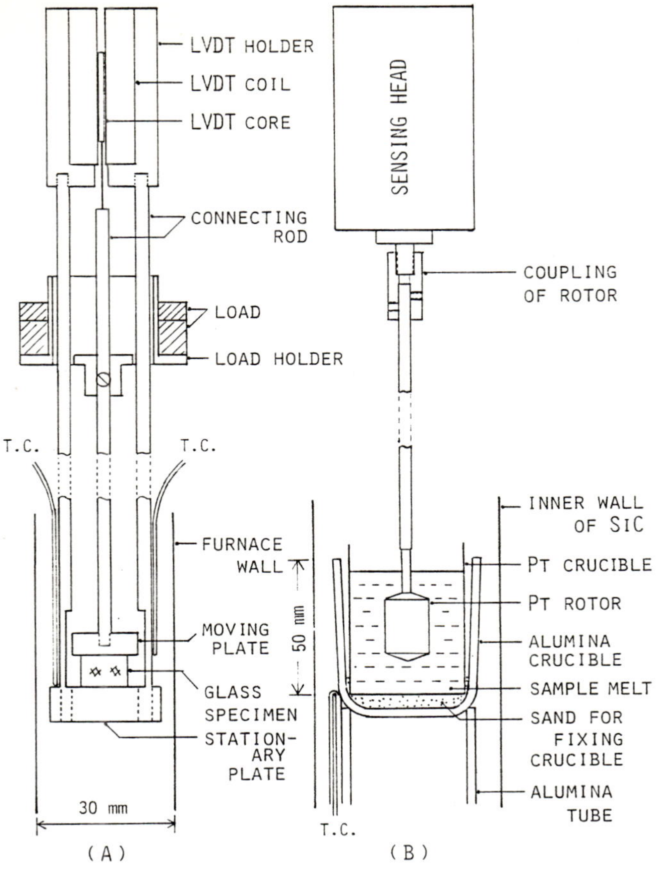

Fig. 1 Schematic diagram of measuring apparatus. (A); Parallel plate viscometer. (B); Rotating cylinder viscometer.

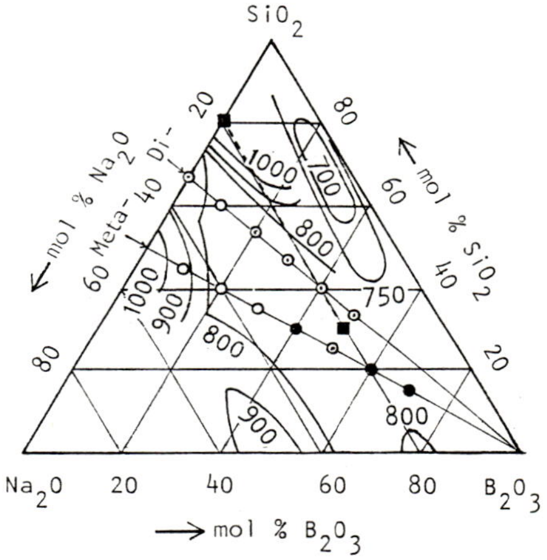

Fig. 2 Composition of the samples. ⊙; measured in both glassy and molten samples, ○; in glassy samples, ●; in molten samples. ■ are those of Fulcher's experimental range.

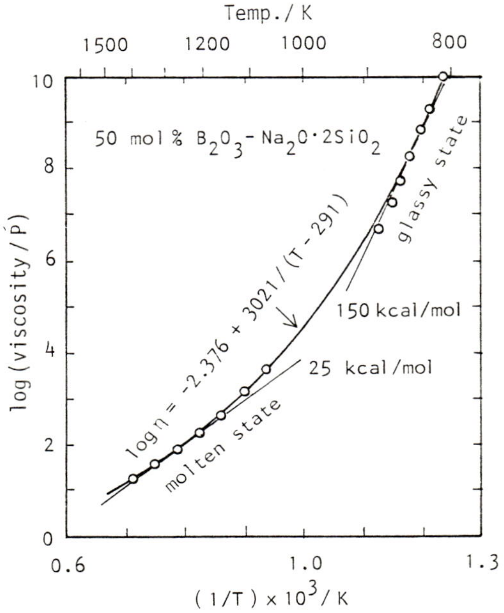

Fig. 3 Relationship between log viscosity and 1/T.

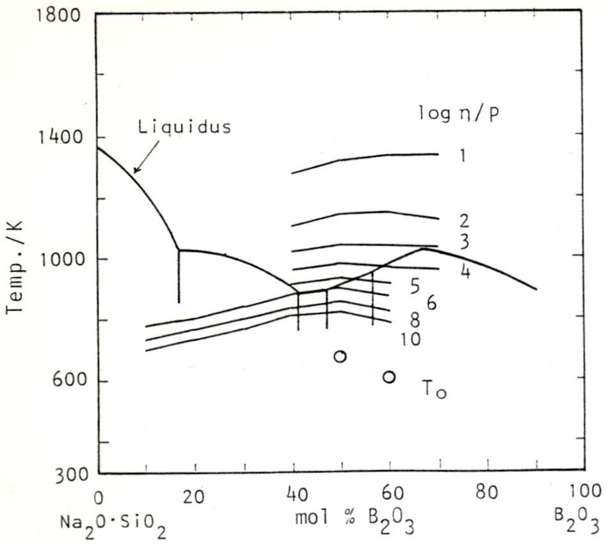

Fig. 4 Iso-viscosity curves in the pseudo-binary system, $Na_2O{\cdot}SiO_2$-B_2O_3; T_o is the Fulcher constant.

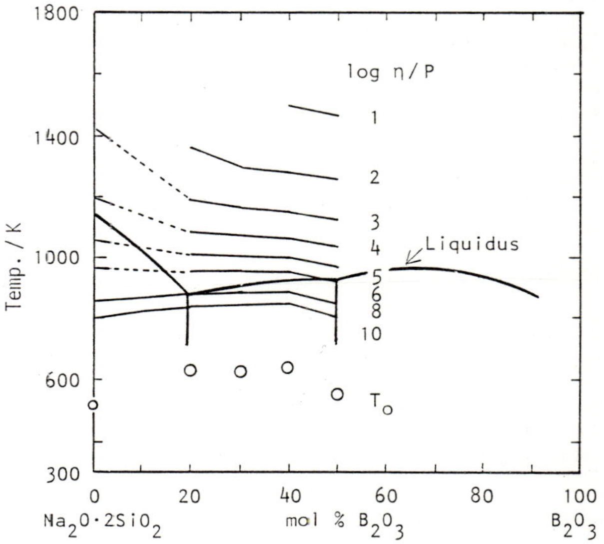

Fig. 5 Iso-viscosity curves in the pseudo-binary system, $Na_2O{\cdot}2SiO_2$-B_2O_3; T_o is the Fulcher constant.

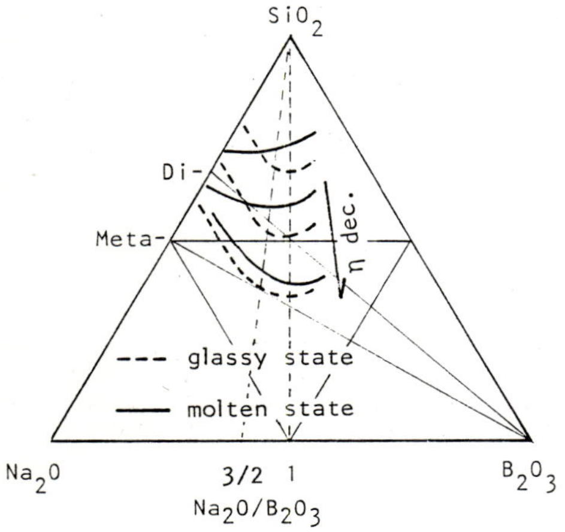

Fig. 6 Schematic representation of iso-viscosity curves in glassy and molten states. Solid and broken lines are the molten and the glassy states, respectively.

Determination of viscosity of B_2O_3, PbO and binary B_2O_3-PbO melts using an oscillating-plate viscometer

T IIDA, Z MORITA and T ROKUTANDA

The authors are in the Department of Materials Science and Processing, Osaka University, Osaka, Japan.

SYNOPSIS

The viscosities of B_2O_3, PbO, and binary B_2O_3-PbO melts have been measured using an oscillating-plate viscometer, constructed by the authors. The viscosities of B_2O_3-PbO melts decrease rapidly with increasing concentration of PbO. The network parameter proposed previously by the authors was found to decrease with increasing PbO content. The network present in molten B_2O_3 is gradually broken down by additions of PbO.

INTRODUCTION

As is well known, the viscosity of a liquid is very dependent upon the structure of the melt. In this paper experimental results for the viscosities of B_2O_3, (a network former) PbO, (a network breaker) and binary B_2O_3-PbO melts, are presented and these are discussed in terms of both the anion-cation attraction parameter and the network parameter proposed by the authors(1).

EXPERIMENTAL

The oscillating-plate viscometer illustrated in Fig.1 has been constructed by the authors, to measure the viscosities of molten slags and fluxes(2). The principle underlying the oscillating-plate method is based on the measurement of the amplitude of plate oscillation in both air and in the melt; the relationship between viscosity and amplitude can be expressed in the following form(3).

$$\rho\mu = K\left[\frac{E_a}{E} - 1\right]^2 \qquad (1)$$

where respectively, ρ and μ are the liquid density and viscosity, E_a and E are the resonant amplitudes of plate oscillation in air and in liquid, K is the apparatus constant, the value of which was determined experimentally by the use of Standard Reference Materials for viscosity measurements.

Compositions of the samples and the materials used for the fabrication of the oscillating plates are given in Table 1. An equal volume of sample ($1.3 \times 10^{-4} m^3$), was used in all experiments. After dehydrating, the sample was further dried by placing it in a magnesia crucible (53 mm diam × 100 mm), where it was heated for 2 hours at 1273K in a $MoSi_2$ furnace. A Pt-30% Rh/Pt-6% Rh thermocouple was then inserted directly into the melt and instantaneous and continuous data were recorded automaticaly for the product ($\rho\mu$) of density and viscosity of the melt by measuring the amplitudes of the plate oscillation with the resonant frequency of 20 Hz. The amplitudes of the plate were observed precisely using a non-contact electro-optical measurement system. These measurements were carried out only during the cooling process with the cooling rate of ca. $2.8 \times 10^{-8} Ks^{-1}$.

RESULTS

The values obtained in the present study for the viscosity of pure molten B_2O_3 are given in Fig.2, together with the data recorded by other investigators. It can be seen from Fig.2, that the results of the three independent measurements are in close agreement.

The experimental viscosity data for B_2O_3, PbO, and binary B_2O_3-PbO melts are shown in Fig.3. The values of density for pure molten B_2O_3 and PbO were taken from the data reported by Shartsis et al(5) and Riebling(6), respectively, and values of the density of the mixtures were computed using the additivity principle. As is shown in Fig.3, the viscosities of B_2O_3-PbO melts decrease rapidly with increasing concentration of PbO. The viscosities of B_2O_3 and binary B_2O_3-PbO melts, with the exception of the data for PbO and B_2O_3-90% PbO, do not obey the general form of the Arrhenius equation. This is especially true for the viscosities of the B_2O_3-70% PbO melt, which shows anomalous temperature variations.

Values of the isothermal viscosities of binary B_2O_3-PbO melts are given in Fig.4, together with the data reported by Ejima and Kameda(7). It was estimated that the experimental uncertainties were approximately ± 5%.

DISCUSSION

The authors have proposed a parameter calculated from viscosity data, (the network parameter, Ψ) can provide a measure of the structure of the melt. This parameter, Ψ, can be defined by equation (2)

$$\Psi = \log\left[\frac{\mu}{\mu_o}\right]^2 \qquad (2)$$

where μ_0 is the viscosity of non-network-forming liquids, which can be expressed in terms of well-known physical quantities(1)(9), and μ is the viscosity of liquids obtained experimentally. Values of the network parameter can thus provide a quantitative description of the network structure of liquids.

The computed values of the network parameter for pure molten B_2O_3 and PbO at 1173K are 8.8 and 1.0, respectively. It can be seen from Figs.5 and 6, respectively, that the network parameter decreases with (i) increasing PbO content and (ii) decreasing anion-cation attraction parameter. This implies that the network structure of molten B_2O_3 is gradually broken down by the addition of PbO.

CONCLUSIONS

(i) The viscosities of binary B_2O_3-PbO melts decreased rapidly with increasing concentration of PbO.

(ii) The network parameter was for pure, molten B_2O_3 and PbO at 1173K and values of 8.8 and 1.0, respectively were obtained.

(iii) The network present in molten B_2O_3 was gradually broken down the addition of PbO.

REFERENCES

(1) Z MORITA, T IIDA, H OKUDA, and M KAWAMOTOR: The 19th Committee, the Japan Society for Promotion of Science (JSPS), Rep. No. 629, (May 1986).

(2) Z MORITA, T IIDA, M KAWAMOTO, and A MORI: Tetsu-to Hagané, 1984 70, 1242.

(3) T IIDA and R I L GUTHRIE: The Physical Properties of Liquid Metals, (1988), Clarendon Press, Oxford.

(4) E T TURKDOGAN: Physicochemical Properties of Molten Slags and Glasses, (1983), p 13, The Metals Society London.

(5) L SHARTSIS, W CAPPS, S SPINNER: J. Am. Ceram. Soc., 1953, 36, 319.

(6) E F RIEBLING: Inorg. Chem., 1964, 3, 958.

(7) T EJIMA and M KAMEDA: J. Jpn Inst. Metals, 1967, 31, 120.

(8) Z MORITA, T IIDA, H OKUDA, and M KAWAMOTOR: The 19th Committee, the Japan Society for Promotion of Science (JSPS), Rep. No. 629, (May 1986).

(8) T IIDA, Z MORITA, and T MIZOBUCHI: Proc. 3rd Inter. Conf. Molten Slags and Fluxes, Glasgow, (1988).

TABLE 1. Compositions of samples and materials of oscillating-plates

Sample	Oscillating-plate (square-cross-section)		
mol%	Material	Length of side/mm	thickness/ mm
Pure B_2O_3	Stainless steel	20	0.5
B_2O_3-5% PbO	Platinum	20	0.2
B_2O_3-10% PbO	Platinum	20	0.2
B_2O_3-15% PbO	Platinum	30	0.2
B_2O_3-20% PbO	Platinum	25	0.2
B_2O_3-70% PbO	Magnesia	30	0.8
B_2O_3-90% PbO	Magnesia	30	0.8
Pure PbO	Magnesia	30	0.8

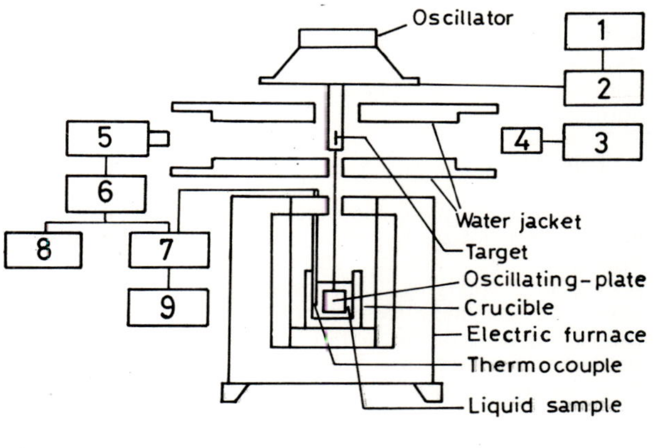

1. Schematic diagram of the oscillating-plate viscometer.
(1) Function generator (2) Amplifier
(3) DC Power supply (4) Light source
(5) Camera unit }
(6) Control unit } amplitude measurement system
(7) Multi-meter (8) Waveform analyser
(9) Microcomputer.

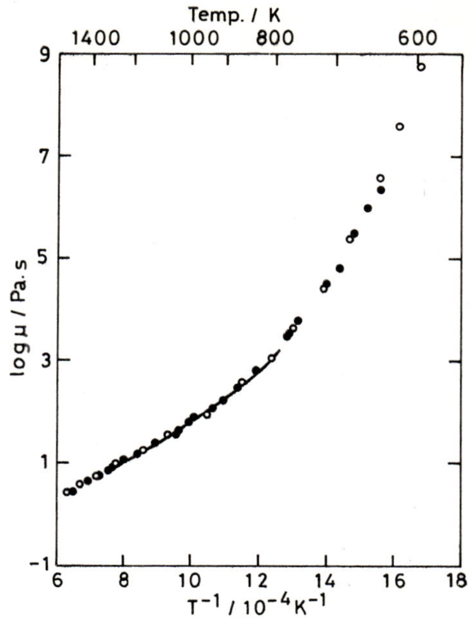

2. The temperature dependence of the viscosity of B_2O_3; o, Napolitano et al; •, Eppler, (as given in ref (4)); ——, present investigation.

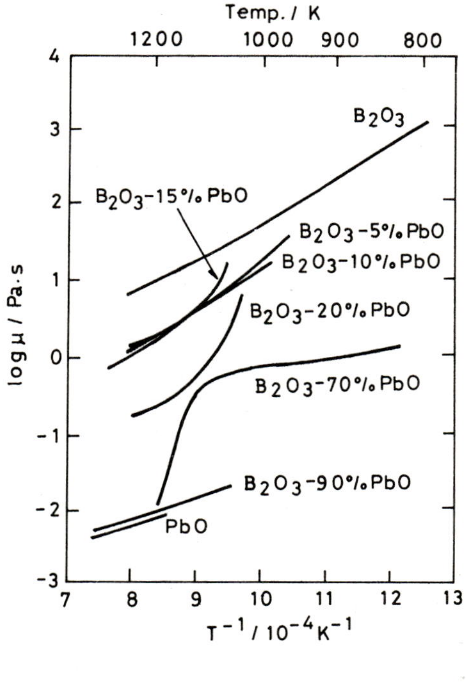

3. The temperature dependence of the viscosity of B_2O_3–PbO melts (numerals refer to mol% PbO)

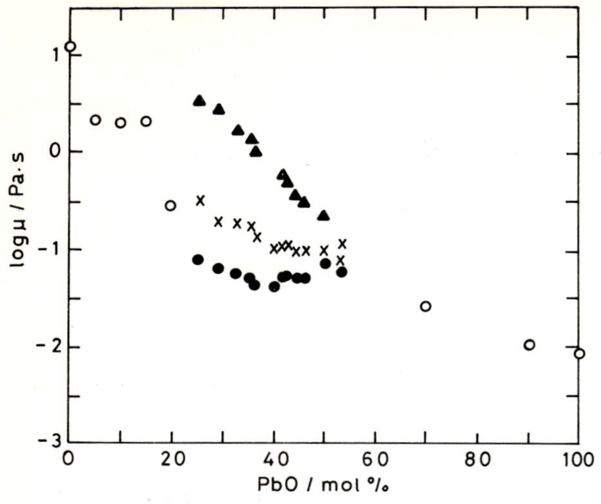

4. The isothermal viscosities of binary B_2O_3–PbO melts. o, Present investigation (o 1173 K); Δ × ●, Ejima and Kameda (▲, 1073 K, ×, 1273 K, ●, 1473 K)

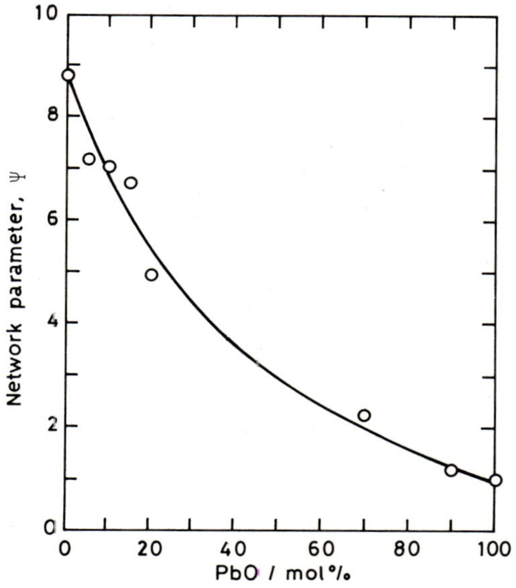

5. The network parameter, Ψ, as a function of the PbO content at 1173K

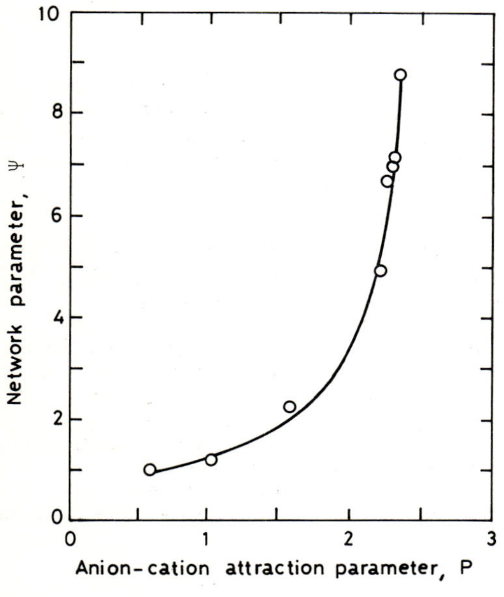

6. The network parameter, Ψ, as a function of the anion–cation attraction parameter at 1173K. $\Psi = 1.92/(2.56-P)$. □

Relationships between a parameter deduced from viscosity and some physico-chemical properties of molten slags

T IIDA, Z MORITA and T MIZOBUCHI

The authors are in the Department of Materials Science and Processing, Osaka University, Osaka, Japan.

SYNOPSIS

In order to provide a quantitative description of the network structure of slags, we have introduced a simple parameter, which can be deduced from the viscosity of melts, since the latter is very sensitive to the structure of the slag. It has been established that relationships exist between this network parameter and (i) the anion-cation attraction parameter, (ii) the sulphide capacity, (iii) the lattice energy, and (iv) the optical basicity of molten slags, respectively.

INTRODUCTION

The physico-chemical properties of molten slags and fluxes are influenced by their network structure. In order to obtain a clear understanding of the physico-chemical properties of slags, it is necessary to have quantitative descriptions of the network structures of the slag. Consequently, we have derived a parameter from viscosity data which can provide reliable and quantitative values of the network structure.

DEFINITION OF NETWORK PARAMETER

The following viscosity equation holds approximately for monatomic liquids, such as metals, which do not form networks[1],

$$\mu_0 = 1.8 \times 10^{-7} \frac{(MT_m)^{\frac{1}{2}}}{V_m^{2/3}} \frac{\exp(H_\mu/RT)}{\exp(H_\mu/RT_m)} \quad \text{(in SI units)} \quad (1)$$

$$H_\mu = 5.1\ T_m^{1.2}$$

where μ_0 is the viscosity of non-network-forming liquids, (ie. hypothetical molten slags) M the formula weight, T_m the melting point, V_m the molar volume at T_m, R the gas constant. In order to obtain a tractable solution, values of μ_0 are computed for mixtures using the additivity principle, despite this assumption being only approximately true.

Since a rigorous quantitative treatment of network structure of molten slags would be, in practice, impossible to derive at the present time, we have intuitively introduced a simple parameter deduced from the viscosity, this property being particularly sensitive to the network structure[2]. The network parameter Ψ is defined as

$$\Psi = \log \left[\frac{\mu}{\mu_0} \right]^2 \quad (2)$$

where μ is the viscosity of molten slag obtained experimentally. It has been assumed that molecules in network-forming liquids behave in such a way that the mass of the molecules is apparently different to those for liquids which contain no network structure. In other words, the mass of the molecules in a network-forming liquid, viz. the apparent mass, is much larger than that in hypothetical liquids which do not form networks. Consequently it is possible to derive a contribution to the viscosity of a liquid which is related to the network structure and which could be calculated solely from the apparent mass (or an apparent formula weight M^*) of the molecules. Thus, from eqn (1), based on this assumption, we are led to the conclusion that

$$\Psi \simeq \log \frac{M^*}{M} \quad (3)$$

In order to facilitate the calculation of Ψ for complex oxide melts, values of μ_0 at various temperatures are listed in Table 1 for the principal constituents of slags for temperatures between 1400 and 2000 K; a worked example is also presented in this table.

RELATIONSHIPS BETWEEN THE NETWORK PARAMETER AND THE PHYSICO-CHEMICAL PROPERTIES OF MOLTEN SLAGS

It could be expected that relationships would exist between the network parameter and the physico-chemical properties of slags. Four examples, demonstrating such relationships, are shown in Figs 1 to 4. It can be seen from these figures, that simple relationships were found to exist between the network parameter and the anion-cation attraction parameter, the sulphide capacity, the lattice energy, and the optical basicity of molten slags, respectively. However in the case of the optical basicity – (Ψ) plot (Fig. 4) each slag would appear to lie on its own curve. This is perhaps not surprising since the optical basicity is associated with the electron donor properties of the slag and consequently, may not be related to the structure of the melt. In the case of oxide melts, the network structure would depend upon the states of oxygen in the melt, eg bridged, non-bridged, or free oxygen. The compositional ranges of the various slags studied are given in Table 2.

It should be noted that the values of the anion-cation attraction parameter, P, and the lattice energy, $\bar{E}_c$, were computed using equations (4) and (5), respectively. The anion-cation attraction parameter for a multi-component system is defined as

$$P = \frac{\sum(Z_i^+ n_i^+ X_i P_i)}{\sum(\bar{Z}_i \bar{n}_i X_i)} \quad (4)$$

$$P_i = \frac{Z_i^+ Z_i^-}{a_i^2}$$

where, respectively, Z^+ and Z^- are the valences of the cation and anion, n^+ and n^- are the numbers of the cation and anion, X is the mole fraction, a is the distance between cation and anion, and the subscript i refers to the component.

The lattice energy per mole of O^{2-} ion for a multi-component system is given by[3]

$$\bar{E}_c = \frac{825X_{CaO} + 919X_{MgO} + 3662X_{Al_2O_3} + 3137X_{SiO_2}}{X_{CaO} + X_{MgO} + 3X_{Al_2O_3} + 2X_{SiO_2}} \quad (5)$$

CONCLUSION

Simple relationships exist between a parameter derived from viscosity data, viz: the network parameter, and some of the physico-chemical properties of slags.

REFERENCES

1. T. IIDA and R.I.L. GUTHRIE: The Physical Properties of Liquid Metals, (1988), Clarendon Press, Oxford

2. Z. MORITA, T. IIDA, H. OKUDA, and K. KAWAMOTO: The 19th Committee, JSPS, Rep.No.629, (May 1986).

3. E. ICHISE and A. MORO-OKA: The 140th Committee, JSPS, Rep.No.195, (Dec. 1986).

Table 1
Calculated values for the viscosity μ_0 of non-network-forming liquids

$\mu_0/10^{-1}$ Pa.s =Poise

Component	1400	1450	1500	1550	1600	1650	1700	1750	1800	1850	1900	1950	2000/K
SiO_2	.082*	.072	.063	.056	.050	.45	.041	.037	.034	.031	.029	.027	.025
Na_2O	.012	.011	.010	.0095	.0090	.0085	.0080	.0076	.0073	.0070	.0067	.0064	.0062
CaO	.82	.67	.55	.45	.38	.32	.28	.24	.21	.18	.16	.14	.13
FeO	.058	.052	.047	.042	.039	.036	.033	.031	.029	.027	.025	.024	.022
Al_2O_3	.21	.18	.15	.13	.12	.10	.091	.081	.073	.066	.060	.055	.050
K_2O	.0071	.0067	.0063	.0060	.0057	.0055	.0052	.0050	.0048	.0047	.0045	.0044	.0042
Li_2O	.077	.067	.059	.052	.046	.042	.038	.034	.031	.029	.027	.025	.023
PbO	.023	.021	.020	.018	.017	.016	.016	.015	.014	.014	.013	.013	.012
B_2O_3	.0037	.0036	.0035	.0033	.0032	.0032	.0030	.0030	.0029	.0028	.0027	.0027	.0026
TiO_2	.12	.11	.094	.082	.073	.065	.059	.053	.048	.044	.040	.037	.035
MgO	1.6	1.2	1.0	.81	.67	.56	.48	.41	.35	.30	.27	.23	.21
MnO	.14	.12	.11	.095	.084	.076	.068	.062	.056	.052	.048	.044	.041
Fe_2O_3	.010	.0098	.0092	.0088	.0084	.0080	.0077	.0074	.0072	.0069	.0067	.0065	.0061

* .082 = 0.082

For example, in the case of $CaO-Al_2O_3-SiO_2$ system at 1800 K, a value of μ_0 is given by:

$$(\mu_0)_{mixture} = 0.21X_{CaO} + 0.073X_{Al_2O_3} + 0.034X_{SiO_2}$$

where X is the mole fraction and

$$X_{CaO} + X_{Al_2O_3} + X_{SiO_2} = 1$$

Table 2
Composition ranges of various slags covered in Figs 1 to 4.

Symbol	Slag	Composition range/mol%
o	$CaO-SiO_2$	CaO=43-70, SiO_2=30-57
●	$CaO-Al_2O_3-SiO_2$	CaO=20-50, Al_2O_3=10-40, SiO_2=20-70
Δ	$CaO-MgO-SiO_2$	CaO=9-51, MgO=7-39, SiO_2=39-57
■	$CaO-MnO-SiO_2$	CaO=0-50, MnO=0-50, SiO_2=0-50
▲	$CaO-Al_2O_3-SiO_2-MgO$	CaO=18-47, Al_2O_3=5-11, SiO_2=29-65, MgO=7-22
□	$Al_2O_3-MgO-SiO_2$	Al_2O_3=5-16, MgO=16-41, SiO_2=54-69
×	SiO_2-BaO	SiO_2=52-85, BaO=15-48

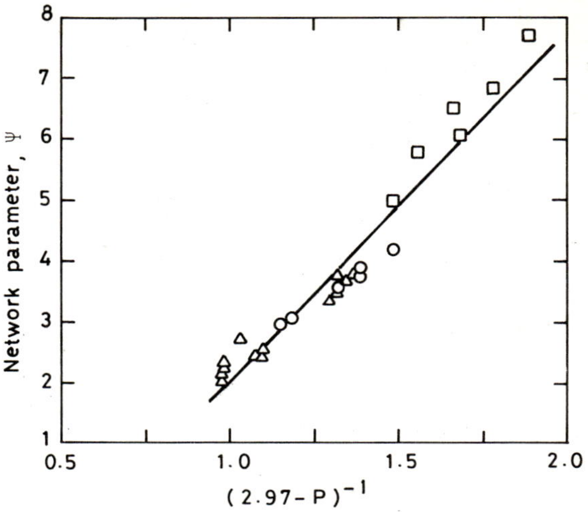

1. Relationship between network parameter, Ψ, and anion-cation parameter, P, for molten slags at 1773 K.

$\Psi = \{5.73/(2.97-P)\} - 3.66.$

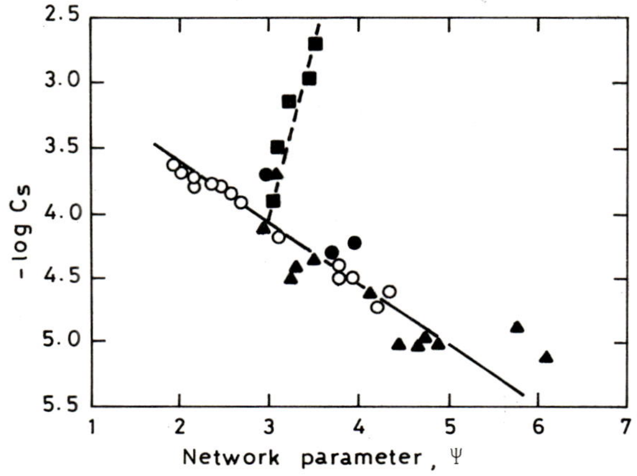

2. Relationship between network parameter, Ψ, and sulphide capacity, C_S, for molten slags at 1773 K.

$-\log C_S = 0.479\Psi + 2.65.$

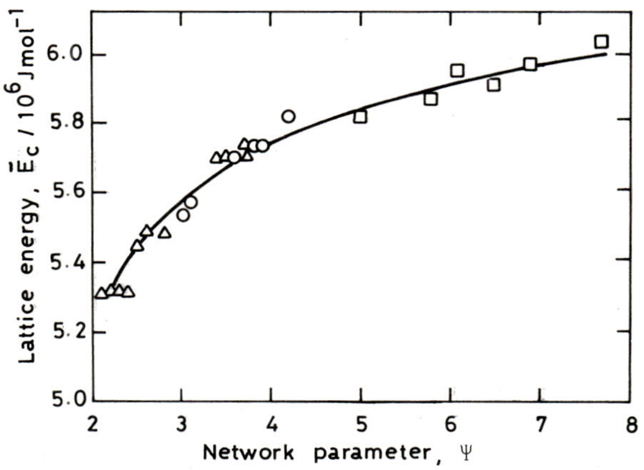

3. Relationship between network parameter, Ψ, and lattice energy, $\bar{E}_c$, for molten slags at 1773 K.

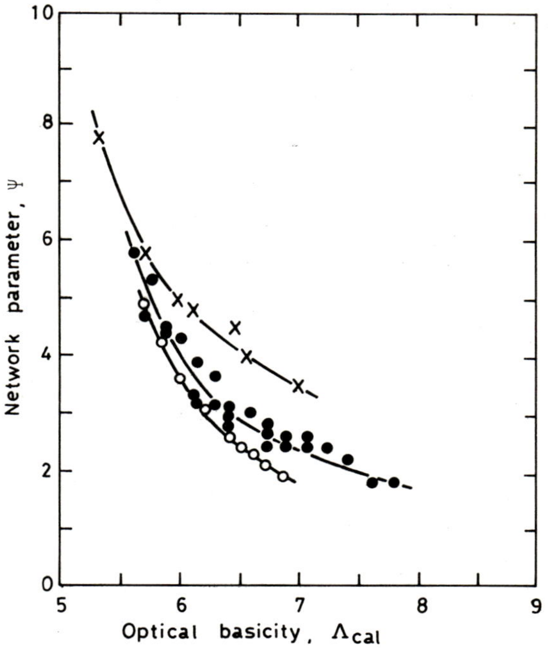

4. Relationship between network parameter, Ψ, and optical basicity, Λ_{cal}, for molten slags at 1973 K.

□

Anionic modification of electrical conductivity in liquid silicates

R A BERRYMAN and I D SOMMERVILLE

The authors are with the Ferrous Metallurgy Research Group, Department of Metallurgy and Materials Science, University of Toronto, Ontario, Canada.

Synopsis

Modification of the anionic structure in calcium silicate liquids was studied by means of electrical conductivity measurements. Up to 10 mass percent of calcium fluoride or aluminum oxide were added to a series of calcium silicate liquids. A series of related experiments was also performed in which a partial substitution of magnesia was made for calcium oxide. Results showed analogous behaviour between fluorine and oxygen and between aluminum and silicon in the anionic structure, and between magnesium and calcium cations.

Introduction

Electrical conductivity data for silicate melts is of both fundamental and technological interest. In considerable early work, such as that of Bockris and co-workers[1,2], the variation of conductivity was examined across binary systems as a means for drawing conclusions regarding melt structure. In work of direct industrial interest, measurements of conductivities of electro-slag remelting fluxes[3,4] are perhaps an obvious example. Similarly, electrical conductivities have been reported for melts encountered in chromium[5] and nickel[6] electric smelting, among others.

In the present work, the effect of adding various third components to calcium silicate liquids was studied using measurements of electrical conductivity. In previous work by Keller et al.[7] measurements of calcium diffusion rates were compared with those of electrical conductivity in melts of the same composition. It was shown that in pure calcium silicates liquid at 1873K (1600 °C) or below, the calcium (Ca^{2+}) ion was responsible for the majority, if not all, of the charge transferred during alternating current electrical conduction. A structural interpretation of their results, consistent with various silicate models[8-11] is that on the silica-rich side of the orthosilicate composition (the only composition range liquid at 1873K), the only species present in significant amounts are calcium cations and silicate anions. The latter do not contribute to conduction owing to their size. Under the influence of an externally applied electric field, then, calcium ions move between positions of local minimum potential energy in the silicate structure. Those positions favourable to occupation by the calcium ions are surrounded by several oxygen atoms bonded singly to parts of silicate anions. The difficulty of calcium ion movement is increased as the mole fraction of silica, and hence the degree of melt polymerisation, increases. Under such conditions, the number of sites which calcium ions can occupy decreases, thus increasing the average distance for each jump, and the structure between sites develops an increasing number of oxygen bridges, raising the activation energy required by calcium ions for motion.

If an additional component is now introduced to the melt, its behaviour in the silicate environment will affect the ease of calcium ion movement. Additions studied in this work were CaF_2 and Al_2O_3, either individually or together. Calcium fluoride dissociates to form calcium and fluoride ions. In acid melts, the fluorine can replace oxygen on silicate sites. Evidence of this behaviour was suggested in early viscosity work[12], and proved subsequently by mass-loss studies[13], in which the increasing loss of SiF_4 gas at increasing silica contents demonstrated an association between silicon and fluorine in the liquid state. In very basic melts, on the other hand, such as CaO-CaF_2-Al_2O_3 fluxes used for ESR, the fluoride ions are found in the free (F^-) state, and contribute substantially to the conduction process[3]. Thus, an equilibrium can be proposed[13]:

$$\text{-Si-O}^- + F^- = O^= + \text{-Si-F}^o \qquad (1)$$

When the basicity is high this equilibrium shifts to the left, freeing the fluoride ions to conduct, while under acidic conditions, the fluorine tends to be bonded to silicon.

Aluminum has been assumed to be capable of substitution for silicon in silicate chains[14], with a tetrahedral coordination. In keeping with the amphoteric character of aluminum ions, however, a six-fold coordination state for the Al^{3+} ion was also postulated. In this case, aluminum acts as a donor, rather than as an acceptor of oxygen, and such behaviour would be expected under conditions of high alumina and/or silica content. In other work[15] a triangular coordination for aluminum has been assumed, in forming ions such as $Al_2O_5^{4-}$. Obviously, the dissolution of aluminum oxide into a melt would consume more oxygen in the case of four-fold coordination, causing a higher degree of polymerisation, than for three-fold coordination. By measuring conductivity, and comparing the results to those for alumina-free melts, it should be possible to determine the coordination number of aluminum.

Finally, the effect of substituting part of the calcium by magnesium was studied. Unlike the other additions, magnesium ions would be expected to behave in a manner similar to calcium, and to conduct. However, since the magnesium ion is smaller than calcium, and hence more highly polarizing, the actual conductivity measured might be somewhat different.

Experimental

Experimental equipment used for measuring electrical conductivity was similar to that used by Victorovich et al[6]. A single molybdenum electrode was located concentrically in a cylindrical molybdenum crucible, and could be moved vertically using a threaded thumbwheel to locate the electrode tip at any point from above the melt surface to the crucible bottom. Resistance measurements were made using a Hewlett Packard 4328A milliohmeter operating at a frequency of 1kHz. A series of 10-11 resistance measurents were made at predetermined depths as the electrode was lowered, and again as it was raised. After repeating the procedure, the four resistance measurements taken at each depth were averaged, to remove the effect of meniscus deflection by the electrode. Since the electrode was traversed between the liquid surface and crucible bottom, the depth of the liquid could be measured, and used in estimating the density at each temperature.

Calibration of each crucible was performed at room temperature, prior to the experiment. Since the oxide melts could be expected to dissolve any molybdenum or other oxides from the surface of the crucible or electrode, thus ensuring good contact between the metallic surface and melt, some care was taken with preparation of the crucible and electrodes prior to calibration, to ensure a similar degree of contact between the aqueous calibration solution and the molybdenum surfaces. Molybdenum surfaces were polished to a 25 micron (#600) finish with silicon carbide paper, and carefully cleaned with detergent solution to remove any grease and loose abrasive. After swabbing with acetone and drying, the crucible was filled to a depth of 40mm with aqueous potassium chloride solution (usually 1 molar) for calibration. Because of the high temperature sensitivity of the electrical conductivity of the calibration solution, the crucible was immersed in a water bath maintained at 293K (20 °C).

Melts were prepared by heating premixed powders of the desired composition in the precalibrated molybdenum crucible, using 10kHz induction heating, with an Ar-15%H_2 atmosphere to prevent oxidation of the molybdenum. After fusion of the powder mixture, a molybdenum paddle-type stirrer was used to ensure homogenization of the melt. Generally, two melting cycles were required to reduce the volume of the initial powder charge (110g) to a liquid having a depth of 35-40mm.

Each experiment was performed under an atmosphere of purified argon, in a closed-end recrystallized alumina tube heated externally by a molybdenum resistance furnace. A series of radiation shields above the crucible helped to maintain a hot zone with less than ± 3 degrees fluctuation along the length of the crucible. A 'B'-type thermocouple was inserted through holes in each of the radiation shields into the top of the crucible, immediately above the melt. Details of the apparatus are shown in Figure 1. Prior to the experiment, the apparatus was lowered into the furnace tube over a period of four hours, using a screw-driven carriage mounted on linear bearings. A period of 1-2 hours was allowed for thermal equilibration, before the first series of measurements was made. At each temperature, traverses of depth with the electrode required about 30 minutes. A further 30 minutes were allowed for re-equilibration at each new temperature, typically 20-30 degrees different from the last. Measurements were made between a maximum of 1910K (1637 °C) and the melting points of the liquids, which ranged from about 1620 to 1820 K (1347~1547 °C).

Data were treated by finding the slope of the straight line obtained when the cell constant G (cm^{-1}) obtained from the calibration trial is plotted as a function of resistance R (ohms) of the melt. Pairs of points (R,G) on each line represent data for a common immersion depth of the electrode. Figure 2 shows several such plots schematically, representing a range of melt temperatures. The slope of each line represents the conductivity k ($ohm^{-1} \cdot cm^{-1}$) of the melt. All calculations were done using a computer program.

Results

CaO-SiO_2

Results for lime-silica melts without additions are shown in Figure 3, showing conductivity as a function of composition at 1823K (1550 °C). Data from other work[1,7,16] are shown for comparison. Agreement is best between this work and the work of Keller et. al[7], while the other data shown tend to suggest higher conductivities for any given composition.

CaO-CaF_2-SiO_2

Points showing ln (conductivity) as a function of amount of calcium fluoride added to different calcium silicate compositions are plotted in Figure 4, for a temperature of 1800K (1527 °C). Calcium silicate compositions with or without fluoride additions shared a series of common CaO/SiO_2 ratios. Curves have been drawn for each of these lime/silica ratios according to a scheme explained below. For all values of lime/silica ratio shown, the effect of calcium fluoride addition is to strongly increase the liquid conductivity. The slopes of the curves are greatest for the most acidic compositions, and become steeper as the amount of fluoride approaches zero.

CaO-Al_2O_3-SiO_2

As in the case of fluoride additions, the data for alumina additions have been plotted to show ln (conductivity) as a function of mole fraction of alumina, at 1800K. These data are shown in Figure 5. Again, curves are drawn representing constant lime/silica ratios, and the association between the points and each curve is indicated on the graph. The data suggest a strong decrease in conductivity with alumina addition, consistent with the consumption of oxygen by alumina upon dissolution. The effect, as reflected in the slopes of the curves, is greater at more acidic compositions, which is consistent with the smaller amount of oxygen available per network-forming atom.

Discussion

Figure 4 showed an increase in conductivity with CaF_2 additions at constant CaO/SiO_2 ratios. Two factors are responsible for this increase. The first, obviously, is that in adding calcium fluoride, the concentration of conducting calcium ions increases. This effect can be accounted for by re-expressing the electrical conductivity as an equivalent conductivity of calcium ion, $A_{Ca^{++}}$:

$$A_{Ca^{++}} = k_{Ca}/n_{Ca} \cdot z_{Ca} \qquad (2)$$

where n_{Ca} is the concentration of calcium ion, in moles per cm^3, and z_{Ca} is the valence of the ion. In using this equivalent conductivity, it is assumed that the transport number of calcium is unity so that k_{Ca} is equal to the measured specific conductivity of the sample. Using equivalent conductivity not only accounts for the effect of increasing the calcium ion content by calcium fluoride addition, but also takes into account the increasing calcium ion concentration as the CaO/SiO_2 ratio increases.

The second effect on conductivity introduced by calcium fluoride is that of the fluorine. As was mentioned previously, association between silicon and fluorine has been shown in acidic calcium fluorosilicates. However, it is possible, depending on melt basicity, that appreciable amounts of free fluoride ion, F^-, could also be present in the melt, contributing to the total charge transported. In order to distinguish between the contribution of fluorine to silicate network-breaking, and the possible contribution to the transport of charge, the conductivity data were plotted to show equivalent conductivity of calcium as a function of a stoichiometric parameter, B:

$$B = \frac{n_O + n_F}{4 n_{Si}} \qquad (3)$$

In physical terms, this represents the number of network breakers (oxygen or fluorine) available, per bond on each silicon atom. The definition of this parameter assumes that all of the fluorine introduced to the melt has behaved in the same manner as oxygen, in depolymerizing silicate structure. Further, by converting measured specific conductivity to equivalent conductivity of calcium, it has been assumed that conducting fluoride ions are not present. Thus, if a plot of experimental data showed different behaviour between melts with and without fluoride additions, this would suggest that fluorine is found, at least to some extent, as a free ion.

By a minor modification to the parameter B, we can also include the effect of alumina, if its behaviour is acidic. Thus, B becomes:

$$B = \frac{n_O + n_F}{4 n_{Si} + x n_{Al}} \qquad (4)$$

where x is the coordination number of aluminum. Figure 6 shows the experimental results for the different melts, plotted as equivalent conductivity of calcium against B. In this plot, a value of three was used for x. The agreement between the points representing alumina-containing compositions, and the remainder of the data, was somewhat better than when a value x=4 was chosen. The line drawn is a least-squares fit. Interestingly, it passes through zero conductivity at a point on the B axis representing a composition, for pure calcium silicate, close to the SiO_2 liquidus. It can be seen that points representing fluoride-containing melts appear to follow the same A_{Ca} vs. B relationship as do those points for calcium silicate melts. Thus, the results suggest that in the range of composition studied, fluorine is bonded to silicon. At high values of the parameter B, the availability of oxygen is greater, and more fluoride could be expected to be found in the free state, consistent with equation (1).

In the case of alumina additions, the good fit of the data to the same line confirms the assumed behaviour. Partial substitution of silicon by aluminum in the silicate network leads to little change in the network structure, resulting in good agreement between the different conductivity data.

To allow the inclusion of results from the experiments where magnesium was substituted for calcium, the equivalent conductivity parameter was modified, to an average (weighted by cation fraction) of the equivalent conductivities for calcium and magnesium. As seen in Figure 6, magnesium substitution for calcium leads to little measurable difference in liquid conductivities. This is, to some extent, consistent with viscosity data[17], where the effect of MgO was indistinguishable from that of CaO. However, a trend in the current data, if one exists, is that the conductivity of melts with MgO substitution falls away from that of melts without MgO, as B increases. In highly acidic melts, the two cations conduct equivalently, but in more basic melts, the higher polarizing ability of the magnesium ion relative to calcium appears to reduce its mobility.

Data from three experiments in which both alumina and calcium fluoride, as well as magnesia in one case, were added, are also shown in Figure 6. Reasonable agreement with the other data is seen, although all three points lie above the line.

Conclusions

Electrical conductivity measurements were made in calcium silicate liquids, with additions of CaF_2, Al_2O_3, and MgO, either singly or in combination. Results were plotted assuming that only Ca and Mg ions transport charge, and that the fluorine and aluminum enter the silicate anion structure. Good agreement between the data for different compositions was shown, suggesting that this interpretation is correct, over the composition range studied.

Acknowledgements

The authors wish to thank Albright and Wilson Americas, Incorporated for financial support of this work. R. Berryman gratefully acknowledges personal support received from the Natural Sciences and Engineering Research Council.

References

1. J. O'M. Bockris, J. A. Kitchener, S. Ignatowicz and J. W. Tomlinson, Disc. Farad. Soc., **4** (1948), pp265-281.
2. J. O'M. Bockris, J. A. Kitchener, S. Ignatowicz and J. W. Tomlinson, Trans. Farad. Soc., **48** (1952), pp75-91.
3. A. Mitchell and J. Cameron, Metall. Trans., **2** (1971), pp3361-3366.
4. S. Hara, H. Hashimoto and K. Ogino, Trans. ISIJ, **23** (1983), pp1053-1058.
5. D. I. Ossin, D. D. Howat and P. R. Jochens, Proc. Electric Furnace Conf., AIME, Toronto, **29** (1971), pp94-100.
6. G. S. Victorovich, C. Diaz and D. K. Vallbacka, Proc. 2nd. Int'l. Symp. Metallurgical Slags and Fluxes, AIME, (1984), pp907-924.
7. H. Keller, K. Schwerdtfeger and K. Hennesen, Metall. Trans. B, **10B** (1979), pp67-70.
8. G. W. Toop and C. S. Samis, Trans. TMS-AIME, **224** (1962), pp878-887.
9. C. R. Masson, Proc. Roy. Soc. (London), **A287** (1965), pp201-221.
10. C. R. Masson, J. Iron Steel Inst., **210** (1972), pp89-96.
11. P. L. Lin and A. D. Pelton, Metall. Trans. B, **10B** (1979), pp667-675.
12. P. Kozakevitch, Rev. Met., **51** (1954), pp569-584
13. D. Kumar, R. G. Ward and D. J. Williams, Disc. Farad. Soc., **32** (1962), pp147-154.
14. P. Kozakevitch, Phys. Chem. of Proc. Metallurgy, G. St. Pierre, Ed., pp97-116, Wiley (Interscience), NY 1961.
15. Shi Xue Dou, J. Phys. Chem., **85** (1981), pp3859-3863.
16. T. Hoster and J. Potschke, Arch. Eisenhuttenw., **54** (1983), pp389-94.
17. E. T. Turkdogan and P. M. Bills, Am. Cer. Soc. Bull., **39** (1960), pp682-687.

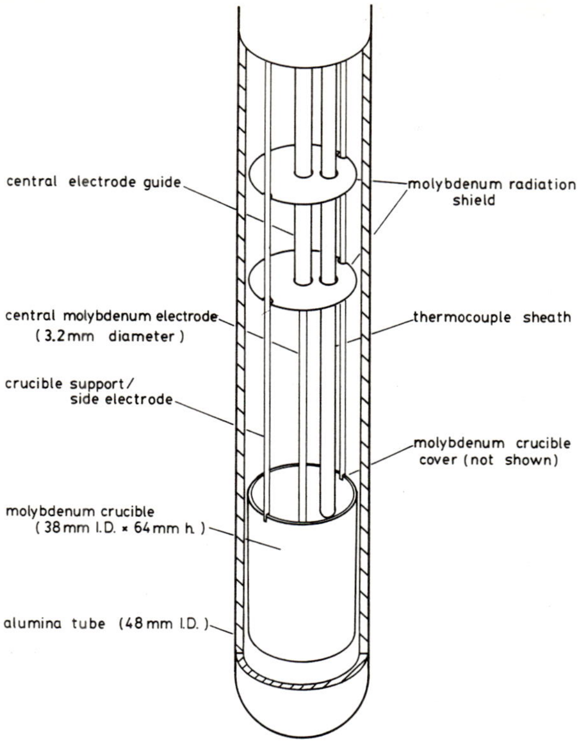

Figure 1 Electrical conductivity apparatus.

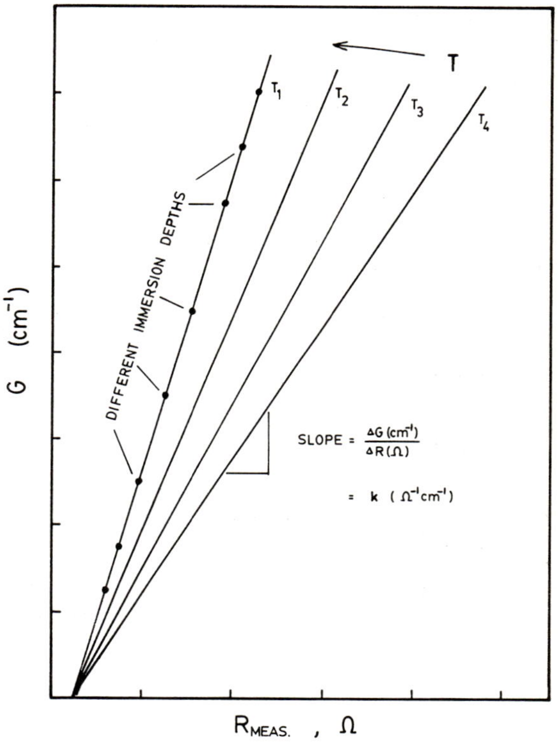

Figure 2 Determination of liquid conductivity by plotting cell constant for each depth as a function of melt resistance measured at the same depth.

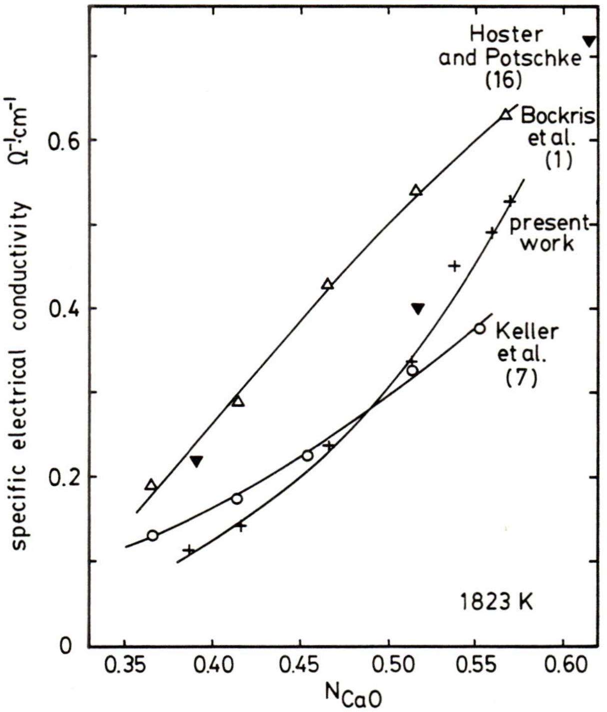

Figure 3 Comparison of this work (CaO-SiO_2 melts) with other studies at 1823K[1,7,16].

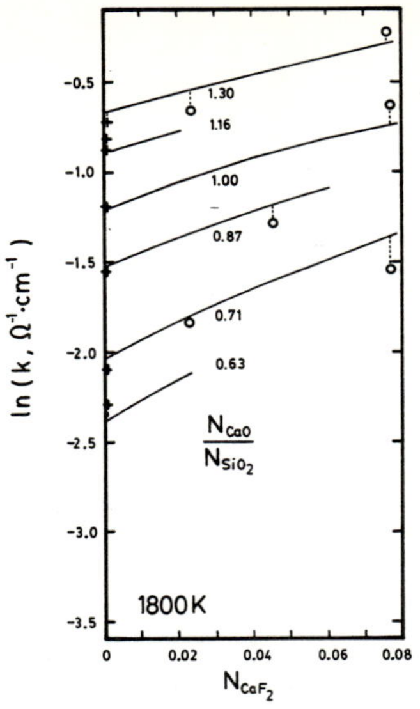

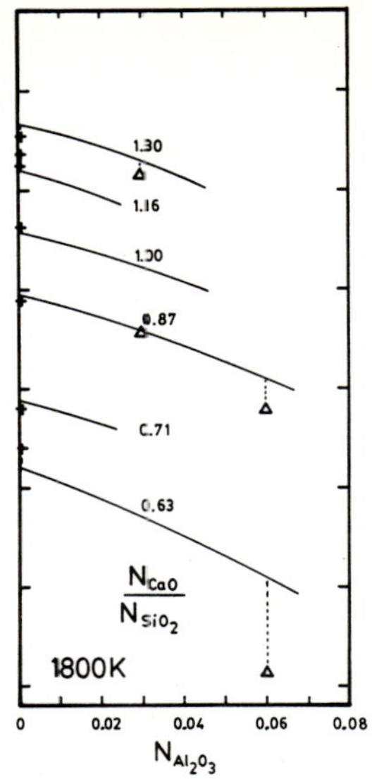

Figure 4 Conductivities of $CaO-CaF_2-SiO_2$ melts.

Figure 5 Conductivities of $CaO-Al_2O_3-SiO_2$ melts.

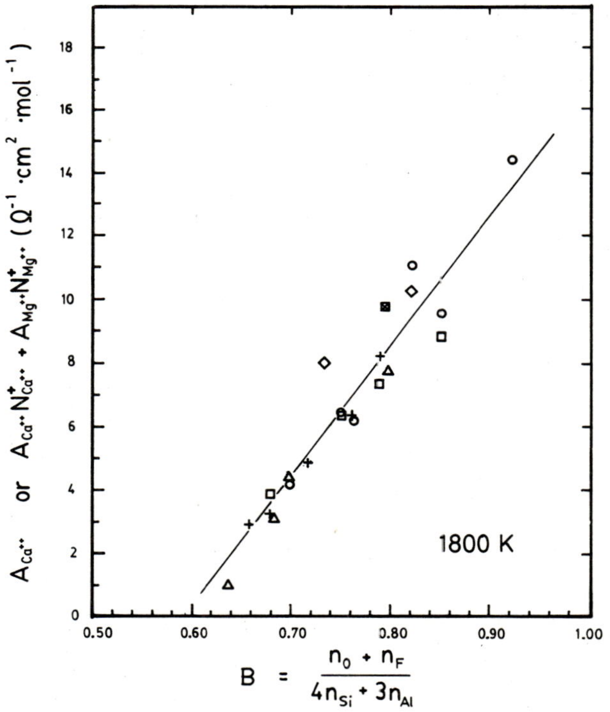

Figure 6 Results plotted showing equivalent conductivity of (calcium+magnesium) ions as a function of stoichiometric parameter B. + $CaO-SiO_2$;
o $CaO-CaF_2-SiO_2$; △ $CaO-Al_2O_3-SiO_2$;
□ $CaO-MgO-SiO_2$; ◇ $CaO-Al_2O_3-CaF_2-SiO_2$;
⊠ $CaO-Al_2O_3-CaF_2-MgO-SiO_2$.

Interfacial phenomena of molten slags

T EL GAMMAL and P STRACKE

Prof Dr-Ing El Gammal is in the Department of Steelmaking, Institute for Ferrous Metallurgy, Aachen University of Technology, FRG;
Dr-Ing Stracke, formerly of the same department, is now with the Daimler-Benz AG, Mannheim, FRG.

SYNOPSIS

The interfacial tension between liquid slags of varying composition and liquid steel has been measured. Additions of Na_2O and CaF_2 have very strong effects. The results show that the structure, and in particular the ionic character of the slags, are the main factors influencing interfacial tension.

INTRODUCTION

Interfacial phenomena between liquid steel and molten slag play fundamental rôles in the control of metallurgical processes. Such processes include:

1. nucleation of deoxidation and desulphurisation reaction products
2. exchange of elements between the 2 phases during slag-metal reactions
3. passage and absorption of non-metallic inclusions from the steel bath
4. metal drop formation in the electro slag remelting process
5. control of surface quality of continuous casting slabs

Due to the difficulty of determining the interfacial tension between liquid steel and molten slag, there is a lack of data, which makes calculations for optimizing metallurgical processes difficult.

With the development of the Drop Detachment Technique[1)] it has been possible to determine, experimentally, interfacial tension data for various liquid steel/ molten slag systems, and to study the effects produced by changes in composition of the slag. These effects, which are closely allied to changes in structure and ionic character of the molten slag, together with the results of interfacial tension determination, form the subject of this paper.

EXPERIMENTAL

The drop detachment technique for determining interfacial tension, which has been described fully elsewhere[1)], consists of allowing a drop of liquid steel to form at the end of a capillary immersed in liquid slag, and recording the changes in weight during formation and detachment of the drop. The principle of the method is shown in fig. 1, and schematically, the recording of the change in weight of the drop up to detachment in fig. 2.

The change in weight caused by the detachment of the drop is related to the interfacial tension of the system as follows:

$$= \frac{m_2 \cdot g}{2 \cdot \quad \cdot r \cdot} \qquad (1)$$

where m_2 = change in weight at drop detachment
g = acceleration due to gravity
r = capillary radius
= correction factor for residual drop at capillary

Preparation of Materials

For this investigation synthetic slags were produced from CaO, SiO_2 and Al_2O_3 of high purity, the CaO, Al_2O_3 contents being varied to assess the influence of these components on interfacial tension. To the 3-component system additions of CaF_2, Na_2O & MgO were made to assess their influence. The compositions of the slags are listed in table 1, and for clarity, these are marked in the corresponding phase diagrams. The composition of the reference steel CK 45 is given in table II.

The structure of liquid slags and their relationship to interfacial phenomena

The ionic nature of liquid slags is now well established. Non-metallic or acid oxides are predominantly covalent in character, giving rise to complex anionic groups. As the silica content of a liquid slag increases, progressive polymerization of the SiO_4^{4-}-ion takes place, to give a continuous structure or network of SiO_4^{4-} tetrahedra. Phosphorus and aluminium oxides can exist also as complex anions - PO_4^{3-} and AlO_3^{3-} - which, with increasing concentration, can associate to give complexes of a higher order.

Basic oxides, with a high difference in electronegativity, do not form stable bondings between ions in the melt. On dissociation, the metal ions tend to break up or transform the networks of anionic complexes in acid slags (network modifiers), as shown for calcium in fig.3[2)].

The strength of the anionic networks has a strong influence on the mobility of free ions in the melt. Where the networks are relatively intact the ease of passage of these ions through the network to the liquid metal/liquid slag interface, will depend on their ionic radius. Smaller ions such as Ca^{2+} (0.96 A°) are more mobile than the larger O^{2-} (1.45 A°) or anionic complexes, and the enrichment of the interfacial layer in the one or the other species can lead to changes in the resultant forces acting at the interface, as shown in fig.4. The presence of surface active species reduces the resultant force by virtue of attraction, as compared with the uncharged condition, leading to a drop in interfacial tension.

The spectrum of changes in interfacial tension caused by ionic species is reflected in the so-called electrocapillarity curve, which records the variation in interfacial tension between the two phases under an applied external EMF. Such a curve is shown in fig.5[3)], which was determined for a steel drop in a 50/50 mixture of CaO and Al_2O_3. Changes in the potential difference across the interface can also be brought about by changes in the concentration or character of the electrolyte i.e. without externally applied charge. This would correspond to a displacement of the electrocapillary curve either to the left or to the right of the zero point, in effect to a recalibration of the zero point. An increase in anionic concentration of the electrolyte at the interface would promote the dissolution of positive ions from the electrode (as in redox reactions) and increase its negative charge (LH-side of zero potential line). This gives low values of interfacial tension (increase in attractive forces). An increase in cationic concentration in the electrolyte, however, would tend to reduce the charge in the electrode (RH-side of zero line), to a point where the potential difference across the interface is at its lowest. This corresponds to a maximum in interfacial tension.

RESULTS AND DISCUSSION

1. Effect of varying CaO content

The effect of increasing the CaO content at constant Al_2O_3 (15 %) from 38.3 % to 44.5 % is shown in fig. 6. After a steady increase in interfacial tension there is a relatively sharp drop at 42.5 % CaO. A glance at the $CaO-SiO_2-Al_2O_3$ phase diagram fig. 7, shows this to correspond to a change in slag structure from pseudowollastonite to gehlenite.

The ionic behaviour of CaO and its effect on the silicate network plays an important rôle here. The increase in CaO content serves initially to weaken the silicate network by partially replacing the 3-dimensional O-Si-O link with the 2-dimensional O-Si-O-Ca-O-Si-O link, but not to such an extent as to release sufficient O^{2-}ions, which could then migrate to the interface. In this range single bonded oxygen, which is not mobile, still predominates. Ca^{2+}ions, on the other hand, because of their smaller ion radius (0.96 A°), can migrate to the interface, where they will increase the interfacial tension due to electrocapillarity.

At the phase boundary (42.5 % CaO) the 3-dimensional network is largely transformed, releasing O^{2-}ions which can migrate to the interface with consequent lowering of the interfacial tension.

2. Effect of varying Al_2O_3 content

The effect of varying the Al_2O_3 between 0 - 15 % at a constant $CaO:SiO_2$ ratio (B = 1) is shown in fig. 8, where it can be seen that up to 5 % Al_2O_3 there is an increase in interfacial tension, which is less pronounced at higher temperatures. This increase is probably due to the network formation of AlO_3^{3-}ions with consequent reduction of the activity of free O^{2-}ions. Rising temperatures would then tend to disrupt the anion complexes. All compositions can be seen to lie within the pseudowollastonite range (fig. 7).

3. Effect of MgO-addition to $CaO-SiO_2-Al_2O_3$ mixtures

The effect of additions of 0-6 % MgO to $CaO-SiO_2-Al_2O_3$ mixtures at fixed contents of SiO_2 and Al_2O_3 is shown in fig. 9. Particularly at the higher temperatures there is a sharp change in interfacial tension approaching 6 % MgO.
This corresponds to the intersection of the pseudowollastonite/anorthite phase boundary, as indicated in the CaO-MgO-SiO_2 phase diagram for the 15 % Al_2O_3 section (fig. 10).
As the MgO addition was made at the

expense of the CaO content, an increase in the interfacial tension was to be expected, due to the less strongly basic character of MgO as opposed to CaO, and consequently, to its weaker effect as silicate network modifier. In terms of the ion-oxygen attraction I[4]:

$$I = \frac{2z}{(r + 1.45)^2} \qquad (2)$$

where z = valency of the cation
r = radius of the cation (A°)
and 1,45 = radius of the oxygen ion (A°)

MgO shows with I = 0.95 a higher ion-oxygen attraction than CaO (I = 0.70). Its readiness to give up oxygen ions is thus correspondingly less. Furthermore, the cationic radius of Mg^{2+} of 0.65 A° is lower than that of Ca^{2+} (0.96 A°), which allows it to migrate more easily to the interface.

At higher temperatures the increase in interfacial tension up to the pseudowollastite/anorthite phase boundary practically disappears, suggesting that these effects lose their significance as the temperature increases.

4. Effect of CaF_2 addition

The effect of CaF_2 addition in the range 0 - 10 % CaF_2 at constant $CaO:SiO_2:Al_2O_3$ ratio (B = 0,82) is shown in fig. 11. There is a sharp drop in interfacial tension between 4and6 % CaF_2, which may be ascribed to the disintegration of the silicate network by the F^- ions, due to their stronger attraction for Si^{4+} cations. It will be noted that at this basicity of 0.82 there is initially a slight rise in interfacial tension at the higher temperature. At low basicities the release of Ca^{2+} cations and F^- anions from CaF_2 addition will result in a preponderance of Ca^{2+} ions over O^{2-} ions at the interface, due to the inferior mobility of the latter through the still largely intact silicate network. As the basicity of the slag increases, the initial increase in interfacial tension disappears to give finally a decrease at a basicity of B = 1.24 (figs. 12, 13). Evidently at this basicity the O^{2-} and F^- ions are sufficient to outweigh the influence of the Ca^{2+} ions at the interface.

The increased drop in interfacial tension with additions of CaF_2 in excess of 4 % is matched by the drop in viscosity, as shown in fig. 14. This would indicate a relationship between interfacial tension and viscosity.

5) Effect of Na_2O addition

The effect of Na_2O addition to a $CaO-SiO_2-Al_2O_3$ slag of basicity B = 1.25 is shown in fig. 15. With increasing Na_2O addition there is a clear drop in interfacial tension, particularly for additions of 0 - 2 %. This is in line with the sharp drop in activation energy for viscous flow reported for $Na_2O + SiO_2$ systems at low concentrations for Na_2O[5], indicating the strong rupturing effect of Na_2O on silicate networks.
Na^+ ions have a relatively large radius (0.95 A°) and consequently a small peripheral field. This results in a low oxygen-ion attraction (I = 0.36), and strong tendency to structure breaking of the silicate network. Also, because of their large radius, the ability of Na^+ ions to migrate to the interface will be lessened, with consequent effects on the electrocapillarity.

CONCLUSIONS

The systematic investigation of the interfacial tension between liquid steel and molten slags of varying composition has shown up the differing effects of some of the individual components of these slags on interfacial phenomena. Whereas changes in the CaO, Al_2O_3, or MgO content of the slags produced , at least in the initial stages, relatively minor changes in interfacial tension, additions of Na_2O and CaF_2 caused marked effects, lowering the interfacial tension considerably.

Applying the ionic theory of slags to the interpretation of the results, it would seem that the interfacial properties are directly affected by the structure of the slags. Changes in interfacial tension may be attributed to alterations in the electric charges, brought about by differences between the accumulation behaviour of the respective ions at the interface. Similarities have been observed in the interfacial and viscous behaviour, which in the case of these slags, are both dependent on the structure of the melt.

References

1. T. El Gammal & R.D. Müllenberg
"The simultaneous determination of density and interfacial tension of liquid slags and metals"
Archiv f. das Eisenhüttenwesen 51 (1980) No. 6 pp. 221-226

2. J. O'M Bockris & A.K.N. Reddy
Modern Electrochemistry Vol. 1 p. 602
Plenum Press New York 1970

3. T. El Gammal, O. Mayer
"Electrocapillarity between molten slags and liquid metals"
Proc. Its Intern. Symp. Molten Salt Chemistry and Technology, Kyoto, 1983
I-213 pp. 239-242

4. G. Hofmaier
"Viscosity and structure of liquid silicates"
Berg- und Hüttenmännische Monatshefte
113 (1968) No. 7 pp. 270-281

5. J. O'M Bockris & A.K.N. Reddy
ibid 3, p. 605

Table 1 Chemical composition of slags

Slag No.	CaO %	SiO_2 %	Al_2O_3 %	CaF_2 %	Na_2O %	MgO %
1	44.5	40.5	15.0	-	-	-
2	42.5	42.5	15.0	-	-	-
3	45.0	45.0	10.0	-	-	-
4	47.5	47.5	5.0	-	-	-
5	50.0	50.0	-	-	-	-
6	51.7	41.3	7.0	-	-	-
7	49.3	39.3	6.7	4.7	-	-
8	48.8	38.9	6.6	5.7	-	-
9	48.6	38.7	6.6	6.1	-	-
10	47.0	37.5	6.4	9.1	-	-
11	50.7	40.4	6.9	-	2.0	-
12	49.8	39.7	6.7	-	3.8	-
13	48.8	38.9	6.6	-	5.7	-
14	40.3	44.7	15.0	-	-	-
15	38.3	44.7	15.0	-	-	2.0
16	36.3	44.7	15.0	-	-	4.0
17	34.3	44.7	15.0	-	-	6.0
18	38.3	46.7	15.0	-	-	-
19	36.3	44.4	14.3	5.0	-	-
20	35.6	43.5	13.9	7.0	-	-
21	34.4	42.1	13.5	10.0	-	-
22	47.1	45.2	7.7	-	-	-
23	44.8	43.0	7.3	4.7	-	-
24	43.6	41.8	7.1	4.6	2.9	-
25	43.6	41.8	7.1	7.5	-	-
26	41.5	39.8	6.7	7.2	4.8	-
27	42.5	40.7	6.9	9.9	-	-
28	41.5	39.8	6.8	11.9	-	-

Table II Chemical Composition of Reference Steel CK 45

C %	Si %	Mn %	P %	S %	Cr %	Ni %	Mo %	Cu %	O ppm	N ppm
0,43	0,17	0,58	0,02	0,038	0,046	0,022	0,006	0,019	32	90

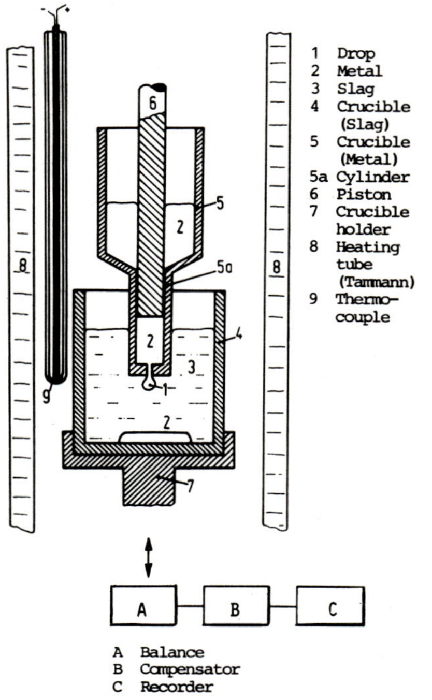

1. Principle of drop detachment method for determining interfacial tension

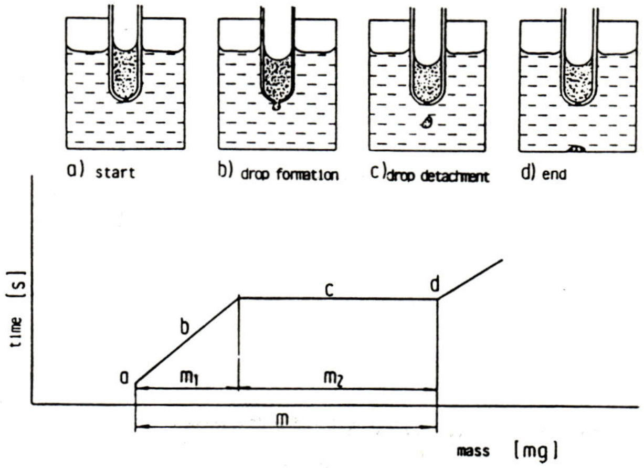

2. Schematic representation of formation and detachment of drop

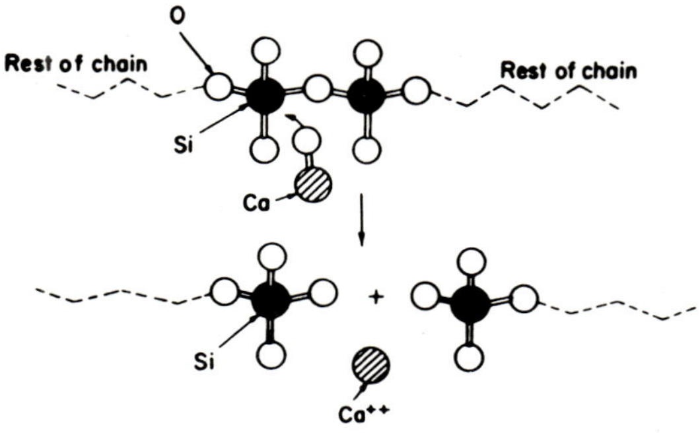

3. Rupture of SiO_4 tetrahedra by CaO[2)]

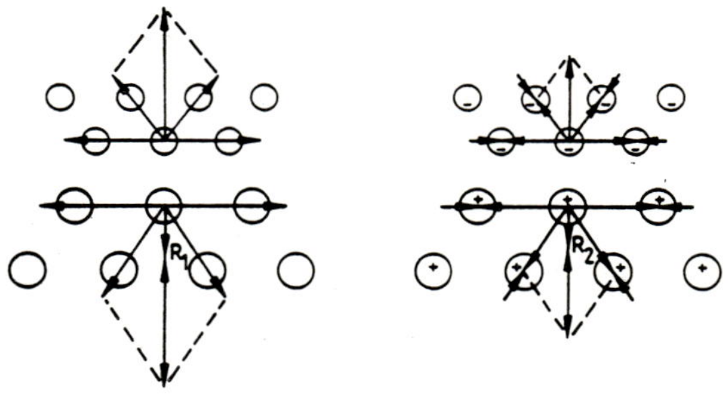

4. Change in force relationships at an interface due to electric charges

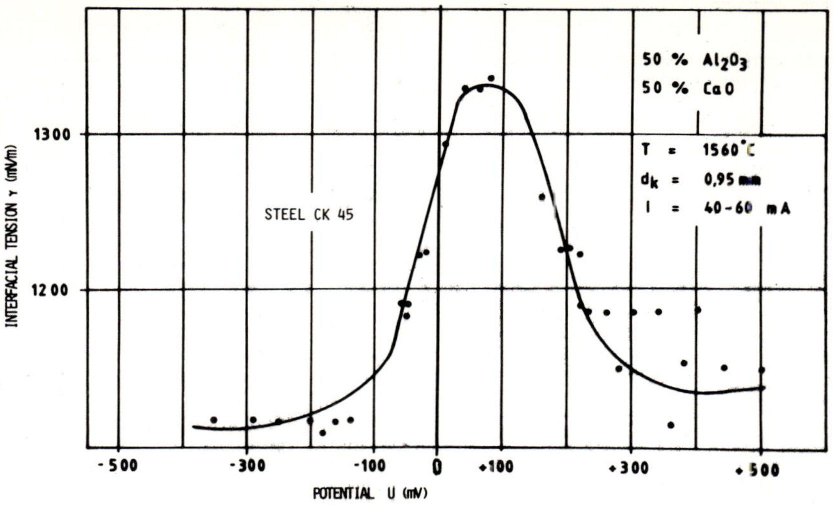

5. Electrocapillarity curve for steel CK 45 in contact with CaO/Al_2O_3 slag[3)]

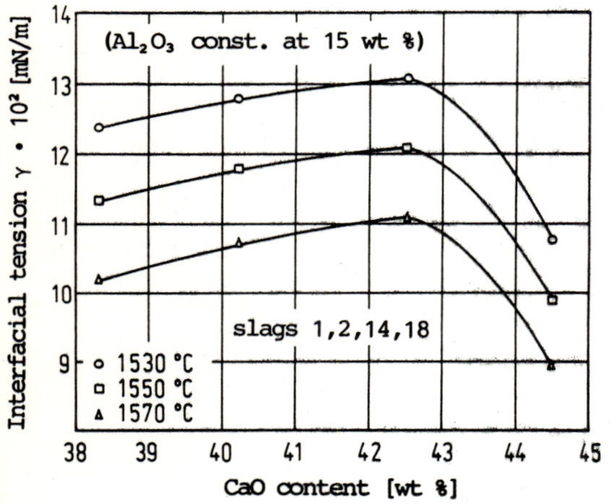

6. Change in interfacial tension on varying CaO content at constant Al_2O_3 content (15 %)

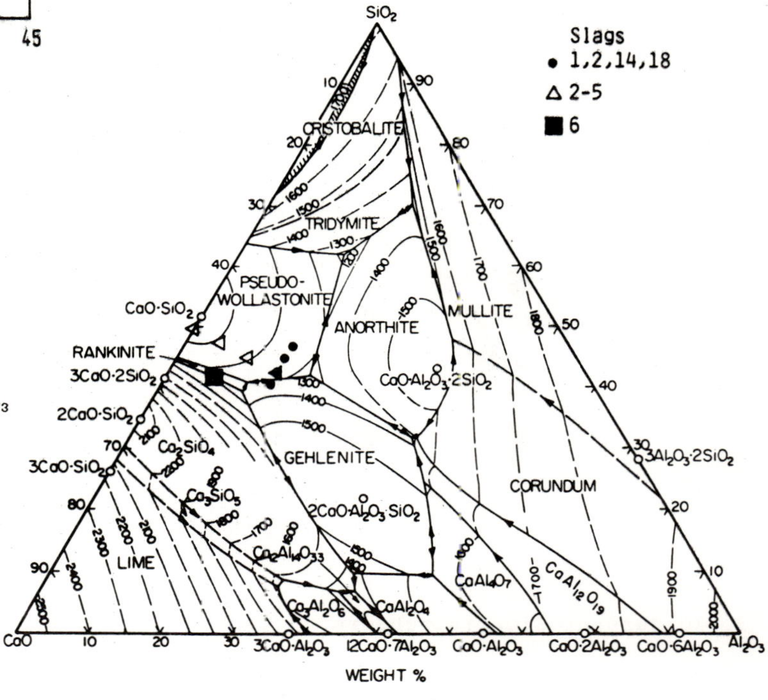

7. Phase diagram for $CaO-SiO_2-Al_2O_3$ system

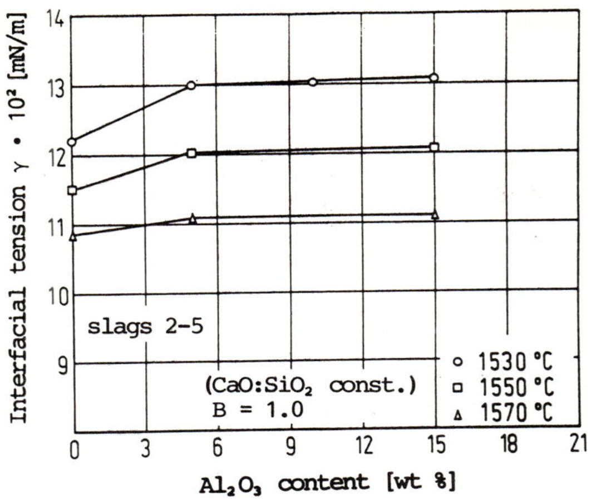

8. Change in interfacial tension on varying Al_2O_3 content at constant CaO-SiO_2 ratio (B = 1)

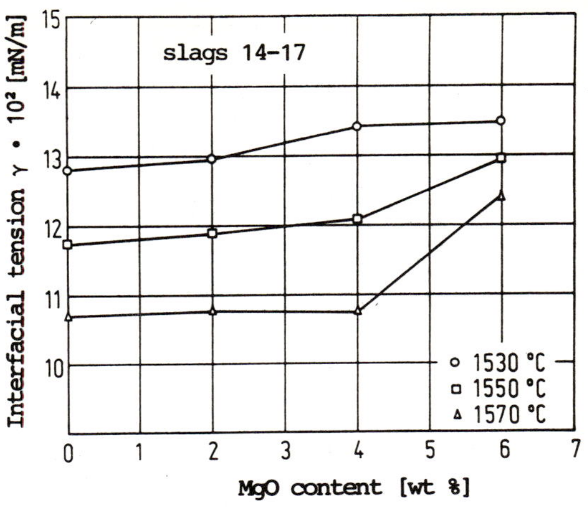

9. Change in interfacial tension on varying MgO content at fixed SiO_2 and Al_2O_3 contents

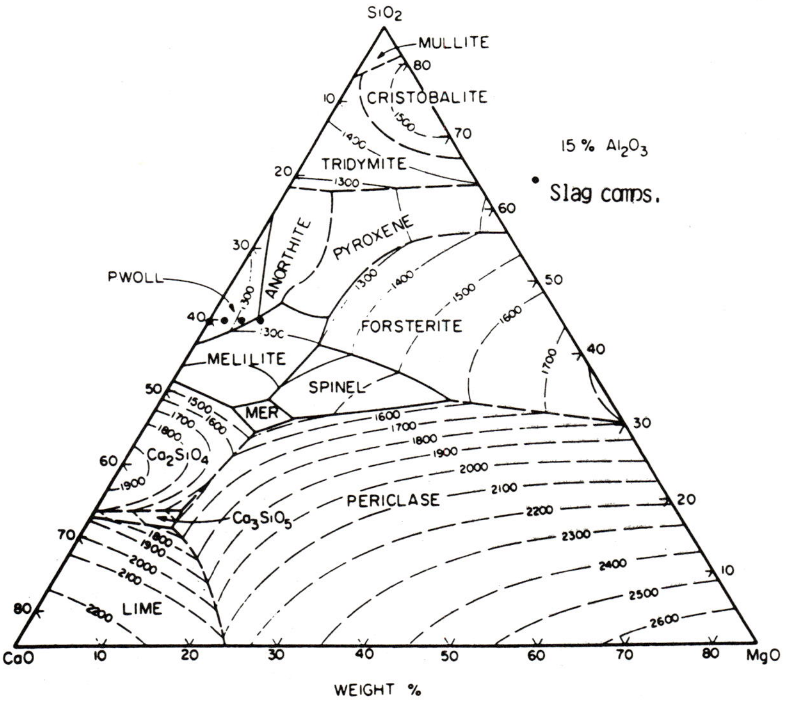

10. Phase diagram for CaO-MgO-SiO_2 system at 15 % Al_2O_3 section

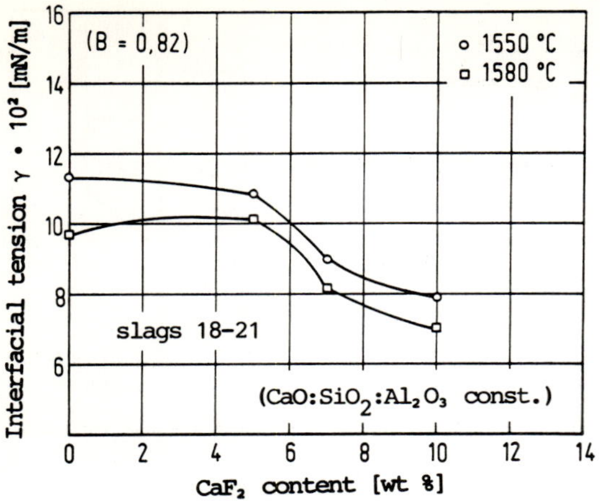

11. Influence of CaF_2 addition on interfacial tension (B = 0.82)

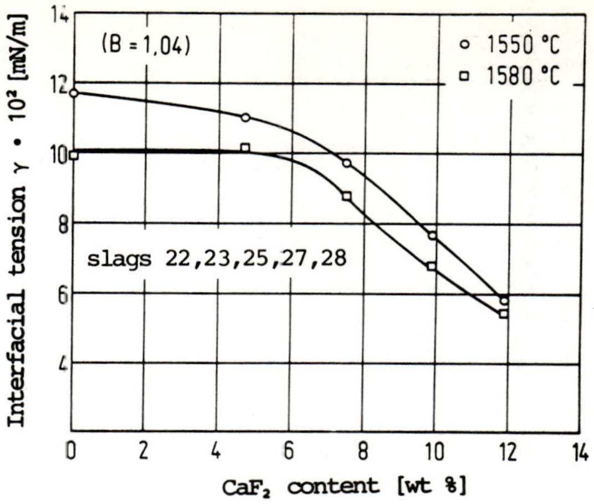

12. Influence of CaF_2 addition on interfacial tension (B = 1.04)

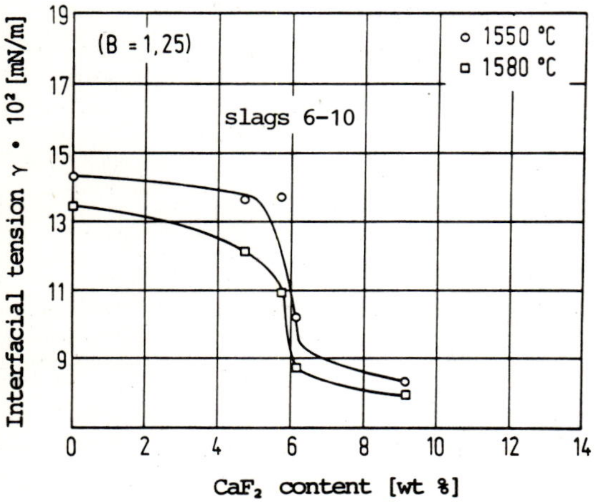

13. Influence of CaF_2 addition on interfacial tension (B = 1.25)

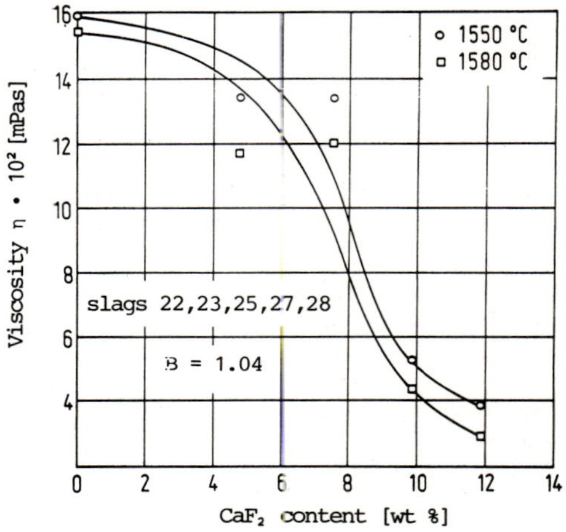

14. Change in viscosity with CaF_2 addition (B = 1.04)

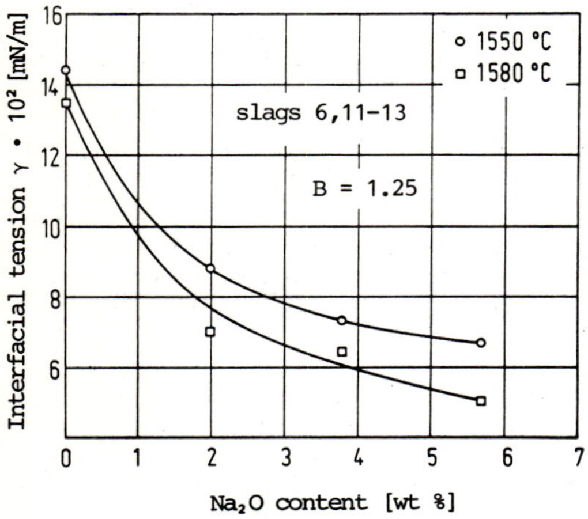

15. Influence of Na_2O addition on interfacial tension (B = 1.25)

Direct observation of local corrosion process of solid oxide at slag-metal interface

K MUKAI, K GOUDA, J YOSHITOMI and K HIRAGUSHI

Dr Mukai is in the Department of Metal Processing Engineering at the Kyushu Institute of Technology; Mr Gouda, Dr Yoshitomi and Mr Hiragushi are in the Technical Research Laboratory at the Kurosaki Refractories Co. Ltd.

SYNOPSIS

A direct observation of the local corrosion of a transparent quartz glass crucible at an interface with a ($PbO-SiO_2$) slag and lead, revealed the formation of a slag film between the inner wall of the crucible and the liquid lead and also the motion of the film caused by Marangoni effect in the local corrosion zone. The motion of the slag film has been found to be the main cause of the local corrosion of this system.

INTRODUCTION

Mukai et al (1) have recently examined the corrosion behaviour of solid oxides at a slag-metal interface and proposed a mechanism to explain the local corrosion of this system. This mechanism is based on the intense motion of the slag film that forms between metal and solid oxide in the local corrosion zone due to an interfacial tension gradient.

The aim of the present study was to substantiate the proposed mechanism by directly observing the process of local corrosion for a system involving a transparent quartz glass crucible with ($PbO-SiO_2$) slag and lead.

EXPERIMENTAL

Direct observation was made of the corrosion process of the inner wall of the transparent quartz glass crucible at the interface of PbO(70 mol%)$-SiO_2$ slag and lead at 1073 K in an argon atmosphere. The respective depths of the metal and slag phases were about 10 mm and 20 mm for all experiments except for those conducted to investigate the influence of the depth of the slag phase on the rate of the local corrosion. The shape of the crucible wall in the local corrosion zone was photographed to determine the corrosion rate, and the intensity of the slag film motion in this zone was calculated by analysis of a colour cinefilm record.

RESULTS AND DISCUSSION

The transparent crucible had an outer diameter of 20 mm and an internal diameter of 11 mm, and by use of appropriate lighting, it was possible to visually observe both the corrosion behaviour, and the motion of the slag film.

With increasing temperature, the metal in the crucible first melts down and then the slag melts. The molten slag begins to penetrate into the boundary between the molten metal and the crucible and reaches the bottom of the crucible within 60 seconds after the beginning of the penetration; this results in the formation of the slag film between the liquid metal and the inner wall of the crucible. The whole slag film in the vicinity of the level of the slag-metal interface flows downwards for the first 0.18 ks after formation of the slag film, as shown in Fig.1a. Then a few narrow upward streams begin to occur in the film (Fig.1b), and these increase in number with increasing time (Fig.1c), whilst the downward flow becomes slower with increasing the time (Fig.1d).

Fig.2 shows the shapes of the local corrosion zone which were photographed during the corrosion process. The local corrosion proceeds mainly below the initial level of the slag-metal interface. Time variation of the inner radius, r_{SM}, of the crucible at the most corroded portion was determined with equation (1) by measuring $r_{SM(n)}$ (as shown in Fig.3) which was recorded on photographic film every thirty seconds during the progress of the local corrosion,

$$r_{SM} = [-r_{SM(n)}^3 + \{r_{SM(n)}^6 + (n^2 r_0^2 - n^2 r_{SM(n)}^2 - r_{SM(n)}^2) \times r_{SM(n)}^2 r_0^2\}^{\frac{1}{2}}]/(n^2 r_0^2 - n^2 r_{SM(n)}^2 - r_{SM(n)}^2) \qquad (1)$$

where n is the index of refraction of the quartz glass and r_0 is the outer radius of the crucible as shown in Fig.3. This figure indicates that as a whole, the corrosion rate, dr_{SM}/dt, remains constant for about the first 0.6 ~ 1.2 ks after the formation of the slag film and then gradually decreases with time. Fig.3 also indicates that the initial period with high corrosion rate decreases with decreasing depth of the slag phase, which is due to the increase in SiO_2 content in the slag caused by the local corrosion, as reported in the previous work (1).

The intensity of the downward flow of the slag film (v_z) for the initial period with high corrosion rate was determined from cinefilm records, by measuring the velocity, v_z, in the vertical direction of suitably illuminated spots in the slag film*. Fig.4

---

*The intensity of the downward flow is equal to the velocity, in the vertical, direction of suitably illuminated spots in the slag film but it cannot be substantiated that v_z is equal to the velocity of the slag film.

indicates that as a whole, the intensity of the downward flow during the period of high corrosion rate is greater than that during the period of low corrosion rate. This result shows that the motion of the slag film promotes the local corrosion rate. The lower corrosion rate and intensity, v_z, observed during the period with the initial high corrosion rate may be due to the development of the upward stream.

Fig.5 shows that the thickness of the solidified slag film in the most actively corroded zone is from 50 to 100 μm. Concentration changes in the slag film were detected along and perpendicular to the inner wall of the crucible by analyzing the slag film with EPMA. The interfacial tension gradient along the vertical direction in the local corrosion zone was evaluated to be about 3.0 N/m^2 from the vertical concentration gradient of the solidified slag film along the interface between the slag film and the metal and the relation between the interfacial tension and the composition of the slag of this system (2).

From the above results the local corrosion of the crucible with this sytem can be explained as follows: The penetration of the slag film indicates that the relation $\gamma_{OM} > \gamma_{SM} + \gamma_{SO}$ is satisfied in this system, where γ is the interfacial tension and 0, S and M indicate solid silica, liquid slag and metal, respectively. Since the contact time of the lower slag film with the crucible wall is longer than that for the upper slag film, the lower slag film will have a higher SiO_2 content than the higher film due to the dissolution of SiO_2 from the crucible wall into the film. The difference in SiO_2 content causes an interfacial tension gradient in the vertical direction, since the interfacial tension between $PbO-SiO_2$ slag and lead increases with increasing SiO_2 content (2). Thus the slag film is continuously pulled down by the interfacial tension gradient, which induces a downward flow of the slag film. The inner wall of the crucible in the vicinity of the metal level is continually washed with fresh slag film, which results in enhanced corrosion in this vicinity and gives rise to the local corrosion zone. The role played in this system by the slag film is essentially the same as the one in the local corrosion of solid oxide at a slag surface (3).

SUMMARY

The present results and discussion strongly support the mechanism proposed to explain local corrosion ie that Marangoni effects caused by the dissolution of the crucible in the slag film promote motion of the slag film and mass transfer which results in local corrosion.

ACKNOWLEDGEMENTS

We wish to express our appreciation to the Japanese Ministry of Education, Science and Culture for providing us with a Grant-in-Aid for Scientific Research. We also wish to express our appreciation to Mr Hisatoshi Sakai, who assisted in performing our experiment.

REFERENCES

(1) K. Mukai, T. Masuda, K. Gouda, T. Harada, J. Yoshitomi and S. Fujimoto, "Local Corrosion of Solid Oxides at the Liquid ($PbO-SiO_2$) Slag-Pb Interface", Journal of the Japan Institute of Metals, 48(7) (1984) pp 726-734.

(2) K. Mukai, J.M. Toguri, I. Kodama and J. Yoshitomi, "Effect of Applied Potential on the Interfacial Tension between Liquid Lead and $PbO-SiO_2$ slags", Canadian Metallurgical Quarterly, 25(3) (1986) pp 225-231.

(3) K. Mukai, T. Harada, T. Nakano and K. Hiragushi, "Mechanism of Local Corrosion of Solid Silica at $PbO-SiO_2$ Slag Surface", Journal of the Japan Institute of Metals, 50(1) (1986) pp 63-71.

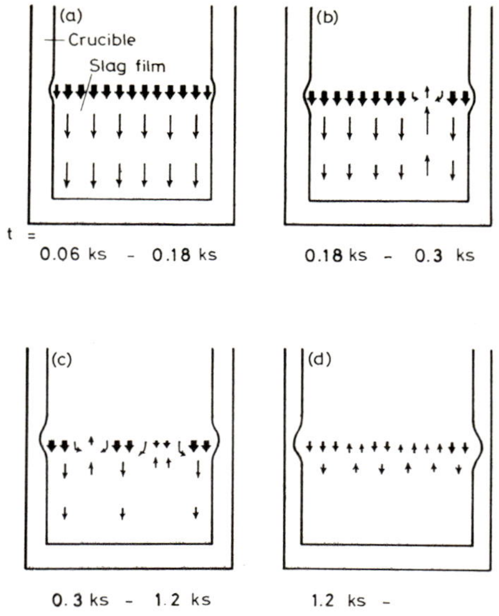

Fig.1. Schematic presentation of the flow pattern of the slag film. t: elapsed time after the formation of slag film.

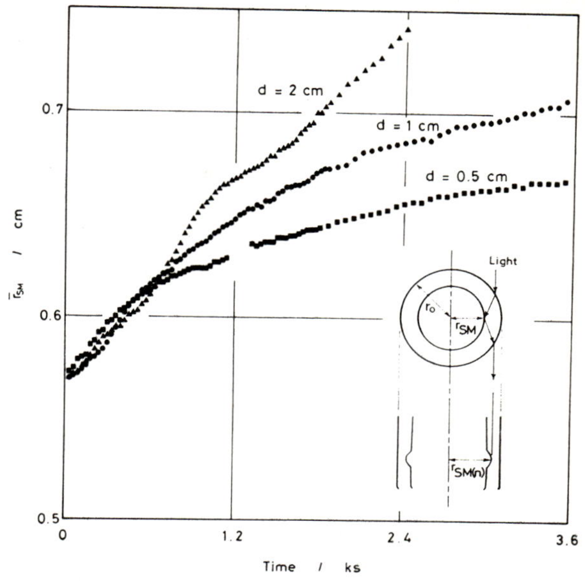

Fig.3. Relation between $\bar{r}_{SM}$ and time. $\bar{r}_{SM}$: the mean of the maximum radii of right hand and left hand sides of the crucible in the local corrosion zone. d: depth of the slag phase.

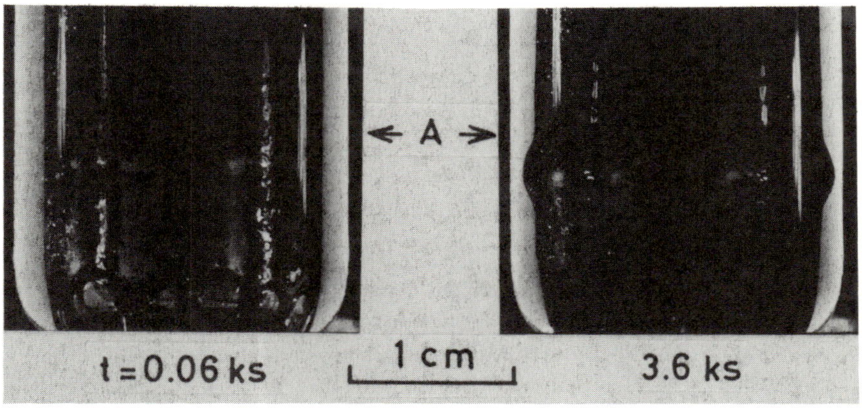

Fig.2. Shapes of the inner wall of the crucible in the local corrosion zone in situ photographed. A: initial level of the slag-metal interface.

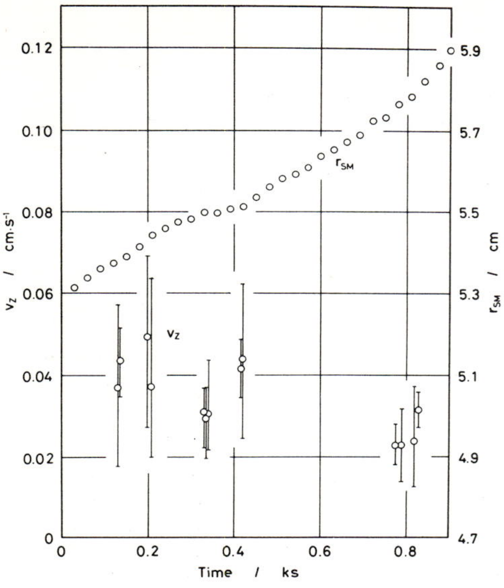

Fig.4. Variations of the maximum radius of the crucible, r_{SM}, and the intensity of the downward flow of the slag film, v_Z, with time.

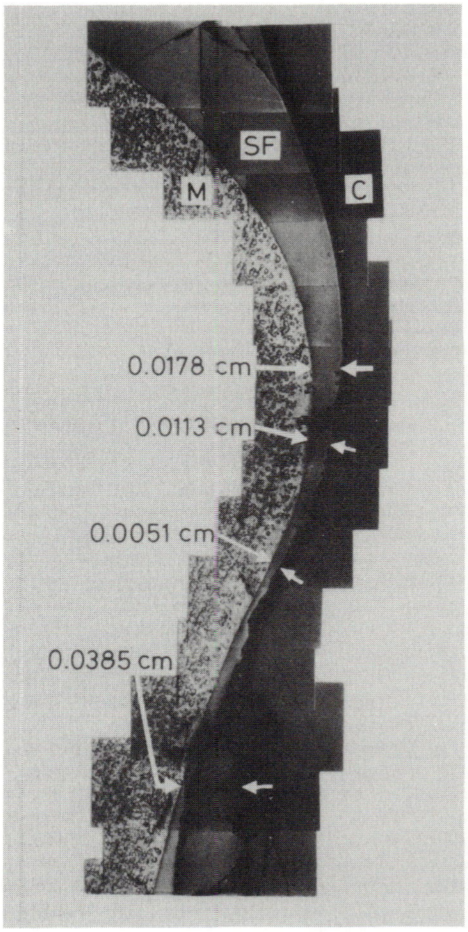

Fig.5. Solidified slag film between the metal and the inner wall of the crucible in the local corrosion zone. M: metal; SF: slag film; C: crucible

Phase separations among oxides and sulphides and their metallurgical implications

A YAZAWA and Y TAKEDA

The authors are with the Research Institute of Mineral Dressing and Metallurgy (SENKEN), Tohoku University, Sendai, Japan.

SYNOPSIS

Phase separation was investigated for the system FeS-FeO-AO, where AO is a strong acidic oxide. With the increasing addition of AO, the homogeneous FeS-FeO liquid tends to separate into two phases of matte and slag. When AO is replaced by basic oxide BO, phase separation is not observed in the FeS-FeO-BO systems, but observed between ferrite slag and high grade matte when FeS is substituted by Cu_2S in high proportion. In the oxide systems consisting of AO-BO-XO, where XO is a neutral metal oxide, immiscibility is often observed between XO and xBO-AO like $2CaO\text{-}SiO_2$. Oxidic dissolution of metal in slag was discussed in relation to this immiscibility trend with special reference to the tin-slag system. Reducing various metallurgical processes into ternary systems, the immiscibilities were discussed thermodynamically based upon the data of each binary system.

INTRODUCTION

Slag is accepted as a top layer above a metal phase and dissolves readily any kind of non-metal components, but sometimes it tends to separate further into two phases. The miscibility or immiscibility in slag systems is likely to be closely related to the formation of matte, the loss of metal in slag or the removability of impurities in slag. This point is not recognised adequately by metallurgists.

In the present paper, some experimental data obtained recently in the authors' laboratory are summarized with special reference to the immiscibility in slag systems. Thermodynamic explanations are given for phase separation, and the metallurgical significance of the immiscibilities among oxides and sulphides are discussed.

EXPERIMENTAL RESULTS

1. Phase Separation in FeS-FeO-AO Systems.

Phase separation behaviour was investigated for the system FeS-FeO-AO where AO is a strong acidic oxide such as B_2O_3, P_2O_5 or SiO_2. The mixture of ternary components was brought into equilibrium in an iron crucible at 1100°C, and the samples obtained were analysed chemically. The phase relations obtained for $FeS\text{-}FeO\text{-}B_2O_3$ are illustrated in Fig. 1 (1). It has been well recognized that FeS and FeO form a homogeneous eutectic liquid, and do not show any immiscible tendency. However, when an acidic component such as B_2O_3 is added, separation into two phases is observed, as shown in Fig. 1. Similar phase separations between matte and slag are observed in $FeS\text{-}FeO\text{-}SiO_2$(2) and $FeS\text{-}FeO\text{-}P_2O_5$(1) systems, although the existence of a non-variant three-liquid equilibrium among matte, slag and acidic oxide was first confirmed experimentally in Fig. 1. In previous reports (2,3), it was confirmed that the addition of Cu_2S into $FeS\text{-}FeO\text{-}SiO_2$ results in clear-cut separation between matte and slag. From the structural viewpoint the addition of acidic oxide AO creates a polymerized, ionic, liquid slag, and separates from a covalent-type liquid matte.

2. Mutual Dissolution between Matte and Ferrite Slags.

It would be interesting to consider the immisibility between matte and slag when acidic oxide AO is replaced by basic oxide BO such as CaO or Na_2O which may form a ferrite slag with FeO_x. However, any immiscible trend was not observed in FeS-FeO-BO systems, and the phase separation into matte and slag was realized only after a considerable amount of Cu_2S was added.

In a magnesia crucible ferrite slag and matte were kept at 1250°C to reach equilibrium under a flow of 10% SO_2-Ar gas, and the samples obtained were analysed chemically. In Fig. 2 (4), the distribution ratio X defined by

$$L_X^{s/m} = (\%X \text{ in slag}) / (\%X \text{ in matte}) \qquad (1)$$

is plotted against copper contents in the matte. The thick lines in the middle illustrate the data for ferrite slag-matte equilibria, while thin curves at the top and bottom of the figure are for silicate slag-matte. The distribution ratios of Ca, Ba, Si and Fe are always more than unity and increase with increasing matte grade. This behaviour is exactly opposite to that observed for S and Cu, although the inflextion points exist in those of Cu at around 70 to 75 %Cu in matte due to the existence of oxidic copper in slag. As shown in the figure, it should be noticed that low grade matte, less than 40 or 48 % Cu, forms a homogeneous liquid phase with calcium or barium ferrite slag. That is, ferrite slag is more miscible with the matte than silicate slag, and as the extreme case, phase separation is hardly observed in the sodium or lithium ferrite and sulphide systems.

3. Oxidic Dissolution of Tin in $FeO-CaO-SiO_2$ Slag.

Immiscible trends are sometimes observed in liquid slag systems consisting merely of oxides, especially in soda slag systems (5,6), as shown later in Fig. 3. Similar trends exist in the $FeO-SiO_2-CaO$ system.

The experimental studies were carried out for the equilibria between $FeO-CaO-SiO_2$ slag and tin metal(7). A binary or ternary slag was brought into equilibrium with liquid tin in a magnesia crucible at 1300°C, and the samples obtained were analysed chemically. In Fig. 4 the tin contents in the slags are plotted against measured oxygen potentials. The slag compositions are plotted in Fig. 5 (f); A and E are binary fayalite ($FeO-SiO_2$) and ferrite (FeO-CaO) type slags, while, B, C and D are ternary slags with different ratios of CaO/SiO_2. In the region of medium oxygen potential, the slopes of the lines of 1/2 suggest divalent dissolution of tin in slag, which are to be expected thermodynamically from the oxide dissolution equilibria of tin as follows:

$$Sn(l) + \tfrac{1}{2}O_2 = SnO(e)$$

$$K = a_{SnO} / P_{O_2}^{1/2} = \gamma_{SnO} N_{SnO} / P_{O_2}^{1/2} \qquad (2)$$

At higher oxygen potential considerable amounts of tin dissolved in the slags, and SnO-Sn equilibria are illustrated by the point S.

It should be noted that the solubility of tin in binary fayalite(A) or ferrite(E) slag is much higher than in the ternary $FeO-CaO-SiO_2$ slag, and a higher ratio of CaO/SiO_2 results in a lower content of tin in the slag at a given oxygen potential, and theoretically the minimum %Sn in slag may be expected at $CaO/SiO_2 = 2$, although ternary slag of this ratio is hardly in the molten state at 1300°C. As discussed later on, such trends are caused by immiscibility in $FeO-CaO-SiO_2$ slags.

Based on the data obtained for various slag compositions, the activity coefficients of SnO in slag were derived from eq. (2), and illustrated in Fig. 6 with solid lines. Because a higher value of the activity coefficient means lower solubility of tin in slag, the minimum solubilities may be expected on the line connecting FeO with $2CaO-SiO_2$, as shown by the estimated, dashed lines.

DISCUSSIONS

It is very convenient to discuss phase separation in various metallurgical processes in terms of ternary system, because immiscibility is easily treated thermodynamically on the basis of binary data, and can be illustrated in ternary diagrams. Thus, referring to various heterogenous pyrometallurgical processes, ternary immiscibilities are illustrated collectively in Fig. 5.

Postulating the regular solution behaviour for X-Y-Z system, ternary activity is derived from a function of each binary as follows(8):

$$\ln\gamma_x = \alpha_{xy} N_y^2 + \alpha_{xz} N_z^2 + (\alpha_{xy} + \alpha_{xz} - \alpha_{yz}) N_y \cdot N_z \qquad (3)$$

The value of α suggests the degree of negative or positive deviation in the binary system. Estimating this α function semi-quantitatively as shown on each binary in Fig. 5, ternary immiscibilities were calculated thermodynamically, and illustrated in Fig. 6(a)-(f).

In Fig. 5(a), a homogeneous and small positive or negative deviation is estimated for the $FeS-FeO_x$ or FeO_x-AO binary, respectively, but because of the immiscibility in the FeS-AO system, a wide miscibility gap is observed in the ternary, which corresponds to matte-slag separation caused by addition of acidic oxide as shown in Fig. 1. Due to the immiscible characters of X(metal), Cu_2S or Ni_3S_2 against both FeO_x and AO, as shown in Fig. 5(b), a clear-cut separation is observed between slag and metal, or high grade matte.

When AO is replaced by basic oxide BO, the ternary is converted from (b) to (d) or (a) to (c) in Fig. 5, where a more miscible ferrite slag is formed instead of iron silicate slag.

Considerable miscibility is expected in a ferrite slag and matte system from figure (c), and this is observed experimentally (Fig. 2).

It is interesting to find miscibility gaps in Fig. 5(e) and (f), where the slag consists merely of oxides, and all binaries are well mixed, show negative deviations and do not show any trends for phase separation. However, thermodynamic calculation shows that if one binary system exhibits extremely negative deviations and forms strong intermediate compounds, like $_x$BO-AO, another rather neutral component, such as FeO_x or some metal oxide, exhibits an immiscible trend against such a strong compound.

Figure 5(f) may correspond to a $FeO-CaO-SiO_2$ slag and a distinct miscibility gap is not necessarily observed, but there is clearly a trend towards immiscibility. Because of such immiscibility, the content of FeO_x or XO in ternary slag must be at a minimum on the line connecting xBO-AO and FeO_x or XO. The experimental results shown in Figs. 4 and 6 can be understood by referring to the thermodynamic model demonstrated in Fig. 5(e) and (f).

Large immiscibility gaps similar to those shown in Fig. 5(e) are well known between FeO and phosphates such as $3CaO-P_2O_5$ or $3Na_2O-P_2O_5$(9,10), see Fig. 3(a). The immiscibility between oxides and carbonates or sulphates, as illustrated in Fig. 3(B)(11) and (C)(5), may be explained by the same principle, since the carbonate, or sulphate, is a compound consisting of strong base and acid such as Na_2O and CO_2 or SO_3. Undoubtedly, the miscibility gap between Na_2O-SiO_2 and Cu_2O in Fig. 3(C) is similar to that in Fig. 5(e). Some examples occurring in alloy systems were shown in a previous report(6).

CONCLUSIONS

A homogeneous liquid phase of FeS-FeO separates into matte and slag by addition of acidic oxides such as B_2O_3 or SiO_2. Clear-cut separation is obtained by addition of Cu_2S. When an acidic oxide is replaced by a basic oxide, the resulting ferrite slag is highly miscible with sulphides, and separation into matte and slag is observed only with high grade mattes. A trend towards immiscibility is also observed between neutral oxides and slags consisting of a strong acid and base, and this has great significance to the oxidic dissolution of valuable metal in the slag.

Postulating regular solution behaviour, ternary immiscibilities are estimated thermodynamically from binary data, and various heterogeneous smelting systems are reasonably explained.

ACKNOWLEGEMENT

The authors wish to express their appreciation to Professor Y.A. Chang, University of Wisconsin, for his valuable advices in calculating the ternary activity data.

REFERENCES

(1) A.K. Espeleta, H. Ujiie and A. Yazawa : To be published in J. Min. Metall. Inst. Japan.

(2) A. Yazawa and M. Kameda : Tech. Rep. Tohoku Univ. 18, No.1 (1953),40.

(3) A. Yazawa : Tech. Rep. Tohoku Univ. 21, No.1, (1956), 31.

(4) C. Acuna and A. Yazawa : Trans. Jap. Inst. Metals 27 (1986), 881.

(5) Y. Takeda, G. Riveros, Y.J. Park and A. Yazawa : Trans. Jap. Inst. Metals 27 (1986), 608.

(6) A. Yazawa and Y. Takeda : Metall. Review MMIJ 4 (1987), 53.

(7) Y. Takeda, Po Po Chit and A. Tazawa : To be published in J. Min. Metall. Inst. Japan.

(8) L.S. Darken : Trans. Met. Soc. AIME 239 (1967), 90.

(9) W. Oolsen and H. Maetz : Arch. Eisenhuttenw. 19, (1948), 111.

(10) G. Troemel and H.W. Fritze : Arch. Eisenhuttenw 28 (1957), 489.

(11) P. Grieveson and E.T. Turkdogan : Trans. Met. Soc. AIME 224 (1962), 1090.

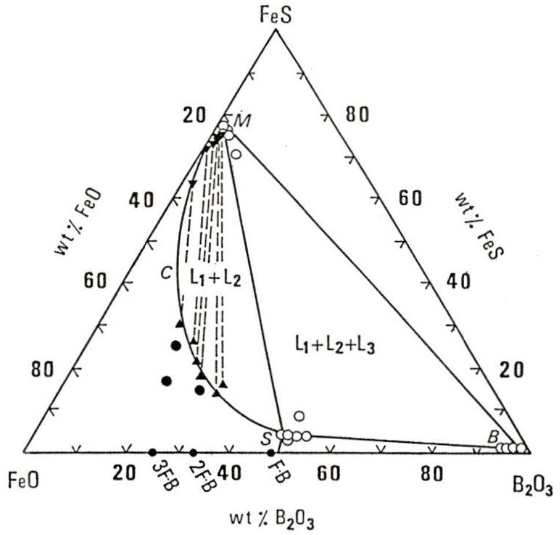

Fig.1 Phase separation in the system FeS-FeO-B_2O_3 at 1100°C.

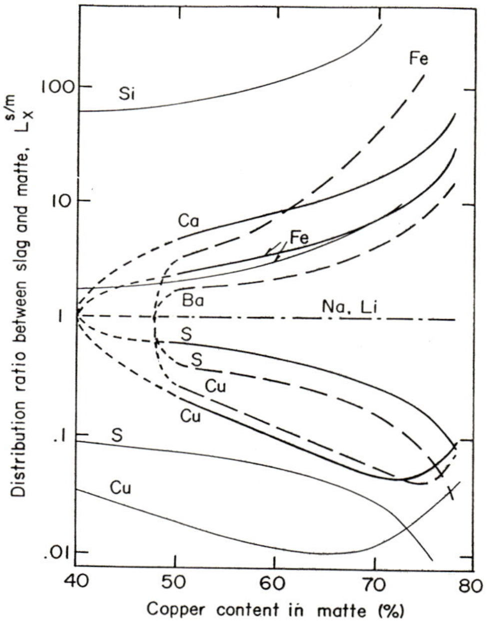

Fig.2 Distribution ratios of various elements between slag and matte plotted against copper content in matte. Solid lines : CaO-FeO_x slag. Dashed lines : BaO-FeO_x slag. Dash-dot lines : Na_2O- or Li_2O-FeO_x slag. Thin lines : FeO-SiO_2 slag.

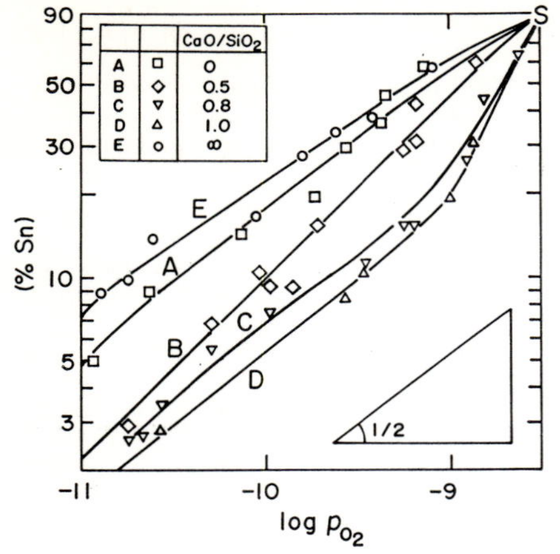

Fig.3 Solubility of tin in FeO-CaO-SiO_2 slag equilibrating with liquid tin at 1300°C. Point S represents liquid SnO.

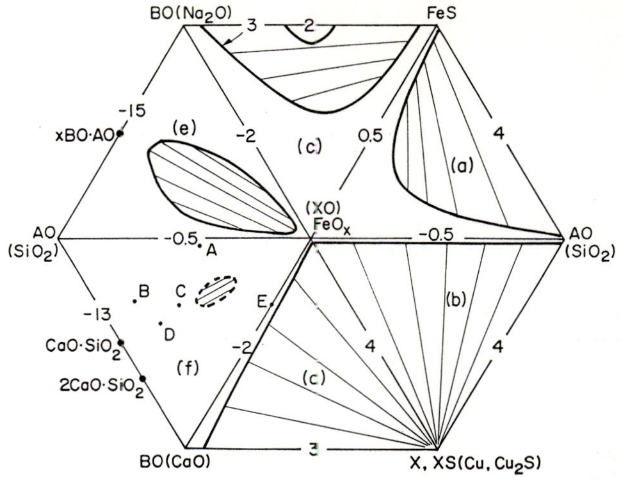

Fig.5 Ternary immiscibility assessments relating with smelting process. The value of binary α-function is given on each binary.

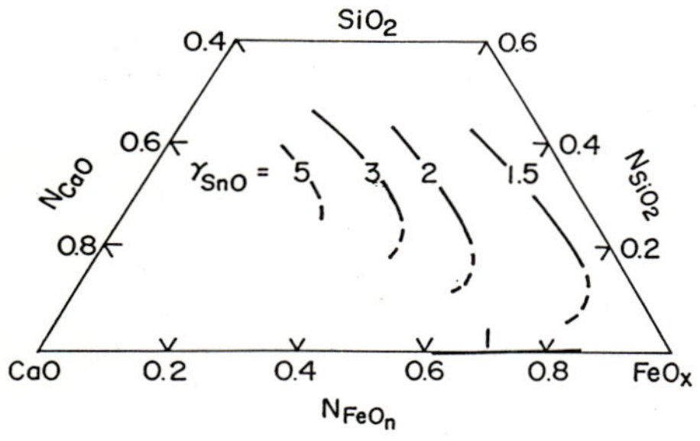

Fig.4 Contour lines of activity coefficient of SnO in FeO-CaO-SiO_2 slag at 1300°C.

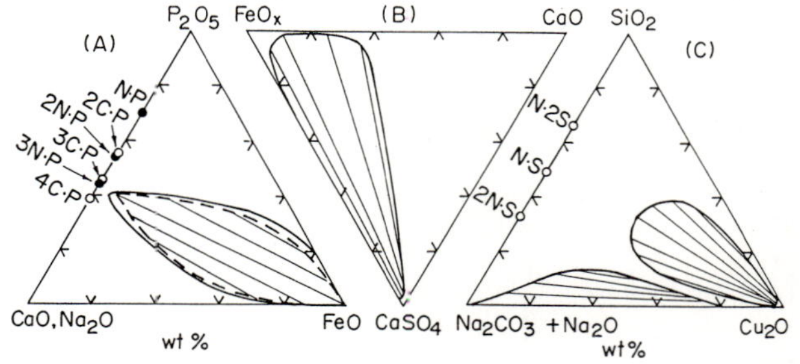

Fig.6 Examples of the miscibility gaps in ternary slag systems.

Thermal diffusivity of molten ferrite slag by laser flash method

H OHTA, IN-KOOK SUH and Y WASEDA

Dr Ohta is in the Department of Engineering at Ibaraki University; Mr Suh and Prof Waseda are with the Research Institute of Mineral Dressing and Metallurgy (SENKEN) at Tohoku University.

SYNOPSIS

The thermal diffusivity of calcium ferrite slags has been determined in the composition range from 33 mol% to 55 mol% CaO at temperatures between 1023K and 1673K by the two layered laser flash method. The sample cell system consists of liquid layer and thin crucible. An initial temperature rise analysis was employed for reducing difficulties due to the conductive heat leak to the crucible wall and the radiative heat leak. The thermal diffusivity of calcium ferrite slags is the order of 3×10^{-3} cm^2/s and its temperature dependence is found to be small.

INTRODUCTION

The importance of physical properties of ferrite slags has been well recognized because their properties indicate the potential capabilities to remove inherent inconvenient elements such as arsenic and phosphorus in the oxidation smelting process of copper or nickel sulfide ores. Recently, density, viscosity and electrical resistivity of these slags have been reported[1]. The thermal diffusivity, however, has not been determined yet.

By measuring the thermal diffusivity of a cell filled with a liquid sample using the laser flash method, several attempts have already been made[2-4]. However, as indicated by Ang et al [5,6], the reservations should be stressed regarding the quantitative accuracy of these results, because of the leakage of heat due to the conduction to the crucible wall and the radiation.

The main purpose of this work is to present our own thermal diffusivity measurements of calcium ferrite slags using the two layered laser flash method coupled with the data processing for the initial time region of the temperature response curve, where the heat leak to the crucible wall is negligibly small.

EXPERIMENTAL PROCEDURE

Figure 1 shows the schematic diagram of the cell assembly for measuring thermal diffusivity of ferrite slag presently employed. A cell assembly is placed on the alumina pipe pedestal. At the desired temperature, a pulsed laser beam is flashed on the top surface of a sample. The temperature response curve was obtained by means of a liquid-N_2-cooled InSb infrared detector focused on back surface of the crucible through silicon lens and gold mirror. The cell system was heated by a platinum wire wound furnace, and the sample temperature was monitored by the Pt/Pt-13% Rh thermocouple. The temperature response obtained by the remote optical sensing device is stored in a digital transient memory connected to the microcomputer. The CaO content was varied in the range of 33-55 mol% and the thermal diffusivity data were determined at temperatures between 1023K and 1673K under air atmosphere.

The dimensions of a Pt crucible are 0.2mm thick, 1.3mm depth and 10mm diameter. The cell preparation was as follows. Pre-melted ferrite powder sample was charged in the crucible to produce the flat surface on melting. The amount of powder is estimated from the density value[7] in advance. After holding at temperature about 100K above the melting point for an hour in order to remove bubbles by the evacuation, the cell system was cooled down to the room temperature. Since the thermal expansion of calcium ferrite slags is the order of 1%, no shrinkage was observed in solid sample and then the thermal expansion of the ferrite layer was neglected in the present work.

RESULT AND DISCUSSION

At initial stage of temperature rise of back surface of a two layered cell, a plot of

$\ln[(\sqrt{t})\Theta(\alpha,t)]$ against $1/t$ show the straight line with slope $-(n_1+n_2)/4$ (see for example, James[8]). Where $n_i = \ell_i/\sqrt{\alpha_i}$, α is the thermal diffusivity, ℓ is thickness of the ith layer, t is time and Θ is the temperature rise.

According to this theoretical basis, the thermal diffusivity of solid samples were estimated for the two layered sample which consists of ferrite slag and thin Pt plate, or single ferrite slag layer alone. The analysis was done in the time region between 10% and 30% of the maximum temperature rise.

As shown in figure 2, the thickness appears not to affect the results in both single layer and two layer systems. This proves that the usefulness of the initial temperature rise analysis is well appreciated and the present sample is optically so thick that the permeation depth of the laser beam is negligible small.

To evaluate the effect of the radial heat flow on thermal diffusivity, the measurements were carried out for three samples with different diameter of 17,13 and 10 mm, and the radial heat flow was confirmed to be insignificant in thermal diffusivity measurements by this method. Thus, one dimensional heat flow approximation is well established in this time region.

The heat leak by radiation is also known to be very small in the initial stage of temperature response curve. Regarding this point, the temperature response curve at 1500K and 1673K was calculated by considering the radiative properties of the present cell system using the heat transfer equation proposed by Heckman[9]. The results show that the uncertainty due to the radiative heat transfer is less than 0.5% at 1500K and 0.6% at 1673K for the present case.

The thermal diffusivity of ferrite slag containing 45% CaO is shown in figure 3 in both solid and liquid states. On the other hand, figure 4 shows the thermal diffusivities of calcium ferrite slags at 1633K in the composition range 33 mol% to 55 mol% CaO. The thermal diffusivity of molten calcium ferrite slags is the order of 3×10^{-3} cm^2/s and its temperature dependence is found to be slightly negative in the temperature range between the melting point and 1673K.

ACKNOWLEDGMENTS

Authors(HO and YW) are grateful to the Ministry of Education, Science and Culture for financial support(#62850122) of a Grant-in-Aid.

REFERENCES

1. J.T.Schriempf, Rev. Sci.Instrum. **43** 781 (1972).
2. S.Sumita, T.Mimori, K.Morinaga and T.Yanagase, J. Jpn. Inst. Metals **44** 94 (1980).
3. J.L.Bates, "Thermal Diffusion and Electrical Conductivity of Molten and Solid Slags", Battelle Pacific Northwest Lab. Rep. (1976).
4. R.E.Taylor and H.J.Lee, Rep. 14th thermal Conductivity Conf. 135 (1974).
5. C.S.Ang H.S.Tan and S.L.Chen, J.App.Phys.**44** 687 (1973).
6. C.S.Ang S.L.Chen and H.S.Tan, J.App.Phys.**45** 179 (1974).
7. S.Sumita, K.Moriga na and T.Yanagase, J. Jpn. Inst. Metals **47** 127 (1983).
8. H.M.James, J.App.Phys. **51** 4666 (1980).
9. R.C.Heckman, J.App.Phys. **44** 1455 (1973).

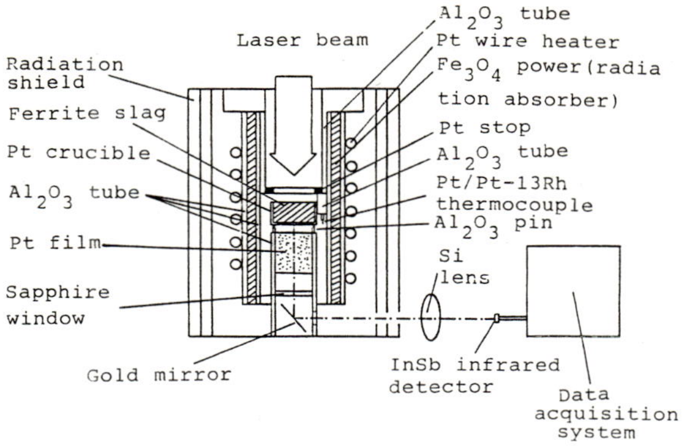

Figure 1 Schematic diagram of the cell assembly used in this work.

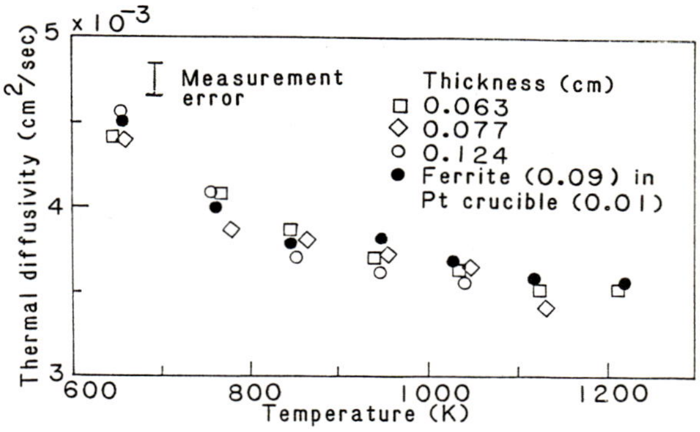

Figure 2 Thermal diffusivity of ferrite slags of different thickness and cell dimension.

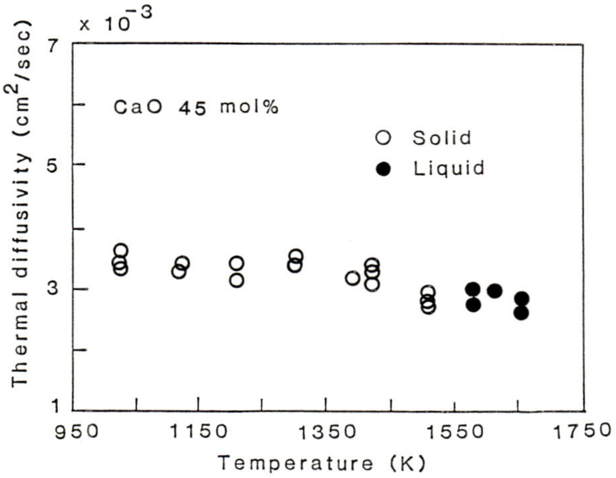

Figure 3 Thermal diffusivity of ferrite slags containing 45 mol% CaO.

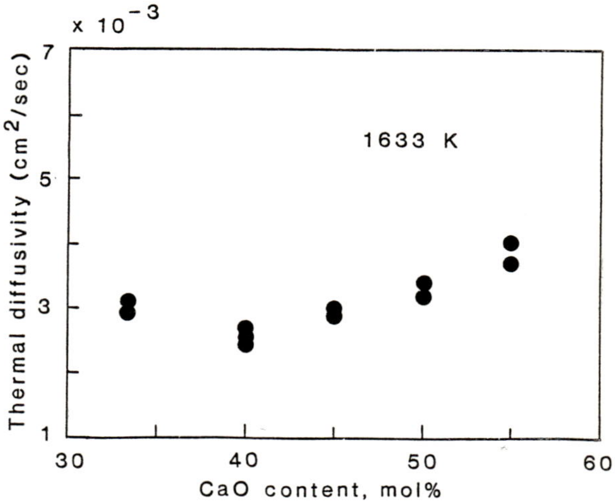

Figure 4 Composition dependence of thermal diffusivity of calcium ferrite slags at 1633K.

Equation for thermal conductivity of molten salts and slags

T IIDA, K C MILLS and Z MORITA

Drs Iida and Morita are in the Department of Materials Science and Processing, Osaka University, Osaka, Japan; Dr Mills is in the National Physical Laboratory, Teddington, Middlesex, UK.

SYNOPSIS

A simple expression is proposed for the thermal conductivities (λ_m) of molten salts and slags at their melting points from a knowledge of their molar volumes (V_M) at this temperature viz.

$$\lambda_m = 1.8 \times 10^{-5} V_M^{-1} \quad \text{(in SI units)}$$

INTRODUCTION

In recent years, the mathematical modelling of heat and mass transfer in high temperature processes has made an important contribution to the improvement of process control. This has led to a demand for reliable thermal conductivity data for the various phases involved in high temperature processes, such as steelmaking, casting and solidification, coal gasification and glassmaking. Unfortunately, there are few accurate data available for the thermal conductivities of molten salts and slags since the measured values frequently contain large and unquantified contributions from both convection and radiation conduction effects. Given both the demand and the inherent difficulties in measuring the thermal conductivity of these melts, it would be extremely useful to have a means of estimating the lattice (or phonon) thermal conductivity from a knowledge of other physico-chemical properties. In this paper, a method is proposed for the calculation of the thermal conductivities of molten salts and glasses from their molar volumes.

Properties such as the velocity of sound in a liquid and the thermal conductivities of non-metallic melts are largely dependent upon the oscillatory motion of the molecules in the fluid. Consequently, expressions can be derived to link these properties with other physico-chemical properties, provided that appropriate values can be assigned to both the effective frequency of oscillation and the relevant proportionality constants[1]. In order to relate the velocity of sound at the melting point (U_m) to other physical properties of the liquid, the authors have previously derived[1] a modified form of the Gitis-Mikailov expression and this is given in equation (1) where, M, V and k_u, respectively, are the formula weight, the molar volume and a constant and the subscript, m, denotes the melting temperature of the liquid.

$$U_m = k_u M^{-\frac{1}{4}} V_M^{-\frac{1}{2}} \tag{1}$$

It should be noted that the value of k_u is dependent upon the type of liquid involved eg for molten metals, k_u has value 5.3 in SI units.

The velocity of sound has previously been related[1] to the thermal conductivity of a dielectric liquid by equn. (2).

$$\lambda_m = 3.3 \times 10^{-7} U_m V_M^{-2/3} \tag{2}$$

Combination of equations (1) and (2) produces equation (3)

$$\lambda_m = (3.3 \times 10^{-7} k_u) M^{-\frac{1}{4}} V_M^{-7/6} \tag{3}$$

Since, $V_m = (M/\rho)$, where ρ is the density of the liquid, equation (3) can be rewritten in the form

$$\lambda_m = 1.8 \times 10^{-6} \rho_m^{7/6} M^{-17/12} \tag{4}$$

The values of $V_m^{-1/6} M^{-\frac{1}{4}}$ are roughly constant for the majority of molten salts and slags and consequently equation (3) can be further simplified to

$$\lambda_m = 1.8 \times 10^{-5} V_M^{-1} \tag{5}$$

In a similar manner equation (5) can be simplified to the form of equation (6), where ρ_N is the number density ie the number of molecules per unit volume.

$$\lambda_m = 3.0 \times 10^{-29} (\rho_N)_m \tag{6}$$

VERIFICATION OF THERMAL CONDUCTIVITY RELATIONSHIPS

In order to check the validity of equations (4) and (5), physical property data for a large number of molten salts and slags have been collated[2-10]. It can be seen from Figs 1 and 2 that linear relationships do exist between λ_m and the parameters ($\rho_m^{7/6} M^{-17/12}$) and V_M^{-1}, respectively. The proportionality constants given in equns. (4) to (6) were obtained from a best fit of the data. The calculated values of λ_m for 29 of the dielectric liquids lay within the bands of ± 30% and only the data for CuCl and AgBr lay outside these bounds. These error limits are probably of similar magnitude to the experimental uncertainties associated with the measurements. It should also be noted that the reported values of λ_m may contain contributions from convection and radiation conduction which result in an erroneously high value for λ_m. This can readily be seen in the reported λ_m values for molten KNO_3 and $NaNO_3$, where recent values are appreciably lower than those obtained previously, due to the steps taken to minimise convection and radiation conduction. Contributions from radiation conductivity (λ_r) are potentially large in molten slags since the melting points of slags are high and λ_r is proportional to $(T)^3$, where T is the thermodynamaic temperature. Consequently for slags only λ_m values which have been determined by transient techniques(7-10), were considered in this study, since errors from these sources are smaller in short-duration experiments. Since the λ_m values of many of the molten salts may be appreciably higher than the true values, the proportionality constants in equations (4) to (6) may need to be revised downwards, when more reliable data become available for these melts.

DISCUSSION

This work has clearly demonstrated that the thermal conductivities of dielectric liquids are inversely proportional to the molar volume. Other studies have also reported that the volume of a substance influences the thermal conductivity. Li et al[11] have reported that for a family of liquids (eg the alkanes) the thermal conductivities are inversely proportional to the 'reduced volume', a parameter which was calculated from viscosity data. Furthermore, Slack[12] has proposed that the decrease in thermal conductivity which occurs when a substance (eg molten salt) melts can be accounted for by the change in volume (V), or density (ρ) which accompanies this transition viz $[\lambda_S/\lambda_L] = [\rho_S/\rho_L]^g$ where subscripts S and L denote the solid and liquid phases, respectively, and $g = [\delta \ln V/\delta \ln T]_T$ Measured values of g fall in the range $1<g<4$.

The relationships shown in equations (4) and (5) obviously refer to melts which contain ionic bonding. However slags also contain strong valence bonds and given the polymeric nature of slags, it was anticipated that the network structure would have a significant effect on the thermal conductivities and that this would lead to the deviations from the linear relationship shown in Fig. 2. Since these departures are apparently small (Fig. 2) it would appear that structural effects have only a secondary effect on the thermal conductivity. One reason could be that the densities of slags are also dependent upon network structure, although for densities 'network effects' are relatively small (ca 5%). In this respect the thermal conductivity is apparently similar to the electrical conductivity of slags, which is largely determined by the size of the cation and the 'network effect' (due to the hinderence of cationic movement) is of secondary importance.

CONCLUSIONS

1. The thermal conductivities of molten salts and slags at their melting points are inversely dependent upon their molar volumes.
2. It would appear that 'network structure' effects on the thermal conductivities of molten slags are secondary since the latter parameter is largely determined by the molar volume of the liquid.

REFERENCES

1. T. IIDA and R.I.L. GUTHRIE. The Physical properties of Liquid Metals (1988) Clarendon Press, Oxford.

2. D.T. JAMIESON, J.B. IRVING and J.S. TUDHOPE. Liquid Thermal Conductivity, a data survey to 1973, HMSO, Edinburgh 1975.

3. T. ASAHINA and M. KOSAKA. Jap. Soc. for Promotion of Science, 140th Committee, Report T5, Dec 1980.

4. V.D. GOLYSHEV, M.A. GONIK, V.A. PETRON and Y.M. PUTILIN. High temperatures, 1984, 21, 684.

5. TAN HE PING and M. LALLEMAND. High Temp. - High Pressure, 1987, 19, 417.

6. J.S. POWELL and K.C. MILLS. National Physical Laboratory, Teddington, UK. unpublished results.

7. M. KISHIMOTO, M. MAEDA, K. MORI and Y. KAWAI. Proc. 2nd Intl. Symp. 'Metallurgical Slags and Fluxes', Lake Tahoe, Nev. Nov 1984, publ. Metall Soc. AIME, Warrendale, Pa. p.891.

8. T. SAKURAYA, T. EMI, H. OHTA, Y. WASEDA. Nippon Kinzoku Gakkaishi (J Jap. Inst. Met.) 1982, 46, 1131.

9. H. OHTA, Y. WASEDA, Y. SHIRAISHI, Proc. 2nd Intl. Symp. 'Metallurgical Slags and Fluxes', Lake Tahoe, Nev. Nov 1984, publ. Metall Soc. AIME, Warrendale, Pa. p.863.

10. K. NAGATA and K.S. GOTO. Proc. 2nd Intl. Symp. 'Metallurgical Slags and Fluxes', Lake Tahoe, Nev. Nov 1984, publ. Metall Soc. AIME, Warrendale, Pa. p.875.

11. S.F. LI, G.W. MAITLAND, W.A. WAKEHAM. High Temp. - High Pressure, 1985, 17, 241.

12. G.A. SLACK. Solid State Physics, 1979, 34, 1.

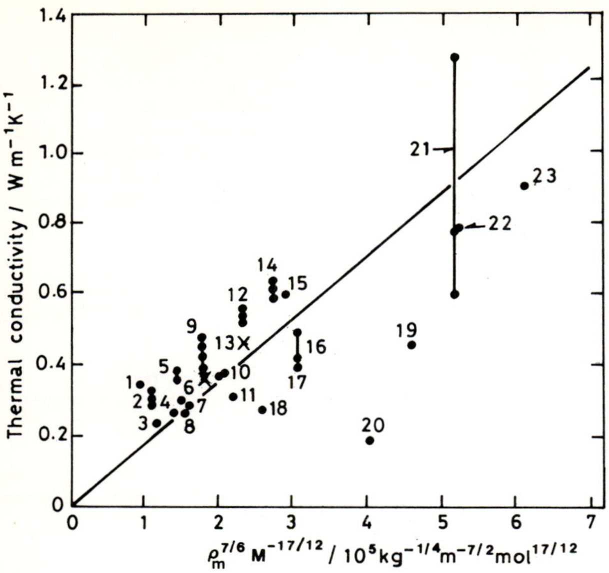

Fig. 1. Thermal conductivity of molten salts and slags as a function of $(\rho_m{}^{7/6} M^{-17/12})$. Points linked by a vertical line represent different experimental values for a single salt or slag, X, values from ref. 6; data for LiF have not been included [λ_m = 1.25, $(\rho_m{}^{7/6} M^{-17/12})$ = 11.2],

(1) $PbCl_2$ (2) C_sNO_3 (3) KI (4) KCSN
(5) $RbNO_3$ (6) $ZnCl_2$ (7) KBr (8) NaI
(9) KNO_3 (10) KCl (11) NaBr (12) $NaNO_3$
(13) $AgNO_3$ (14) $LiNO_3$ (15) $NaNO_2$
(16) NaCl (17) LiBr (18) AgBr (19) LiCl
(20) CuCl (21) 40 CaO–40 SiO_2–20Al_2O_3
(22) 35 CaO–45 SiO_2–20Al_2O_3 (23) NaOH
(24) LiF (25) 40 CaO–45 SiO_2–15 Al_2O_3
(26) 50 CaO–35 SiO_2–15 Al_2O_3
(27) 30 Na_2O – 70 SiO_2
(28) 40 Na_2O – 60 SiO_2
(29) 40 K_2O – 60 SiO_2

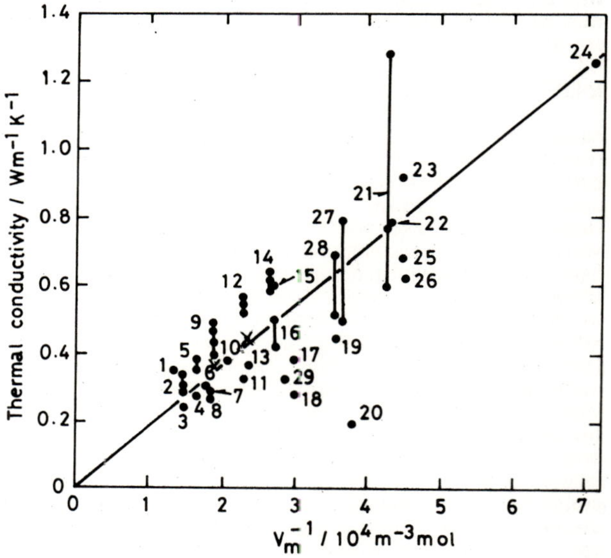

Fig. 2. Thermal conductivity of molten salts and slags as a function of the reciprocal of molar volume. Points linked by a vertical line represent different experimental values for a single salt or slag.

Thermal conductivities of liquid slags

K C MILLS

The author is in the Division of Materials Applications, National Physical Laboratory, Teddington, UK.

SYNOPSIS

In this study, recent thermal conductivity (k) data for molten silicates have been examined and shown to be dependent upon the chain length of the silicate anions. Values of the thermal conductivity derived by attributing thermal resistances to the valence and ionic bonds are in good agreement with the experimental data. A method is proposed for the estimation of the thermal conductivities of complex molten silicates. The change in slope of the k-temperature relationship has been shown to be due to the change in thermal expansion coefficient which accompanies the glass transition.

INTRODUCTION

In recent years mathematical modelling of the heat transfer mechanism occurring in high temperature processes, such as steelmaking, coal gasification and non-ferrous metal production, has resulted in improvements in process control. However, reliable values for the thermophysical properties of the phases involved are required for the succesful operation of these models. Thermal property data have been measured for many of the materials but it has proved very difficult to obtain reliable values for the lattice (phonon) conductivity (k_c) of liquid slags. When conventional, steady state techniques are used the thermal conductivity values are obtained from measurements of the temperature gradient and the heat flux density (Q) [equation (1)] which can contain large contributions from convection and radiation conduction. Consequently, in recent years transient methods have been developed to minimise the contributions from convection. However, conductivity values derived in this manner will contain a small, but unquantifiable, contribution from radiation conduction. Nevertheless, data obtained by transient techniques are significantly more reliable than those obtained by steady state techniques, especially for liquid slags since their melting temperatures are high and the radiation conductivity is proportional to (T^3), where T is the thermodynamic temperature.

$$Q = -k(dT/dx) \qquad (1)$$

Recently, Iida *et al*[1] have proposed that the thermal conductivities of molten salts at their melting points (denoted k_m) can be estimated to within ± 30% from data for their molar volumes (V_m) by the use of equation (2)

$$k_m = 1.8 \times 10^{-5} V_m \quad \text{(in SI units)} \qquad (2)$$

Molten slags are also ionic conductors and these materials were found to obey equation (2). Since slags are polymeric, this finding implies that any structural contributions to the thermal conductivity are relatively small in magnitude. Consequently, in this study the results of recent measurements of thermal conductivity in liquid slags have been analysed to determine the effect of the network structure on the values obtained.

RESULTS AND DISCUSSION

Recent thermal conductivity data[2-5] for slags from the $Na_2O + SiO_2$ and $CaO + Al_2O_3 + SiO_2$ systems are shown in Figs 1 and 2, respectively, and for slags involved in the continuous casting of steel[6] and coal gasification[7] in Fig 3. The following observations can be made from these data:

(i) in all cases there was a maximum in the k_c-temperature relationship;

(ii) the thermal conductivities of liquid slags were appreciably lower than those for the solid state;

(iii) the thermal conductivities of liquid slags increased with increasing silica content.

The origin of the maximum in the k_c-T relationship

Kishimoto *et al*[2] suggested that the maximum (at T_{max}) represented the Debye temperature for the slag. However similar maxima have been reported for continuous-casting[6] and coal slags[7] and in these cases the maxima were attributed to the glass transition temperature (T_g). When a glass is heated through T_g, it is transformed into a supercooled liquid. This transition if accompanied by (i) an abrupt (step-type) change in the heat capacity, C_p, thermal expansion coefficient and velocity of sound (u) and (ii) changes in slope of the temperature dependence of the enthalpy and molar volume (or density). Consequently, it would seem probable that the maxima in the k_c-T plots are associated with the glass transition, and more specifically with the increase in "free volume" which occurs above the glass transition temperature, T_g, when vacancies (of the magnitude of interatomic distances) become free to expand. Support for this view is provided by the fact that, (i) the T_g value derived from C_p values was virtually identical to T_{max}(2), (ii) similar maxima in k_c–T curves have been observed at T_g for amorphous polymers(8) and Ge+Te glass(9) and (iii) Nagata *et al*(10) observed a change in the velocity of sound in silicates at temperatures around T_g.

In polymeric silicates there are three types of bonding, viz; (i) primary valence bonds between the atoms in any one chain, (ii) ionic bonds formed between the end members of the chain (anions) with cations and (iii) van der Waals forces between the chains. An explanation for the change in slope of the k_c–T relationship in organic polymers, which occurs at T_g, was put forward by Eiermann(11). He assumed that the polymeric chains could be sub-divided into segments (with dimensions of intermolecular proportions) and these defined the phonon free path length (L). Thermal resistances (1/k) were attributed to primary valence bonds and van der Waals forces, with the values for the latter being much greater. A network of thermal resistances was assumed and equation (3) was derived from calculations of the forces associated with both types of bond, where Δ, α and β represent the change in property occurring at T_g, the linear and volume thermal expansion coefficients, respectively. Eiermann(11) found that equation (3) could accurately account for both the change in slope and the value of thermal conductivity of the polymer*.

$$\frac{\Delta\left[\frac{1}{k}\frac{dk}{dT}\right]}{\Delta\beta} = \frac{\frac{1}{2}\left[91-28\left[1+{}^{3}/_{2}\int_{0}^{T}\alpha dT\right]^{6}\right]}{\left[13-7\left[1+{}^{3}/_{2}\int_{0}^{T}\alpha dT\right]^{6}\right]} \qquad (3)$$

Ignoring the effect of the ionic bonding in the slags, values for the change in slope were calculated using equation (4), which was derived by introducing the value of 0.009 for $\int_0^T \alpha dT$ in equation (3), this being a typical value for silicates. Although equation (4) does predict a change in slope at T_g, the values of $\Delta[dk/dT.k]$ are lower than the experimental results by a factor of 1.5 to 5. Some of the discrepancies can be attributed to the difficulty in obtaining accurate results for Δβ. However, in slags, ionic and valence bonds predominate and consequently the most probable cause of the discrepancy is that the constant (5.60) in equation (4) is inappropriate for the ionic bonding present. Nevertheless, equation (4) does show that the change in slope is almost certainly associated with the change in expansion coefficient which occurs at the glass transition, since the interionic distances would be unaffected by temperature in glasses but would increase markedly above T_g with the increased 'free volume' available in the supercooled liquid.

$$\Delta\left[\frac{1}{k}\frac{dk}{dT}\right] = -5.60\,\beta \qquad (4)$$

The effect of structure on the thermal conductivity of molten silicates

In order to determine the magnitude of any structural contributions to the thermal conductivities by comparing extant data for molten silicates, it is necessary first to define a reference temperature for these comparisons; the melting point has been used in this study and this is denoted by the subscript, m, eg. k_m. Several methods have been used to derive a quantitative description of the network structure, viz;

(i) the average chain length† of the $(Si_nO_{2n+3})^{6-}$ rings formed(12); (ii) the ratio of non-bridging oxygen (NBO) atoms/(Si+Al atoms)†(13) and (iii) the network parameter, ψ, recently proposed by Iida *et al*(14). The latter was calculated from estimated values of the viscosity using the methods reported by Riboud *et al*(15) and Urbain *et al*(16). Only the thermal conductivity data obtained by transient techniques(2–5) have been used in this examination. These data may contain a small but unquantified contribution from the radiation conductivity but errors from this source will be considerably smaller than those associated with values obtained using steady state techniques. Despite the experimental errors involved in both the measurements (which must be at least ± 0.1 $Wm^{-1}\,K^{-1}$), and the various structural parameters, it can be clearly seen from Figs 4–6 that a relationship does exist between k_m and these parameters. It can also be seen that k_m increases with increasing chain length (ie. increasing silica content).

The strength of the van der Waals forces in slags can be assumed to be negligible compared with those of the primary valence and ionic bonds. This view is supported by the fact that the thermal conductivities of slags (0.3–1.5 $Wm^{-1}\,K^{-1}$) are considerably higher than those of organic liquids (ca. 0.1 $Wm^{-1}\,K^{-1}$) where van der Waals forces perdominate. Thus the thermal resistance (R_m) of molten slags can be considered to be composed of thermal resistances associated with primary valence and ionic bonds (R_{PVB} and R_{IB}, respectively).

Bockris *et al*(12) proposed that molten silicates were composed of discrete $(Si_nO_{2n+3})^{6-}$ rings, the value of n in these rings being determined by the mole fraction (x) of CaO or Na_2O eg. $(Si_3O_9)^{6-}$ at x = 0.5, $(Si_6O_{15})^{6-}$ at 0.33, $(Si_9O_{21})^{6-}$ at 0.25 and $(Si_{12}O_{27})^{6-}$ at 0.2 etc. Schematic representations of the structure of the first three anions are given in Fig. 7. It was also proposed that two of these anions coexisted in the intermediate compositional ranges.

In Fig. 4 the relative amounts of these coexisting anions were taken into account by deriving an 'average chain length' which was obtained by interpolation of the curve relating n to x_{CaO} (or x_{Na_2O}).

This approach was also used in Fig. 8 which shows the thermal resistance, R_m, of the slags as a function of the 'average chain length' of the anion ring.

It would be anticipated that the resistance to thermal conduction would be much greater between chains than that along the chain ie. $R_{IB} >> R_{PVB}$. The sharp decrease with increasing chain length shown in Fig 8 supports this view. Silica can be considered to contain only primary valence bonds since it has a high degree of cross-linkage and thus a value can be obtained for R_{PVB} from the measured values of the thermal conductivity of silica. In a similar manner values of R_{IB} can be derived from the thermal resistances of the monomer (n=1) where ionic forces

---

* This confirms the importance of the effect of volume on the thermal conductivity (1)

† Al_2O_3 was assumed to be fully integrated into the silicate structure.

predominate, and this can be achieved by extrapolating the curve in Fig. 8 to n=1 (R_m=2.6 mKW^{-1}).

Eiermann[11] assumed that the thermal resistance of organic polymers could be treated as a three-dimensional network of resistances arising from van der Waals forces (R_{VdW}) and primary valence bonds (R_{PVB}). The Debye, path-length theory was applied to elements (of interatomic dimensions) in the long polymeric chains, both types of resistances being connected in series ($R \triangleq R_{PVB} + 2R_{VdW}$). It can be seen from Fig 7 that $(Si_6O_{15})^{6-}$ and $(Si_9O_{21})^{6-}$ rings are not planar like the organic polymers and thus application of the three-dimensional network approach would be difficult. Consequently, an alternative approach has been adopted to calculate the thermal resistance of slags, in which the thermal resistances arising from the primary valence and ionic bonds for each silicon atom are derived for the various $(Si_nO_{2n+3})^{6-}$ ring configurations. Assuming series connections of R_{PVB} and R_{IB}, the thermal resistances associated with each silicon atom in the ring have been derived below and related to R_m, where N is the total number of silicon atoms.

$(Si_3O_9)^{6-}$: $R_m = N(2R_{PVB} + 2R_{IB})$

$(Si_6O_{15})^{6-}$: $R_m = N(3R_{PVB} + 1R_{IB})$

$(Si_9O_{21})^{6-}$: $R_m = N(3.33\ R_{PVB} + 0.67\ R_{IB})$

$(Si_{12}O_{27})^{6-}$: $R_m = N(3.5R_{PVB} + 0.5R_{IB})$

The thermal conductivity of SiO_2 at temperatures around 1200 K is about 2 $Wm^{-1}K^{-1}$ and corresponds to four primary (Si-O) valence bonds, thus R_m = 0.5 mKW^{-1} = $4(NR_{PVB})$ and thus NR_{PVB} = 0.13 mKW^{-1}. The thermal resistance of the SiO_4^{4-} anion formed by the monomer was derived from Fig. 8, ie. R_m = 2.6 $mK^{-1}W^{-1}$; this pertains to four ionic bonds involving O^- and thus (NR_{IB}) = (2.6/4) = 0.65 mKW^{-1}. Values of R_m were calculated for the various $(Si_nO_{2n+3})^{6-}$ rings using these values of R_{PVB} and R_{IB} and the results are presented in Fig. 8. It can be seen that the calculated values are in good agreement with the experimental data and also reproduce the curvature of the R_m-'average chain length' relationship. Thus the increase in thermal conductivity with increasing chain length can be accounted for in terms of the relative proportions of valance and ionic bonds.

Bockris *et al*[12] reported that monovalent ions, like Na^+, were incapable of forming the 'bridges' between anion rings which divalent Ca^{2+} ions produce. Thus R_{IB} would be expected to be higher in sodium silicate melts than in calcium silicates and it can be seen from Fig 8 that the open circles do tend to lie above the averaged curve. However, it is also possible that this behaviour is due to the weakening of the Si-O bond with the introduction of Al into the chain.

The estimation of thermal conductivities of molten silicates

Thermal conductivities (k_m) can be estimated using Figs 4-6. The relationship between the network parameter (ψ) and thermal conductivity (Fig. 6) is particularly useful in this context since it can be used to estimate the thermal conductivities of complex slags from a knowledge of the melting point and chemical composition. First the chemical composition can be used to calculate the viscosity at the melting point[15,16] (μ_m) and then ψ can be derived from equation (5) and the tabulated data for μ_o[14] and finally k can be derived from Fig 6 using the calculated value for ψ.

$$\psi = \log [\mu_m/\mu_o]^2 \qquad (5)$$

The mean-free path length in molten slags

Recently, Rivers *et al*[17] and Nagata *et al*[10] have reported values for the velocity of sound (u) in molten slags. This information can then be used along with heat capacity (C_p)[2,3] and density (ρ) values to calculate the mean free path length (L) from equation (6)

$$L = 3k/C_p u.\rho \qquad (6)$$

Calculated values for the mean free path length were around 0.15 nm but varied between 0.09 and 0.30 nm. The following values have been reported for the various relevant interionic distances in molten silicates, Si-O, 0.16 nm; O-O, 0.27 nm; Si-Si, 0.31 nm; Na-O and Ca-O, 0.24 nm. Thus the experimental values for the mean free path length correlate best with the Si-O interionic distance.

CONCLUSIONS

1. The thermal conductivities of molten silicates contain a substantial contribution from the structure of the slag, this contribution increasing with increasing chain length of the silicate anions.

2. Thermal conductivity values calculated by: (i) attributing thermal resistances to the ionic and valence bonds; and (ii) taking the structural aspects of $(Si_nO_{2n+3})^{6-}$ anion rings into account are in good agreement with experimental data.

3. Thermal conductivities of complex, molten slags can be estimated from a knowledge of the chemical composition and melting point of the slag, by use of the relationship between thermal conductivity and the network parameter, ψ.

4. The change in slope of the thermal conductivity-temperature relationship is due to the change in thermal expansion coefficient which occurs at the glass transition.

ACKNOWLEDGEMENTS

The author acknowledges the valuable discussions held with his colleagues Dr A. Olusanya, Mr J. S. Powell and Dr R. Morrell and with Prof. P. Grieveson [Imperial College, London].

REFERENCES

1. T. IIDA, K.C. MILLS and Z. MORITA, Proc. of this Conference.

2. M. KISHIMOTO, M. MAEDA, K. MORI and Y. KAWAI. Proc. of 2nd Intl. Symp. 'Metallurgical Slags and Fluxes', Lake Tahoe, Nev. Nov. 1984 publ. Metall. Soc. AIME, Warrendale, Pa. p.891.

3. H. OHTA, Y. WASEDA, Y. SHIRAISHI, ibid p.863.

4. K. NAGATA, K.S. GOTO, ibid p.875.

5. T. SAKURAYA, T. EMI, H. OHTA and Y. WASEDA. Nippon Kinzoku Gakkaishi (J Jap. Inst. of Met.) 1982, 46, 1131.

6. R. TAYLOR and K.C. MILLS, Ironmaking Steelmaking, 1988, 15, 187.

7. J.M. RHINES and K.C. MILLS. High Temp – High Pressure, 1985, 17, 173.

8. D. HANDS. 'The thermal transport of polymers', RAPRA Members Report 1977, 11.

9. Q. XU, K. ICHIKAWA, J. Phys. C; (Solid State Phys.) 1985, 18, L 985.

10. K. NAGATA, K. OHIRA, H. YAMADA and K.S. GOTO. Metall. Trans. B, 1987, 18B, 549.

11. K. EIERMAN. Rubber Chem. Technol. 1966, 39, 841.

12. J. O'M BOCKRIS, J.D. MACKENZIE and J.D. KITCHENER. Trans. Farad. Soc. 1954, 51, 1734.

13. B.O. MYSEN, D. VIRGO and C.M. SCARFE. Am. Mineralogist 1980, 65, 690.

14. T. IIDA, Z. MORITA and T. MIZOBUCHI. Proc. of this Conference.

15. P.V. RIBOUD, Y. ROUX, L.D. LUCAS and H. GAYE. Fachber. Huttenprax. Metallweiterv. 1981, 19, 859.

16. G. URBAIN, F. CAMBIER, M. DELETTER and M.R. ANSEAU. Trans. J. Brit. Ceram. Soc. 1981, 80, 39.

17. M.L. RIVERS and I.S. CARMICHAEL. J. Geophys. Res. 1987, 92, 9247.

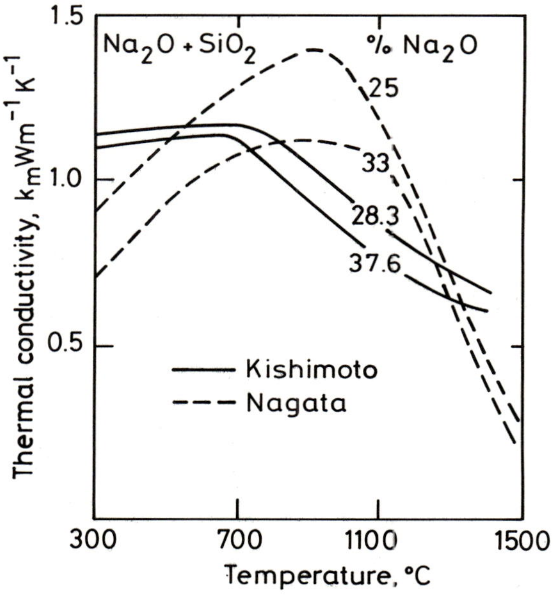

1. The thermal conductivity of Na_2O + SiO_2 slags as a function of temperature.

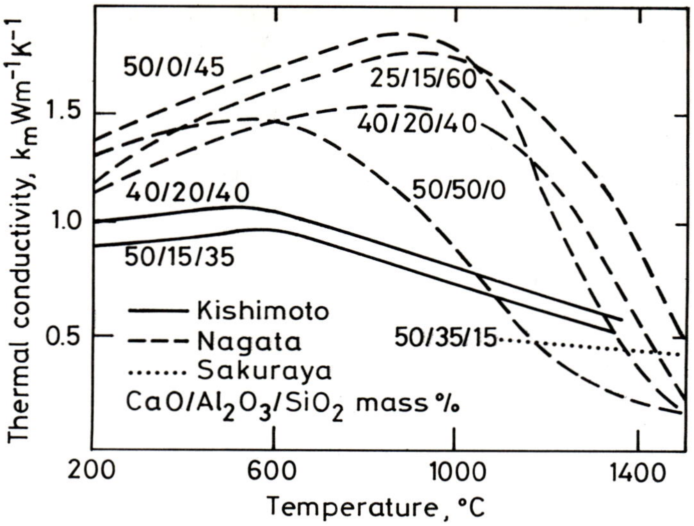

2. The thermal conductivity of CaO + Al_2O_3 + SiO_2 slags as a function of temperature.

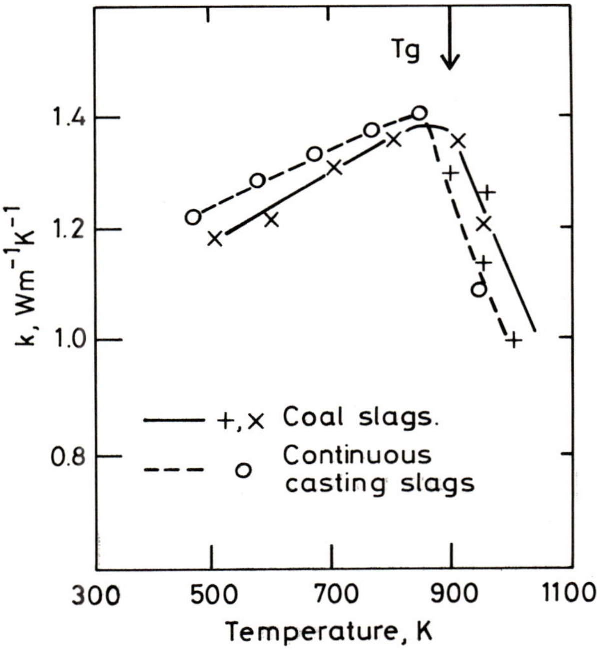

3. The thermal conductivity of continuous casting and coal slags as a function of temperature.

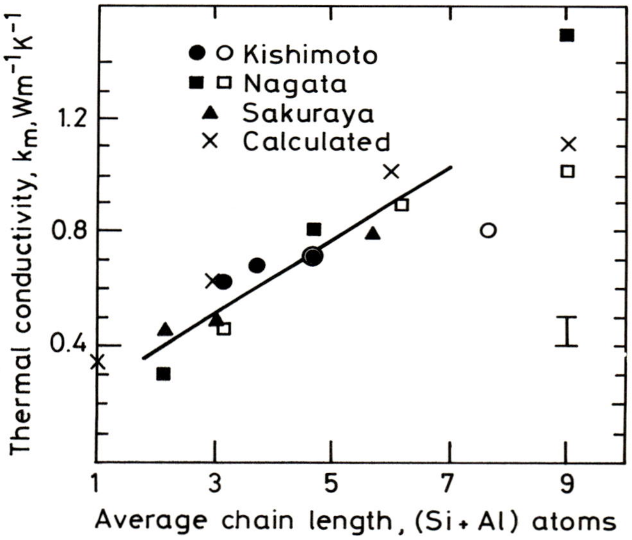

4. Thermal conductivity of slags at their melting point (k_m) as a function of the average chain length of the $(Si_nO_{2n+3})^{6-}$ rings(12); the open circles represent data for the Na_2O + SiO_2 system and closed circles the CaO + Al_2O_3 + SiO_2 system; I represents the minimum experimental uncertainty.

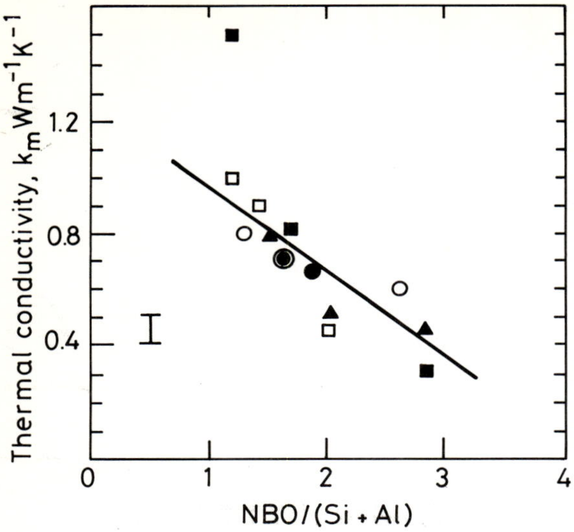

5. Thermal conductivity of slags (k_m) as a function of the ratio of non-bridging oxygen (NBO) atom/(Si + Al atoms); symbols as in Fig 4.

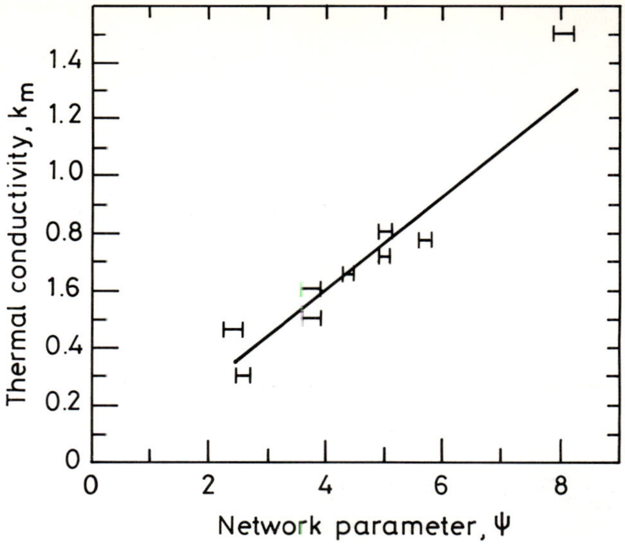

6. Thermal conductivity of slags, (k_m) as a function of the network parameter ψ; H represents the spread of ψ resulting from viscosoity estimations[15,16].

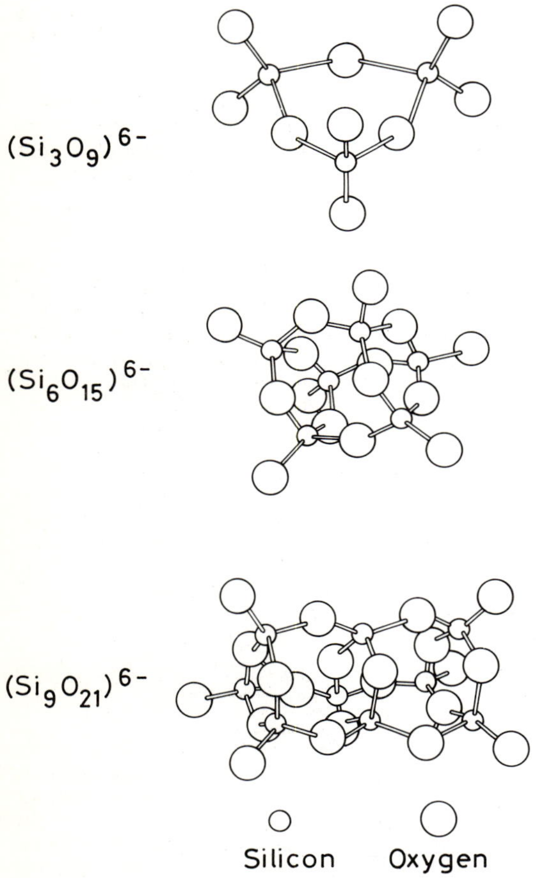

7. Schematic representations of various $(Si_nO_{2n+3})^{6-}$ rings, after Bockris *et al*[12].

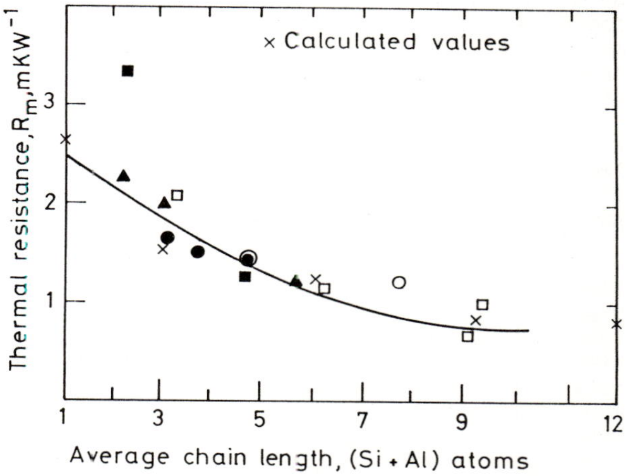

8. The thermal resistance (R_m) as a function of the average chain length of the $(Si_nO_{2n+3})^{6-}$ rings[12]; symbols as in Fig 5.

Heat transfer simulation for continuous casting

S BAGHA, N C MACHINGAWUTA and P GRIEVESON

Dr Bagha, formerly Research Student at Imperial College, is now with Davy McKee, Stockton; Dr Machingawuta is a Research Associate in the Department of Materials, Imperial College; Professor Grieveson is in the Department of Materials, Imperial College, London.

SYNOPSIS

A simple technique have been used to simulate heat transfer in the mould during continuous casting. Commercial casting fluxes and synthetic slags have been studied. In addition, the effect of cooling water flowrate on heat transfer has been investigated. The overall heat transfer is shown to be a very important function of the heat transfer coefficient between the infiltrated flux and the mould. It is demonstrated that the value of this heat transfer coefficient is dependent on the composition of the flux and the cooling water flowrate.

With a knowledge of the solidus and recrystallisation temperatures, the effective thermal conductivities of the glassy, partially recrystallised and liquid regions of the infiltrated flux are estimated.

INTRODUCTION

Whereas continuous casting fluxes have a variety of functions during the casting process, undoubtedly one of the most important is to regulate heat transfer between the solidifying shell and the water-cooled mould. In order to gain information on the factors which influence this heat transfer, a simple simulation technique was developed for use in the laboratory. It is well known that the mechanism of heat transfer is complex. The molten flux penetrates into the gap between mould and strand and exerts a resistance to the withdrawal of heat from the strand. It has been established that the infiltrated slag layer consists of:

a) a solid slag layer formed on the mould wall and with most fluxes this layer travels with the mould during oscillation.

b) a thin liquid layer (0.1-0.3mm) is formed adjacent to the strand which moves downward with the strand, and,

c) the solid layer comprises both a glassy layer adjacent to the mould and a partially recrystallised layer between the glassy and liquid layers.

Using the simulation test with a variety of industrial fluxes, values of thermal flux obtained across the infiltrated slag layer between the solidifying strand and the mould have been determined to be consistent with those observed in industrial practice. Our initial work showed that the adhesion of the flux to the copper mould varied significantly with the chemical composition of the flux at constant water flow rate. It was also noted that the high heat fluxes were observed when this adhesion was strong. In general, the experimental heat of fluxes decreased with increasing sodium oxide and alumina additions to the flux.

The present work describes observations of the effect of variations in the cooling water flow rate on the heat flux. A more sophisticated analysis of the heat flux through the infiltrated flux is attempted.

EXPERIMENTAL TECHNIQUE

Thermal flux measurements were made using the apparatus shown in Figure 1. About 3kg of electrolytic iron were melted in an alumina-graphite crucible (8cm inside diameter and 11cm high) which acted as a susceptor activated by a 15kW induction generator. The temperature of the liquid iron was monitored using a Pt-Pt/6% Rh thermocouple. When the required operating temperature was reached ~75g of casting powder was placed on the molten metal to produce a layer 7-10mm deep. The water-cooled, copper finger was lowered mechanically into the molten iron to ensure that a uniform layer of slag was formed on the copper surface. Variation in the rate of lowering the finger allowed infiltrated flux thickness values of ~1 to 3mm. The flow rate of the cooling water was monitored and its temperature at entry and exit of the mould were recorded using a mercury thermometer and a Cu/Cu-Ni thermocouple respectively. The temperature rise of the water (ΔT) reached a maximum value after approximately 30 seconds when iron started to solidify and then subsequently decreased with time. After completion of the experiment, the copper finger was removed from the iron, with the flux layer surrounded by a layer of solidified iron. The infiltrated slag layer was detached from the copper finger and solidified iron. The thickness of the slag layer was measured and the ease of removal from the copper finger assessed. Samples of the infiltrated slag were taken for examination by optical microscopy and X-ray diffraction studies.

Optical metallographic examination revealed the presence of a smooth glassy layer adjacent to the mould wall, a partially recrystallised region and a previously liquid layer next to the iron surface. The thickness of each layer was determined.

Treatment of thermal flux data
The following assumptions were made in calculating the rate of heat transfer across the slag layer during simulation experiments:

a) the heat flow across the slag layer was unidirectional

b) conduction was the primary mode of heat transfer and,

c) the heat losses to the surroundings were negligible.

The heat flux per unit area, Q, can be calculated using equation [1]

$$Q = \frac{m.Cp.\Delta T}{A} \qquad [1]$$

where m, Cp and ΔT are the mass flow rate, heat capacity and temperature rise of the water and A is the area of the copper finger in contact with the molten metal.

The heat flux across the slag can also be expressed in the form of equation [2]:

$$Q = \frac{T_1 - T_2}{\left\{ \left[\frac{t_1}{k}\right]_{Cu} + \left[\frac{t_2}{k_{eff}}\right]_{slag} \right\}} \qquad [2]$$

where T_1 and T_2 are the temperature of the molten slag in contact with iron and the water at the wall of the copper finger, t_1 and t_2 are the thicknesses of the copper wall and slag layer respectively and k and k_{eff} are the thermal conductivities of copper and the slag.

A value of the effective thermal conductivity of the slag can be determined from a plot of

$$\frac{\Delta T}{Q} - \left[\frac{t_1}{k}\right]_{Cu}$$

against t_2.

As mentioned earlier our previous work had shown a good correlation between the adhesion of the slag to the copper finger and the heat flux e.g. high heat fluxes are associated with good adhesion and therefore strong glass/cold finger bonds whilst low heat fluxes relate to poor adhesion and weak glass/cold finger bonding. These observations are consistent with the presence of a significant heat transfer coefficient existing at the slag/mould interface. An analysis is attempted by dividing the

$$\left[\frac{t_2}{k_{eff}}\right]_{slag} \text{ term into two terms - } \left[\frac{1}{h_s} + \frac{t_2}{k_{av}}\right]$$

where h_s represents the heat transfer coefficient at the slag/copper interface and k_{av} is the average thermal conductivity of the slag phase. The value of h_s can be calculated from the intercept of a graph of

$$\frac{\Delta T}{Q} - \left[\frac{t_1}{k}\right]_{Cu}$$

against t_2 and the value of k_{av} from the slope.

Variation of heat transfer with cooling water flowrate for commercial fluxes
To investigate the effect of cooling water flow rate on heat transfer, simulation tests have been carried out on several commercial casting powders at cooling water flowrates of 0.05, 0.10, 0.15, 0.175, and 0.20 l/s. The analyses of some of the fluxes investigated are presented in Table 1. Some results including the thermal fluxes measured are given in Table 2.

The effect of cooling water flow rate on the heat transfer was found to depend upon the composition of the powder. For the commercial fluxes investigated, there appears to be an optimum cooling water flow rate to produce perfect heat transfer at the slag/mould interface which is characteristic for a particular casting powder. For powder SPH-SL409-801/A, the conditions for perfect heat transfer are produced in the simulation experiments when the cooling water flowrate is close to 0.10 l/s, i.e. under these conditions, the experimental data lie on the maximum thermal flux line where the heat transfer coefficient approaches infinity. At flowrates which are lower or higher than this optimum value, there is a reduction in heat transfer coefficient at the interface, such that a value of ~3kWm^{-2} K^{-1} is observed at water flowrates of 0.05 and 0.2 l/s. For the powders H and Scorialit C 178-95 III, the conditions observed for best heat transfer in the simulation experiments are produced when the water flowrate is between 0.10 and 0.15 l/s and between 0.15 and 0.175 l/s respectively.

The results obtained for the average thermal conductivity through the slag layers are remarkably similar for all powders and give a value of ~1.2 Wm^{-1} K^{-1}.These results indicate the importance of the bond between the glass and mould in determining the heat transfer. At high cooling water flowrates, the quenching effect is severe and may lead to contraction of the glass away from the copper surface due to thermal expansion. At low flowrates the bond is also poor due to the higher temperature at the copper/slag interface. Hence it appears that moderate flowrates produce optimum conditions for a good bond, a high heat transfer coefficient and a high thermal flux across the infiltrated slag.

Effect of chemical constitution on heat transfer
The observation that the chemical constitution of commercial casting fluxes play an important role in heat transfer led to a preliminary study of the effect of chemical composition on heat transfer. Commercial continuous casting fluxes are essentially mixtures of CaO, Al_2O_3, SiO_2, Na_2O and fluoride (usually CaF_2) with minor additions of MgO, K_2O, MnO and FeO together with carbon which is added to modify the melt agglomeration at the mould surface. With such a large number of constituents, it is impossible to compare the performance of powders on a compositional basis. A series of synthetic powders were prepared with CaO/SiO_2 ratios of 0.8, 1.0 and 1.2 respectively and 5% CaF_2 to which various quantities of soda was added. These powders were investigated using the simulation test and the results are presented in Figure 2.

The heat flux values determined were consistent with those obtained using a wide variety of commercial powders. Once again, high heat flux values were associated with good adhesion between the slag and the copper finger and low Q values with poor adhesion.

Heat transfer coefficients at the copper/slag interface were estimated and are shown in Figure 3. The results show that both soda content and CaO/SiO_2 ratio affect the heat transfer coefficient at the slag/finger interface for a water flowrate of 0.2 l/s.

These results indicate the complex nature of heat transfer in continuous casting and confirms the problems experienced in industrial practice of the choice of flux for the wide variety of operational conditions. However, more detailed information may ease casting operators' problems in the future.

Estimation of thermal conductivities of the glass, partially crystalline and liquid layers of infiltrated flux

An attempt is made to estimate the effective thermal conductivities of the glassy, partially crystalline and liquid layers from considerations of the thermal fluxes and the boundary conditions across the infiltrated slags. For the two commercial slags SPH-SL409-801/A and C-178 95-III, the solidification and crystallisation temperatures were estimated from results of differential thermal analysis of the powders. Using this information, the temperatures at the boundary between liquid and partially recrystallised layers and partially recrystallised layer and the glassy layer are known and allow the estimation of the effective thermal conductivities of these layers to be computed from equation [3]

$$Q = k_G \frac{T_R - T_M}{t_G} = k_P \frac{T_S - T_R}{t_P} = k_L \frac{T_F - T_S}{t_L} \qquad [3]$$

where k_G, k_P and k_L are the thermal conductivities of the glassy, partially recrystallised and liquid layers respectively, T_R, T_S, T_M, and T_F ar the temperatures at boundaries of the glassy/recrystallised, recrystallised/liquid, mould/glass and liquid/iron interfaces respectively; and t_G, t_P and t_L are the thicknesses of the glassy, partially recrystallised and liquid layers respectively.

The results of these calculations are presented in Table 3. From the experimental results, the average thermal conductivity for the glassy region of SPH-SL409-801/A is 0.58 $Wm^{-1}K^{-1}$ and for C-178 95-III is 0.78 $Wm^{-1}K^{-1}$.

The thermal conductivities calculated for the partially recrystallised regions are observed to increase with increasing thickness of the layer, as shown in Figure 4.

The extent of recrystallisation is expected to vary across the layer. At high temperatures, the thermodynamic driving force for recrystallisation is small but diffusion is fast whilst at low temperatures the thermodynamic force will be large but diffusion will be slow. It is anticipated that recrystallisation will be more developed in the middle of the layer. In addition, the thermal conductivity of crystalline solids are known to be greater than glasses. The variation of thermal conductivity is expected to be a complex relationship as will the temperature profiles through the layer. It is perhaps not surprising that the apparent thermal conductivity of the partially recrystallised region increases with increasing thickness. However, the true nature of variation of thermal conductivity across this layer is beyond our present knowledge.

It is interesting to note that the average liquid thermal conductivity is lower than that observed for the glassy section. A careful perusal of the average liquid thermal conductivity values derived from these experiments indicates that the values increase with increasing thickness of the liquid layer. As stated by Mills in the previous paper, heat conduction through liquid slags is very complex and radiation conduction is expected to become more significant as the thickness of the slag layer increases. If we extrapolate the experimental values to zero thickness, a value of $\sim 0.3 Wm^{-1}K^{-1}$ is obtained. It is suggested that the value of 0.3 $Wm^{-1}K^{-1}$ is a reasonable one for phonon conductivity of liquid casting fluxes.

CONCLUSIONS

The values of thermal fluxes determined in simulation experiments are consistent with those experienced in industrial practice.

It has been confirmed that heat transfer characteristics of the infiltrated slag are related to adhesion between the slag and mould. High heat transfer coefficients at the slag/mould interface are associated with good adhesion.

There is an optimum cooling water flow rate for maximum heat transfer coefficient at the slag/mould interface and this optimum value varies with the chemical composition of the casting flux.

An estimation of the thermal conductivities of the glassy, partially recrystallised and liquid portions of the infiltrated slag is made. A value of $0.3 Wm^{-1}K^{-1}$ is suggested for the phonon conductivity of liquid continuous casting fluxes.

ACKNOWLEDGEMENTS

The authors are grateful to the National Physical Laboratory for the full financial support of this work. We wish to express our appreciation to Dr. K.C. Mills of the National Physical Laboratory for his interest and inspiring discussions during the course of the work.

TABLE 1

Chemical compositions of powders studied in the simulation experiments

Powder	%SiO_2	%CaO	%MgO	%Al_2O_3	%Na_2O	%K_2O	%Fe_2O_3	%MnO	%F	%C_{free}
A	33.6	35.0	4.9	2.7	9.7	0.5	3.9	0.1	5.5	3.9
H	33.8	33.1	2.3	5.4	12.4	0.3	0.3	0.3	7.3	3.0
SPH409	34.34	35.77	0.69	6.12	U	0.19	U	0.019	U	5.7
		%CaO/MgO			**%Na_2O/K_2O**					
SPHC-178	33.5-35.5	32.0-34.0		5.0-6.0	6.5-8.0		< 0.5	< 0.1	5.0-6.0	10.0-11.0

Note: U = unknown
SPH-409 is short for SPH-SL 409-801/A
C-178 is short for SPH C-178-95III

TABLE 2

Thermal flux measurements for various fluxes using different cooling water flowrates

Powder	T_{water} °C	m l/s	t_{slag} mm	Q MWm^{-2}
A	11.8	0.10	0.77	1.16
A	9.7	0.15	0.74	1.43
A	7.8	0.16	0.76	1.23
A	9.3	0.165	0.62	1.51
A	6.9	0.175	0.66	1.19
H	12.8	0.10	0.18	1.26
H	7.1	0.15	1.03	1.05
H	9.3	0.16	0.70	1.47
C-178	12.8	0.10	0.70	1.26
C-178	8.5	0.15	0.95	1.26
C-178	7.5	0.175	0.92	1.29
C-178	6.7	0.175	1.01	1.16
SL-409	18.2	0.05	1.04	0.86
SL-409	18.9	0.05	0.82	0.97
SL-409	22.5	0.05	0.77	1.11
SL-409	20.8	0.05	0.81	1.03
SL-409	12.2	0.10	0.78	1.19
SL-409	13.0	0.10	0.78	1.28
SL-409	14.2	0.10	0.73	1.40
SL-409	5.5	0.20	0.76	1.08
SL-409	24.5	0.05	0.55	1.21
	then	0.20	0.60	1.26
SL4-0	6.4	0.20	0.60	1.26
	then	0.05		

TABLE 3

Estimation of thermal conductivities of the glassy, crystalline and liquid sections of the slag layer

Powder	m l/s	t_{slag} mm	t_{glass} mm	t_{cryst} mm	t_{liq} mm	t_g/t_s	k_{glass} $Wm^{-1}K^{-1}$	k_{cryst} $Wm^{-1}K^{-1}$	k_{liq} $Wm^{-1}K^{-1}$	h_{slag} $kWm^{-2}K^{-1}$
SPH-SL 409-80I/A	0.05	0.83	0.32	0.42	0.13	0.39	0.68	2.18	0.43	3.79
"	0.10	0.78	0.40	0.29	0.09	0.51	0.60	1.74	0.35	infinity
"	0.10	0.78	0.40	0.24	0.15	0.51	0.64	1.55	0.62	"
"	0.20	0.72	0.17	0.42	0.14	0.24	0.39	2.24	0.49	2.98
C-178 95-III	0.10	0.70	0.28	0.28	0.14	0.40	0.90	1.42	0.42	3.70
"	0.15	0.95	0.30	0.34	0.32	0.32	0.51	1.73	0.96	infinity
"	0.175	0.92	0.46	0.33	0.13	0.50	0.81	1.72	0.40	"
"	0.175	1.01	0.57	0.28	0.16	0.56	0.92	1.31	0.44	"

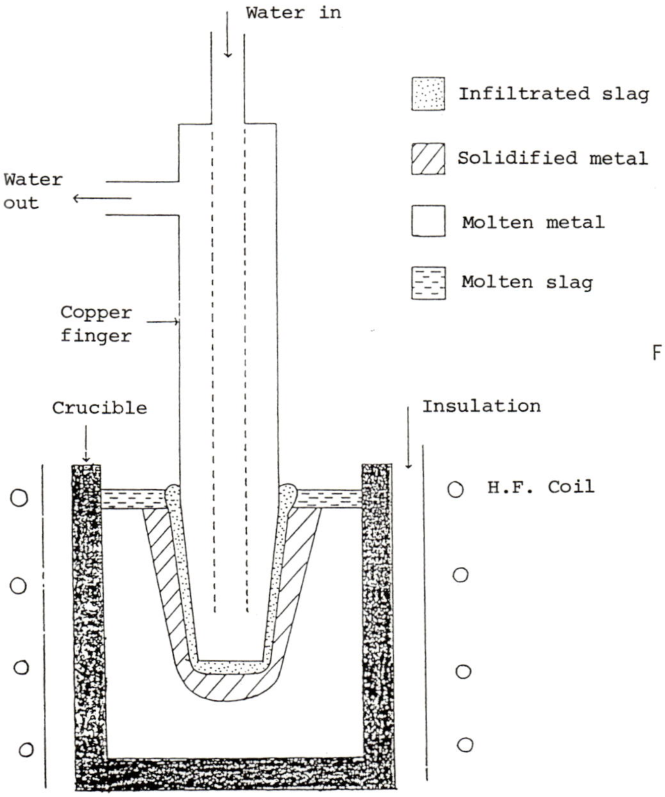

Figure 1 Diagram of the simulation test for continuous casting

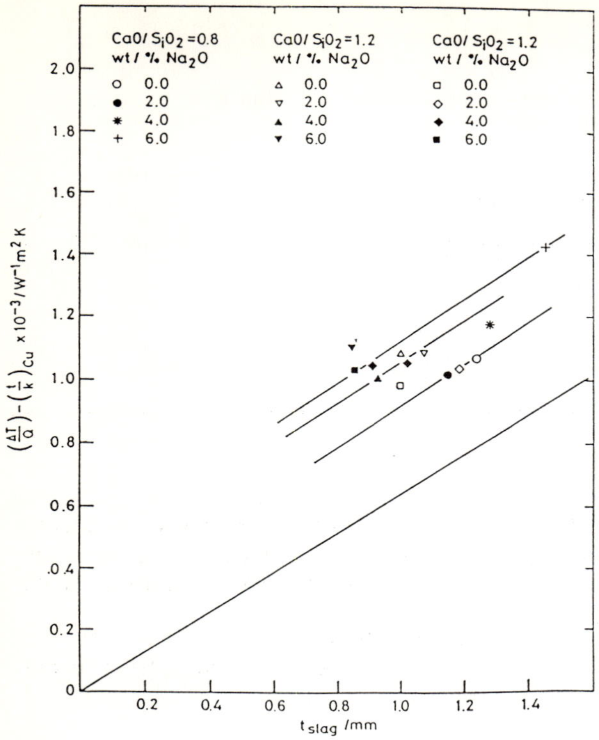

Figure 2 Estimation of heat transfer coefficients of synthetic casting fluxes (water flow rate = 0.2 l/s)

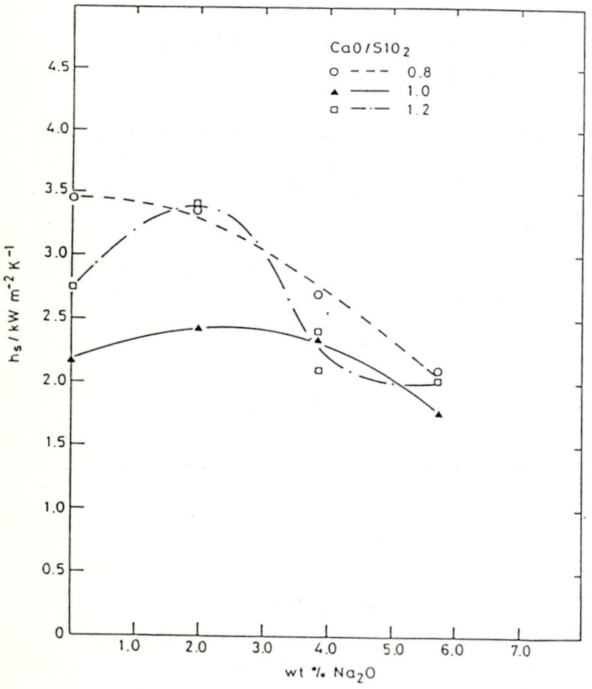

Figure 3 Variation of heat transfer coefficient with soda content of synthetic slags

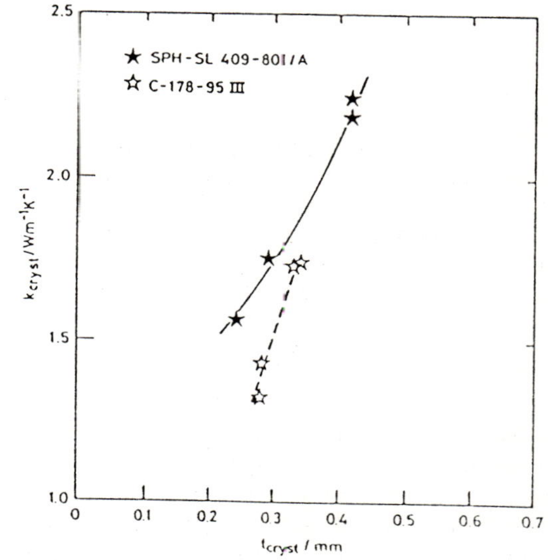

Figure 4 Variation of thermal conductivity of partially recrystallised layer with layer thickness

Extension of polymer theory to ternary silicate melts

C R MASSON, I D SOMMERVILLE and D J SOSINSKY

Dr Masson was formerly with the Atlantic Research Laboratory, National Research Council of Canada, Halifax, NS, Canada; Dr Sommerville is in the Department of Metallurgy and Materials Science, University of Toronto, Ontario, Canada; Mr Sosinsky, formerly at the University of Toronto, is now with Lambda Technology Corporation, Columbia, MD, USA.

SYNOPSIS

Although conventional polymer theory has, in the past, been used with success to evaluate ionic distribution in binary melts $MO\text{-}SiO_2$, its extension to ternary and higher systems has been delayed by our inability to derive values of K_{11} for such systems where K_{11} is the equilibrium ratio for the reaction:

$$2SiO_4^{4-} = Si_2O_7^{6-} + O^{2-}$$

An empirical correlation observed previously between K_{11} and the optical basicity of the oxide MO allows K_{11} to be estimated for such systems. Ion fractions N of monomer SiO_4^{4-}, dimer $Si_2O_7^{6-}$, trimer $Si_3O_{10}^{8-}$ and "free" oxide ion O^{2-} were calculated as functions of composition for basic melts of the systems 'FeO'-CaO-SiO_2 and 'FeO'-MnO-SiO_2 at 1600°C. The results are illustrated with the aid of iso-N diagrams, which yield some interesting insights into the detailed ionic constitution of these melts.

INTRODUCTION

Conventional polymer theory, modified to allow for the presence of "free" oxide ions O^{2-} in an array of discrete silicate ions $Si_nO_{3n+1}^{(2n+2)-}$ at equilibrium, has been used with success[1] to describe the thermodynamic properties of binary silicate melts. In this theory, knowledge of the equilibrium ratio K_{11} for the reaction

$$2\,SiO_4^{4-} = Si_2O_7^{6-} + O^{2-} \qquad (1)$$

allows ion fractions N of monomer SiO_4^{4-}, dimer $Si_2O_7^{6-}$, trimer $Si_3O_{10}^{8-}$ and higher ions as well as of "free" oxide ion O^{2-} to be evaluated as functions of composition in melts more basic than the metasilicate.

For binary melts $MO\text{-}SiO_2$, K_{11} may be evaluated by assuming that the activity of MO is equal to the ion fraction of oxide ion:

$$a_{MO} = N_O \qquad (2)$$

It should be stressed that, in the context of this work, equation (2) does not require the validity of Temkin's Law but may be derived from the theory[2] on the basis that the activity coefficients of the anionic constituents are in geometric series:

$$\gamma MO/\gamma M_2SiO_4 = \gamma M_2SiO_4/\gamma M_3Si_2O_7 =$$
$$\gamma M_3Si_2O_7/\gamma M_4Si_3O_{10} = \ldots . \qquad (3)$$

This is a reasonable assumption which is supported by the good agreement between the observed and calculated activities for systems with a common cation.

From this assumption it also follows that for reactions such as

$$2SiO_4^{4-} = Si_2O_7^{6-} + O^{2-} \qquad (1)$$

or

$$2M_2SiO_4 = M_3Si_2O_7 + MO \qquad (1')$$

activity coefficients vanish in the expressions for equilibrium constants so that:

$$K_{11} = N_2N_O/N_1^2 \qquad (4)$$

where K_{11} is the equilibrium constant of reaction (1) or (1') and N is the ion fraction of the indicated species. Ion fractions may thus be used in place of activities. This allows values of K_{11} and hence ionic distributions to be evaluated from thermodynamic data.[2,3]

It is now recognised[1] that for melts with $K_{11} >$ 1 the data are best represented in terms of a linear chain model, in which the monomer SiO_4^{4-} is regarded as pseudo-bifunctional whereas, for $K_{11} < 1$ the treatment corresponds to that for a pseudo-trifunctional condensation. Reasons for this have been discussed elsewhere[1].

For ternary and higher systems, however, equations (2) and (3) cannot be applied due to the competitive interactions which arise between the anions and the various cations. For this reason, it has not been possible so far to derive values of K_{11} from thermodynamic data and hence to extend conventional polymer theory to these systems.

Optical Basicity

The concept of optical basicity, pioneered by Duffy and Ingram in the field of glass chemistry[4-9], is being applied with increasing interest to the study of metallurgical slags. It has now been successfully applied to correlate data on sulphide capacity[10-12], phosphate capacity[13,14], water and carbonate capacities[15], nitride, cyanide and carbide capacities[16], arsenic capacity[17], the activity coefficient of P_2O_5[14], the slag-metal distribution of phosphorus[13,14,18], the activity coefficient of Cr_2O_3[19], the activities of CaO and Na_2O[20] and redox equilibria such as As^{3+}/As^{5+}[21].

Although such correlations are empirical and may be valid only for the systems and ranges of composition for which they were derived, they are useful in allowing large amounts of data to be correlated on the basis of a single parameter and for making estimates when experimental data are not available.

It has recently been shown[1] that for silica in its binaries with CaO, MgO, MnO and FeO a smooth correlation exists between log K_{11} and the optical basicity Λ_{MO} of the corresponding oxide as calculated from the equation

$$\Lambda_{MO} = {}^{1}/1.36(x - 0.26) \qquad (5)$$

where x is the Pauling electronegativity of the element M. This correlation provides a means of estimating K_{11} for ternary melts AO-BO-SiO_2 along lines of fixed AO/BO ratio. Such melts may be regarded as pseudo-binaries of fixed cationic composition. One of the most useful aspects of the treatment developed by Duffy and Ingram is the ability to evaluate average values of Λ for mixtures of oxides by means of the expression

$$\Lambda = X_A\Lambda_A + X_B\Lambda_B + X_C\Lambda_C + \ldots.. \qquad (6)$$

where Λ is the optical basicity and X the equivalent ion fraction of the indicated component, based on the fraction of negative charge neutralized by the charge on the cation concerned. The most convenient method of calculating values of X for components in slags has been outlined elsewhere[11,12].

PROCEDURE

Equation (5) was used to calculate optical basicities of 'FeO', MnO and SiO_2 with reference to $\Lambda_{CaO} = 1$, while equation (6) was employed to calculate the optical basicities of mixtures in the two ternary systems, 'FeO'-CaO-SiO_2 and 'FeO'-MnO-SiO_2. Values of log K_{11} corresponding to these values of Λ for the mixed oxides were then obtained by interpolation from the correlation between log K_{11} and Λ described previously (ref I, Fig. 10). The results of such calcuations are shown in Fig. I as iso-K_{11} lines in the 'FeO'-CaO-SiO_2 system. Values of K_{11} derived in this way along pseudo-binary compositions were used to calculate ion fractions of oxide ion at various silica contents ($0 < X_{SiO_2} < 0.50$) by means of the equation:

$$\frac{1}{X_{SiO_2}} = 2 + \frac{1}{1-N_O} - \frac{3}{1 + N_O(\,3/K_{11} - 1\,)} \qquad (7)$$

These, in turn, were used to evaluate ion fractions of discrete silicate ions by use of the expression[2]:

$$N_x = \frac{(3x)!}{(2x+1)!\,(x)!}\left[\frac{1}{1+\dfrac{3N_O}{K_{11}(1-N_O)}}\right]^{(x-1)} \times \left[\frac{1}{1+\dfrac{K_{11}(1-N_O)}{3N_O}}\right]^{(2x+1)} (1-N_O) \qquad (8)$$

where N_x is the ion fraction of a silicate ion of chain length x.

Results and Discussion

A typical set of distribution curves for melts in the system 'FeO'-CaO-SiO_2 with $X_{FeO} / X_{CaO} = 2/3$ (i.e. along the line indicated by $K_{11} = 0.00933$ in Fig. 1) is shown in Fig. 2.

Ionic distributions obtained in this way were used to construct lines of constant N_O, N_1, N_2 and N_3 for the systems 'FeO'-CaO-SiO_2 and 'FeO'-MnO-SiO_2 at 1600°C. Figures 3 and 4 show lines of constant ion fraction of oxide ion, O^{2-}, and monomer, SiO_4^{4-}, respectively in the system 'FeO'-CaO-SiO_2. The distribution patterns for the dimer, $Si_2O_7^{6-}$ and trimer, $Si_3O_{10}^{8-}$ were similar in form to that of the monomer, except that the "saddle" occurred at the pyro- and tripoly- compositions ($X_{SiO_2} = 0.40$ and 0.429, respectively), instead of the ortho- composition ($X_{SiO_2} = 0.333$), as exhibited by the monomer in Fig. 4. This is illustrated in Fig. 5 for the dimer in the system 'FeO'-MnO-SiO_2.

Ionic distributions along various sections of the ternaries were also constructed. An example is shown in Fig. 6 for compositions along the line joining Fe_2SiO_4 with the CaO apex. On adding CaO to a melt of composition Fe_2SiO_4, the ion fraction of oxygen at first decreases, due to the strong tendency of CaO to induce depolymerization, with a consequent uptake of "free" oxide ions by the silicate ions. This is accompanied by a corresponding increase in the ion fraction of monomer, as shown in Fig. 6. With further additions of CaO, however, the ion fraction of oxide ion increases at the expense of the silicate ions.

Conclusions

1. The correlation between the value of log K_{11} for a binary or pseudo-binary silicate system and the optical basicity of the end-member other than silica provides a basis for the extension of polymer theory into ternary silicate melts.

2. Values of K_{11} obtained in this way were used to calculate ion fractions of O^{2-}, SiO_4^{4-}, $Si_2O_7^{6-}$ and $Si_3O_{10}^{8-}$ in the two ternary systems 'FeO'-CaO-SiO_2 and 'FeO'-MnO-SiO_2, for compositions more basic than the metasilicate.

3. The results are illustrated by means of plots of th‘ distribution of the various ionic species across pseudo-binary systems, and by iso-N plots in the ternary systems.

References

1. C.R. Masson, Proc. 2nd Int. Symp. Met. Slags and Fluxes (ed. H.A. Fine and D.R. Gaskell), Met. Soc. of AIME, (1984), pp. 3-44.
2. C.R. Masson, I.B. Smith and S.G. Whiteway, Can.J. Chem., 48, (1970), pp. 1456-1464.
3. C.R. Masson, J. Iron and Steel Inst., 210, (1972), pp. 89-96.
4. J.A. Duffy and M.D. Ingram, J. Amer. Chem. Soc., 93, (1971), pp. 6448-54.
5. J.A. Duffy and M.D. Ingram, J. Chem. Phys., 54, (1971), pp. 443-444.
6. J.A. Duffy and M.D. Ingram, J. Inorg. Nucl. Chem., 36, (1974), pp. 43-47.
7. J.A. Duffy and M.D. Ingram, Phys. Chem. Glasses, 16, (1975), pp. 119-123.
8. J.A. Duffy and M.D. Ingram, J. Inorg. Nucl. Chem., 37, (1975), pp. 1203-1206.
9. J.A. Duffy and M.D. Ingram, J. Non-Cryst. Solids, 21, (1976), pp. 373-410.
10. J.A. Duffy, M.D. Ingram and I.D. Sommerville, J. Chem. Soc., Faraday Trans. I, 14, (1978), pp. 1410-19.
11. I.D. Sommerville and D.J. Sosinsky, Proc. 2nd. Int. Symp. Met. Slags and Fluxes (ed. H.A. Fine and D.R. Gaskell), Met. Soc. of AIME, (1984), pp. 1015-26.
12. D.J. Sosinsky and I.D. Sommerville, Met. Trans. B, 17B, (1986), pp. 331-7.
13. H. Suito and R. Inoue, Trans. I.S.I. Japan, 24, (1984), pp. 47-53.
14. T. Mori, Trans. Japan Inst. Metals, 25, (1984), pp. 761-771.
15. D.J. Sosinsky, I.D. Sommerville and A. McLean, Proc. Sixth Process Technology Conference, ISS-AIME, (1986), pp. 697-703.
16. I.D. Sommerville, Proc. Internat. Symposium on "Foundry Processes: Their Chemistry and Physics", General Motors Research Laboratories, Warren, Mich., (1986).
17. T. Lehner, Personal Communication, (1986).
18. D.R. Gaskell, Trans. I.S.I. Japan, 22, (1982), pp. 997-1000.
19. T. Sakuraya, Personal Communication, (1986).
20. D. R. Gaskell, Proc. Internat. Symposium on "Foundry Processes: Their Chemistry and Physics", General Motors Research Laboratories, Warren, Mich., (1986).
21. G. Jeddeloh, Phys. Chem. Glasses, 25, (6), (1984), pp. 163-164.

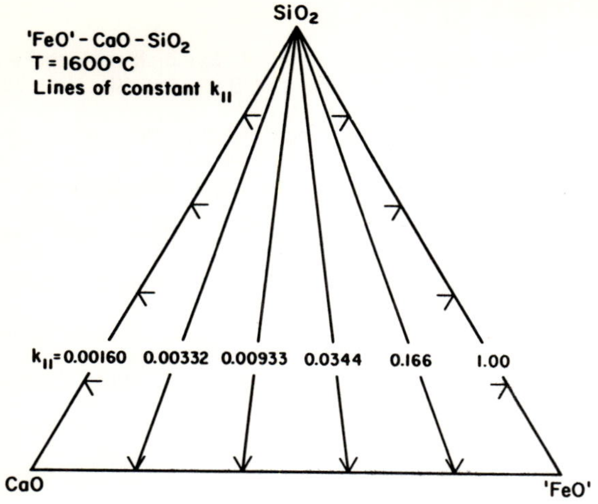

Fig. 1 Lines of constant K_{11} in the system 'FeO'-CaO-SiO_2 at 1600°C.

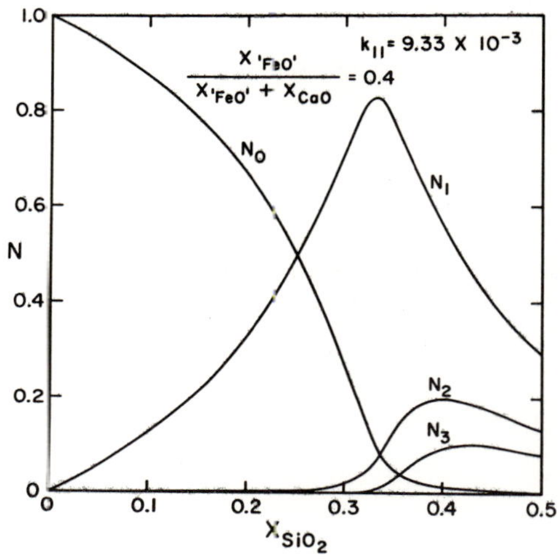

Fig. 2. Calculated ion fractions of oxide ion N_O, monomer N_1, dimer N_2 and trimer N_3 in the system 'FeO'-CaO-SiO_2 at 1600°C for compositions with $X_{'FeO'} / X_{CaO} = 2/3$ (K_{11} = 0.00933 in Fig. 1).

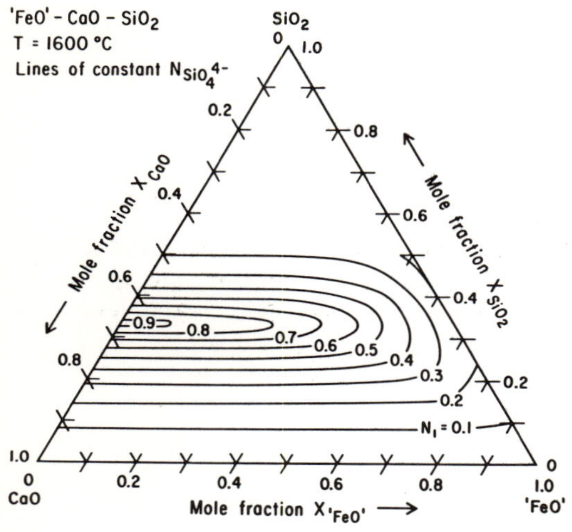

Fig. 3. Lines of constant oxide ion fraction in the system 'FeO'-CaO-SiO_2 at 1600°C.

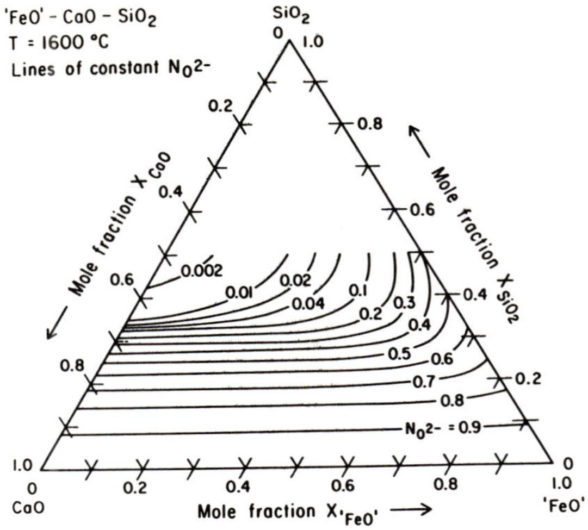

Fig. 4 Lines of constant ion fraction of monomer SiO_4^{4-} in the system 'FeO'-CaO-SiO_2 at 1600°C.

Fig. 5. Lines of constant ion fraction of dimer $Si_2O_7^{6-}$ in the system 'FeO'-MnO-SiO_2 at 1600°C.

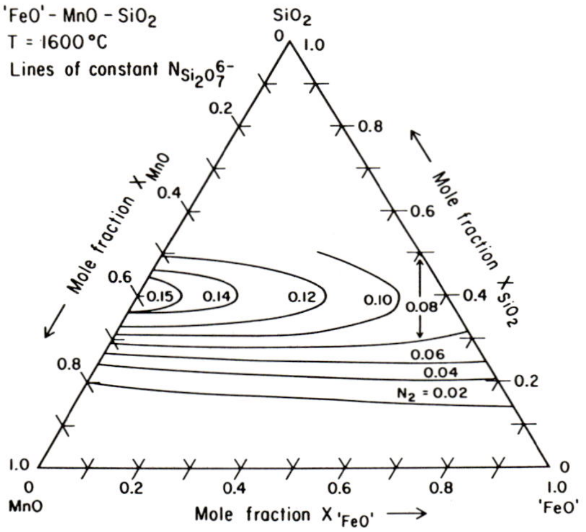

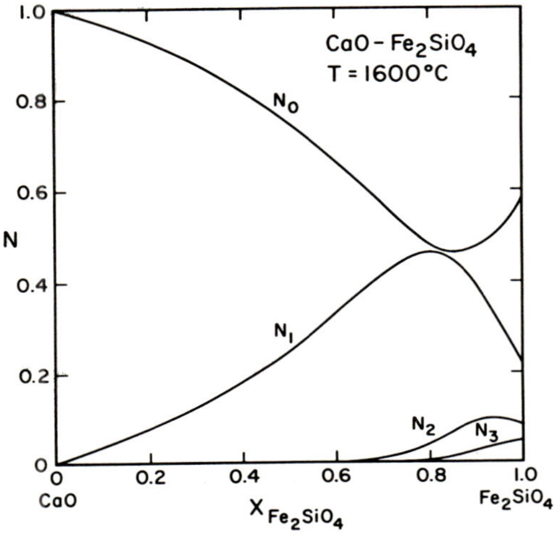

Fig. 6. Calculated ion fractions of oxide ion N_O, monomer N_1, dimer N_2 and trimer N_3 along the CaO-Fe_2SiO_4 join in the system 'FeO'-CaO-SiO_2 at 1600°C.

Application of MTDATA to modelling of slag, matte, metal, gas phase equilibria

A T DINSDALE, S M HODSON and J R TAYLOR

Dr Dinsdale is in the Division of Materials Applications and Mrs Hodson in the Division of Information Technology and Computing at the National Physical Laboratory, Teddington; Dr Taylor is at the Johnson Matthey Technology Centre, Reading, UK.

SYNOPSIS

Computer calculation of phase equilibria provides a very powerful and cost-effective tool in understanding complex chemical and metallurgical problems. The results of some calculations made with MTDATA relating to extractive metallurgy are described which model phase equilibria involving slag, matte, metal and gas phases.

INTRODUCTION

Over the last 15 years great advances have been made in the calculation of phase equilibria for multicomponent systems from critically assessed thermodynamic data for the subsystems. Three key steps have made this possible:

(a) the development of reliable software to calculate equilibria between a large number of different types of phases (e.g. slags, mattes, alloys, gases, aqueous solutions, molten salts) for systems with many components.

(b) the much deeper understanding of how thermodynamic data for these phases can be modelled accurately as a function of temperature and composition.

(c) the growth of a substantial body of critically assessed data.

In this paper the results are described of some calculations relating to extractive metallurgy made with MTDATA (1), the Metallurgical and Thermochemical Databank developed by the National Physical Laboratory.

MTDATA is a very flexible system allowing a wide range of problems to be analysed for a variety of phases. Of particular interest is the module MULTIPHASE (2) which is designed to determine the phases present and the amounts of each constituent within each phase of a multicomponent, multiphase chemical or metallurgical system at equilibrium. The overall composition, temperature and pressure or volume of the system must be defined prior to the calculation. Various models have been included, e.g. ideal mixing of gas phase species, non-ideal associated solutions, Redlich-Kister polynomials (3) modified when necessary to include sublattices (4) with either fixed or variable site ratios (5) and an extended Kapoo-Frohberg model for liquid slags (6). These reliable mathematical models form the basis of the predictive approach used by MTDATA. They enable critically assessed data for simpler systems, for which experimental data are usually well established, to be extrapolated into higher order systems in order to calculate multicomponent phase equilibria. This approach has proved to be extremely successful over the years especially for alloys system (7). Thus using the facilities within MTDATA it is possible to explore compositions and conditions for which no direct experimental data exist. MTDATA features menu driven input, access to databases (including that derived by SGTE, Scientific Group Thermodata Europe, (9,10) of which NPL is a member), tabulation of thermodynamic functions and allows output in graphical form if required.

Provided that a complete and consistent set of thermodynamic properties of the various phases which could exist in a given chemical or metallurgical system is available, any equilibrium state may be determined within the system by minimising the Gibbs energy. Two distinct methods are used by MULTIPHASE. In the first method the species amounts are taken as variables, the linear equations forcing the given component concentrations are always satisfied and a lower value of the non-linear Gibbs energy function is found at each iteration of the process until a minimum is reached. At this point the projection of the gradient vector onto the set of constraints is zero which is equivalent to equating the chemical potentials. The procedure is extremely reliable and experience has shown that, except where a given dataset gives rise to a miscibility gap, convergence is guaranteed. However, the nature of the minimisation imposes a limit of approximately 10^{-6} on the accuracy with which the amounts of phases or the compositions of the constituents can be attained. This certainly is satisfactory for most applications.

In the second method the logarithms of the mole fractions are taken as variables together with the amount of each phase. The projection of the gradient is forced to be zero and then the equations to impose the component concentrations are solved by an interactive process. The accuracy with which the mole fractions can be determined is much higher than the first method and convergence can be quicker but it is found that the method is reliable only when a good initial guess is provided for the compositions and phase amounts. In practice the first method is often used to provide an excellent starting point for the second method. The second method has many similarities to procedures used in other comparable computer programs, for example that described by Eriksson for SOLGASMIX (11-13).

EXTRACTION FROM SULPHIDE ORES

One of the areas in which MTDATA has been applied is in the pyrometallurgical extraction of values from sulphide ores. Pellets made from the ore are smelted with suitable fluxes and the matte, which is basically a Cu-Fe-Ni-S liquid, is blown with oxygen. It is then slow cooled after which different metal phases can be separated magnetically. Phase equilibrium calculations have been made to throw light on the different aspects of this process. The thermodynamic and phase diagram data for the Cu-Fe-Ni-S system were critically assessed in order to provide a dataset which could represent all the thermodynamic information for that system(14). Figure 1 is a calculated isothermal section of the phase diagram for the Cu-Ni-S system for 973 K and it shows some of the many phases involved in the system. Of particular importance is the liquid sulphide phase which, for this work, was modelled as a two sublattice phase using the model of Hillert et al. (15,16) but extended by allowing the ratio of sites between the sublattices to vary (14,17). Excellent agreement was obtained between the experimental and observed properties. Such a dataset allows the partition of elements between the matte and the alloy phase to be calculated.

The liquid phase dataset can also be used to model the equilibrium between the matte and the slag phases during the smelting operation. In order to test the level of agreement expected a series of equilibrations of silica saturated iron silicate slags were carried out under controlled conditions of temperature and partial pressure of SO_2. Figure 2 shows the excellent agreement which can be obtained between calculation and experiment. A more detailed thermodynamic analysis of the process will be described elsewhere (18).

FERROMANGANESE PRODUCTION

Another example which shows the utility of the calculation of phase equilibria is in the production of ferromanganese. This normally takes place through the silicothermic process in which silicon, contained in silicomanganese, reduces manganese oxide in the form of high grade ore. Silicomanganese itself is produced through the smelting of the manganese ore, silica and coke. Both of these processes are very dependent on the equilibrium between the $MnO-SiO_2$ slag and the essentially metallic Mn-Si melts. Lime and/or magnesia is/are added to lower the SiO_2 activity and encourage the partition of manganese to the metallic phase and silicon to the slag phase. The equilibria have been studied experimentally by Kor (19).

Calculations were carried out to explore the agreement with the experimental information. The thermodynamic data for the Nm-Si system had been recently assessed by Gisby and Dinsdale (20). Figure 3 shows the calculated phase diagram with experimental data superimposed and is a good example of the agreement expected. The data for the slag phase were modelled using an extended Kapoor-Frohberg model (6) and taken from the SGTE database (9,10), the work of Gaye and Welfringer (21) and the assessment of the $CaO-SiO_2$ system by Taylor and Dinsdale (22). It is common practice to represent these types of equilibria by plotting the ratio k against the slag basicity B where:

$$k = \frac{[\text{wt\% Mn}]^2}{(\text{wt\% MnO})^2} \frac{(\text{wt\% SiO}_2)}{[\text{wt\% Si}]} \qquad B = \frac{(\text{wt\% (CaO+MgO)})}{(\text{wt\% SiO}_2)}$$

where the quantities in round brackets refer to the slag phase and those in square brackets to the metal phase. Kor (19) suggested, as a result of his experimental work, that there is a linear relationship between the logarithm of the equilibrium ratio and the slag basicity. The calculations however, show that the relationship is more complex. In agreement with Kor it was found that the ratio is only mildly dependent on the temperature and on the relative concentrations of CaO and MgO. In contrast it was found that the relative amounts of Mn and Si in the system had a much greater effect. Figure 4 shows the logarithm of the calculated equilibrium ratio plotted against the basicity of the slag;for overall ratios of Mn to Si of 200:1 and 10:1 in terms of weight, the ratios correspond approximately to 0.3 and 5 wt% silicon respectively in the metal phase. Superimposed on these curves are the experimental results of Kor (19), which relate to a metal phase containing typically 1-2 wt% silicon, and they agree well with the calculations presented.

MODELLING OF BLAST FURNACE EQUILIBRIA

So far in this paper it has been shown how critically assessed thermodynamic data can be used to calculate equilibria between a matte and a slag phase or between a slag and a metal phase and it has been shown that the calculations agree extremely well with what is observed experimentally. MTDATA and in particular the MULTIPHASE module is particularly useful for studying very complicated equilibria involving many components and a large number of phases. As an example a series of calculation shave been performed to demonstrated how aspects of equilibria within an iron blast furnace could be modelled. The dataset used for the calculations comprises 6 components, Fe, Si, C, O, Ca and N and a total of 29 phases including the liquid slag, assumed to contain Ca, Fe, Si and O, various metallic phases modelled as containing Fe, C and Si, the gas phase and various solid oxide phases. The data were drawn from a number of sources: the gas phase data from the SGTE database (9,10), the data for the Fe-C-Si system from a recent assessment of Dinsdale et al. (23) and the data for the liquid slag and solid oxide phases from the work of Taylor and Dinsdale (22) and Gaye and Welfringer (21). For a given temperature and pressure and amounts of the individual components MULTIPHASE will calculate the amount and compositions of the phases in equilibrium. Table 1, for example, shows an example of the calculated compositions in weight percent of the phases in equilibrium for a temperature of 1600^oC and a pressure of 101325 Pa. The overall composition is also shown in the table. The percentage of each component within the various phases has been tabulated. It should be emphasised that these calculations are for the equilibrium state and may not relate directly to the conditions in a blast furnace where kinetic factors may limit the attainment of equilibrium.

A series of calculations of this sort can be carried out, for example by varying the temperature or the amount of one of the components such as oxygen. The temperature, pressure and other compositions are the same as used in the calculations and the results show the mass of the various phases predicted to be in equilibrium as oxygen is blown through the furnace. Of particular interest is the appearance of the slag phase as the amount of oxygen exceeds 1.4 mol and the disappearance of graphite as it is burnt off. Other graphs of a similar kind can be plotted based upon these calculations. Figure 6 shows how the composition of the metallic liquid phase varies as the oxygen is added. Initially the carbon and silicon levels remain roughly constant. However, after a certain amount of oxygen has been added the level of silicon declines rapidly as the liquid slag is formed, and subsequently the carbon level decreased gradually towards a final approximately constant value. Figures 5 and 6 are examples of diagrams which can be obtained very easily. Other properties, such as the composition of the slag and distribution of components between the various phases, can all be plotted giving great insight into conditions in the furnace.

SUMMARY

In this paper it has been shown how phase equilibria between slags, mattes, metallic phases and a gas phase modelled using MTDATA can have powerful applications for extractive metallurgy. The method is, however, limited through the rather sparse database available for oxide systems. The uses of such calculations within a package for process control could have great potential.

ACKNOWLEDGEMENTS

Part of this work was sponsored by Matthey Rustenberg Refiners.

REFERENCES

1 MTDATA Handbook - Documentation for the NPL Metallurgical and Thermochemical Databank, October 1987, National Physical Laboratory.
2 Hodson S.M., NPL Report, To be published.
3 Redlich O., Kister A.T., Ind. Eng. Chem. (1948) 40, 345.
4 Sundman B., Agren J., J. Phys. Chem. Solids (1981) 42, 297.
5 Hillert M., Jansson B., Sundman B., Agren J., Royal Institute of Technology Report TRITA-MAX-0218 (Revised), Feb 1984.
6 Kapoor M.L., Mehrotra G.M., Frohberg M.G., Proc. Aust. Inst. Mining Metall. (1975) 254, 11.
7 Barry T.I., in 'Chemical Thermodynamics in Industry: Models and Computations', Ed. T I Barry, Publ. Blackwell Scientific Publications, Oxford, 1985.
8 Kapoor M.L., Frohberg G.M., Proc. Symp. 'Chemical Metallurgy of Iron and Steel', Sheffield, 1971.
9 Dinsdale A.T., NPL News, No 365, Summer 1986, 26.
10 Ansara I., Sundman B., CODATA Report 'Computer Handling and Dissemination of Data' (1987) 154-158.
11 Eriksson G., Rosen E., Chem. Scr. (1973) 4, 193.
12 Eriksson G., Acata Chem. Scand. (1971) 25, 2651.
13 Eriksson G., Chem. Scr. (1975) 8, 100.
14 Dinsdale A.T., Thesis, "The Generation and Application of Metallurgical Thermochemical Data", Brunel University, 1984.
15 Hillert M., Staffansson L.I., Metall. Trans. (1975) 6B, 37.
16 Fernandez Guillermet A., Hillert M., Jansson B., Sundman B., Metall. Trans. (1981) 12B, 745.
17 Dinsdale A.T., Chart T.G., Barry T.I., Taylor J.R., High Temperatures - High Pressures (1982) 14, 633.
18 Dinsdale A.T., Hodson S.M., Barry T.I., Taylor J.R., presented at the 27th Annual Conference of Metallurgists, Montreal, Canada, August 1988.
19 Kor G.J.W., Metall. Trans. B, (1979) 10B, 367.
20 Gisby J.A., Dinsdale A.T., To be published.
21 Gaye H., Welfringer J., Second International Symposium on Metallugical Slags and Fluxes, Ed. H.A. Fine and D.R. Gaskell, Lake Tahoe (1984) 357-375, Publ. Metall. Soc. AIME., New York, 1984.
22 Taylor J.R., Dinsdale A.T., To be published.
23 Dinsdale A.T., Chart T.G., Hassall G.J., Unpublished work.

Table 1 Results from a MULTIPHASE calculation

Temperature = 1873.00 K
Fixed gas pressure = 1.013250E+05 Pa
Calculated gas volume = 5.243074E-02 m³

	C	Ca	Fe	O	Si	N	
	Overall component amounts (mol)						
	0.3000	0.6000	1.1000	1.7000	0.4000	0.1000	
Phase	Weight fraction of component within phase						Total weight
LIQUID	0.0017	0.0000	0.9982	0.0000	0.0001	0.0000	0.0611
GAS	0.3646	0.0000	0.0002	0.4890	0.0001	0.1461	0.0096
Ca2SiO4	0.0000	0.4654	0.0000	0.3715	0.1631	0.0000	0.0134
SLAG	0.0000	0.3976	0.0092	0.3913	0.2019	0.0000	0.0448
	Percentage distribution of components between phases						
LIQUID	2.963	0.000	99.326	0.000	0.057	0.000	
GAS	97.037	0.001	0.003	17.241	0.004	100.000	
Ca2SiO4	0.000	25.996	0.000	18.351	19.497	0.000	
SLAG	0.000	74.003	0.671	64.408	80.442	0.000	

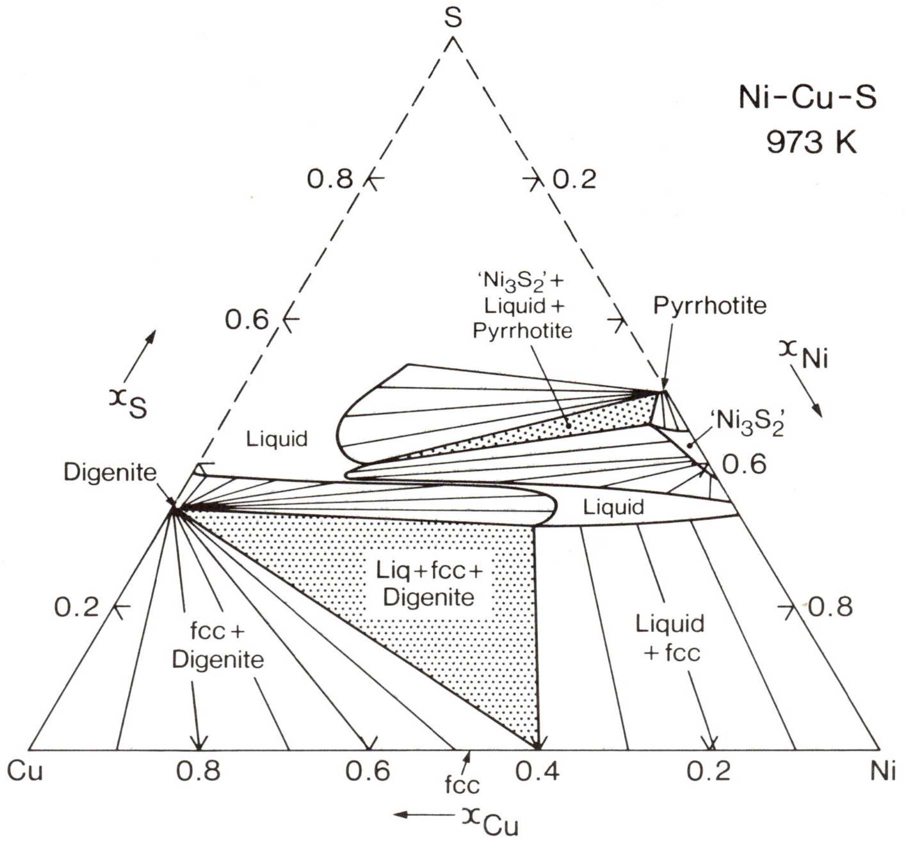

Figure 1. Calculated phase diagram for the Cu-Ni-S system for 973 K.

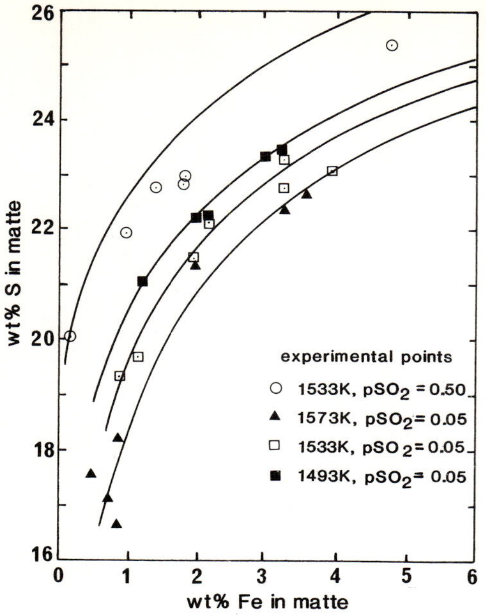

Figure 2. Calculated matte compositions in equilibrium with Fe-O-SiO_2 slags at different temperatures with constant partial pressures of SO_2.

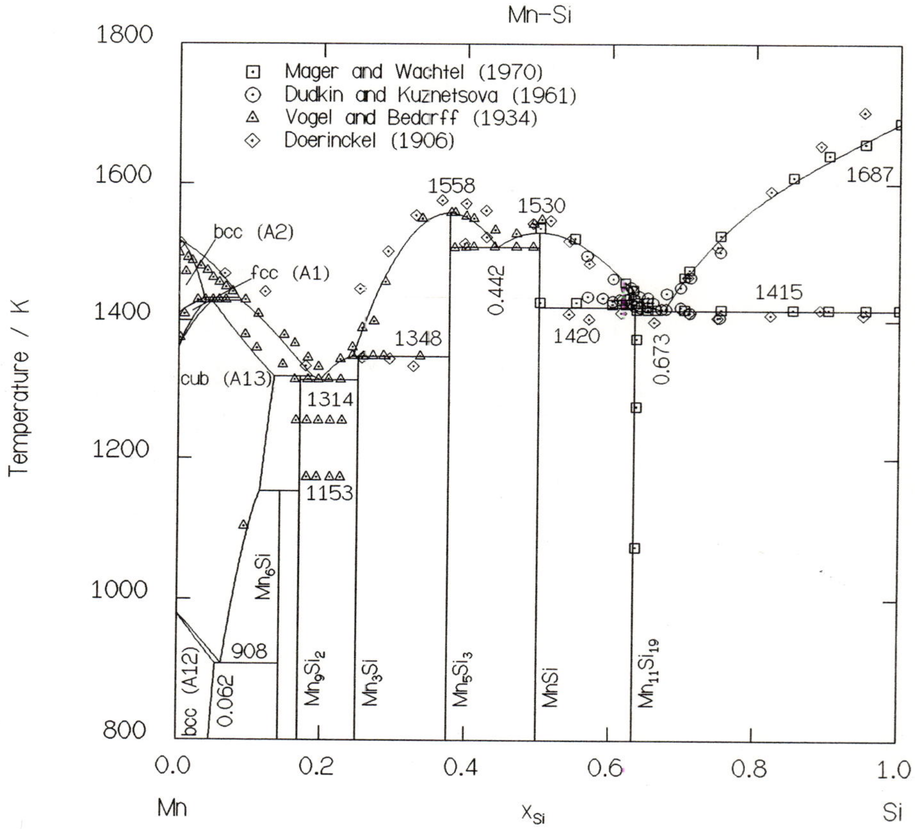

Figure 3. Calculated phase diagram for the Mn-Si system (20). The experimental information is superimposed.

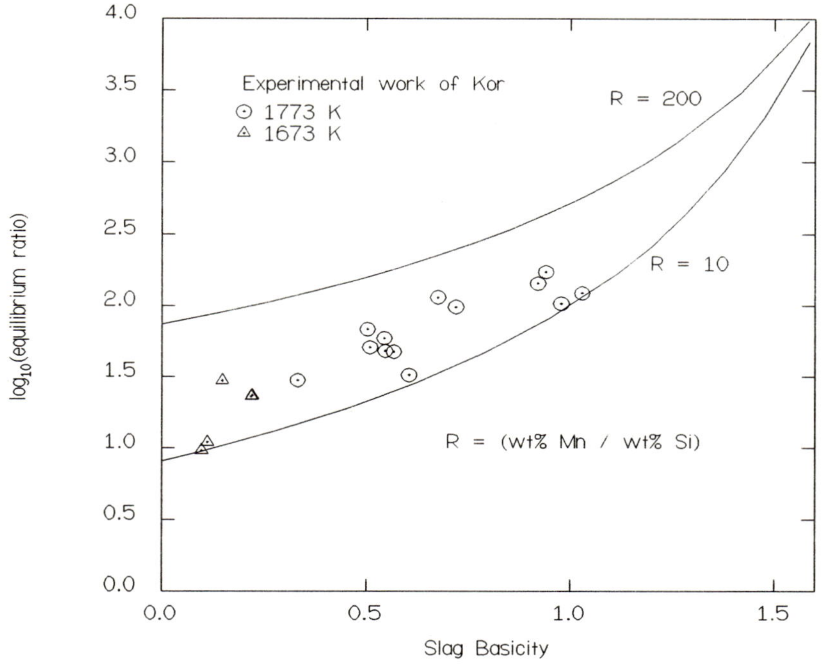

Figure 4. The partition of manganese and silicon between the slag phase and the metal phase plotted as a function of the slag basicity. The two curves relate to different relative amounts of Mn and Si. Experimental data of Kor (19) are superimposed.

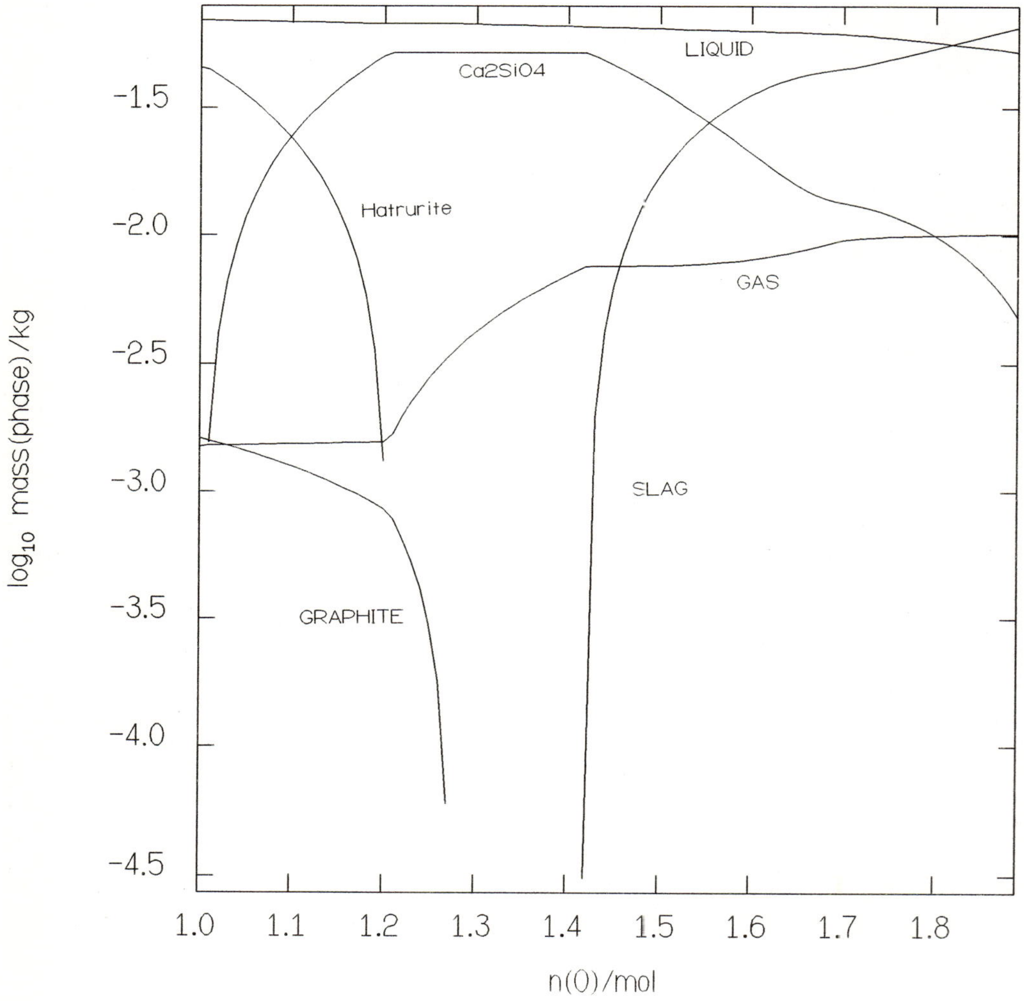

Figure 5. The amount of various phases calculated to form at 1600°C as oxygen is added to a prototype blast furnace mixture.

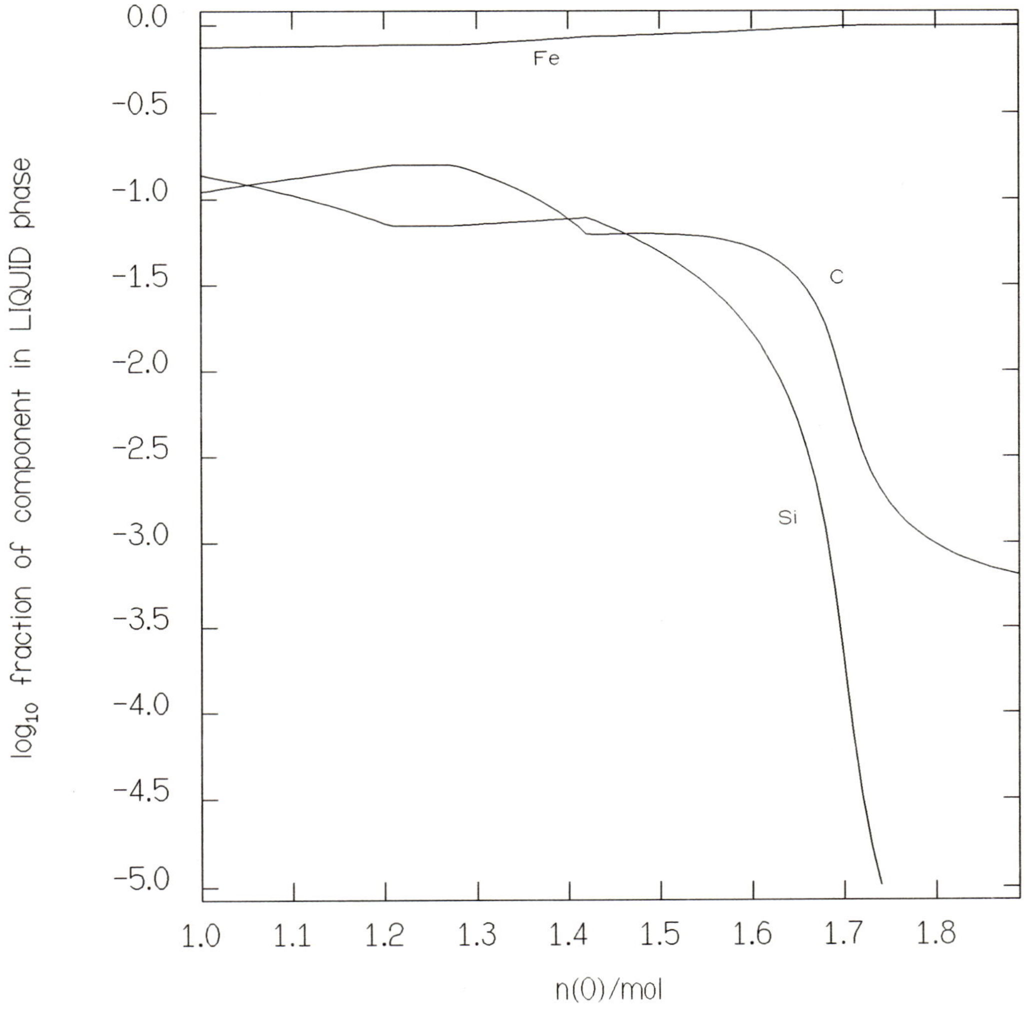

Figure 6. The change in the composition of the metallic liquid phase as oxygen is added to the prototype blast furnace mixture.

Thermodynamic activity predictions for molten slags and salts, IX

J W HASTIE, D W BONNELL and E R PLANTE

The authors are in the Metallurgy Division of the Institute for Materials Science and Engineering at the National Bureau of Standards, Gaithersburg, MD, USA.

SYNOPSIS

A predictive model has been developed for calculating detailed equilibrium phase compositions and activities. The model has been demonstrated for such non-ideal mixtures as molten salts and polymeric oxide slag, glass, and ceramic systems.[1,2,3] The model is particularly applicable to high-order systems and has been successfully applied to oxide and halide compound mixtures of up to eight atomic components, and containing elements selected from the set Al, B, Ca, Cd, Cs, Fe, K, Li, Mg, Na, Pb, Si, O, Cl and Br. Examples considered in the present study include the systems "FeO"-Al_2O_3, an MHD (magnetohydrodynamic) coal slag, a steel blast furnace slag, and CsCl-$PbCl_2$. The model is based on assigning, in a formal thermodynamic sense, solution complex-components, such as alkali-silicates or halide-associates, which account for the non-ideal solution interactions. These complex-components, together with the non-complexed constituent components, are explicitly included in an extensive database of standard Gibbs energies of formation at high temperatures. Multicomponent equilibrium codes (such as SOLGASMIX)[1] are then used to determine the equilibrium concentration of each component in any phase, subject to such retrictions as the phase rule. These concentrations are found to be equivalent to activities through comparisons of model and experimental values. Model validation tests have usually been made by mass spectrometric determination of species partial pressures and activities using special techniques described elsewhere.[4] Useful activity information can also be determined from phase boundary locations using established thermodynamic principles.[5]

INTRODUCTION

Polymeric oxide slag and certain molten salt mixtures are known to exhibit substantial deviations from ideal thermodynamic mixing behavior, e.g. see the texts of Richardson[5] and Sundheim[6]. While some experimental activity and phase information is available for such systems it is sparse and difficult to obtain. Also, the expressing of activities, as a function of system composition or temperature, has proven to be unreliable for extrapolation beyond the range of available experimental data. Detailed structural models of polymeric oxide or salt mixtures require considerable experimental input for each system of interest and also have limited utility as extrapolation tools. To avoid these difficulties, and to extend the application of sparsely available experimental data to practical conditions not easily obtained in the laboratory, we have developed a generic predictive model as discussed in detail earlier.[1,2] The model also has the advantage that non-expert users can apply it without any knowledge about the system thermochemistry.

ACTIVITY MODEL

The rationale and theoretical basis for the mixing model used was presented earlier in Parts I and II of this research activity.[1,2] To summarize, the key features of the model are as follows. We attribute the experimentally observed large negative deviations from ideal thermodynamic activity behavior to the formation of stable complex liquid components (and solids) such as K_2SiO_3, $KAlSiO_4$, ... The terminology of components and complex-components refers to the usual reference state oxides, e.g. K_2O or SiO_2, and to their compounds (liquid or solid), e.g. K_2SiO_3, respectively. The component and complex-component oxides formed are assumed to mix ideally with each other, in accordance with Raoult's law. Hence thermodynamic activities and <u>effective</u> (as distinct from the gross or nominal) mole fractions are equivalent quantities with this model. It should be stressed that explicit structural assumptions are not necessary in the present treatment. The component liquids considered here are distinctly different from the hypothetical associated species used in other modeling approaches. For the most part, the component liquids are established neutral, stable, thermodynamically defined compounds, appearing in phase diagrams in equilibrium with congruently melting solids, and also in reference tables of thermodynamic functions. The standard Gibbs energies of formation are either known or can be estimated for these complex-component liquids (and solids). By minimizing the total system free energy one can calculate the equilibrium composition with respect to these components. Thus, for instance, the <u>effective</u> mole fraction of K_2O present in equilibrium with K_2SiC_3, and other complex liquids (and solids) containing K_2O, can be calculated. As we have shown previously for the potassia, and other alkali systems, the component activities can, to a good approximation, be equated to these effective mole fraction quantities.[1,2] Thus physical interactions among the complex components are considered negligible compared with the strong chemical interactions leading to their formation.

Similar models can be traced to the early work of Schenck and others.[7] However, the model was not actively pursued in the interim owing to its apparent non-applicability to systems exhibiting positive deviations in mixing activities and also immiscibility. We have not pursued this aspect as the systems studied by us have shown large negative departures from ideal activity behavior. However, we feel that for an appropriate choice of component species, anti-mixing systems could be accomodated by the model. It is clear that the reliability of the model, given that the mixing assumption is valid, depends closely on the inclusion of all appropriate complex-components, and on the accuracy of their Gibbs energies of formation.

Table 1 lists the condensed phase components present in the oxide model database. Standard state gass or vapor phase species, also present, such as O_2, K, Na, etc are not listed. The database consists of polynominal-term coefficients expressing ΔG_f vs T for each component, solid(S) or liquid(L). Gaseous and vapor species are added as needed. Note that the components in parentheses are hypothetical liquids with estimated ΔG_f's. Also the Li database is not comprehensive, e.g. silicates have not yet been included. It is clear from Table 1 that many complex liquid and solid components are required for the model to have general utility. In polymeric silicate systems, the presence of three- and four-component complexes are key to the model's ability to reproduce experimental activity data. This behavior contrasts with that found so far for salt and metal systems where two-component complexes usually suffice to reproduce the mixing thermochemistry. A present limitation of the model are the relatively large error limits in the ΔG_f values for some of the complex-liquid components, which typically may be $\pm$ 10 kJ/mol. A listing of the ΔG_f values is given elsewhere.[2,3]

RESULTS

"FeO"-Al_2O_3 System

Many industrial slags contain FeO_x in various forms. In earlier work we have developed, and successfully tested, oxide slag models where FeO_x was absent.[1,2] The presence of FeO_x adds considerable additional complexity, both in the experimental activity determinations and in the construction of a model database. In the development and testing of the FeO_x database the "FeO"-Al_2O_3 system was investigated where "FeO" is the non-stoichiometric Wustite component (i.e. $Fe_{0.947}O$). Figure 1 shows a representative case where the phase boundaries predicted by the model are in good agreement with the recommended literature experimental phase diagram.[8] The departures between the model and experimental phase data are comparable with those found for various experimental determinations. The most sensitive comparison areas are the eutectics. Here the uncertainties in the known Gibbs energies are much larger than the free energy differences between the various phases and the nearly parallel nature of the phase free energy surfaces magnifies the model sensitivity to even small uncertainties.

MHD-Channel Potassia-Enriched Coal Slag

For the relatively simple binary system shown above, the key equilibria which determine the activities and phase boundaries involve the solid and liquid phases of "FeO," Al_2O_3 and Al_2FeO_4. However, for typical industrial slag systems, many solid and liquid complex-components are required by the model. For instance, $Na_2Si_2O_5$, $KAlSiO_4$, and $KCaAlSi_2O_7$ are found by the model to be key complex-components in many mineral-derived slags.[2] Note that these liquid forms effectively represent binary, ternary and quaternary interactions, respectively, between the constituent oxide compounds.

A representative example of a complex slag system is given by the MHD potassia-enriched coal (Illinois number 6) slag with the (room temperature) composition in wt %: SiO_2 (47.7), K_2O (19.8), Fe_2O_3 (14.8), Al_2O_3 (12.3), CaO (3.9), MgO (1.0), and Na_2O (0.5).[9] In this system, the K_2O activity is controlled primarily by the presence of $KAlSi_2O_6$ and $KCaAlSi_2O_7$ as stable liquids. Figure 2 shows good agreement between experimental and model predicted K-pressures, as determined from the model K_2O activities and the reference state pressures for pure $K_2O(\ell)$. Note that the agreement between the model and experimental data is within the experimental uncertainty. The predicted phase boundaries are also indicated in the figure. Above 1500 K the slag is predicted to be completely liquid and below 1300 K completely solid, with $KAlSi_2O_6$ and Fe_3O_4 being the main solid phases.

Steel Blast Furnace Slag

Pressures of sodium, potassium, and oxygen have been measured over a composition approximating a typical blast furnace slag but with slightly more Na_2O and K_2O present to facilitate the partial pressure determinations. The slag initially contained, SiO_2 (41), CaO (41), Al_2O_3 (10), MgO (4), Na_2O (2), and K_2O (2) with the wt % composition given in parentheses. Although one might anticipate that K would be the more volatile component of the glass, Na and K were vaporized at nearly the same rate. This may indicate that Na and K are strongly associated in the condensed phase. The non-additive behavior of mixed-alkalies in glass is well known[10] and because of the similar concentration of Na_2O and K_2O in this slag such effects require consideration. The model database (Table 1) represents this mixed-alkali association with two tentative estimated liquids, $KNaSiO_3(\ell)$ and $KNaSi_2O_5(\ell)$. Figure 3 shows curves of least-squares data fits for a typical experimental run, together with comparison model curves. As the alkali is depleted by vaporization, the condensed phase composition changes and data were thereby obtained over a range of alkali content. The pairs of model curves shown in figure 3 bracket the initial and final composition conditions of the experimental run. Figure 4 shows a typical example for variable K_2O-content (gross) and fixed temperature. Note in figures 3 and 4 that the agreement between the model and experimental curves is not very satisfactory although similar trends with temperature and composition are indicated. Possible non-equilibrium conditions in the experimental results are under investigation as a possible cause of this difference. Also, the model influence of the mixed-alkali components requires further investigation as no independent data are available for their stability. It is noteworthy that this is the only complex oxide case studied by us where such a serious difference between model and experiment has been found.

Molten Salts

Molten salt mixtures, particularly where the cations are dissimilar (e.g. alkali metal plus heavy metal), mix non-ideally with respect to the component activities. Complex-ion formation has often been indicated as the cause of non-ideal mixing.[6] With the present model, the formation of complex liquids, which most likely also contain complex ions, can be used to account for the experimental activity data. For the salt systems investigated to date we have found that one or two binary complexes of the form $AX \cdot BX_2$ (where AX and BX_2 are component metal halides) are sufficient to account for the observed activity data. Figure 5 shows a typical comparison between model and experimental (Bloom and Hastie[11]) activity curves. Note that unlike the oxide case, where the Gibbs energy data could usually be obtained independent of the activity data, for the halides no Gibbs energy data are available for the complex liquid components and hence the values used were obtained by an iterative fit of the model to the experimental activities. The solid model curve in figure 5 is seen to fit the experimental data (within experimental error) when ΔG_f for the complex is given by the ΔG_{rn} value of -5.9 kcal/mol for the reaction

$$PbCl_2(\ell) + CsCl(\ell) = CsPbCl_3(\ell).$$

The dashed curve shows the models' sensitivity to a 2 kcal/mol uncertainty in the ΔG_f $CsPbCl_3(\ell)$ value.

CONCLUSIONS

The "ideal mixing of complex components" model presented here appears to reliably predict activity and related data for polymeric oxide slag and salt systems. For systems where the mixing approaches ideality, the model may be limited by the significant error limits associated with the complex liquids Gibbs energies of formation. In principle, a more precise database and model can be developed using coupled thermochemical phase equilibria least-squares optimization approaches and such work is in progress.[12] It should be emphasized that the current model is predictive, rather than optimizing, and that information regarding phase boundaries, activities, or eutectic compositions is usually not used in designing the model database. Since the model is equally capable of handling high- or low-order systems, where essentially no phase stability or activity information is available, for the former, we feel that the degree of agreement found represents a significant improvement in the prediction of activity and phase relations for complex systems.

REFERENCES

1. Hastie, J.W., W.S. Horton, E.R. Plante and D.W. Bonnell, High Temperature-High Pressure, 14, 569 (1982).

2. Hastie, J.W. and D.W. Bonnell, High Temp. Science, 19, 275 (1985).

3. Plante, E.R., D.W. Bonnell and J.W. Hastie, "Experimental and Theoretical Determination of Oxide Glass Vapor Pressures and Activities," Proc. Int. Conf. on Fusion of Glass, Alfred Univ., June 14-17, 1988, Am. Ceramic Soc., Columbus (in press).

4. Hastie, J.W., Pure and Applied Chemistry, 56, 1583 (1984).

5. Richardson, F.D., "Physical Chemistry of Melts in Metallurgy," Academic Press, (1974).

6. Sundheim, B.R., "Fused Salts," McGraw Hill, (1964).

7. Schenck, H., "Introduction to the Physical Chemistry of Steelmaking," The British Iron and Steel Research Association, (1945).

8. Levin, E.M., C.R. Robbins, and H.R. McMurdie, "Phase Diagrams for Ceramists," 1969 Supplement, Diagram 2070, Am. Cerman. Soc., Columbus (1969).

9. Hastie, J.W., D.W. Bonnell and E.R. Plante, Proc. 23rd Symp. Engineering Aspects of Magnetohydrodynamics, June 25-28, 1985, Hidden Valley.

10. Day, D.E., J. Non-Cryst. Solids, 21, 343 (1976).

11. Bloom, H. and J.W. Hastie, J. Phys. Chem. 72, 2361 (1968).

12. Spear, K., Penn State, PA, Private Communication, (1988).

Table 1. Thermodynamic Database for Oxide Phase Equilibria/Activity Model

Name	Formula	Phases		Name	Formula	Phases	
Hercynite	Al_2FeO_4	S	L	Feldspar	$KAlSi_3O_8$	S	L
Alumina	Al_2O_3	S	L	(potassium melilite)	$KCaAlSi_2O_7$		L
Mullite	$Al_6Si_2O_{13}$	S	L		$KFeO_2$	S	
	$CaAl_2O_4$	S	L	Potassium beta-alumina	$K_2Al_{18}O_{28}$	S	L
Anorthite	$CaAl_2Si_2O_8$	S	L		$K_2Fe_{12}O_{19}$	S	
(calcium leucite)	$CaAl2Si_4O_{12}$		L	Potassium monoxide	K_2O		L
	$CaAl_4O_7$	S		Potassium metasilicate	K_2SiO_3		L
Calcium ferrite	$CaFe_2O_4$	S	L	Potassium disilicate	$K_2Si_2O_5$		L
	$CaMgSi_2O_6$		L		$K_2Si_4O_9$		L
Calcium oxide	CaO	S	L		$KNaSiO_3$		L
Pseudo-wollastonite	$CaSiO_3$	S	L		$KNaSi_2O_5$		L
Gehlenite	$Ca_2Al_2SiO_7$	S	L	Lithium aluminate	$LiAlO_2$	S	L
Di-calcium ferrite	$Ca_2Fe_2O_5$	S	L		$LiAlO_8$	S	L
Akermanite	$CaMgSi_2O_7$	S	L	Lithium monoxide	Li_2O	S	L
Larnite	Ca_2SiO_4	S	L		Li_5AlO_4	S	L
Diopside	$Ca_3Al_2O_6$	S		Spinel	$MgAl_2O_4$	S	
Grossular	$Ca_3Al_2Si_3O_{12}$	S		Magnesia	MgO	S	L
Merwinite	$Ca_3MgSi_2O_8$	S		Clinoenstatite	$MgSiO_3$	S	L
	$Ca_{12}Al_{14}O_{33}$	S	L	Forsterite	Mg_2SiO_4	S	
Ferrous oxide	FeO	S	L	Nepheline	$NaAlSiO_4$	S	L
Wustite	$Fe_{0.947}O$	S		Jadeite	$NaAlSi_2O_6$	S	L
Hematite	Fe_2O_3	S		Albite	$NaAlSi_3O_8$	S	L
Fayalite	Fe_2SiO_4	S	L		$NaFeSi_2O_6$	S	
Magnetite	Fe_3O_4	S	L	Sodium monoxide	Na_2O		L
Potassium aluminate	$KAlO_2$		L	Sodium metasilicate	Na_2SiO_3		L
Kaliophilite	$KAlSiO_4$	S	L	Sodium disilicate	$Na_2Si_2O_5$	S	L
Leucite	$KAlSi_2O_6$	S	L	Cristobalite	SiO_2	S	L

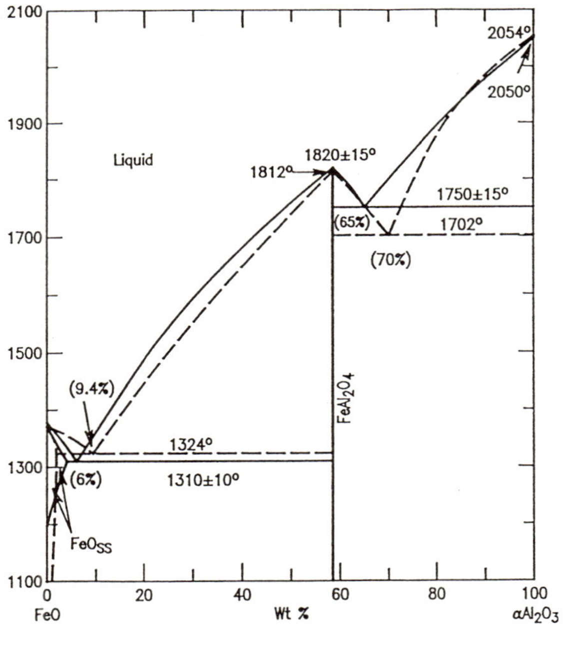

1. Comparison of literature (solid curve) and calculated (dashed curve) phase diagram of the "FeO"-Al_2O_3 system. Abscissa is degrees Celcius.

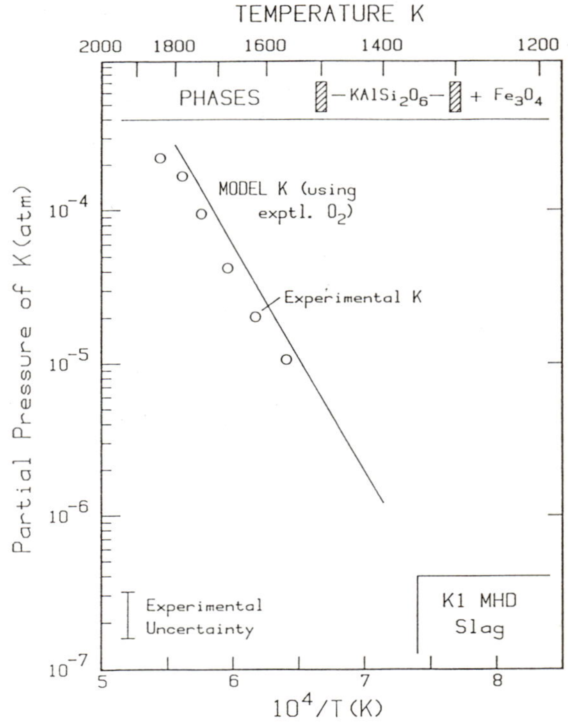

2. Model and experimental partial pressure data for K over an actual MHD slag.

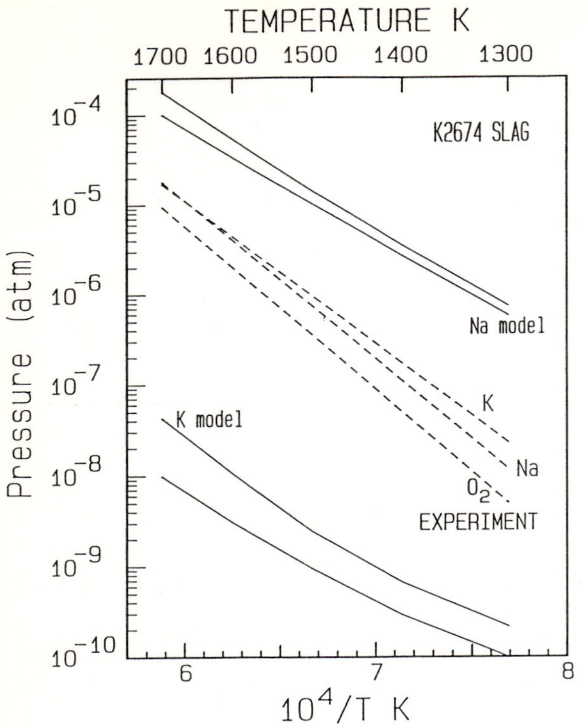

3. Experimental (dashed curves) and model (labeled solid curves) K, Na, and O_2 partial pressures over an alkali-enriched blast furnace slag.

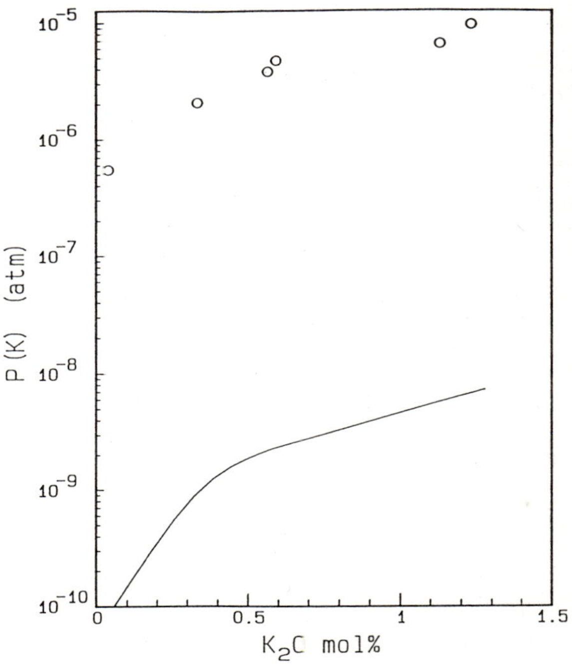

4. Experimental (data points) and model (solid curve) K partial pressures as a function of condensed phase K_2O content (gross) at 1600 K.

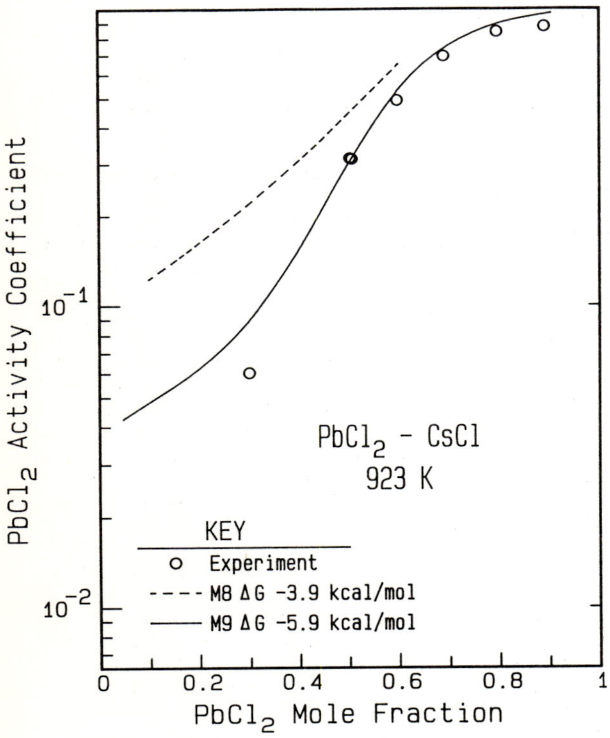

5. Experimental (data points) and model (solid and dashed curves) activity coefficients for $PbCl_2$ in the $PbCl_2$-CsCl system at 923 K.

Use of slag model to describe slag-metal reactions and inclusions precipitation

H GAYE, P V RIBOUD and J WELFRINGER

Drs Gaye and Riboud are in the Department of Physical Chemistry at IRSID, Maizieres les Metz, France; Dr Welfringer, formerly in the same department, is now in the Steelmaking Department at IRSID.

SYNOPSIS

A very efficient statistical thermodynamics "cell model" has been developed which allows an accurate representation of the properties (phase diagrams and component activities) of complex iron and steelmaking slags over wide ranges of temperature and compositions. It has been inserted in various multiphase equilibrium calculation programs and kinetic models for the description and monitoring of metal-gas-slag treatments and for the control of oxide and sulphide inclusions in steels.

INTRODUCTION

The basics of the slag model developed at IRSID were presented at the 2nd International Conference on Slags and Fluxes (1). Since then, our efforts have been devoted to expanding the range of systems covered, but also to inserting the slag model in specific procedures for the study of iron and steelmaking reactions.

Although the model could be used by expert thermodynamicists in standard multiphase equilibria algorithms (e.g. Solgasmix), it appeared very soon that a routine use of these concepts by the technical staff of our plants and research centres required the device of self-sufficient, efficient and reliable packages aimed at describing specific reactions (slag-metal-gas reactions, inclusions precipitation in steels...).

This presentation will outline both facets of our activities in the domain.

STATE OF THE MODEL

The statistical thermodynamics 'cell model' is based on an idea originally proposed by Kapoor and Frohberg (2) for describing binary and ternary liquid oxide systems, which has been extended to represent systems containing an arbitrarily large number of components. It calculates simultaneously slag component activities and phase diagrams.

It makes use of a limited number of binary parameters (2 to 3 per binary system) and, owing to the fact that it requires no ternary or higher order parameters, constitutes a fairly safe means of extrapolating data into unexplored compositions or temperature domains. Highly efficient numerical techniques for the statistical thermodynamics treatment have been implemented, so that the model can be used as much in complex multiphase equilibria calculations for which repetitive calls of the program are required, and it can also be inserted into reaction-kinetics models.

At the present time, this model gives a fairly accurate description of the properties (phase diagram, component activities, slag-metal equilibrium partition ratios) for systems consisting of

$$SiO_2\text{-}Al_2O_3\text{-}Fe_2O_3\text{-}FeO\text{-}CaO\text{-}MgO\text{-}MnO$$

over wide compositions (basic and acid slags) and temperature ranges.

As an illustration, Figure 1 shows a comparison of computed and experimental phase diagram and silica activity in the $CaO\text{-}SiO_2\text{-}Al_2O_3$ system at 1550°C. The agreement for domains of composition representative of blast-furnace slags, ladle metallurgy slags and most actual inclusions precipitation domains is quite satisfactory. The only serious discrepancy affects the silica-rich side of the binary $SiO_2\text{-}Al_2O_3$ system. It is our opinion that this discrepancy cannot be corrected by fiddling with the values of the parameters, but a slight change of the model would be required in order to reflect the amphoteric nature of alumina. Attempts to include Na_2O and P_2O_5 have not yet been fully successful. The description of phase diagrams (in particular $CaO\text{-}SiO_2\text{-}P_2O_5$- iron oxides) is rather satisfactory, as shown on Figure 2 which represents the experimental and computed phase diagram of the $CaO\text{-}P_2O_5$- iron oxides system in equilibrium with iron at 1600°C. Note that the line a(FeO)=1 does not represent the contour of the computed miscibility gap, but it is inside this contour; it was drawn to show that it is indeed possible to predict the existence of the miscibility gap without ternary parameters. In order to describe the very sharp change in lime and phosphorus pentoxide activities in the binary $CaO\text{-}P_2O_5$ at the stoichiometry of tricalcium phosphate, it was necessary to insert P_2O_5 in the model as $(PO)_2-O_3$ (which is not totally surprising as, in the PO_4 tetrahedron, one of the oxygens is doubly linked to phosphorus). Although the oxygen activity is also accurately represented, there remains for the phosphorus equilibrium slag-metal partition ratio, with the constant parameters retained so far, a mismatch of about 5 between the domain of basic low phosphorus refining slags and that of acid slags (basicity around 1) of the same system. This is quite inadequate for our needs; for high and low phosphorus basic refining slags, estimates using a FLOOD's type model (3) are still more accurate.

Up to now, the behaviour of sulphur dissolved in the slag is represented using sulphide capacity - optical basicities correlations (4-5). This is not fully satisfactory and we are in the process of inserting sulphur directly into the model. This has already been attempted, using this very model (6) or a slightly modified one (7), and seems very promising.

CRYSTALLIZATION PATH

In many steelmaking situations the oxide system (slag or inclusions) may happen to be only partly liquid. In order to apply thermodynamic relationships between oxides and metal, it is necessary to determine, from sample analysis, the composition of liquid and proportions of solids at treatment temperature.

To that effect, the crystallization path of the slag is computed: starting from the liquidus temperature, the equilibrium state is calculated for decreasing temperatures, until treatment temperature is reached. At each step, the composition of equilibrium liquid is computed using Taylor series expansions in terms of temperature decrement, the coefficients of which are obtained by application of the slag model. This is of course coupled with the mass balance equations, and the simplicity of the method relies on the approximation that most solid phases are strictly stoichiometric. Provision is made, however, for the existence of solid solutions (e.g. spinel phases, melilite ...) which are described as ideal solutions of end members.

It is thus possible to obtain the equilibrium state of the slag at any temperature, down to the solidus temperature. An example of computer printout is shown in Table 1.

This procedure for crystallization path calculation is a central piece of all the exploitation programs of the slag model. It is of course useful to determine the equilibrium state of a system at a given temperature, but also to study the evolution of slags or inclusions upon cooling. For instance, it has been used to rationalize studies of slag-refractories reactions. It is also of very practical interest to investigate a new complex system: the calculations made for a single composition yield a large amount of information.

SLAG-METAL-GAS REACTIONS

Various applications of the slag model for studying industrial treatments have been made. They include the distribution of deoxidants (Al, Si and Mn) and steel desulphurization in ladle metallurgy (5), the evaluation of the oxygen potential of refining slags at equilibrium with either iron or a gas, the study of iron or manganese blast furnace reactions, and the description of microslags formed during sintering of iron ores. Only two examples will be given here.

The first example, pertaining to the partition of deoxidants (al, Si and Mn) between steel and slag during ladle metallurgy treatments, corresponds to the treatment under vacuum in a well stirred 30 ton ladle with a slag of composition around 13%SiO_2 - 25%Al_2O_3 - 50%CaO - 10%MgO - .05%MnO. From the analysis of metal and slag samples, the oxygen activities corresponding to the $\underline{Al}/(Al_2O_3)$, $\underline{Si}/(SiO_2)$ and $\underline{Mn}/(MnO)$ elementary equilibria at various stages of the treatment were computed. The first sample was taken shortly after an Al addition: at that time, $\underline{Mn}$ and $\underline{Si}$ were close to mutual equilibrium, whereas $\underline{Al}$ was much too high. Aluminium is then progressively oxidized ($\underline{Al}$ drops from 0.16 to 0.023%), $\underline{Si}$ increases due to silica reduction (from 0.18 to 0.24%) and the small amount of reduced MnO does not significantly alter $\underline{Mn}$ content in the steel. At end point, the contents in the three deoxidants were quite coherent with the computed equilibrium. On the basis of this study, it was possible to standardize the sequence of additions and time of stirring before sampling in order to obtain a meaningful sample and achieve a better process control.

The second example corresponds to reactions involving slags containing iron oxides. For these slags, the model allows calculations for a fixed thermodynamic constraint: equilibrium with iron, or with a given oxygen potential. The results summarized in Table 2 compare, for samples taken at the end of refining in a well-stirred 6 ton EAF, the values of iron oxides distribution, equilibrium oxygen activity and sulphur partition predicted with the slag model, with those obtained using a Flood type model (3). The latter has been shown to give very accurate estimates in this limited range of basic refining slags but can only describe the conditions at slag-metal equilibrium. The agreement between the two models is quite fair. The equilibrium sulphur was computed, respectively, from direct correlations of partition coefficient and, for the present model, from the calculated oxygen activity and sulphide capacity correlations. It appears that the new correlation proposed by Sosinky and Sommerville (8) is quite inaccurate at the high temperature of this experiment (1654°C). The last line of Table 2 gives the iron oxides distribution predicted in the bulk slag, corresponding to the oxygen potential measured with an oxygen probe: the large difference in oxygen potential between bulk slag and interface is accommodated by a very slight difference in Fe_2O_3 content.

INCLUSIONS PRECIPITATION IN STEELS

The calculation consists, on the basis of the global analysis of a metal sample (that is, total Al, Mn, Si, Ca, Mg, ..., O, S contents), in evaluating the composition and amounts of residual inclusions (oxides and sulphides) which may have been formed at treatment temperature and during the subsequent cooling until the steel starts solidifying. It assumes equilibrium between steel and inclusions as they precipitate. Since the first presentation of this procedure (5), it has been largely improved in order to increase its reliability and take into account the precipitation of solid oxides.

The method makes use of a recurrent procedure for complex equilibria calculation, combining material balances and equilibrium constraints. Starting with an approximate composition of the liquid oxide inclusions (which may be quite different from the equilibrium composition), the scheme consists of a "search pattern" which leads to the equilibrium composition of the oxides, the possible precipitation of sulphides (CaS or (Ca-Mn)S) being checked at each step. At each step of the pattern, the slag model is used to compute the oxide activities, so that, although the method has been used mostly for Ca-treated steels, it can be applied in a wide range of situations (e.g. for silica-rich inclusions, Mn alumino-silicates ...). Quite obviously, this "heavy method" is feasible only thanks to the efficiency of the slag model. Typical calculation times are a few seconds on IRSID's VAX 11/780 for a single composition and temperature.

The results of these calculations have been checked against the analysis of inclusions extracted from steel samples by non-aqueous electrolytic dissolution of the metal matrix. A detailed description of the extraction method can be found in Refs (9-10); selective chemical analyses of these inclusions not only provide their global composition, but also allow identification and quantification of the various phases present: CaO and ($CaO-Al_2O_3$), SiO_2, Al_2O_3, CaS, (Ca-Mn)S, AlN.

In Figure 4 is shown the comparisons of analysed and calculated amounts of oxidized aluminum fixed as Ca alumino-silicates were prepared for several steel grades with varied sulphur levels and amounts of oxide inclusions, and total Al contents larger than 150 ppm. The agreement is rather satisfactory: the differences between the two values do not exceed $\pm$ 5ppm. The comparison for calcium present as oxides or sulphides is very satisfactory; for all samples in which calcium sulphide was analysed, and only for these samples, the calculation predicted the stability of these inclusions, and in about the right amount.

CONCLUSIONS

The vast experience we have acquired using the IRSID slag model to evaluate plant data on slag-metal reactions and inclusions precipitation shows that, although well needed extensions (e.g. addition of S, F, P_2O_5, Cr_2O_3, ...) and possible improvements are expected, it constitutes a very useful tool for process control studies and possibly for product testing purposes.

REFERENCES

(1) H. GAYE and J. WELFRINGER: 2nd International Conference on Molten Slags and Fluxes. Lake Tahoe, 1984.

(2) M.L. KAPOOR and M.G. FROHBERG: Symposium on Chemical Metallurgy of Iron and Steel. Sheffield, 1971.

(3) H. GAYE, J.C. GROSJEAN and P.V. RIBOUD: Congres PCS 78, Versailles, France, 1978.

(4) J.A. DUFFY, M.D. INGRAM and I.D.SOMMERVILLE: J. Chem. Soc. Faraday Trans. I 74, (1978), pp. 1410-1419.

(5) H. GAYE, C. GATELLIER, M. NADIF, P.V. RIBOUD, J. SALEIL and M. FARAL: Clean Steels III, Balatonfured, 1986, pp. 137-147.

(6) W. YAMADA, T. MATSUMIYA and N. NAKAMURA: CAMP-ISIJ Vol.1 (1988) - 250, p. 168.

(7) C. SAINT-JOURS and M. ALLIBERT: This Conference, paper 23.

(8) D.J. SOSINSKY and I.D. SOMMERVILLE: Metall. Trans. B, 1986 17B, pp. 331-337.

(9) D. HENRIET and P. de GELIS: Revue de Metallurgie CIT, (Sept. 1984), pp. 703-715 and (Oct. 1984), pp. 799-807.

(10) O. BOUCHER, L. CARROT, G. GIROUD, J. SALEIL: Revue de Metallurgie CIT (Feb. 1986), pp. 117-126.

Table I: Example of Computed Crystallization Path.

	SiO_2	Al_2O_3	FeO	CaO	MgO	MnO	% Phase
Global	33.50	12.00	0.50	45.00	8.00	1.00	
	Liquidus Temperature: 1459 Celsius						
	Primary Phase: $2CaO-SiO_2$						
At 1450 Celsius:							
Liquid	33.42	12.65	0.53	43.92	8.43	1.05	94.9% liquid
							5.1% $2CaO-SiO_2$
At 1400 Celsius:							
Liquid	33.28	10.45	2.35	42.07	7.15	4.70	21.3% liquid
							3.9% $2CaO-SiO_2$
							27.9% Merwinite
							46.9% Melilite

Table II: Comparison, for basic refining slag, of present model with a Flood type model (3). For sulphide capacity correlations, (a) after Ref.(5) and (b) after Ref.(8).

	%FeO	$\%Fe_2O_3$	a(O)ppm	S ppm	Equilibrium S	
Slag-metal	27.20	7.65	1210	160	175	Flood model
Interface.	26.84	8.07	1290	"	187(a)/40(b)	This model
Bulk slag.	25.70	9.33	2002			

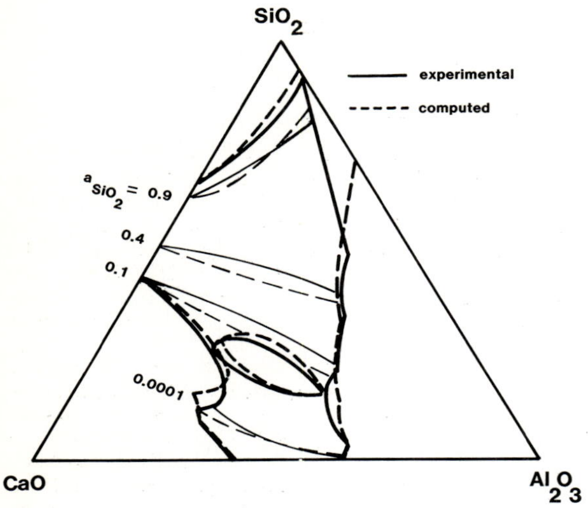

Figure 1

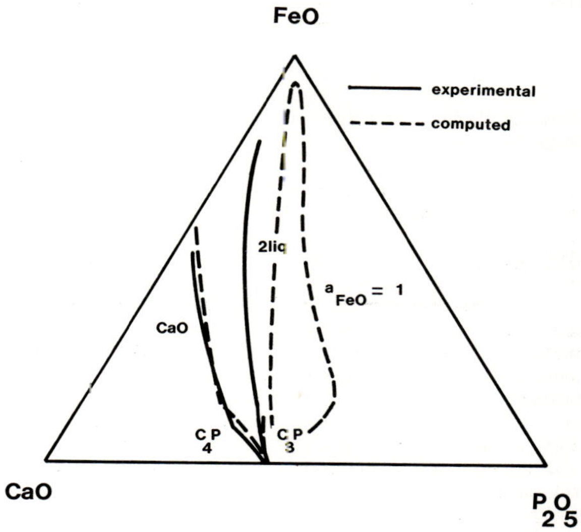

Figure 2

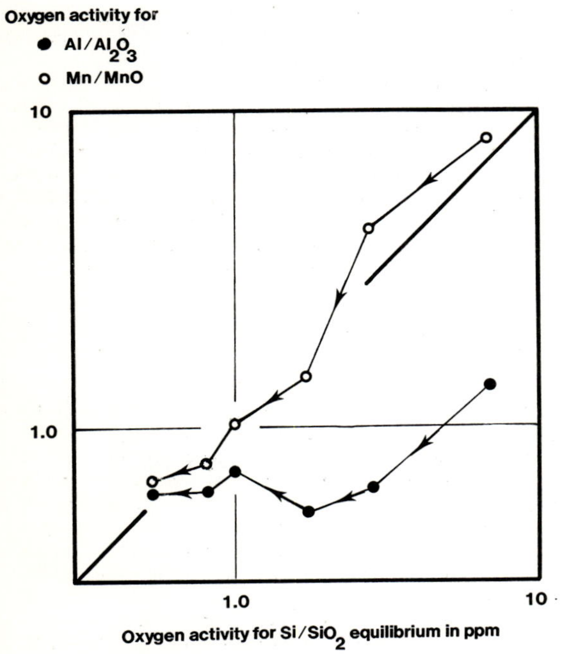

Figure 3

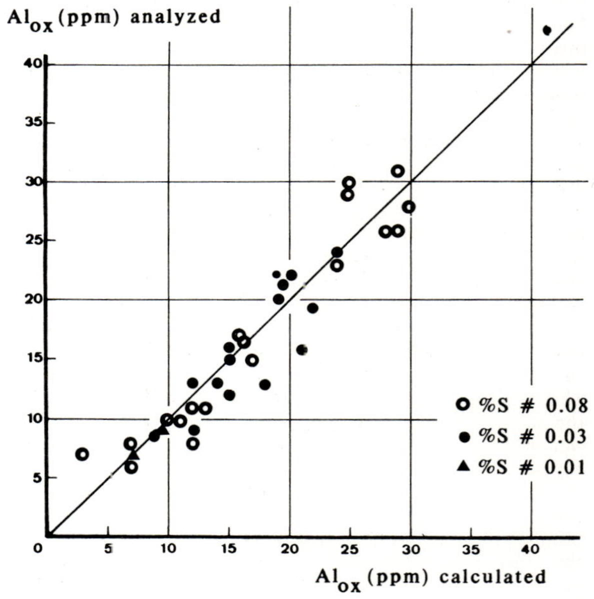

Figure 4

Slag models and inclusion engineering in steelmaking

D A R KAY and JIANG JUNPU

Professor Kay is with the Department of Materials Science and Engineering, McMaster University, Hamilton, Ontario, Canada; Professor Jiang is with the University of Iron and Steel Technology, Beijing, China.

SYNOPSIS

A major objective of modern ladle steelmaking technologies is to control the volume fraction of oxide and sulphide inclusions and to fix their compositions so that they have the least deleterious effects during subsequent casting, working, fabrication and service.

The possible use of oxygen probe measurements and slag models in the control of this 'inclusion engineering' for the production of steels of improved machinability is described in this paper.

INTRODUCTION

The mechanical properties of a steel of a given matrix, are largely dependent on the volume fraction, size, composition, distribution and morphology of inclusions and other second phase particles. Inclusions greater than about 1 μm in size have a markedly deleterious effect on steel ductility and toughness. For this reason, modern ladle steelmaking technologies involving calcium treatment are designed to minimize the volume fraction of inclusions and to control the composition of those remaining so that they have the least deleterious effects during subsequent casting, working, fabrication and service. The production of free-machining steels provides the exception to this rule, in that dispersions of oxide and sulphide inclusions are used to promote chip formation. In this case, ladle steelmaking is used to control the composition of these inclusions so that they are 'softer' than the matrix under machining conditions and do not accelerate tool wear processes. The control of steel deoxidation and desulphurization in ladle steelmaking is therefore an integral part of 'inclusion engineering'.

In order to put inclusion engineering into practice, it is essential that the equilibrium relationship between the liquid steel and the corresponding inclusion should be determined. These relationships must be established using data available to the steelmaker, i.e. steel composition, temperature and the soluble oxygen content as measured by commercially available oxygen probes.

In this paper, equilibrium liquid steel/inclusion composition relationships are presented for the $MnO-Al_2O_3-SiO_2$ and $CaO-Al_2O_3-SiO_2$ systems and discussed in relation to their use in deoxidation and calcium treatment control. Particular attention is given to the control of inclusions in the $CaO-Al_2O_3-CaS$ system where the precipitation of liquid inclusions is important in the prevention of nozzle clogging during billet continuous casting.

OXIDE INCLUSION COMPOSITIONS

The composition domains of 'soft' silicate inclusions, as defined by Bernard et al.(1), in the $MnO-Al_2O_3-SiO_2$ and $CaO-Al_2O_3-SiO_2$ systems are shown in Figures 1a) and 1b), respectively. The shaded areas at high silica contents give the compositions of 'hard' supercooled silicate liquid(s) while the shaded areas at lower silica contents represent the compositions of 'hard' inclusions which have either partially or wholly recrystallized when subjected to the specific thermal history of the study.

The production of 'soft' inclusion dispersions is the key to the development of steels of improved machinability in terms of tool wear. Deoxidation practices can be controlled to produce 'soft' spessartite ($3MnO-Al_2O_3-SiO_2$) inclusions in the $MnO-Al_2O_3-SiO_2$ system, for example. For inclusions in the $CaO-Al_2O_3-SiO_2$ system, standard deoxidation practices are modified by calcium ladle treatments to give either anorthitic/gehlenitic ($CaO-Al_2O_3-2SiO_2$)/($2CaO-Al_2O_3-SiO_2$) inclusions in the case of silicon/manganese deoxidised grades, or duplex calcium aluminate/calcium sulphide inclusions in aluminium deoxidised grades. In the latter case, abrasive alumina inclusions are modified to calcium aluminates on which 'soft' calcium sulphide can precipitate and the advantages of aluminium deoxidation retained. Where the soluble aluminium content is too low for conventional grain refinement (<0.01%), use is made of established microalloying technology.

The prerequisite of steelmaking inclusion engineering is a deoxidation control system in which liquid steel analyses can be related to corresponding equilibrium inclusion compositions using measured temperature and oxygen activity data obtained from commercially available oxygen probes. These thermodynamic predictions can be incorporated into a deoxidation control system and/or used to determine the significance of any kinetic effects during liquid steel treatment, and the extent of reoxidation during transfer operations.

THERMODYNAMIC DATA BASES

For most oxide systems relevant to steelmaking deoxidation, the experimental thermodynamic data base relating inclusion compositions to the corresponding oxide activities is incomplete, and use has to be made of slag (inclusion) models to give a complete data base over the range of temperatures and inclusion compositions encountered in industrial steelmaking. The $CaO-Al_2O_3$ system is an exception, in that there is an adequate thermodynamic data base (2) which can be applied to the modification of alumina inclusions by calcium treatment.

In developing a deoxidation control model for inclusion engineering in the $MnO-Al_2O_3-SiO_2$ and $CaO-Al_2O_3-SiO_2$ systems, a statistical thermodynamic slag model developed by Gaye(3) can be used to calculate component oxide activities from the corresponding weight percent inclusion compositions at a given temperature. The equilibrium Henrian activities of elements in solution in the steel are then calculated for a given Henrian oxygen activity using the thermodynamic data given in Table I. The weight percent concentrations of elements in solution in steel are calculated from the corresponding Henrian activities, using the free energy interaction parameter data at 1600°C given in Table II. The use of these data is considered to be valid over the temperature range of steelmaking deoxidation and ladle treatment.

INCLUSIONS IN THE $MnO-Al_2O_3-SiO_2$ SYSTEM.

The importance of soluble oxygen content and temperature control in inclusion engineering, is illustrated by the calculation of weight percent steel compositions in equilibrium with 'soft' inclusions in the $MnO-Al_2O_3-SiO_2$ system, using an AISI 1045 plain carbon steel base. Iso-weight percent curves for silicon and manganese at 1600°C are plotted as a function of inclusion composition on the ternary phase diagrams in Figures 2a) and 2b). The curves are calculated for Henrian oxygen activities of 0.0070 in Figure 2a) and 0.0080 in Figure 2b). From the data given in Figure 2a), an AISI 1016 steel containing 0.60% soluble manganese and 0.17% soluble silicon, with a soluble Henrian oxygen activity of 0.0070 at 1600°C, would be in equilibrium with an inclusion containing 36%MnO, 31%SiO_2 and 33%Al_2O_3 by weight, i.e., the inclusion composition is given by the intersection of the two iso-weight percent curves. The corresponding equilibrium inclusion weight percent composition of 45%MnO, 37%SiO_2 and 18%Al_2O_3 is obtained from Figure 2b) when the Henrian oxygen activity is increased by 0.0010 to 0.0080. The soluble oxygen content is therefore doubly important since it also controls the volume fraction of indigenous oxide inclusions forming during cooling and solidification.

In cases where the soluble silicon and manganese contents can be approximated to the corresponding total contents obtained from spark emission spectroscopy, steel and exogenous inclusion compositions can be related using temperature and soluble oxygen contents obtained from oxygen probe measurements taken during steelmaking. Aluminium iso-weight percent curves may also be calculated for Figures 2a) and 2b) but they are not required to establish inclusion compositions. In any event, the soluble aluminium contents are too small, in this case (<10 ppm by weight), to be used as control variables during steelmaking.

The effect of temperature on the composition of an AISI 1016 steel in equilibrium with an inclusion containing 30%MnO, 50%SiO_2 and 20%Al_2O_3 is shown in Figures 3a) and 3b) where the manganese and silicon contents, respectively, are shown as a function of the Henrian activity of oxygen. Contents of 0.28%Mn and 0.10%Si by weight at 1550°C are to be compared with contents of 0.42%Mn and 0.30%Si at 1600°C for a given Henrian oxygen activity of 0.0070.

The prediction of the composition of exogenous inclusions in the $MnO-Al_2O_3-SiO_2$ system also provides the base for subsequent inclusion modification by aluminium and/or calcium additions.

INCLUSIONS IN THE $CaO-Al_2O_3-SiO_2$ SYSTEM

For the inclusion engineering of anorthitic and gehlenitic inclusions in the $CaO-Al_2O_3-SiO_2$ system, the analyzed total aluminium and calcium contents are low and are not known with sufficient accuracy during liquid steel processing to be used as estimates of their respective soluble concentrations. Although the soluble silicon content can be approximated to the analyzed total silicon content, an inclusion composition cannot be determined solely from the iso-weight percent silicon curves calculated for a measured soluble oxygen content and temperature.

For liquid steels in equilibrium with inclusions in the $CaO-Al_2O_3$ system - see Figure 4, the total analyzed aluminium contents are higher (typically 0.010 - 0.060%) and are reasonable estimates of the soluble aluminium contents. Using the thermodynamic data base for calcium aluminates given in the literature (2), coexisting aluminate phases can be identified at a given aluminium activity using oxygen activity and temperature readings obtained from an oxygen probe. Hence, oxygen probe readings can be used to control the modification of alumina inclusions in the calcium treatment of aluminium killed steels.

The conditions for aluminate and solid calcium sulphide co-precipitation can also be calculated from the thermodynamics of the reaction:

$$CaO(s) + [S] = CaS(s) + [O] \qquad (1)$$

The results of such calculations are shown in Figures 5a) and 5b) where the minimum [wt.%S] required for the co-precipitation of calcium sulphide with two coexisting aluminates is shown as a function of the Henrian activity of aluminium at 1600 and 1550°C, respectively. The co-existing aluminate phases shown in these diagrams are $CaO.Al_2O_3/CaO.2Al_2O_3$ (CA/CA_2), liquid aluminate/$CaO.Al_2O_3$ (L/CA) and CaO/liquid aluminate (C/L).

As can be seen from these diagrams, calcium sulphide precipitation in calcium treated liquid steels can be avoided, even in those steels with a relatively high sulphur content, if the soluble aluminium content is low and the calcium treatment is just sufficient to form liquid aluminate inclusions. The data in Figure 5b) indicate that a sulphur content of <30ppm is required to avoid calcium sulphide formation at 1550°C for the CaO/liquid aluminate co-existence.

The importance of controlling calcium treatment is again illustrated in Figure 6 where a relative calcium activity is plotted as a function of the Henrian activity of aluminium. A relative calcium activity is obtained by arbitrarily equating one Henrian calcium activity to 10.0 and adjusting the other activities accordingly. In this way, problems associated with the calculation of absolute calcium activities, due to the uncertainty in the standard free energy of formation of lime, are avoided. As can be seen from Figure 6, calcium treatment can be minimized if the Henrian activity of aluminium is low and the steel is not overtreated. As a first approximation, correct treatment at $h_{Al} = 0.010$ as opposed to overtreatment at $h_{Al} = 0.060$ can reduce calcium consumption by as much as a factor of five.

CONTROL OF CALCIUM TREATMENT

The use of oxygen probe measurements in the control of calcium treatment is illustrated by the data given in Table III, where the wt% aluminium and oxygen contents have been approximated to their corresponding Henrian activities. These data are generated using the thermodynamic data base given in the literature (2).

In Example 1, an oxygen probe measurement prior to calcium treatment establishes the Henrian oxygen activity window ($0.00018 < h_O < 0.00047$) for the production of liquid aluminate inclusions at constant temperature. The oxygen probe measurement after calcium treatment confirms the formation of liquid aluminates and establishes the minimum sulphur content (0.0083%) at which the precipitation of calcium sulphide will occur at that temperature.

Example 2 illustrates the effect of increased aluminium deoxidation on subsequent calcium treatment. The size of the oxygen activity window for the formation of liquid aluminates has decreased ($0.00008 < h_O < 0.00022$), as have the actual activity values . In addition, although the probe readings after calcium treatment are the same as those given in Example 1, the equilibrium oxide inclusions are now $CaO.Al_2O_3$ and $CaO.2Al_2O_3$ solid aluminates.

ACKNOWLEDGEMENTS

The financial support of the Department of energy Mines and Resources (CANMET) is gratefully acknowledged. One of us (J.J.) should also like to acknowledge financial support from the Canadian International Development Agency (CIDA).

REFERENCES

1. G. Bernard, P.V. Riboud and G. Urbain: Rev. Met. CIT, vol.78, no. 5, 1981, pp. 421-434.
2. D.A.R. Kay, S.V. Subramanian and R.V. Kumar: 'Inclusions in Calcium Treated Steels', Proceedings of the Second International Symposium on the Effects and Control of Inclusions and Residuals in Steels, 25th. Conference of Metallurgists, Toronto, 1986. CIM
3. H. Gaye and D. Coulombet: IRSID PCM- RE.1064, March, 1984.
4. H. Gaye, C. Gatellier, M. Nadif, J. Soliel and M. Faral: IRSID PCM-RE.1277, May, 1986.
5. D.Ghosh and D.A.R. Kay: J. Electrochem. Soc., vol. 124, 1977, pp. 1836-1845.
6. J.F. Elliott: 'Electric Furnace Steelmaking', 1985, p. 265. AIME
7. G.K. Sigworth and J.F. Elliott: Metal Science, vol. 8, 1974, pp. 298-310.
8. M. Kohler, H-J. Engell and D. Janke: Steel Research, vol.56, no.8, 1985, p. 419.
9. J.F. Elliott, M. Gleiser and V.Ramakrishna: 'Thermochemistry for Steelmaking', vol.II, 1963. Addison-Wesley Publishing Co. Inc.

TABLE I. Thermodynamic Data Base

Reaction	Log K	Reference
$[Si]_{1w/o} + 2[O]_{1w/o} = SiO_2(s)$	(31040/T) - 12.0	4
$[Mn]_{1w/o} + [O]_{1w/o} = MnO(s)$	(15050/T) - 6.7	4
$2[Al]_{1w/o} + 3[O]_{1w/o} = Al_2O_3(s)$	(61304/T) - 20.37	5,6

TABLE II. Interaction Parameters.

	Al	C	Si	O	Mn	S	Ca
Al	0.045(4)	0.043(7)	0.058(4)	−1.170(4)	−	0.035(7)	−0.072(4)
C	0.091(4)	0.140(7)	0.180(4)	−0.130(4)	−0.070(4)	0.110(7)	−0.337(4)
Si	0.056(4)	0.078(7)	−0.107(4)	−0.140(4)	0.060(4)	0.063(7)	−0.097(4)
O	−1.980(4)	−0.098(9)	−0.250(4)	0.0	−0.083(7)	−0.270(7)	−
Mn	−	−0.012(7)	0.033(4)	−0.030(4)	−0.003(4)	−0.026(7)	−0.010(8)
S	0.030(4)	0.046(7)	0.056(4)	−0.133(4)	−0.004(4)	−0.028(7)	−
Ca	−0.047(7)	−0.097(7)	−0.068(7)	−	−	−	−

Note: Figures in parentheses refer to references in the literature.

TABLE III. Calcium Treatment Control Using Oxygen Probe Data.

Example 1.

INPUT		OUTPUT
Before Ca/Si Treatment		
T(F):	2912	
O_2 Probe (mV)	−120	[Al] = 0.027
		For Liquid Aluminates
		Max. [O]ppm: 4.7
		Min. [O]ppm: 1.8
After Ca/Si Treatment		
T(F)	2875	
O_2 Probe (mV)	−200	LIQUID ALUMINATES
		Max. [S]: 0.0083
		to avoid CaS ppn.

Example 2.

INPUT		OUTPUT
Before Ca/Si Treatment		
T(F):	2912	
O_2 Probe(mV)	−180	[Al] = 0.082
		For Liquid Aluminates
		Max. [O] ppm: 2.2
		Max. [O] ppm: 0.8
After Ca/Si Treatment		
T(F):	2875	
O_2 Probe(mV)	−200	CA/CA_2
		SOLID ALUMINATES

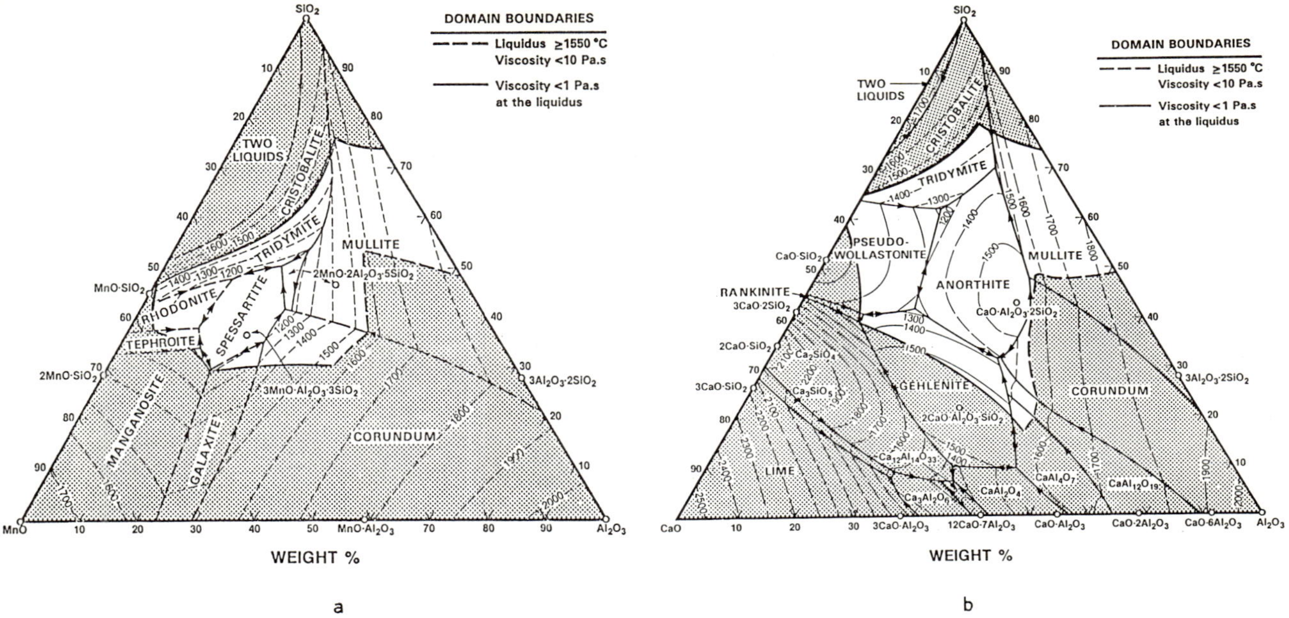

Figure 1. — a) The $MnO-Al_2O_3-SiO_2$ phase diagram and b) the $CaO-Al_2O_3-SiO_2$ phase diagram, showing the composition domains of 'deformable' silicate inclusions.

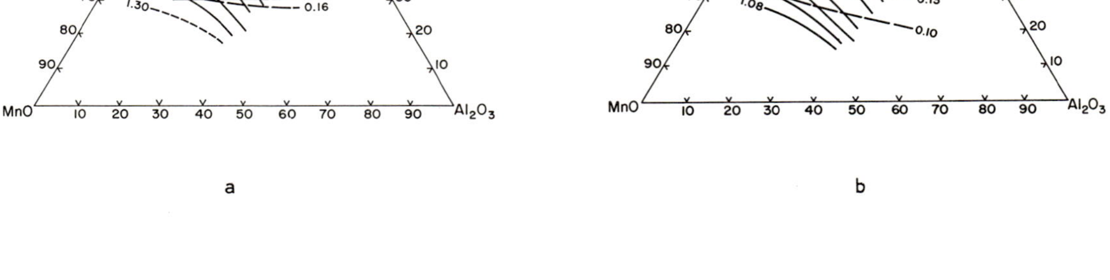

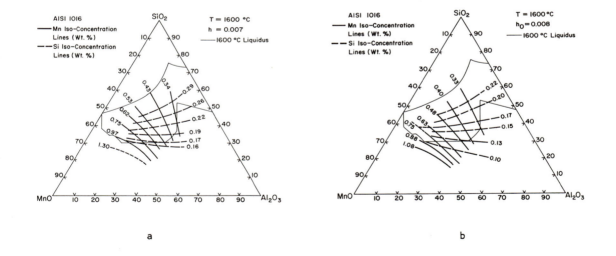

Figure 2. — Iso-weight percent curves for silicon and manganese in the $MnO-Al_2O_3-SiO_2$ for an AISI 1016 steel at 1600°C. a) $h_O = 0.007$ and b) $h_O = 0.008$

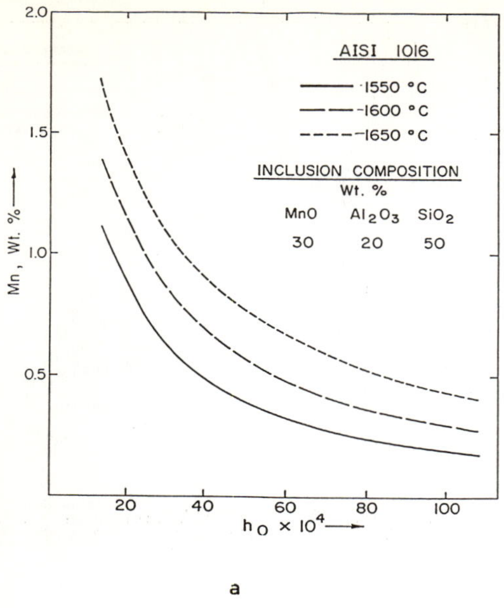

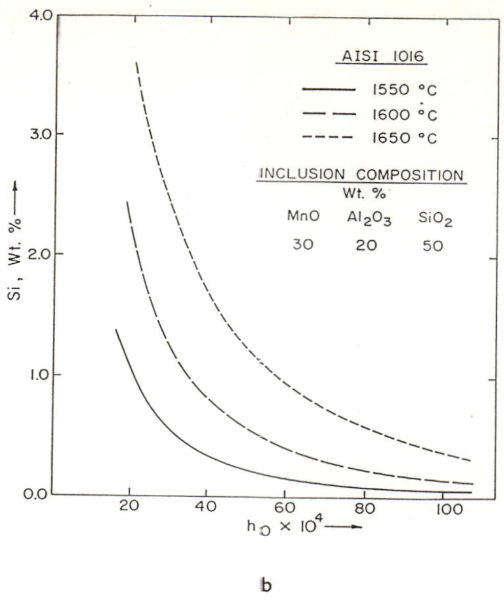

Figure 3. — The effect of oxygen activity and temperature on a) the manganese content and b) the silicon content of an AISI 1016 steel in equilibrium with an inclusion containing 30%MnO, 20%Al_2O_3 and 50%SiO_2 by weight.

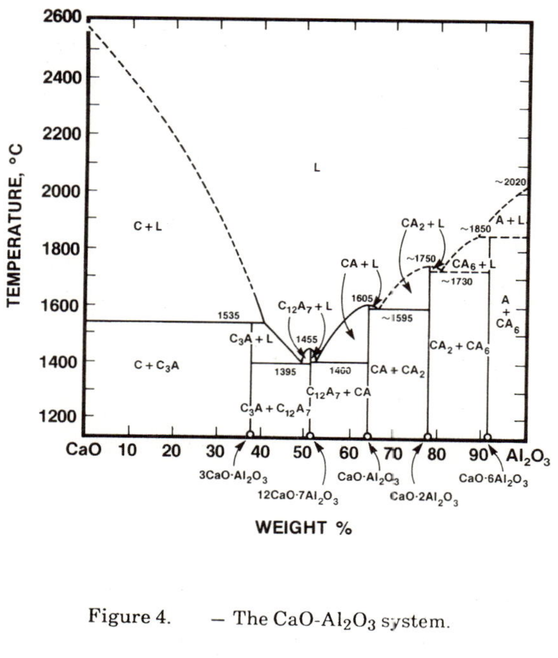

Figure 4. — The CaO-Al_2O_3 system.

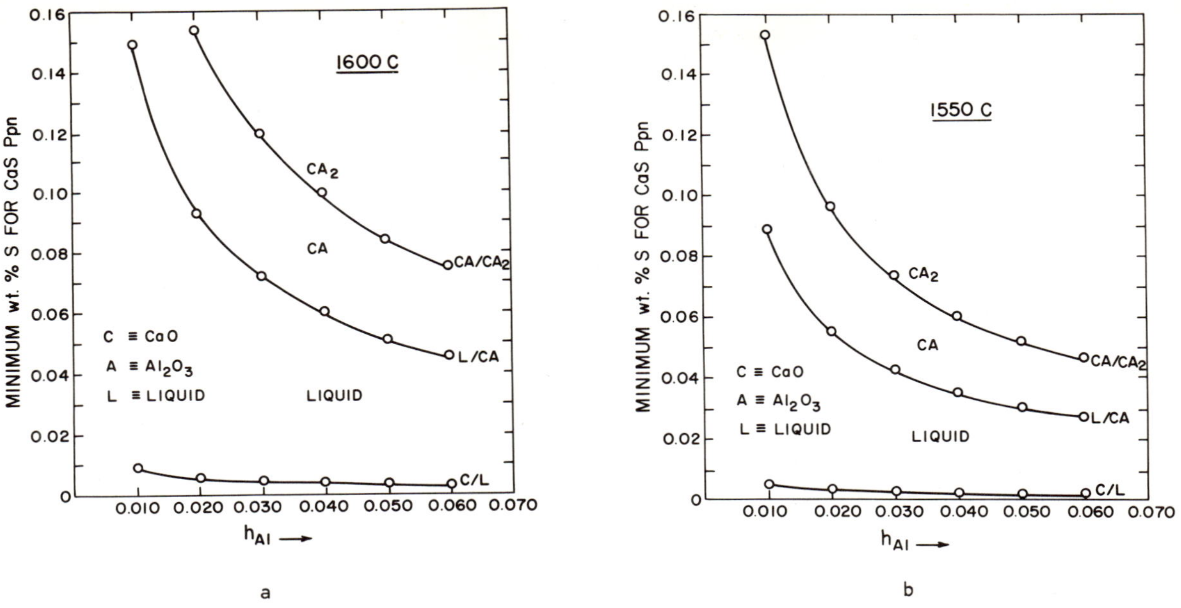

Figure 5. – The minimum [w/oS] required for the coprecipitation of calcium sulphide and two coexisting aluminate phases as a function of the Henrian activity of aluminium, h_{Al}, a) 1600°C and b) 1550°C.
C = CaO, A = Al_2O_3 and L = Liquid

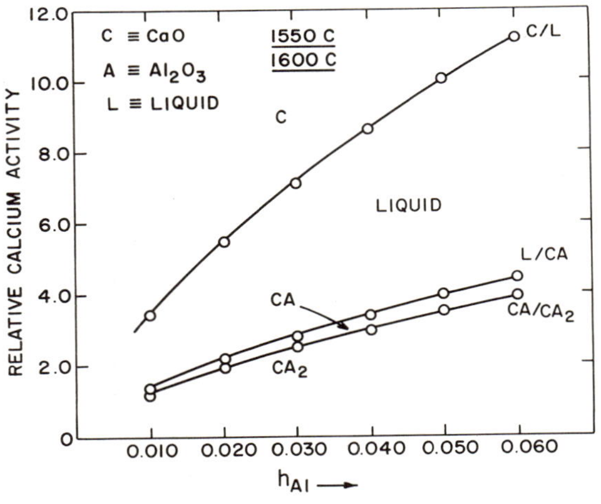

Figure 6. – A relative calcium activity in equilibrium with two aluminate phases as a function of the Henrian activity of aluminium, h_{Al}.
C = CaO, A = Al_2O_3 and L = Liquid

Sulfide capacities of CaO-FeO-SiO_2 slags

B CHEN, R G REDDY and M BLANDER

B Chen is a Graduate Student in the Department of Chemical and Metallurgical Engineering, University of Nevada-Reno, Reno, Nevada; Dr Reddy is an Associate Professor in the Department of Chemical and Metallurgical Engineering, University of Nevada-Reno; Dr Blander is a Senior Scientist and Group Leader in the Chemical Technology Division/Materials Science Program at the Argonne National Laboratory, Argonne, Illinois.

SYNOPSIS

Reliable predictions of the sulfide capacities of binary silicate slags can be made *a priori* using a fundamental solution model based on polymer theory. Predictions of sulfide capacities for the ternary CaO-FeO-SiO_2 system were calculated using the Flood-Grjotheim interpolation equation and are in excellent agreement with measurements.

INTRODUCTION

Over the years, a number of empirical correlations have been proposed between sulfide capacity and basicity of slags. Recently, we showed that sulfide capacities of some binary silicate slags could not be predicted *a priori*, based on the concept of optical basicity, and we presented a new and simple fundamental solution model for making such predictions, based on polymer theory.[1,2,3] In this paper we discuss predictions of the sulfide capacities of CaO-FeO-SiO_2 slags at 1773 K, based on our model.

CALCULATIONS

The sulfide equilibrium reaction for the MO-SiO_2 system can be written as

$$MO_{(\ell)} + 1/2S_{2(g)} = MS_{(\ell)} + 1/2O_{2(g)} \qquad (1)$$

From equation (1) we deduced a relation for the sulfide capacity, C_S, which was defined by Fincham and Richardson[4]

$$C_S = K_M a_{MO} \frac{(wt\%S)}{a_{MS}} \qquad (2)$$

where K_M is the equilibrium constant for reaction (1).

Values of C_S for binary MO-SiO_2 systems were calculated *a priori* using equation (2) with known input data on K_M and a_{MO} together with a method for deducing the activities of MS described in detail.[1,2,3]

For basic melts in the composition range $O<X_{SiO_2}<0.33$ in the MO-SiO_2 system, the derived equation for C_S is

$$C_S = 100\ W_S\ K_M\ a_{MO} \left(\frac{1-2X_{SiO_2}}{\overline{W}}\right) \qquad (3)$$

where $\overline{W}$ is the average molecular weight of the solution, and W_S is the atomic weight of sulfur. For the composition range $0.33<X_{SiO_2}<0.5$ in the MO-SiO_2 system, the equation below for C_S was derived from polymer solution theory[5]

$$C_S = 100\ W_S\ K_M\ a_{MO} \frac{X_{SiO_2}}{\overline{W}} \left(\frac{\Phi_S}{a_{MS}}\right) \qquad (4)$$

were Φ_S is the volume fraction of S^{2-} sites and, in dilute solution, is approximated by $\Phi_S = N_S/N_{Si}$, N_S is the number of sulfide sites, and N_{Si} is the number of silicon sites. The ratio between Φ_S and a_{MS} was obtained from the consideration of Flory's approximation[5] for polymer-monomer mixtures in silicate melts and lies between 1 for only monomeric silica species and 1/e for infinite polymer chains.

In a ternary MO-NO-SiO_2 system, one can apply the Flood-Grjotheim approximation,[6] to deduce the sulfide capacities at constant silica mole fractions

$$\ln C_S = N_M \ln C_{S,M} + N_N \ln C_{S,N} \qquad (5)$$

where $C_{S,M}$ and $C_{S,N}$ are sulfide capacities in the MO-SiO_2 and NO-SiO_2 binary systems, respectively, and C_S is the sulfide capacity of the ternary MO-NO-SiO_2 system.

From calculated values of $C_{S,CaO}$ and $C_{S,FeO}$ (Ref. 1), C_S values were calculated for the CaO-FeO-SiO_2 system using Eq. (7) up to $X_{SiO_2} = 0.5$. The calculated values of C_S and experimental data of Bronson and St. Pierre[7] are compared in Table I and Figure 1. As can be seen in Fig. 1, the experimental data fall between the two limits of m = 1 (where m is the average chain length of the polymer so that only monomers in the slag are considered in a neutral solution at $X_{SiO_2} = 0.33$) and m>>1 (complete polymerization of the silicate melt at $X_{SiO_2} = 0.5$). At 1773 K, the sulfide capacity increases with an increase in the FeO content in the melt at constant $X_{SiO_2} = 0.4525$. Linear interpolation of $\ln C_S$ between 0.33 and 0.50 mole fraction SiO_2 leads to the dashed line in Fig. 1 for X_{SiO_2}

= 0.4525 in the CaO-FeO-SiO_2 system at 1773 K. Since C_S is not a very strong function of m, the dashed line is a reasonable approximation. The results exhibit excellent agreement between the calculated and experimental sulfide capacities with an average percentage deviation of 16.4%. This deviation corresponds to uncertainties in a free energy of about 600 cal, which is less than the uncertainties in the input data. The plotted values are given in Table I where the largest difference between experimental and calculated values of C_S can be seen to be 0.98×10^{-4} and the largest percentage deviation from experiment is −33%, which corresponds to an error in the thermodynamic input data of 1400 calories. This is smaller than the combined uncertainties in the input data for any one point. It should also be noted that differences between measured values of C_S at the same composition are as large as 0.61×10^{-4} so that differences between calculated and measured values may not be significant. Thus, it appears that our model makes predictions of C_S *a priori*, which are about as precise as is permitted by the data needed for the calculation. In Fig. 2 we exhibit our predictions of C_S as a function of composition in the CaO-FeO-SiO_2 ternary at 1773 K.

CONCLUSIONS

The method we propose for calculating C_S in CaO-FeO-SiO_2 slags *a priori* is in good agreement with the available experimental data and it is likely to be valid in higher order multicomponent slags where one can use the general form of the Flood-Grjotheim equations. Further work on the calculation of C_S in other multicomponent slag systems is in progress.

Although it applied accurately in this case, the Flood-Grjotheim approximation does not apply universally with the same accuracy for certain kinds of solution interactions.[6,8] We plan to improve both the polymer models for silicates and the approximation for multicomponent solutions in order to deduce a more general and accurate description.

ACKNOWLEDGMENTS

This work was supported by the U.S. Department of Energy, Division of Materials Science, Office of Basic Energy Sciences, under Contract W-31-109-ENG-38 with the Argonne National Laboratory. One of the authors (B.C.) expresses his appreciation to the Department of Chemical and Metallurgical Engineering and Program Development Funds, University of Nevada-Reno for financial support.

REFERENCES

1. R. G. Reddy and M. Blander, Metall. Trans. **18B**, 591-596 (1987).
2. R. G. Reddy and M. Blander, Metall. Trans., in press.
3. R. G. Reddy, M. Blander, and B. Chen, Proc. of the Joint Int. Symp. on Molten Salts, 172nd Electrochem. Soc. Meeting, Honolulu, HI, October 18-23, 1987, Vol. 87-7, pp. 156-164 (1987).
4. C. J. B. Fincham and F. D. Richardson, Proc. Royal Soc., London, **223A**, 40-62 (1964).
5. J. H. Hildebrand and R. L. Scott, The Solubility of Non-Electrolytes, Third Edition, Reinhold Publishing Corporation, New York, NY, pp. 347-351 (1950).
6. H. Flood and K. Grjotheim, JISI **171**, 64 (1952).
7. A. Bronson and G. R. St. Pierre, Metall. Trans. **12B**, 729-731 (1981).
8. M.-L. Saboungi and M. Blander, Can. Metall. Quarterly **20**(1), 31-36 (1981).

Table I. Sulfide Capacities of CaO-FeO-SiO_2 System at 1773 K*

X_{SiO_2}	X_{FeO}	X_{CaO}	$C_S \times 10^4$ Model	Expt.[7]
0.453	0.040	0.507	1.17	1.45
0.453	0.078	0.469	1.55	1.59
0.453	0.078	0.469	1.55	1.70
0.453	0.115	0.432	2.04	3.02
0.453	0.115	0.432	2.04	2.69
0.453	0.150	0.397	2.66	2.14
0.453	0.150	0.397	2.66	2.75

*Sulfide capacities were experimentally determined at 1776 K for $X_{SiO_2} = 0.4525$.

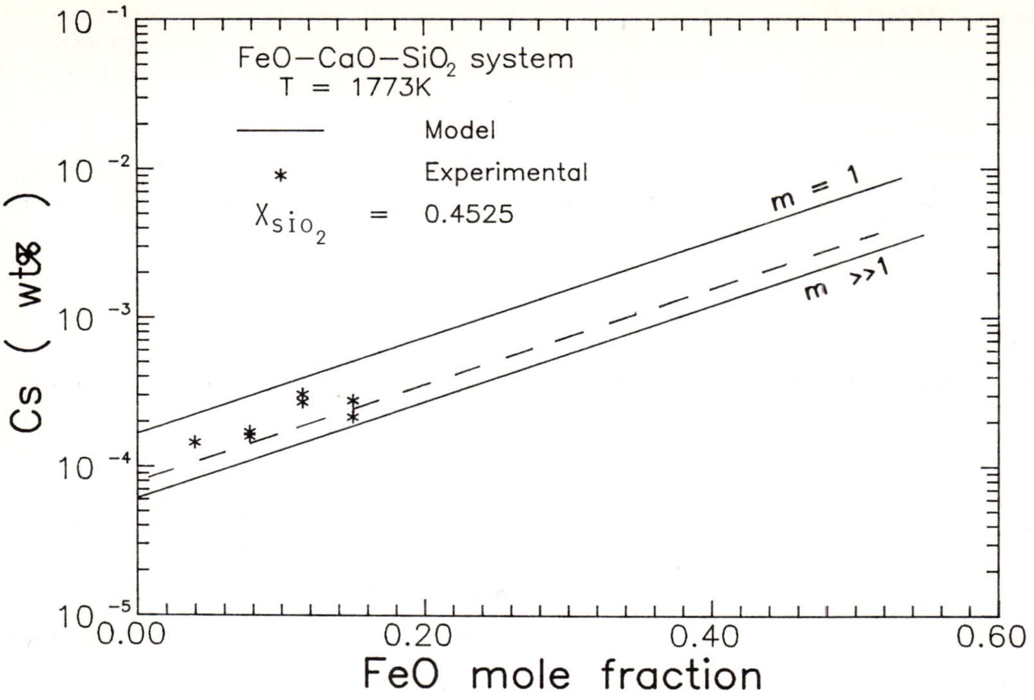

Fig. 1. Comparison of Calculated (Dashed Line) and Measured (Asterisks) Values of (C_S) in the CaO-FeO-SiO2 System at 1773 K for $X_{SiO_2} = 0.4525$.

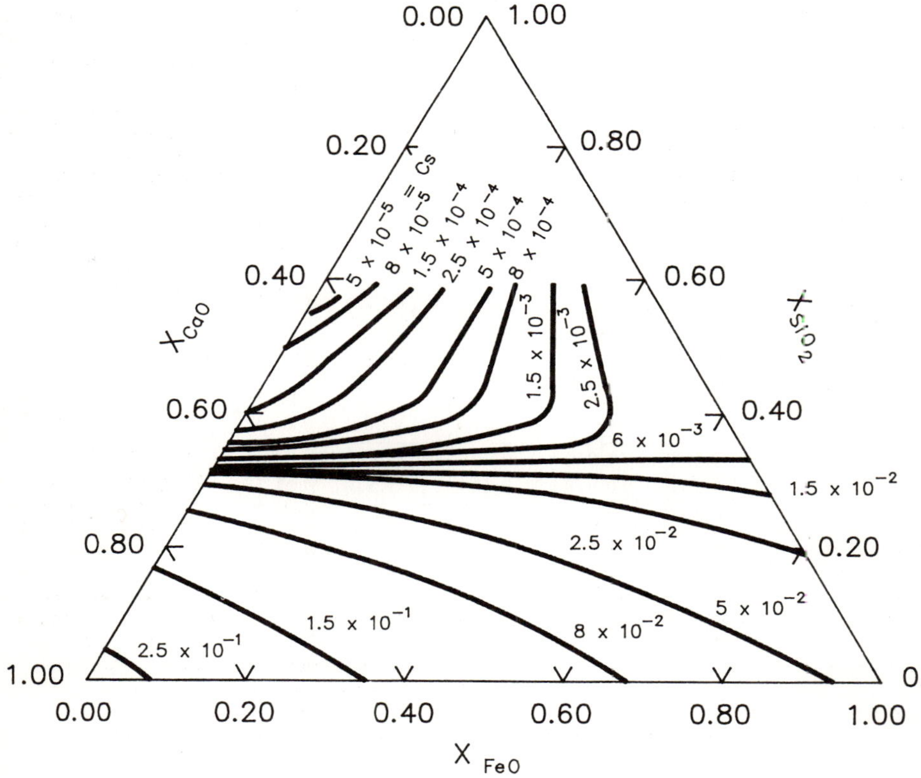

Fig. 2. Calculated Values of C_S in the Ternary CaO-FeO-SiO_2 System.

Application of models for slag formation in basic oxygen steelmaking

R BOOM, B DEO, W VAN DER KNOOP, F MENSONIDES and G VAN UNEN

The authors are in the Research and Works Laboratories, Hoogovens IJmuiden, The Netherlands. B Deo is Visiting Professor from IIT, Kanpur, India.

SYNOPSIS

Evolution of slag composition during the blow in 315 t converters has been studied with the aid of special sublance sensors developed at hoogovens. Effect of addition of soft burnt dolomite on liquidus temperatures of slag during the blow and on phosphorus distribution at end point was investigated with the help of models. Dissolution rate of soft burnt dolomite was estimated from mass balance calculations. Iron oxide content of the slag during the blow was found to be influenced by slag fluidity.

INTRODUCTION

Slag formation research at Hoogovens has a longstanding tradition, whereby theoretical modelling, laboratory experiments and operational results on large-scale oxygen steelmaking are combined [1-4]. In the past decade, steelmaking at Hoogovens has been in a state of transition from full ingot casting to full continuous casting of both slabs and billets. As a consequence ladle treatment, including RH-OB, has been introduced. Many changes have occurred in the operating practices as well as in quality requirements. Silicon, manganese and phosphorus in hot metal decreased respectively from 0.6, 0.7 and 0.11% in 1978 to 0.45, 0.50 and 0.055% in 1988. All compositions are reported in weight per cent. Despite an increase in tapping temperatures from 1600 to 1655°C, problems with phosphorus control were overcome by measures such as improving burnt lime quality and the introduction of bottom stirring. Lime consumption has even decreased from 68 kg/t in 1976 to 44 kg/t in 1978 and to 42 kg/t of steel at present. Sublance and associated mathematical process control models have helped to cope with quality demands.

It is well recognized that an optimal slag formation path helps in diminishing the distance between equilibrium and practice or, in other words, improves the predictabilities of slag and metal compositions. In this work systematic investigations have been made to study the evolution of slag composition under the present steelmaking conditions at Hoogovens. Models based on ionic and molecular theories have been applied to analyse the results.

EXPERIMENTAL

Slag samples, maximum four per heat, were collected during the progress of the blow in 315 t converters by means of a sublance system. Special sensors were developed in cooperation with Electro-Nite. A steel chain was attached to the outer part of a standard sublance sensor and lowered into the bath.

Liquid slag solidified on the cold chain and came with the sensor.

Two series of experiments were conducted in order to study the evolution of slag composition during the blow. The first involved the use of calcined dolomite and lime whereas in the second series only lime was used as a flux. In both series a constant blowing regime (lance height, flow rate, etc.) was used. The experiments conducted in 1978 were used as a reference. In these experiments, however, slag samples were taken by interrupting the blow and lowering a sample pot through the lance hole. The interruption could have affected (to an extent not known) the evolution of the slag composition during the blow.

RESULTS AND DISCUSSION

The evolution of the slag composition during the blow is plotted in Fig. 1. It shows that, although the trends of slag paths are similar for all experiments, the heats containing soft-burnt dolomite had, on the average, higher silica in slag in the early part of the blow, up to about 2/5 of blowing time.

For heats with dolomite a lower FeO fraction in slag was found in general and especially in the later part, i.e. during 2/5 to 4/5 of blowing time. This behaviour can be analysed in terms of

(1) evolution of the iron content of the slag,
(2) liquidus temperatures,
(3) relative dissolution rates of lime and dolomite,
(4) the rates of slag metal reactions and their deviation from equilibrium at the end point.

1) Evolution of the total iron content of slag. It is possible to estimate the weight of slag from the chemical composition of slag at any stage of the blow by setting up a mass balance. Considering the turbulent and dynamic nature of the oxygen steelmaking process and the errors in sampling, some uncertainty in the calculated slag weight is to be expected.

Fig. 2 shows the estimated weight of iron in slag. In spite of the uncertainty there is a clear pattern, similar to that in Fig. 1, for the change in total iron content in slag with time for all the three series of experiments and especially for the heats in which dolomite was used.

During the trials it was observed that taking slag samples from the heats containing dolomite was easier compared with heats containing only lime. The slag without dolomite tended to become dry after 3/5 of the blow, implying that liquidus temperatures were higher in this case.

The implications of total iron content in slag will be discussed later.

2) Liquidus temperatures during the blow. It was not possible to measure the temperature of slag and metal during the blow. From the chemical composition of the slag samples collected at different stages of the blow the liquidus temperatures can be theoretically calculated by applying the Kapoor and Frohberg model [5]. The calculated liquidus temperatures are plotted in Fig. 3.

By means of using reported phase diagrams and G-values the model was adapted to Hoogovens' conditions by incorporating TiO_2 present in the steelmaking slags. The effect of P_2O_5 could not be incorporated; its concentration in the slag was less then 3%.

The calculated average liquidus temperatures were lower by approximately 100^oC in the heats containing dolomite. It may be deduced that, because of the lower liquidus temperatures, the slag was more fluid and thus dissolved a larger proportion of SiO_2 in the early part of the blow as evident in Fig. 1. Furthermore, in a relatively fluid slag, the contribution of the slag-metal reaction

$$(FeO) + [C] = [Fe] + CO_g \qquad (1)$$

is also expected to be high, leading to a low FeO content of slag in the period of 2/5 to 4/5 of the blow. Lower iron contents of the slag in heats with dolomite additions do indicate that metal and slag compositions approached closer equilibrium.

3) Dissolution rates of lime and dolomite. For heats containing dolomite a dimensionless ratio R, defined as weight of MgO in slag/weights of (MgO + CaO) in slag, can be calculated. In the case where both dolomite and lime dissolve at equal rates, the value of this ratio would be 0.18. The observed values were found to be lower than the calculated values during the major part of the blow, indicating that the dissolution rate of dolomite itself was slower than that of lime. This may be attributed partly to the differences in the physical and thermodynamical characteristics of dolomite versus lime, and partly to the driving force for dissolution of MgO in slag. The slag appeared to be nearly saturated in MgO (7%) just about midway through the blow. It indicates that, under the conditions existing at Hoogovens, the amount of dolomite added could have been less so as to optimize the specific advantage of dolomite for fluid slag formation in the period of 1/5 to 4/5 of the blow.

It is difficult to comment, on the basis of the experiments described in the present work, on the quantitative effects of the foaming behaviour and the variation of total iron in slag (Fig. 2) on the relative dissolution rates of lime and dolomite. With lime only the slags tended to be dry at 3/5 of the blow. In dry slags dissolution rates will be slow.

Due to the kinetics of reactions and the associated heat effects, a larger contribution of the endothermic reaction (1) will slow down the temperature rise of slag (see flat region of liquidus temperature in Fig. 3) and consequently retard the dissolution rates of both lime and dolomite. A larger contribution of reaction 91) could have been verified by exhaust gas analysis. However, during the time the experiments were conducted the off-gas analysis data were not available due to operational difficulties.

For the solid (dry) part of the slag it may be commented that MgO and FeO are known to form a larger range of solid solubility than CaO and FeO.

4) Prediction of phosphorus at end point. One of the principal interests in end point predictions is the phosphorus content of the bath.

Figure 4 shows the plots of the predicted phosphorus distribution ratios (P)/[P] against experimentally observed values using (a) Healy's [6] equation, (b) the modified equation used at Hoogovens

$$\log (P)/[P] = 10800/T(K) + 0.026 * (CaO) + 1.62 * \log (Fe) + const. \qquad (2)$$

and (c) the modified Flood and Grjotheim (F.G.) theory [7]. It is clear from Fig. 4 that equ. (2) leads to a better prediction ($o = \pm 15$) than the other two relationships. Predictions from the original Healy's equation are off by a factor 2 to 3 under the conditions existing at Hoogovens. It may be possible to obtain better results by tuning the F.G.-model to Hoogovens' conditions.

In general the end point prediction of (P)/[P] for heats containing dolomite are closer to the actually observed values than for heats in which only lime was used ($0 = \pm 17$). It indicates that metal and slag approached a steady state, closer to equilibrium, in the presence of dolomite.

CONCLUSIONS

(1) Under the present conditions at Hoogovens, the path of slag during the blow has changed drastically since 1978 because of many changes.

(2) In the experimental heats with dolomite the following special features were observed:

- Dissolution rate of dolomite was lower than that of lime.

- Fluid slag with lower liquidus temperatures during the blow was formed enabling higher dissolved silica amounts in the early part of the blow.

- Total iron content of slag was lower in the period 2/5 to 4/5 of the blow indicating closer approach to equilibrium.

- The phosphorus distribution ratio at end point was more predictable than without dolomite.

REFERENCES

1. A.I. van Hoorn, J.T. van Konijnenburg and P.J. Kreijger, pp.2.1-2.6 in Proc. McMaster Symposium on Iron and Steelmaking no.7, May 1976, Hamilton Canada.

2. A. van der Velden and P.J. Kreijger, Iron and Steel International 52 (8) 1979, pp. 237-243.

3. P.J. Kreijger and R. Boom, Canadian Met. Quart. 21 (4) 1982, pp. 339-345.

4. R. Boom, R.R. Beisser and W. van der Knoop, pp. 1041-1060 in 2nd Int. Symposium on Metallurgical Slags and Fluxes, november 1984, Lake Tahoe, Nevada.

5. H. Gaye and J. Welfringer, pp. 357-375 in Second Int. Symposium on Metallurgical Slags and Fluxes, november 1984, Lake Tahoe, Nevada.

6. G.W. Healy, J. Iron and Steel Inst. 208 (1970), pp. 664-668.

7. H. Flood and K. Grjotheim, J. Iron and Steel Inst. 171, (1952), pp. 64-70.

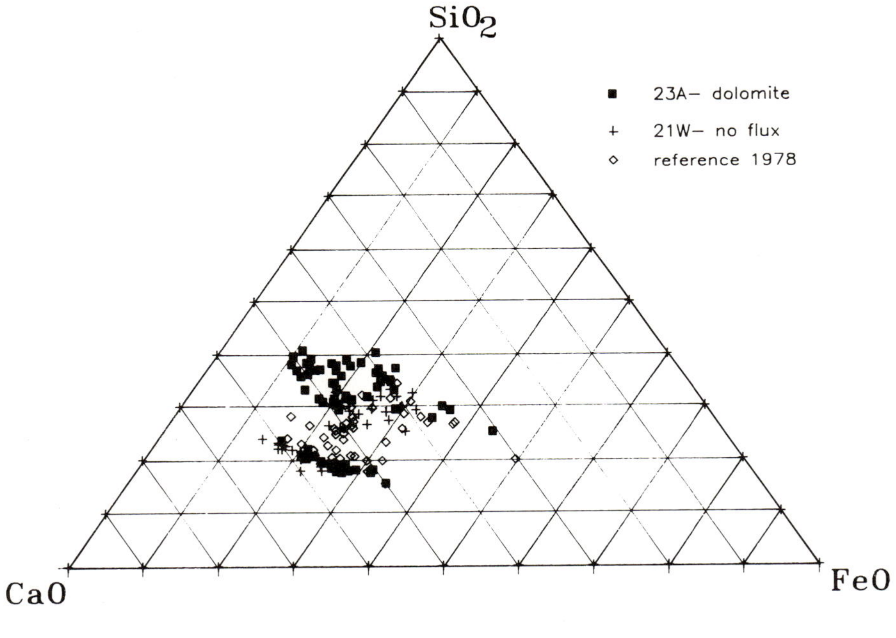

Fig. 1
Evolution of slag composition using a ternary $CaO-SiO_2-FeO$ system.

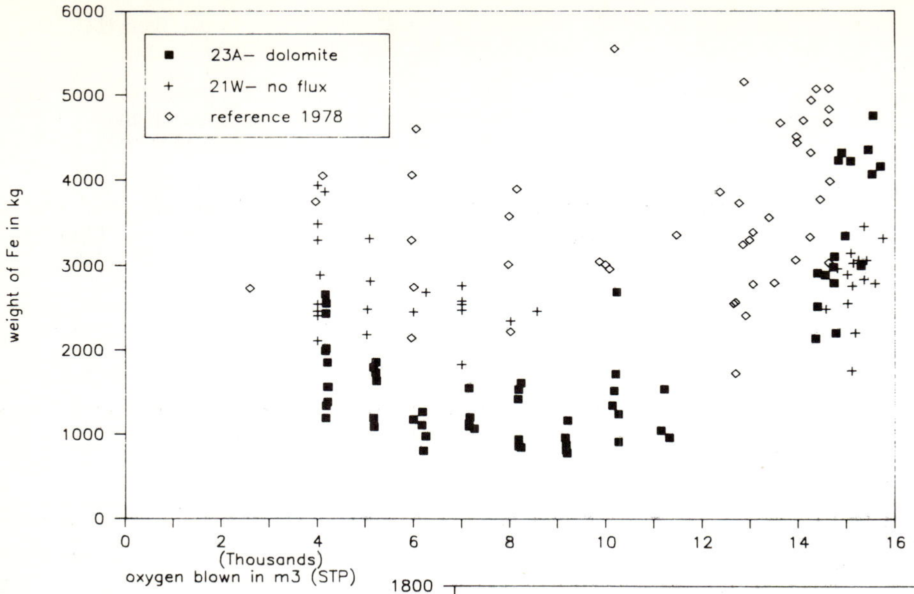

Fig. 2
Evolution of iron weight in the slag

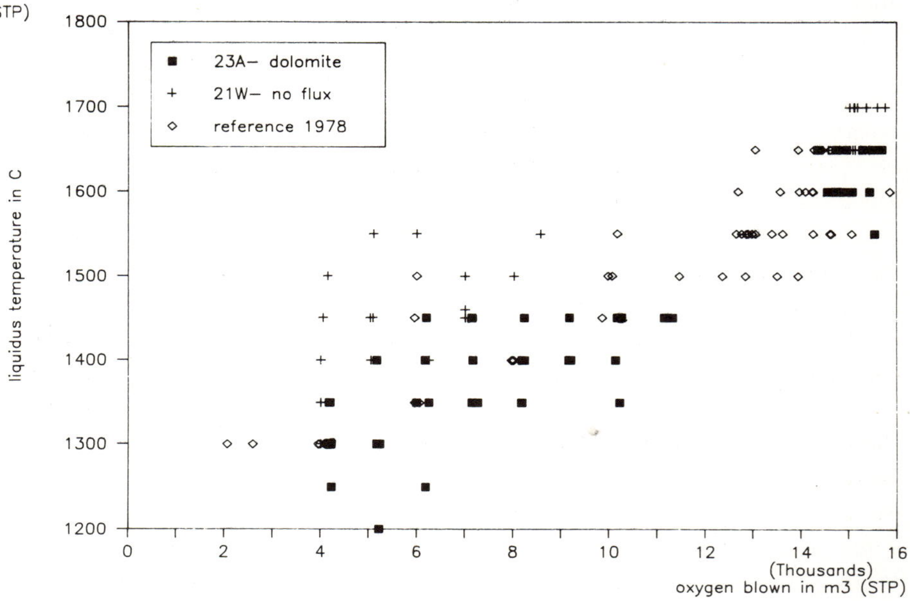

Fig. 3
Evolution of slag liquidus temperature

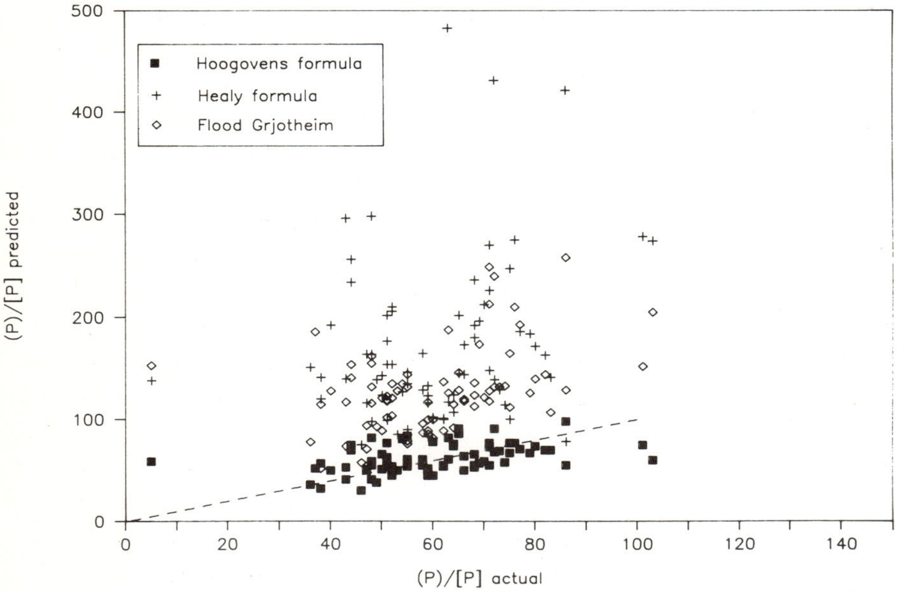

Fig. 4
Prediction of P-distribution

Some applications of slag chemistry to metal refining

H B BELL

The author is in the Division of Metallurgy and Engineering Materials, University of Strathclyde, Glasgow, UK.

In order to keep this paper within a reasonable length and to look at applications in a variety of parameters it is confined to iron and steel and phosphorus and sulphur slag-metal partitions.

During the process of iron and steel manufacture the metal is subjected to a range of oxygen potentials and these have a marked effect on the slag-metal partitions of all the impurities normally present in iron and steel, i.e. carbon, manganese, silicon, sulphur and phosphorus. There have been extensive studies concerning the latter two elements and the most useful slag parameters which have been developed are the sulphide and phosphate capacities. In this paper these will be defined as

$$C_S = (\%S)\sqrt{\frac{pO_2}{p_{S_2}}}$$

$$C_{pO_4} = \frac{(\%PO_4^{3-})}{p_{P_2}^{1/2}\, p_{O_2}^{5/4}}$$

Other expressions have been used but all are interrelated. The advantage of these parameters is that they can be related to thermodynamic data for metal melts to give slag-metal partitions in terms of weight concentrations and these are of more practical interest to metal producers.

Both expressions contain a term in oxygen pressure and it is worth looking first at the range of oxygen pressures encountered in iron and steelmaking. Fig. 1 gives some examples of the oxygen pressure for conditions during the process. In the iron blast furnace we have strongly reducing conditions in that there is iron with a high carbon content and a very reducing gas in the hearth of the furnace, if the metal was carbon saturated and the carbon monoxide pressure was one atmosphere the oxygen pressure would be in the range 10^{-16} to 10^{-17} atmos depending on the temperature. This is often the case in laboratory melts made in graphite crucibles. The conditions in the blast furnace hearth are more correctly described by the line representing a CO partial pressure of 2 atmos and a carbon activity of 0.8, this represents a higher oxygen potential than carbon saturation.

Conditions in steelmaking are much more oxidising and we have a slag containing appreciable concentrations of iron oxides. The data for an iron and ferrous oxide activity of unity show oxygen potentials in the range -8 to -11.5 depending on temperature. Data for lower activities of FeO are shown in Fig. 1. The probable oxygen pressure during Basic Oxygen Steelmaking is greater still and is shown as a range 10^{-6} to 10^{-7} atmos in Fig. 1.

In ladle treatment of iron and steel there can be a range of oxygen pressures depending on the conditions. In the ladle treatment of hot metal the oxygen pressure will be low. Watanabe and Umezawa have studied the oxygen potentials during ladle dephosphorisation of hot metal by Na_2CO_3 and estimate it to be -13.5 at 1450°C falling to -14.3 at 1300°C. This is a factor of 100 times greater than in the blast furnace hearth. After deoxidation steel will have a lower oxygen potential and the ranges for different concentrations of silicon and aluminium are shown in Fig. 1. Vacuum treatment will also lower the oxygen potential of the steel. The effect of lowering the temperature under any given conditions is to lower the oxygen potential.

If the expressions for sulphide and phosphate capacity are examined then two parameters related to the partitions of phosphorus and sulphur can be defined

$$\frac{(\%S)}{\sqrt{p_{S_2}}} = \frac{C_S}{\sqrt{p_{O_2}}}$$

$$\frac{(\%PO_4^{3-})}{\sqrt{p_{P_2}}} = C_{PO_4^{3-}}\, p_{O_2}^{5/4}$$

These show that oxygen potential has opposite effects on sulphur partition and phosphorus partition, i.e. in the case of phosphorus high oxygen potentials favour dephosphorisation while with sulphur the opposite is the case.

In order to quantify the partition coefficient other data are required. The sulphur and phosphorus pressures must be related to the concentrations in the metal, i.e. data on the free energy of solution of phosphorus and sulphur in iron are required together with the effects of other elements on the activities of these.

$$\tfrac{1}{2} P_2 \rightarrow \underline{P}_{\%} \quad \Delta G_S = -29\ 200 - 4.6T \text{ cal/g atom}$$

$$\tfrac{1}{2} S_2 \rightarrow \underline{S}_{\%} \quad \Delta G_S = -32\ 280 + 5.6T \text{ cal/g atom}$$

The effect of alloying elements is represented by an interaction coefficient, in this paper only carbon will be considered.

$$e_P^C = 0.12$$

$$e_S^C = 0.11$$

For high carbon contents plots of f_S against carbon concentration are more satisfactory. The large positive effect of carbon on the activity coefficients of phosphorus and sulphur in solution in iron are important in assessing slag metal partitions. The concentration of carbon is of course also linked to the oxygen potential, high carbons normally having a low oxygen potential whereas low carbons can have a range of oxygen potentials depending on the conditions. Fig. 2 shows the effect of carbon content in the range one to five percent on phosphorus and sulphur partitions for two slags; the effect of oxygen pressure is predominant.

In the subsequent discussion data for a number of slags is used in conjunction with different conditions to illustrate the important factors.

Considering first the effect of temperature on C_S and $C_{PO_4^{3-}}$. All the available data for C_S show that it increases with increase in temperature for a given slag composition. This increase is of the order of 1.25 times for each 50°C increase in temperature. In the case of $C_{PO^{3-}}$ the opposite is the case and this can be seen from the equilibrium

$$3CaO + P_2 + 2.5O_2 = 3CaO.P_2O_5$$

$$\Delta G^\circ = -553\,000 + 133T \text{ cal/g mol.}$$

where log k at 1300°C is 47.9 while at 1600°C it is 35.6; this represents a large change in the reaction with temperature. Published data on the effect of temperature on $C_{PO^{3-}}$ bear out this effect of temperature. The magnitude of the effect varies with the slag composition and appears to be greater for lime based slags. For a $CaO\text{-}P_2O_5$ slag the change is from 10^{24} at 1300°C to 10^{18} at 1600°C while for a $Na_2O\text{-}SiO_2\text{-}P_2O_5$ slag it is 10^{29} at 1200°C and 10^{25} at 1400°C.

The result of this temperature effect is shown in Fig. 3 for an $Na_2O\text{-}SiO_2$ slag in contact with hot metal. The phosphorus partition increases by a factor of 16 when the temperature decreases from 1400°C to 1200°C while the sulphur partition decreases by a factor of 6 over the same range.

There has been considerable discussion of optical basicity as a method of correlating sulphide capacities and this has also been applied on a more limited scale to phosphate capacities.

It is useful to put these correlations into perspective with the slag metal sulphur partitions under various conditions. Fig. 4 shows this for a range of oxygen potentials using the correlation of Sommerville. For a given optical basicity the range of sulphur partitions is very wide. For blast furnace conditions there is a large partition coefficient while for the more oxidising conditions it is low. The p_{O_2} of

10^{-13} atoms is that of Watanabe et al for ladle treatment of hot metal. The correlation of phosphate capacity has not been studied so extensively but some correlations are available for Na_2O-SiO_2 slags and these have been used to compile Fig. 5 for several conditions of oxygen pressure. It is worth noting the influence of temperature, i.e. at an oxygen potential corresponding to ladle treatment of hot metal at 1300°C the phosphorus partition is greater than in the much more oxidising conditions at 1610°C.

There are extensive data for sulphide and phosphate capacities of lime-alumina melts and it is interesting to use these data to examine sulphur and phosphorus partition. Fig. 6 takes as an example an oxygen pressure equivalent to an activity of FeO of 0.1 and shows the effect of increasing the concentration of calcium oxide in the slag. Under these conditions the phosphorus partition is much higher than the sulphur partition. In Fig. 7 a CaO-Al_2O_3 slag of N_{CaO} = 0.63 is used to show the effect of oxygen potential on phosphorus and sulphur partition using the activity of FeO as a measure of the potential. This shows the sensitivity of the partitions to oxygen potential. If a slag to metal ratio of 0.02 is considered with an initial phosphorus of 0.05% and sulphur of 0.03% then the effect of oxygen potential on the equilibrium sulphur and phosphorus contents is shown in Fig. 8.

Much lower oxygen pressures are encountered in hot metal treatment and soda based slags are being used for dephosphorisation and desulphurisation; the effect of oxygen potential under these conditions is illustrated in Fig. 9. This shows that these slags are capable of high degrees of desulphurisation and dephosphorisation under very reducing conditions.

In steelmaking conditions the oxygen content of the steel is a measure of the oxygen potential of the process. Fig. 10 considers a CaO-Al_2O_3 slag of N_{CaO} = 0.6 over the temperature range of 1500°C to 1650°C for two different oxygen contents of 0.001% and 0.01%. The effect of the higher oxygen content is to increase phosphorus partition and decrease the sulphur partition. Fig. 10 again illustrates the divergent effect of temperature on phosphorus and sulphur partition.

A technique widely used to measure the oxygen content of metal melts is the use of solid electrolytes. Fig. 11 takes as an example a CaO-ZrO_2 solid electrolyte with a reference oxygen potential of chromium with Cr_2O_3. Two slags are considered, one based on CaO-Al_2O_3 and the other on CaO-CaF_2, these being typical of those used in practice. Both show the typical effects of oxygen potential on sulphur and phosphorus partition. It is interesting to note that CaO-CaF_2 has a much higher desulphurising capacity than CaO-Al_2O_3 whereas the increase in dephosphorising capacity is not so marked.

After deoxidation with aluminium the oxygen potential of the steel can be related to the aluminium content of the steel through the solubility product $[\%Al]^2[\%O]^3$. Fig. 12 shows the effect of aluminium content for the equilibrium phosphorus and sulphur partitions for two slags: one in the CaO-CaF_2 system and the other in the CaO-Al_2O_3 system. Fig. 13 illustrates the effect of treating a steel containing 0.05%P and 0.02%S with a slag to metal ratio of 0.02 using a CaO-CaF_2 slag.

Figs. 11, 12 and 13 show that in order to attain low sulphur contents the metal must be deoxidised, whereas to attain low phosphorus contents it should be treated before deoxidation.

It is hoped that this paper has illustrated that in refining the slag chemistry is only one parameter to be considered in assessing the best conditions for treatment. Phosphorus and sulphur were chosen to illustrate the situation since in many ways they react in opposite ways to a change in conditions.

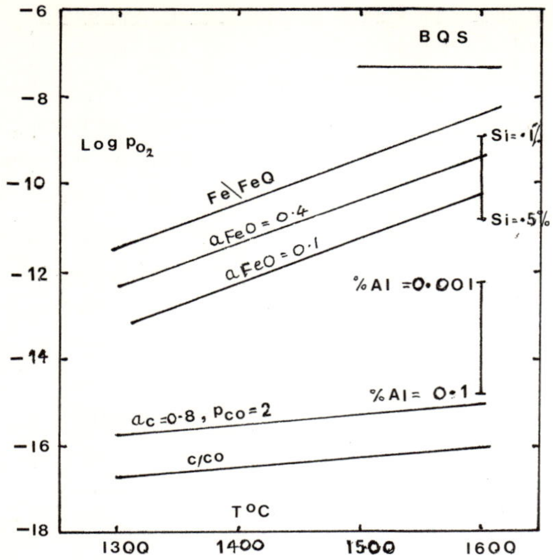

Fig.1 The effect of temperature on the oxygen potential for a range of conditions.

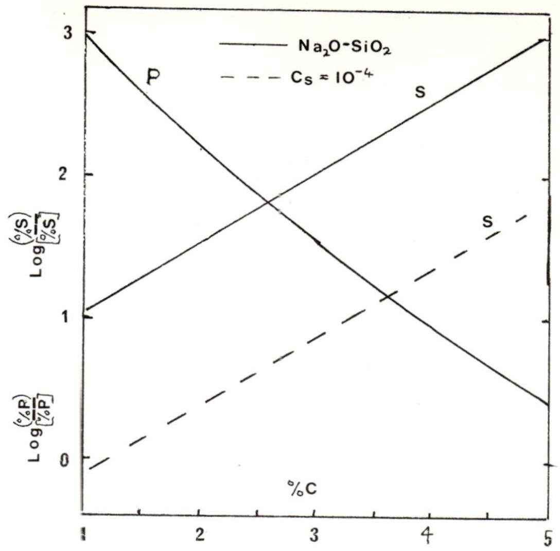

Fig.2 The effect of carbon content of iron on the phosphorus and sulphur partition coefficients for an Na_2O-SiO_2 slag. (Temp.1500°C, p_{CO} = 1, N_{Na_2O} = 0.46).

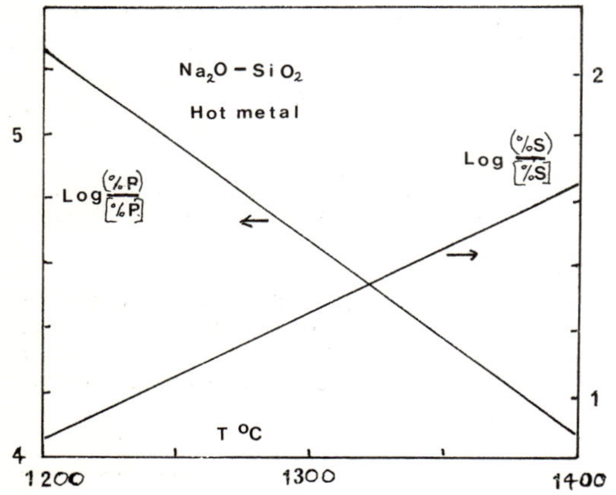

Fig.3 Effect of temperature on the sulphur and phosphorus partition ratios for hot metal treatment. (Na_2O-SiO_2 slag with N_{Na_2O} = 0.46 and log p_{O_2} from Watanabe and Umezawa).

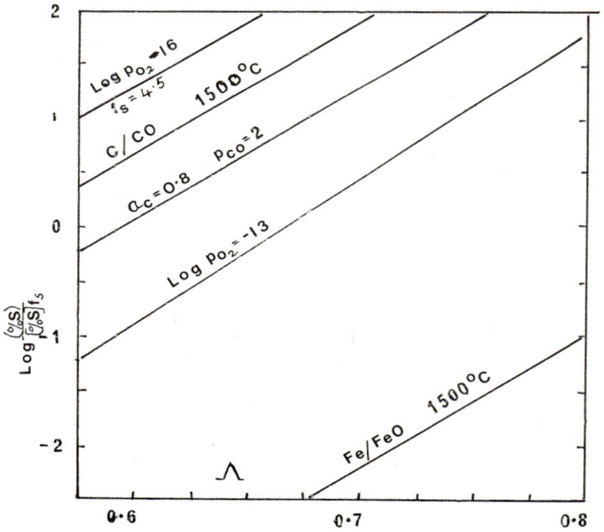

Fig.4 Correlation of cptical basicity with sulphur partiticn for a range of conditions.

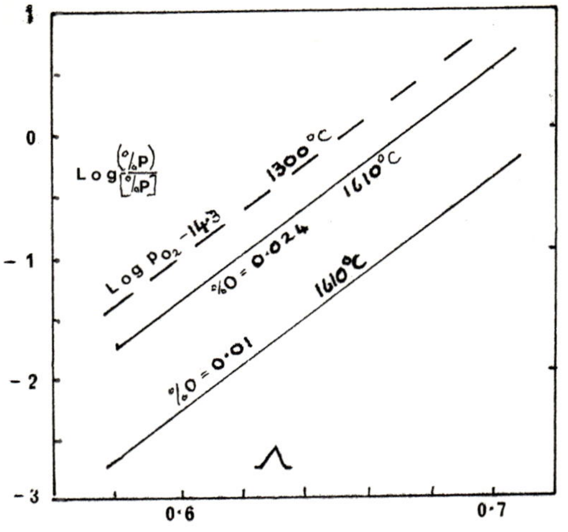

Fig.5 Correlation of optical basicity with phosphorus partition for Na_2O-SiO_2 slags for different condition.

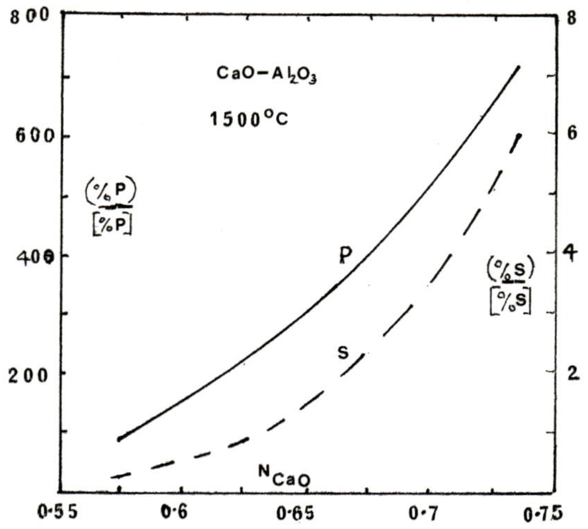

Fig.6 Effect of composition of CaO-Al_2O_3 melts on phosphorus and sulphur partition (Temp. = 1500°C, a_{FeO} = 0.1).

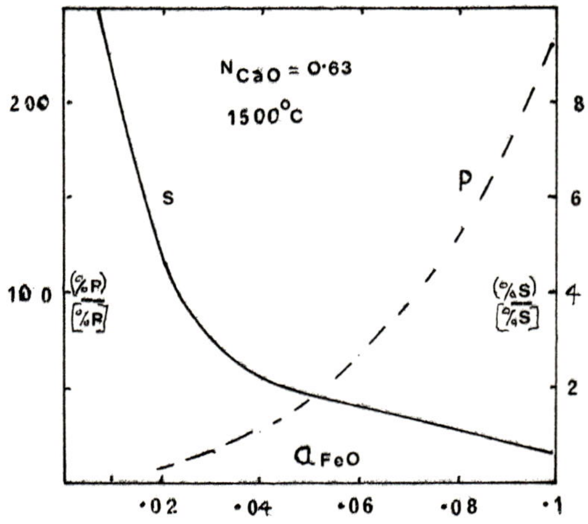

Fig.7 Effect of oxygen potential on the sulphur and phosphorus partition ratios for a CaO-Al_2O_3 melt.

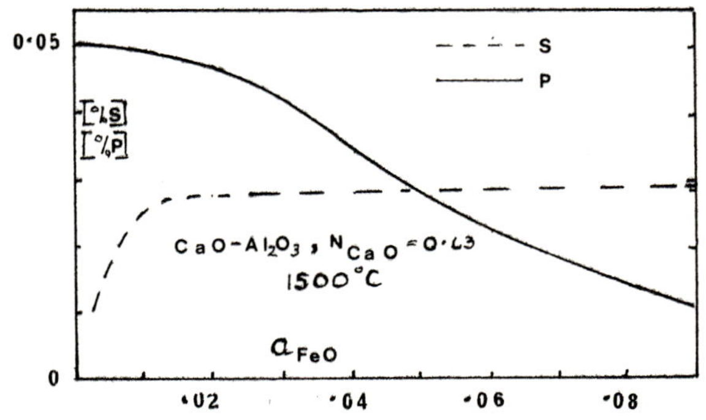

Fig.8 Effect of oxygen potential on the phosphorus and sulphur contents of liquid iron with an CaO-Al_2O_3 slag (Initial %P=0.05, %S=0.03, slag-metal ratio=0.02).

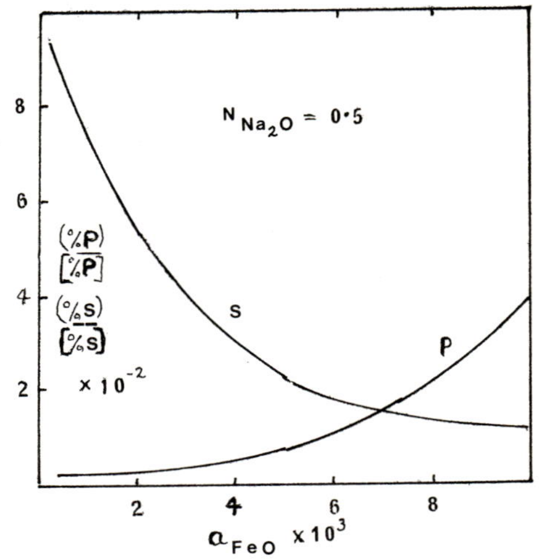

Fig.9 Effect of oxygen potential on phosphorus and sulphur partition ratios for hot metal treatment using Na_2O-SiO_2 slags at 1500°C.

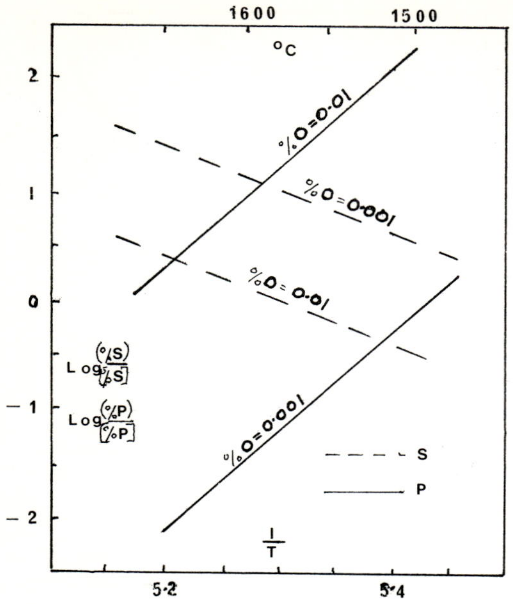

Fig.10 Effect of temperature and oxygen content of steel on the phosphorus and sulphur partition ratios with a $CaO-Al_2O_3$ slag. (N_{CaO} = 0.6).

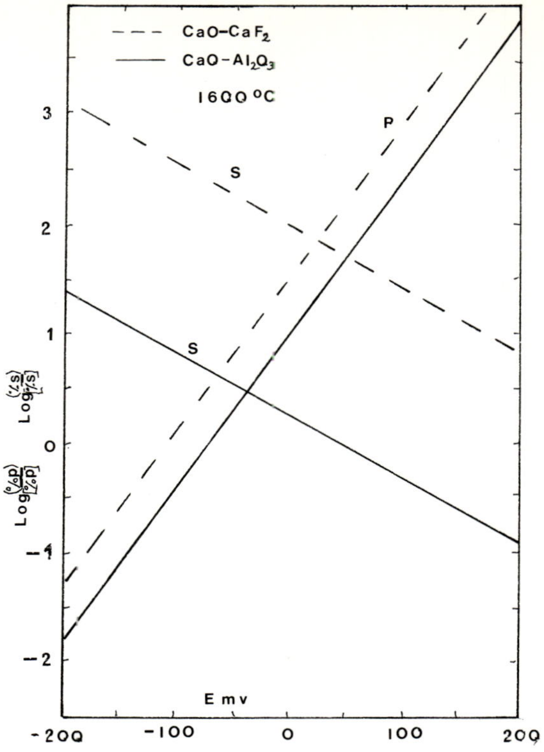

Fig.11 Effect of oxyger potential as measured by an emf cell on the phosphorus and sulphur partition ratios at 1600°C ($CaO-Al_2O_3$ slag, N_{CaO} = 0.63; $CaO-CaF_2$ slag, N_{CaO} = 0.2).

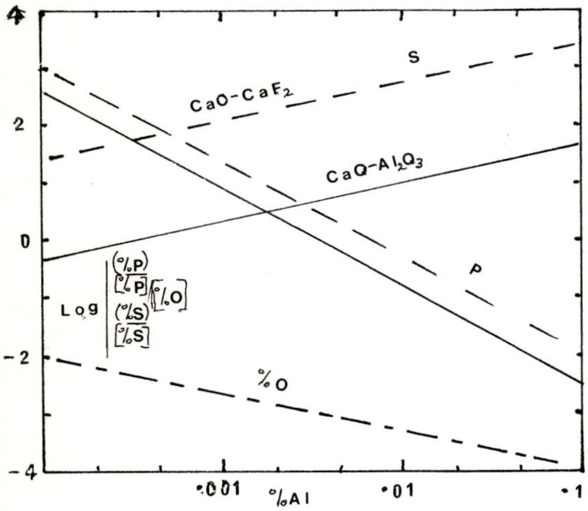

Fig.12 Effect of aluminium content of steel on the phosphorus and sulphur partition ratios for two slags (conditions as in Fig.11).

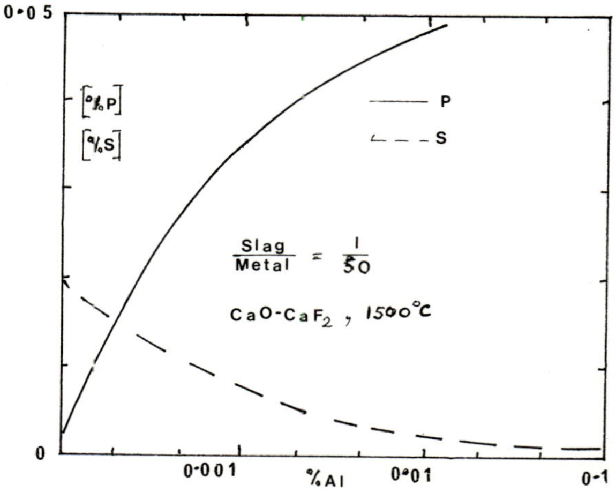

Fig.13 Phosphorus and sulphur contents of steel at equilibrium with a $CaO-CaF_2$ slag (Initial metal composition %P = 0.05, %S = 0.02, slag-metal ratio = 0.02).

POSTERS

Simultaneous dephosphorization and desulphurization of molten Fe-Cr and Fe-Mn alloys by using high basic fluxes

S SIMEONOV, Iv IVANCHEV and B DRAKALIYSKY

Dr Simeonov and Assoc Prof Dr Ivanchev are in the Department of Metallurgy at the Higher Institute of Chemical Technology, Sofia, Bulgaria; Prof Dr Drakaliysky is a General Manager of the Iron and Steel Research Institute, Sofia, Bulgaria.

SYNOPSIS

The investigations have been carried out for simultaneous dephosphorization and desulphurization of molten Fe-Cr and Fe-Mn alloys by using 50%Si-25%Ba-9%Ca and CaF_2-CaC_2 fluxes. The results could be summarized as follows:

1. Molten Fe-Cr alloys, containing up to 20% Cr could be successfully dephosphorized and desulphurized by using slags from the system CaF_2-BaO-$BaCl_2$ and Si-Ba-Ca fluxes. The obtained degrees of dephosphorization and desulphurization were more than 60% at 1550°C.
2. Molten Fe-Mn alloys, containing up to 20% Mn and 0.3-0.7%C could be successfully dephosphorized and desulphurized by using fluxes containing CaF_2 - CaC_2. The levels of dephosphorization and desulphurization achieved in the experiments are over 60%.

INTRODUCTION

In the last several years the demand for high quality steels especially, with low and extra low concentrations of phosphorus and sulphur rapidly increased. Reducing the sulphur and phosphorus contents in steel results in substantional improvement in several mechanical properties and therefore efforts to remove them as low as possible have increased in ironmaking and steelmaking practice as well as in secondary metallurgy.

In conventional refining methods phosphorus is removed from iron alloys under oxidizing conditions, but as it is well known thermodynamically such conditions are not favourable for sulphur removal. However, even for phosphorus this approach is not applicable when the alloys contain less noble elements than iron such as Cr or Mn , because of preferential oxidation of those elements to phosphorus. One alternative to solve this problem is dephosphorization under strongly reducing conditions, which are also favourable for desulphurization. In accord with this view a new dephosphorization process has been proposed and many investigations have been carried out to optimize the process.[1,2,3,4,5].

This work was supported by the Bulgarian State Committee for Science under contract 501.

Thermodynamics consideration

It was experimentaly proved that under strongly reducing conditions $P_{O_2} < 10^{-19}$ Mpa the phosphorus from iron melts could be removed successfully.[1,2] In the same time the reducing atmosphere is favourable for desulphurization. One way to achieve such low oxygen potential is to use material containing alkali earth elements, for example Ca, Ba, etc. Because of strong affinity of Ca and Ba with oxygen it is reasonable first to decrease the oxygen potential of the melts by using Al. In this case the equilibrium P_{O_2} can be calculated from the reactions 1,3.

$$(Al_2O_3)_s = 2\underline{Al} + 3\underline{O} \qquad (1)$$

$$LogK_1 = -\frac{62680}{T} + 19.85 \quad ,[6] \qquad (2)$$

$$1/2\ O_2 = \underline{O} \qquad (3)$$

$$\Delta G^o_3 = - 117152 - 2.89.T \ ,[7] \qquad (4)$$

If Al = 0.05% and at 1823°K the P_{O_2} = $=1.072.10^{-16}$Mpa.

For the Ca addition the equilibrium P_{O_2} can be calculated from reactions 3,5.

$$(CaO)_s = \underline{Ca} + \underline{O} \qquad (5)$$

$$\log K_5 = -\frac{34679}{T} + 7.68 \quad ,[8,9] \qquad (6)$$

The limit solubility of Ca in iron melts is about 0.03% at 1873 K and the P_{O_2} = $=2.24.10^{-28}$ Mpa at 1823 K.

In case of Ba the available thermodynamics data show considerable discrepancy for the reaction (7)

$$(BaO)_s = \underline{Ba} + \underline{O} \qquad (7)$$

$$\log K_7 = -\frac{31331}{T} + 8.27 \ , \quad [10] \qquad (8)$$

$$\log K_7 = -10.07 \quad \text{at } 1873^oK \ , \ [11] \qquad (9)$$

The limit solubility of Ba according to [11] is 0.0004 at 1873 K and the equilibrium P_{O_2} is $1.31.10^{-19}$Mpa from the data [10], while from data of [11] it is $6.56.10^{-22}$Mpa. From above calculations it is obvious that with Ba and Ca it is possible to decrease the oxygen potential below 10^{-19}Mpa so the conditions are favourable for dephosphorization.

Experimental results

The experiments have been carried out in a Tammann furnace. One kg of mother alloys, containig 5 to 20% Cr or 5 to 20% Mn was melted in a MgO crucible. After the melt was kept at temperature of 1823 K, 50g. of BaO-40%$BaCl_2$60% for experiments with Fe-Cr and CaO80%-CaF_2 20% for experiments with Fe-Mn alloys was added in order to protect the melt from oxidation and collect the products formed during experiments.

Temperature was controlled by using a PtRh30/PtRh6 thermocouple. Argon gas for stirring was blown in to the melt using mullite tube. After stabilising the temperature at 1823 K, the melt was deoxidized with Al to obtain about 0.05% Al. After deoxidation by Al, Si-Ba-Ca alloys in proportion of 30g. in case of Fe-Cr and 30g. of CaF_2-CaC_2in case with Fe-Mn alloys were added and melt was stirred mechanically by tungsten rod. During the experiments metal samples were sucked every three minutes, by using silica tubes and quenched in water. After experiments, slag and iron samples were taken. The metals and slags were analized by chemical methods.

The degree of dephosphorization and desulphurization 9 min. after top addition was evaluated in the experiments. The degree of dephosphorization and desulphurization were calculated according to Eq 10

$$\eta_M = \frac{[M\%]_{in} - [M\%]_f}{[M\%]_{in}} . 100 \qquad (10)$$

where $[M\%]_{in}$ and $[M\%]_f$ are the initial and final content of phosphorus and sulphur.

The results are shown at Fig.1 and Fig.2. As it is seen from Fig.1 the degree of dephosphorization increases with increasing chromium content, which could be explained from thermodynamic point of view with the influence of chromium content on the activity of phosphorus. As it is well known the addition of chromium remarkably auguments the activity of phosphorus, hence the rate of dephosphorization increases.

In the case of Fe-Mn alloys the influence of manganese on the dephosphorization is covered by increasing the carbon content from 0.3 to 0.7 from the fluxes containing CaC_2. Because of that activity of phosphorus increases and degree of dephosphorization increases too.

Similar considerations can be used to explain the experimental results in Fig.2. The degree of desulphurization decreases for Fe-Cr melts because activity coefficient of sulphur decreases when chromium content increases.

By analogy with dephosphorization of Fe-Mn alloys the degree of desulphurization is more strongly effected by carbon than manganese and it slightly decreases.

Conclusions

1. Molten Fe-Cr alloys, containing up to 20% Cr could be successfully dephosphorized and desulfurized by using slags from the system BaO-$BaCl_2$ and by injecting in the metal Si-Ba-Ca fluxes. The obtained degrees of dephosphorization and desulphurization were more than 60% at 1823 K.
2. Molten Fe-Mn alloys, containing up to 20%Mn and 0.3-.7%C could be successfully dephosphorized and desulphurized by using fluxes containg CaF_2 - CaC_2 .

REFERENCES

1. H. Monokawa and N. Sano; Met. Transc.,1982, vol. 13B, pp.643-644.
2. S. Tabuchi and N. Sano: Met. Transc.,1984, vol. 15B, pp.351-356.
3. K. Kitamura et al.: Trans. ISIJ,24,1984, pp.631-638.
4. T. Arto et al.: Trans. ISIJ,25,1985, pp.326-332.
5. C. Leal and K. Torssell: Scand. J. Metallurgy 15,1986, pp.265-272.
6. D. Janke and W. A. Fisher: Arch. Eisenhuttenwes.47,1976,pp.195-198.
7. T. P. Floridis and J. Chipman: Trans. Met. Soc. AIME,1958, vol.212,pp.549-553.
8. E. T. Turkdogan: Physical Chemistry of High Temperature Technology, Academic Press, New York,1980,5.
9. D. L. Sponseller and R. A. Flinn: Trans. AIME,230,4,1964,pp.876-888.
10. И. С. Куликов: Металлы, No 6, 1985, с.9-15.
11. M. Breitzmanmn, M. Kohler, D. Janke, H. J. Engell: The 7 Japan-Germany Seminar, May 5-6,1987, pp.95-116.

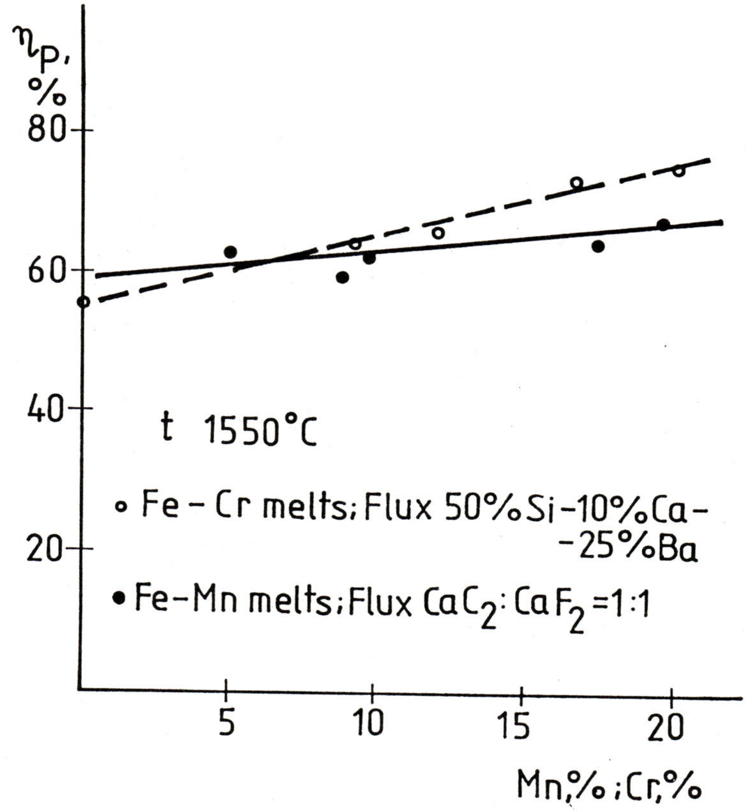

FIG.1 EFFECT OF Cr AND Mn CONTENT ON THE DEGREE OF DEPHOSPHORIZATION

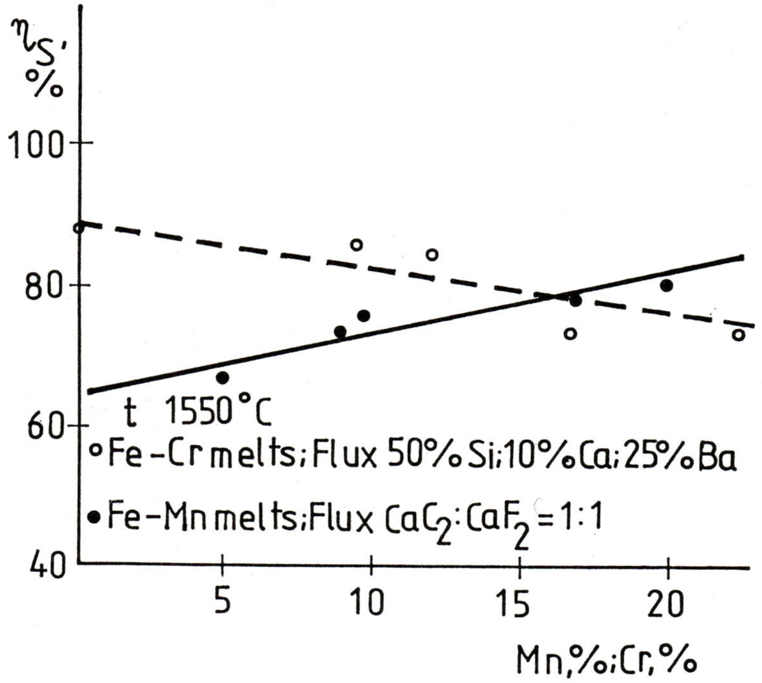

FIG.2 EFFECT OF Cr AND Mn CONTENT ON THE DEGREE OF DESULFURIZATION

Kinetic study of reaction between solid Pt or solid Fe and $CaO-SiO_2-Al_2O_3$ slags by AC impedance method

M HINO, T NITTA, S CHIDA and S BAN-YA

Drs Hino and Ban-Ya are in the Department of Metallurgy at Tohoku University, Sendai, Japan; Mr Nitta, formerly Graduate Student at the Tohoku University, is now in Suwa Works, Tokin Corporation; Mr Chida, formerly Graduate Student at the Tohoku University, is now in Research and Development Lab 1, Nippon Steel Corporation.

INTRODUCTION

An application of A.C. impedance method(1) was tried for the reaction between solid Pt and $CaO-SiO_2-Al_2O_3$ slag to clarify the individual kinetic factors concerning the electrode reactions at the interface between slag and metal, in succession to the successful application(2) of the method for the reaction between solid Pt and Na_2O-SiO_2 slag.

This technique was tried to learn the oxidation-reduction reaction between solid Fe and $CaO-SiO_2-Al_2O_3$ slag as the first application for slag-metal reaction in iron- and steelmaking process.

PRINCIPLE OF THE MEASUREMENT ON A.C. IMPEDANCE METHOD

The major factors affecting the total impedance, $Z(\Omega \cdot m^2)$, consist of electrolyte resistance, $R_{sol}(\Omega \cdot m^2)$, overvoltage of charge transfer, $R_C(\Omega \cdot m^2)$, and preceding or following chemical reaction$[X \underset{k_{-p}}{\overset{k_{+p}}{\rightleftharpoons}} R]$ with diffusion, $Z_R(\Omega \cdot m^2)$, and capacitance of electric double-layer, $C_{dl}(F \cdot m^{-2})$. Since the response for A.C. frequency is different by each factor affecting the total impedance, individual factors in this case can be separately determined by comparing the frequency response observed with that of electrically equivalent circuit of the electrode.

The total impedance, Z, of the electrically equivalent circuit is shown by Eq. 1 and each component is presented by Eqs. 2 to 8.

$$Z = R_{sol} + [j\omega C_{dl} + 1/(R_C + Z_R)]^{-1} \quad (1)$$

$$Z_R = \sqrt{(R_R)^2 + (1/\omega C_R)^2} \quad (2)$$

$$R_R = A[\{(\omega^2 + k^2)^{0.5} + k\}/(\omega^2 + k^2)]^{0.5} \quad (3)$$

$$1/\omega C_R = A[\{(\omega^2 + k^2)^{0.5} - k\}/(\omega^2 + k^2)]^{0.5} \quad (4)$$

Where, $$A \equiv (1/2^{0.5})(RT/n^2F^2)(1/c^oD^{0.5}K_C) \quad (5)$$

$$c^o \equiv c^o_X + c^o_R \quad (6)$$

$$K_C \equiv c^o_X/c^o_R \quad (7)$$

$$k \equiv k_{+p} + k_{-p} \quad (8)$$

and, c^o_X, c^o_R : molarity of reductant or oxidant $(mol \cdot m^{-3})$

D : diffusivity of reductant or oxidant $(m^2 \cdot s^{-1})$

k : rate constant of preceding or following chemical reaction (s^{-1}).

EXPERIMENTALS

The synthetic slag of $CaO-SiO_2-Al_2O_3$ ternary (38-42-20 or 24-61-15 by wt%) was melted in a Pt or a Fe crucible($3-2x10^{-5}$ m^3), which is the counter electrode in the present work, for 8 hr at 1 573 and 1 673 K under Ar, H_2-H_2O-Ar or H_2-Ar atmosphere, and then a Pt or a Fe wire($0.5-1x10^{-5}$ m) of the working electrode was placed at $1-6x10^{-5}$ m deep in the center of the liquid slag.

On this experimental condition, the frequency response for complex impedance was measured in the range from 5 Hz to 10 kHz by overlapping the sinusoidal A.C. overpotential of $\tilde{g}_{max} = 1x10^{-2}$ V on the constant D.C. overpotential from +1 to -1 V between the electrodes. Since the interfacial ratio of a working electrode for a reference electrode was so very small — as 1/130 to 1/180 that the results obtained were directly concerned with a single electrode reaction on the working electrode.

The D.C. polarization curve and current efficiency in potentiostatic electrolysis were also measured in the same apparatus to clarify the kind of the electrode reaction.

RESULTS

Figure 1 shows some experimental results of the polarization curves obtained in the present work. They are the results for the systems of Pt(s)- and Fe(s)-38wt%CaO·42wt%SiO_2·20wt%Al_2O_3 at 1573 and 1673 K by means of a potentiostatic method. The value on the axis of abscissas shows the overpotential, which is corrected by reducing the IR drop due to solution resistance of slag, from the mixed potential of a working electrode against a counter electrode, and the value of interfacial area on the axis of ordinates is evaluated in consideration of slag creeping up the working electrode.

In Pt(s)-slag system, it was confirmed from EPMA that SiO_2 was reduced to Si, which resulted in alloying with Pt(s) in cathode, and oxygen evolved at higher overvoltage in anodic reaction from the observation of fluctuation of current value in polarization curve, but it was not clear what happened at lower overvoltage in anodic reaction. On the other hand, it was confirmed from the result of potentiostatic coulometry that iron was oxidized in the anodic reaction and iron oxide might be reduced in cathodic reaction in Fe(s)-slag system.

In lower figures of Fig. 1, open circles show the corresponding examples of impedance loci, which are showing the relation between the real part of complex impedance, R, and the imaginary part of complex impedance, $1/\omega C$, observed in the Pt(s)- and Fe(s)-38wt%CaO·42wt%SiO_2·20wt%Al_2O_3 systems, in comparison with the controlled overpotential on their polarization curves. The present impedance loci were comparable with kinds of electrode reaction and temperature.

DISCUSSION

The individual kinetic factors R_{sol}, C_{dl}, R_C, A and k, which are concerned in the electrode reactions, were able to be determined by parameter fitting in each impedance loci. Figure 2 shows some results of the individual factors calculated from the results obtained in the present work. Squares in Fig. 1 show the estimated impedance loci by those factors, and they showed good agreement with observed impedance loci.

Two maxima appeared in factor A caused by mass transfer resistance in Pt(s)-slag system as shown in Fig. 2. Their dependency on overvoltage corresponded exactly with two residual currents in anodic field on the polarization curve in Fig. 1. It is considered by extremely low values of A in Fe(s)-slag system, compared with that in Pt(s)-slag system as shown in Fig. 2, that mass concerning the process of oxidation of Fe(s) and reduction of iron oxide transfers easier than that concerning evolution of oxygen and reduction of SiO_2. A value in each system decreased with increasing temperature and slag basicity.

The factor k indicating rate constant of oxidation of Fe(s) and reduction of iron oxide was larger than that of oxygen evolution and SiO_2 reduction. Each k value increased with increasing temperature and overvoltage.

The magnitude of electrical double-layer capacitance between slag and Pt(s) decreased with increasing temperature. The dependency of the capacitance on D.C. overpotential might be caused by diffusion double layer. The capacitance between slag and Fe(s) was larger than that between the slag and Pt(s). Therefore, the thickness of electrical double-layer in Fe(s)-slag system is thiner than that in Pt(s)-slag system.

Charge-transfer resistance in Pt(s)-slag system showed a constant value for each kind of electrode reaction, and decreased with increasing temperature. R_C in Fe(s)-slag system was negregible small.

The conductivity of slags obtained by application of an A.C. impedance method increased with increasing Fe_tO contents, and showed good agreement with those of previous works.

CONCLUSIONS

It was concluded that the A.C. impedance method might be quite useful to determine the individual kinetic parameters concerning the slag-metal reactions in iron- and steelmaking process.

REFERENCES

(1) K. J. Vetters, Electrochemical Kinetics, Academic Press, New York, London, 1967.

(2) M. Hino and S. Ban-ya, Proc. Int. Sympo. on Metall. Slags and Fluxes, ed. by H. A. Fine and D. R. Gaskell, TMS of AIME, Lake Tahoe (1984), p. 669; J. Japan Inst. of Metals, 48(1984), p.595.

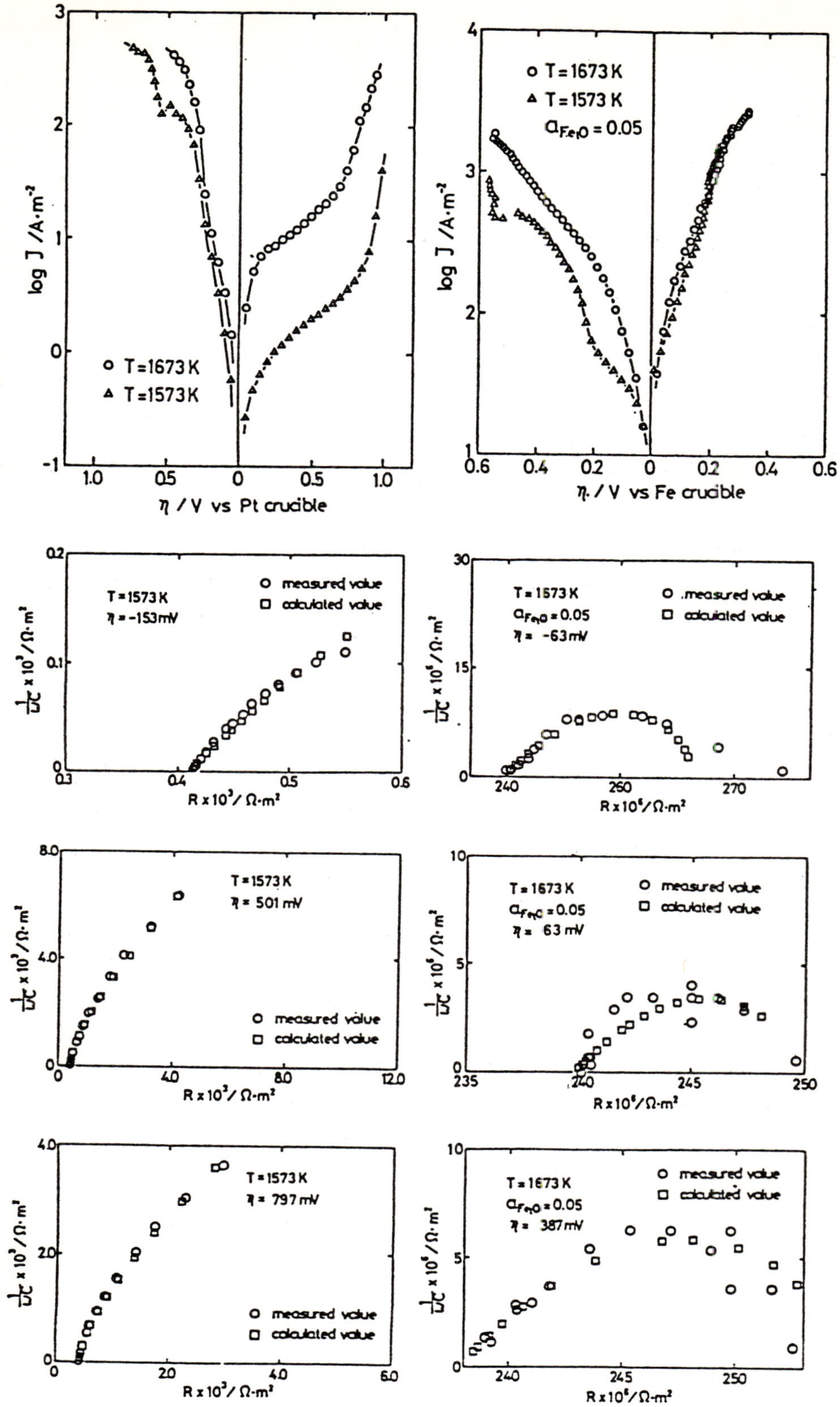

Pt(s)/CaO-SiO_2-Al_2O_3(l) Fe(s)/CaO-SiO_2-Al_2O_3(l)

Fig. 1. Polarization curves and Cole-Cole plot of impedance in Pt(s)/38wt%CaO-42wt% SiO_2-20wt%Al_2O_3(l) and Fe(s)/38wt%CaO-42wt%SiO_2-20wt%Al_2O_3(l) systems.

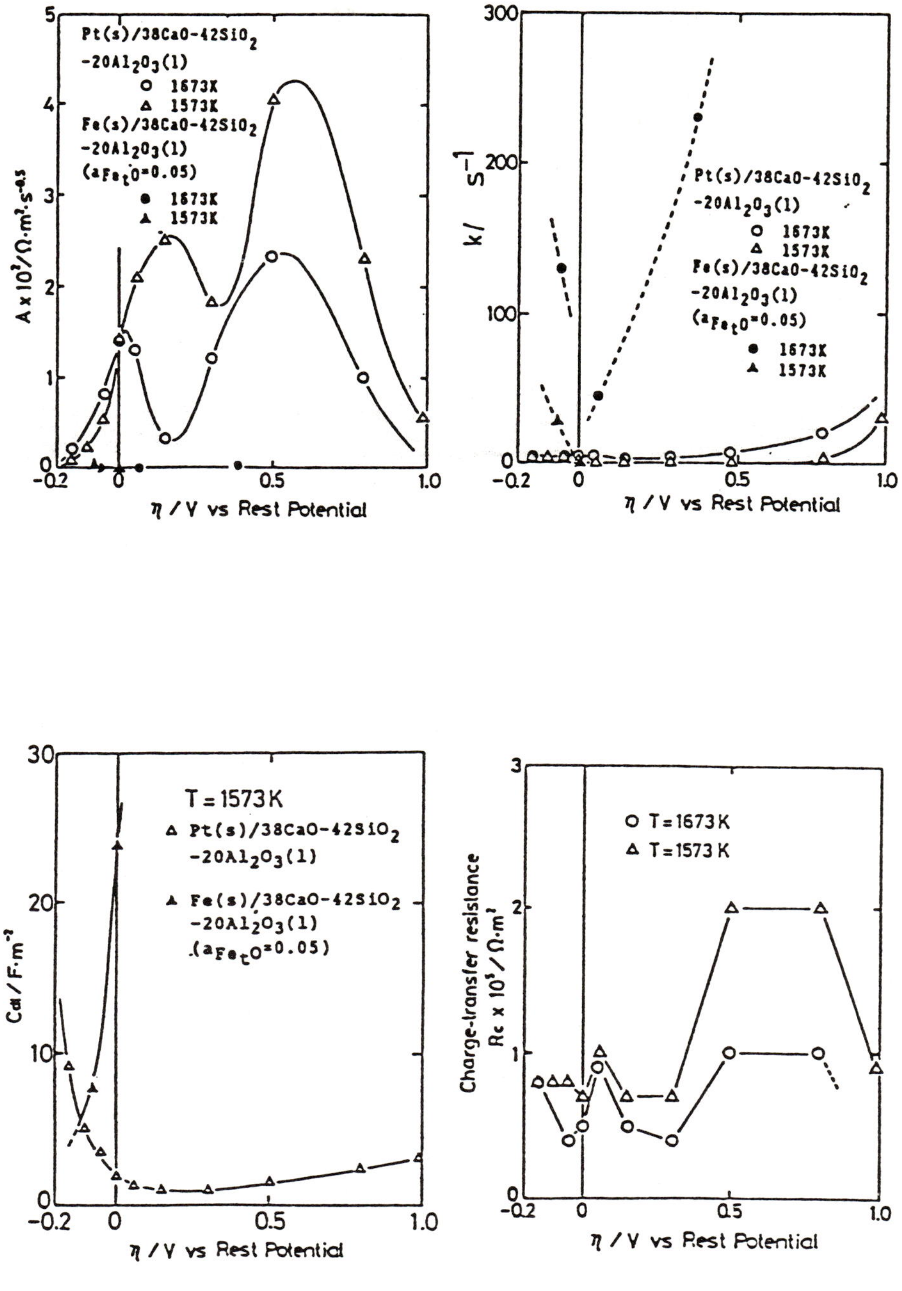

Fig. 2. Overvoltage dependence of factors A, k, double-layer capacitance and charge-transfer resistance in Pt(s)/38wt%CaO-42wt%SiO_2-20wt%Al_2O_3(l) and Fe(s)/38wt% CaO-42wt%SiO_2-20wt%Al_2O_3(l) systems.

Interrelation between phosphate and carbonate capacities

H FUJIWARA, H IRITANI and M IWASE

Mr Fujiwara and Mr Iritani, Graduate Students, and Dr Iwase, Associate Professor, are with the Department of Metallurgy, Kyoto University, Kyoto, Japan.

SYNOPSIS

For MO + MCl_2 melts (M represents Ca, Sr and Ba), phosphate capacities can be predicted from carbonate capacities. For other slag systems, however, such a prediction does not seem to be promising.

PHOSPHATE AND CARBONATE CAPACITIES

The present paper is aimed at testing the validity of the hypothesis made by Wagner[1] through the authors' systematic experiments for MO + MCl_2[2,3]; M represents Ca, Sr and Ba.

Consider the dissolution of carbon dioxide and gaseous diatomic phosphorus, respectively, in MO + MCl_2 binary slags as expressed by equations (1) and (2), respectively[1];

$$CO_2 + O^{2-} = CO_3^{2-} \quad \text{.....(1)}$$

$$(1/2)\ P_2 + (5/4)\ O_2 + (3/2)\ O^{2-} = PO_4^{3-} \quad \text{.......(2)}$$

$$K(1) = \frac{f(CO_3^{2-})\ X(CO_3^{2-})}{f(O^{2-})\ X(O^{2-})\ P(CO_2)} \quad \text{.......(3)}$$

$$K(2) = \frac{f(PO_4^{3-})\ X(PO_4^{3-})}{P(P_2)^{1/2}\ P(O_2)^{5/4}\ [f(O^{2-})\ X(O^{2-})]^{3/2}} \quad \text{.......(4)}$$

Carbonate and phosphate capacities, respectively, can be defined as[1];

$$C(CO_3^{2-}) = X(CO_3^{2-})\ /\ P(CO_2) \quad \text{.......(5)}$$

$$C(PO_4^{3-}) = X(PO_4^{3-})\ /\ P(P_2)^{1/2}\ P(O_2)^{5/4} \quad \text{.......(6)}$$

By substituting equations(5) and (6) for (3) and (4), respectively, one obtains;

$$\log C(CO_3^{2-}) = \log X(O^{2-})$$
$$+ \log [K(1)\ f(O^{2-})\ /\ f(CO_3^{2-})] \quad \text{.......(7)}$$

$$\log C(PO_4^{3-}) = (3/2) \log X(O^{2-})$$
$$+ \log [K(2)\ f(O^{2-})^{3/2}\ /\ f(PO_4^{3-})] \quad \text{.......(8)}$$

With a tentative assumption[1] that the quotients of the activity coefficients in equations(7) and (8) are independent of the molar ratio of X(MO) / X(MCl_2), it can be anticipated that a plot of log C(CO_3^{2-}) against log X(O^{2-}) should be a straight line with a slope of unity, while a logarithmic plot of C(PO_4^{3-}) against X(O^{2-}) should also be linear with a slope of 3/2. It is to be noted here at this point that for MO + MCl_2 melts oxygen anion fractions, X(O^{2-}), can be calculated by using the Temkin model.

The carbonate and phosphate capacities in MO + MCl_2 slags have been determined by the present authors[2,3]. Hence, now we are able to test the validity of Wagner's hypothesis by using our own experimental data. Figure 1 shows the logarithmic plot of C(CO_3^{2-}) against X(O^{2-}) for CaO + $CaCl_2$ slags; the relation can be expressed by a single straight line, being independent of X(CaO)/X($CaCl_2$), and the slope is close to unity, corresponding to equation(7);

$$\log C(CO_3^{2-}) = \log X(O^{2-}) - 0.64 \text{ at } 1473 \text{ K} \quad \text{....(9)}$$

Figure 2 shows the phosphate capacities as the function of X(O^{2-}); the relation can also be expressed by a single straight line, again being independent of the molar ratio of X(CaO)/X($CaCl_2$), in conforming to equation(8);

$$\log C(PO_4^{3-}) = (3/2) \log X(O^{2-}) + 27.2 \quad \text{.......(10)}$$

Now it can be stated that the hypothesis made by Wagner would be valid for MO + MCl_2 melts.

Next, we consider possible correlation between carbonate and phosphate capacities. At infinitely dilute concentrations of CO_3^{2-} and PO_4^{3-} ions, equations(7) and (8) become, respectively;

$$\log C°(CO_3^{2-}) = \log X(O^{2-})$$
$$+ \log [K(1)\ f(O^{2-})\ /\ f°(CO_3^{2-})] \quad \text{.......(11)}$$

$$\log C°(PO_4^{3-}) = (3/2) \log X(O^{2-})$$
$$+ \log [K(2)\ f(O^{2-})^{3/2}\ /\ f°(PO_4^{3-})] \quad \text{.......(12)}$$

where superscript "°" corresponds to zero concentration. By comparing carbonate capacities at infinitely dilute concentrations in different molar ratios of CaO/$CaCl_2$ denoted by prime and double prime subscripts, one obtains;

$$\frac{C^\circ(PO_4^{3-})'}{C^\circ(PO_4^{3-})''} = \left\{\frac{C^\circ(CO_3^{2-})'}{C^\circ(CO_3^{2-})''}\right\}^{3/2} \quad \text{...........(13)}$$

Thus if the carbonate capacities at zero concentrations in two different molar ratios of $CaO/CaCl_2$ have been determined, one may predict the ratio of the phosphate capacities.

The system under consideration is the $CaO + CaCl_2$ binary system. Nevertheless, for the measurements of carbonate and phosphate capacities, experiments have to be conducted for $CaO + CaCl_2 + CO_2$ and $CaO + CaCl_2 + P_2O_5$ ternary melts; otherwise experimental determinations of these capacities are impossible. Hence, in practice, for the determinations of $C^\circ(CO_3^{2-})$ and $C^\circ(PO_4^{3-})$, extrapolation of experimental data for $C(CO_3^{2-})$ and $C(PO_4^{3-})$, obtained for ternary melts, to $X(CO_3^{2-}) = 0$ and $X(PO_4^{3-}) = 0$, respectively, would inevitably be involved. Equations(9) and (10), respectively, would facilitate such extrapolation procedures.

Consider, for example, $CaO + CaCl_2 + CO_2$ ternary slags of $X(CaO)/X(CaCl_2) = 20/80$. The oxygen anion fraction for a $CaO + CaCl_2$ binary melt of $X(CaO)/X(CaCl_2) = 20/80$ is equal to $20/(20 + 80 \times 2) = 0.111$. This value of 0.111 corresponds to the oxygen anion fraction in $CaO + CaCl_2 + CO_2$ ternary slags of the same $CaO/CaCl_2$ mole ratio at zero concentration of CO_3^{2-} ions. Hence by using equation(9), one can obtain the $C^\circ(CO_3^{2-})$ value for $CaO + CaCl_2 + CO_2$ ternary melts of $X(CaO)/X(CaCl_2) = 20/80$ as

$$\log C^\circ(CO_3^{2-}) = \log (0.111) - 0.64 \quad \text{..........(14)}$$

By using the same procedure, values for $C^\circ(PO_4^{3-})$ can also be obtained.

Now by combining equations(9) and (10), we have

$$\log C^\circ(PO_4^{3-}) = (3/2) \log C^\circ(CO_3^{2-}) + 28.2 \quad \text{..............(15)}$$

By using equation(15), for $CaO + CaCl_2$ binary melts, one may predict phosphate capacities from carbonate capacities.

$SrO + SrCl_2$ and $BaO + BaCl_2$ slags

Further, we consider the possiblity for predicting the phosphate capacities for $SrO + SrCl_2$ and $BaO + BaCl_2$ binary slags through equation(15), which was obtained for $CaO + CaCl_2$ melts.

One may tentatively assume that the quotients of the activity coefficients in equation(11) and (12) do not vary with the replacement of Ca^{2+} ions for Sr^{2+} and Ba^{2+} ions.

For $SrO + SrCl_2$ and $BaO + BaCl_2$ melts, linear relations between $\log C(CO_3^{2-})$ and $\log X(O^{2-})$, and $\log C(PO_4^{3-})$ and $\log X(O^{2-})$ were also obtained. By comparing two capacities at infinitely dilute concentrations of CO_3^{2-} and PO_4^{3-} ions, for two different slag systems, denoted by Ca and M subscript, one obtains;

$$\frac{C^\circ(PO_4^{3-})_M}{C^\circ(PO_4^{3-})_{Ca}} = \left\{\frac{C^\circ(CO_3^{2-})_M}{C^\circ(CO_3^{2-})_{Ca}}\right\}^{3/2} \quad \text{......................(16)}$$

or

$$\log C^\circ(PO_4^{3-})_M = \log \frac{C^\circ(PO_4^{3-})_{Ca}}{C^\circ(CO_3^{2-})_{Ca}^{3/2}} + (3/2) \log C^\circ(CO_3^{2-})_M$$

$$= 28.2 + (3/2) \log C^\circ(CO_3^{2-})_M \quad \text{..............(17)}$$

In figure 3, values for $C^\circ(PO_4^{3-})$ are given as the function of $C^\circ(CO_3^{2-})$ for $CaO + CaCl_2$, $SrO + SrCl_2$ and $BaO + BaCl_2$, while the straight line corresponds to equation(17). From this figure, now it can be stated that the phosphate capacities of even $SrO + SrCl_2$ and $BaO + BaCl_2$ slags can be predicted from equation(17), which was obtained for $CaO + CaCl_2$ slags.

Equation(17) seems to be very promising; extension of equation(17) to other slag systems is of strong interest. If this equation is valid for all of the slag systems, then there is no need to measure the phosphate capacities; by measuring carbonate capacities, prediction of phosphate capacities might be possible. It is noted that in general measurements for carbonate capacities are much easier than those for phosphate capacities.

Unfortunately, experimental data that can be used to test the validity of equation(17) are very limited. It is evident, however, from the foregoing discussion that in order to predict phospahte capacities from carbonate capacities, linear relation between $\log C(CO_3^{2-})$ and $\log X(O^{2-})$ must be obtained; otherwise prediction is impossible.

Figure 4 shows the relation between $\log C(CO_3^{2-})$ and $\log X(O^{2-})$ for $R_2O + RX$ melts; R and X denote, respectively, alkaline metals and halogens). Experimental data used in plotting this illustration were taken from Kuwahara et al[4]. As shown in this figure, there is no such a linear relation as observed in figure 1. Hence, for $R_2O + RX$ melts, prediction of phosphate capacities from carbonate capacities would not be possible. The authors eventually felt that for the majority of molten slags used in steelmaking, phosphate capacities might not be estiamted by carbonate capacities.

ACKNOWLEDGEMENTS

Helpful suggestions, discussion and encouragements were given by Professor Alex McLean, University of Toronto, while financial support to this work was given by Nippon Kokan; and these are gratefully acknowledged.

REFERENCES

[1] C. Wagner, Met. Trans., vol.6B, 1975, pp.405/09.

[2] M. Iwase et al., Iron and Steelmaker, vol.15 no.5, 1988, pp.69/80.

[3] M. Iwase et al., Iron and Steelmaker, to be published.

[4] T. Kuwahara, Y. Yamagata and N. Sano, Steel Research, vol.57, 1986, no.4, pp.160/65.

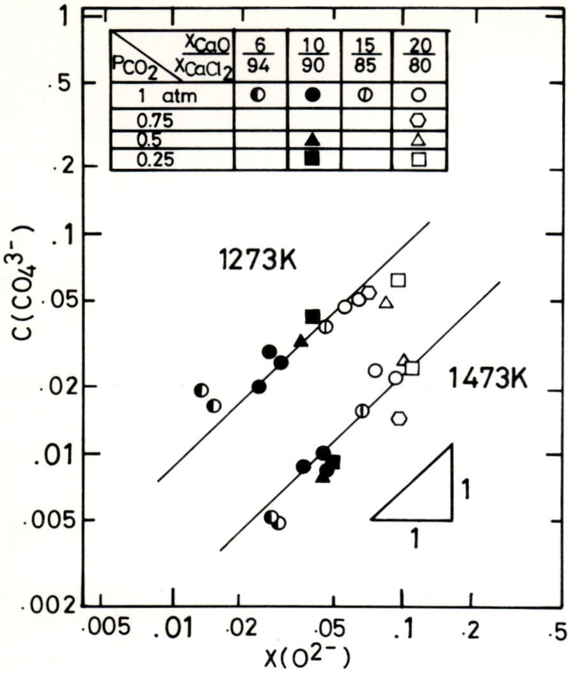

Figure 1 Relation between log $C(CO_3^{2-})$ and log $X(O^{2-})$ for CaO + $CaCl_2$ melts at 1473 K.

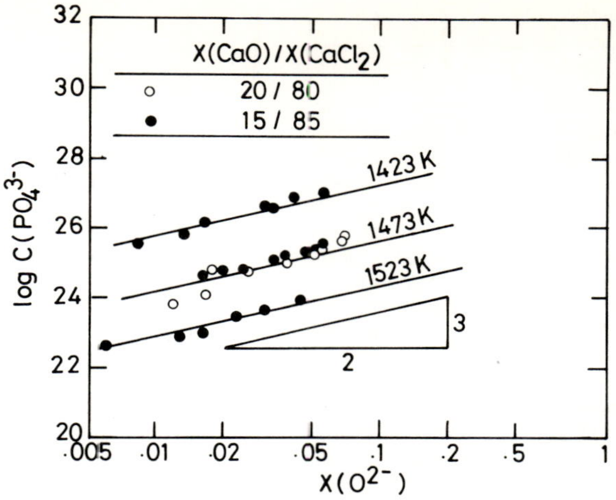

Figure 2 Relation between log $C(PO_4^{3-})$ and log $X(O^{2-})$ for CaO + $CaCl_2$ melts at 1473 K.

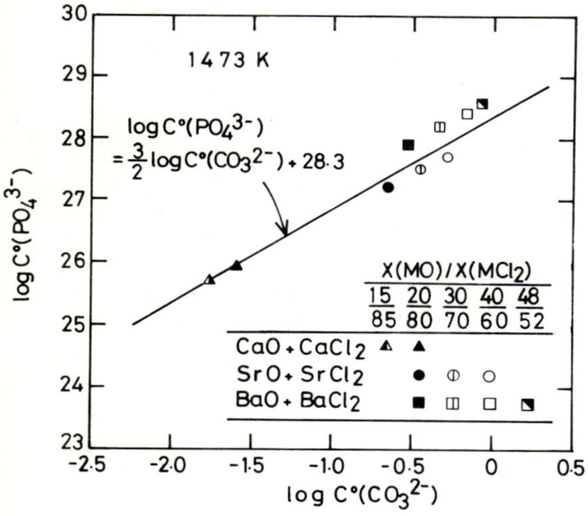

Figure 3 Relation between log $C°(PO_4^{3-})$ and log $C°(CO_3^{2-})$ for CaO + $CaCl_2$, SrO + $SrCl_2$ and BaO + $BaCl_2$ melts at 1473 K.

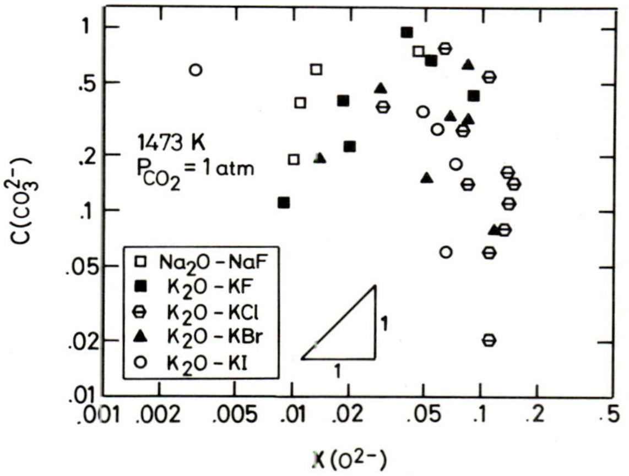

Figure 4 Relation between log $C(CO_3^{2-})$ and log $X(O^{2-})$ for R_2O + RX melts; R and X represent, respectively, alkaline metals and halogens. Experimental data for plotting this figure were taken from Kuwahara et al[4].

Rutile welding fluxes

K G LEEWIS and C J PINHEY

K G Leewis, Welding Institute of Canada, Toronto and C J Pinhey, Dofasco, Hamilton, formerly of the Technical University of Nova Scotia.

INTRODUCTION:

Many of the welding consumables which provide ease of operation and high productivity contain significant amounts of titania[1]. These rutile systems in SMAW (shielded metal arc welding) and FCAW (flux cored arc welding) have found wide application including critical structures[2]. A large body of data has been developed for welding consumables but few involve investigations of the flux system itself[3,4,5,6]. Flux changes have always been related to the mechanical performance of the deposits. Melting temperatures, specific phase diagrams, viscosity and interfacial surface tension values for the major welding titanate systems are sparse. In any case, welders prefer rutile systems because of the ease in controlling the weld pool, good wetability, and hence final weld profile[7].

This enhanced performance has never been properly explained. Titania in fluxes is thought to behave as an amphoteric oxide though this aspect is still in question[8,9,10]. Titania has an optical basicity similar to alumina at 0.61, suggesting that it is a weak network former[11]. Others suggest that TiO_2 acts only as a network modifier since additions act to lower the viscosity at elevated temperatures[12]. It also usually increases the viscosity at the melting temperature[13].

Previous work has shown rutile FCAW consumables to fuse at temperatures above 1600°C[14]. They were extremely corrosive to all but stabilized zirconia containers presumably because of the agglomerated nature of the separate mineral components, and the large volume of iron compounds present. To better understand the role of TiO_2 in metallurgical fluxes a model system was required. Extensive experience with the PbO-SiO_2 system[15,16], the well defined nature of this binary, the much lower working temperatures, and the availability of suitable containers, lead to the choice of the PbO-SiO_2-TiO_2 ternary. Unfortunately the viscosity measurements were not able to solve conclusively the network modifier/former controversy but a very interesting phenomenon was observed for the higher and the lower temperature regions[17].

DISCUSSION:

The fluxes were prepared as in Table 1, with TiO_2 being substituted for SiO_2 and the PbO content of the slags remaining constant at 60 mol/o. The fluxes were crushed and remelted at least three times in platinum to ensure homogeneity. The Brookfield viscometer measured the torque exerted on an alumina cylinder turning in the liquid flux at a known rotational velocity and temperature. Prior calibration of this geometry in fluids of known viscosity enabled the viscosity to be read directly from the chart recorder. Each flux was loaded into a 100cc, stabilized zirconia crucible, then melted under air. Once the system stabilized at 1400°C, the rotating spindle was lowered to a fixed depth in the slag and the furnace turned off. As the furnace continually cooled, a viscosity-temperature profile was plotted, (Fig 1)[17].

The PbO-SiO_2 binary provided the calibration, and as shown by Fig. 2, provided good agreement with published literature[18,19,20]. The substitution of 5% TiO_2, decreased the viscosity and also the activation energy for viscosity, Table 2. However substitution of 10% or more of the SiO_2 by TiO_2 showed a very strong increase in viscosity approaching the fusion temperature of these slags. Two separate activation energies were observed. A lower value for the high temperature portion and a much higher one closer to the melting temperature. There was a definite step of a decade in value between the two, Fig 3. The activation energies for these and a number of welding fluxes are presented in Table 2. These earlier investigators reported their results as viscosity vs temperature. Recalculation as log viscosity vs inverse temperature shows that this step change in flux viscosity behaviour was present, only unreported.

The precipitation of compounds or the addition of solids has been shown to dramatically increase the viscosity of liquids[21]. Rutile slags are known to form precipitates in a number of systems[13]. Unfortunately this apparatus did not enable these ternary slags to be quenched; however the furnace cooled fluxes were sectioned and polished for SEM-EDS investigation. Precipitates were present, Fig 4, and identified as lead-titanates. The higher the titania content of the slag the greater was the volume percentage of precipitates. Judging by the EDS intensities the precipitate composition was essentially constant suggesting that with additional titania the remaining liquid slag became increasingly more acid. As expected the activation energies for the lower temperature segment also increased (Table 2).

This sharp increase in viscosity and change in liquid flux composition could explain some of the favourable operating performance of titania welding fluxes. Both changes would occur simultaneously at the back of the liquid weld pool near the metal solidification front. An increase in viscosity would help stabilize the higher temperature, flatter weld profile and could contribute to a better out of position performance. The decrease in basicity would cause the oxygen activity to soar, lower the interfacial tension, increase wetability, and help maintain a flatter, well profiled, weld geometry. This oxygen would not be dragged into the weld pool if solidification occurred immediately, but would be confined essentially to the weld surface. Thus weld mechanical properties would not be seriously reduced. Further investigation is warranted.

CONCLUSIONS:

The precipitation of titanates on cooling increased the activation energy for viscosity and resulted in at least a tenfold increase in the viscosity values.

ACKNOWLEDGEMENT:

The authors would like to thank Dr. C. R. Masson for his helpful discussions. This work was possible through a contract with the Defense Research Establishment Atlantic. The second author was grateful for the financial support of NSERC.

REFERENCES

1) Dorling D. et al, CANMET/EMR REPORT
2) Anon, Metal Construction, pp 491-4, August, 1986.
3) Jackson C.E., WRI Bulletin No 190, Dec. 1973
4) Davis M.L.B., and Coe F.R., Welding Institute 39/1977/M May 1977
5) Mills K.C., National Physical Laboratory Report, Chem 65,1977, London, HMSO.
6) Urbain G., Steel Reseach 58 (1987, No. 3, p.111
7) Bala S.R. and Santur S., CANMET/EMR DSS Report 7-184-096/F March 1986.
8) Froberg M. and Weber R., Archiv fur das Eisenhuttenwesen,pp 477-80, Vol 7, July, 1965.
9) Van Bemst A. and Delaunois C., Verres et Refractories, pp435-47, Vol 20, No 6, Nov-Dec, 1966.
10) Van der Colf J.and Howat D.D., J. South African Inst Mining and Metallurgy, pp 255-263, April 1979,
11) Sommerville I. and Sosinsky D., Second Int. Symp. on Metallurgical Slags and Fluxes, pp 1015-26, Lake Tahoe, AIME, Nov. 1984.
12) Ohno A. and Ross H., Can. Met. Quart., pp 259-79, Vol 2, No 3, July, 1963.
13) Handfield G. and Charette G.G., Can. Met. Quart., pp 235-40 Vol 10, No 3, 1971.
14) Leewis K.G., DREA DSS Report 09SC97707-4-6146, July, 1987.
15) Caley W.F. and Masson C.R., Can. Met. Quart., pp 359, Vol 15, 1976.
16) Caley W.F. and Masson C.R., "Metal Slag Gas Reactions and Processes", (ed Z.A. Foroulis and W.W. Smeltzer) Electrochemical Society, Princton, pp 140, 1975.
17) Pinhey C.J., M.A.Sc. Thesis, Technical University of Nova Scotia, Halifax, 1987.
18) Nakamura T. et al, Nippon Kinzoku Gakkaishi, pp 1300-4, Vol 41, No 12, 1977.
19) Urbain G., Rev. Int. Hautes Temp. Refract., pp 107-111, Vol 21, 1984.
20) Campbell P., M.A.Sc. Thesis, Technical University of Nova Scotia, 1983.
21) Eliott J.F., Second Int. Symp. on Metallurgical Slags and Fluxes, pp 45-61, Lake Tahoe, AIME, Nov 1984.

FLUX COMPOSITIONS FOR VISCOSITY MODEL

TABLE 1

MELT	AIM (MOL %) PbO	SiO2	TiO2	ACTUAL (MOL %) PbO	SiO2	TiO2	Al2O3
1.1	60	40	0	54.6	45.4	0	7.43
2.1	60	35	5				
2.2	60	35	5	60.8	38.2	1	6.59
3.1	60	30	10				
3.2	60	30	10	56.4	35.8	7.8	6.6
4.1	60	25	15	60.05	24.95	15	6.45
4.2	60	25	15				
5.1	60	20	20				
5.2	60	20	20	57.71	22.73	19.56	8.96

ALUMINA FROM VISCOMETRY SPINDLE

ACTIVATION ENERGIES FOR VISCOSITY IN WELDING FLUXES

TABLE 2

ID	CLASS	V RATIO	ACTIV'N ENERGY LOW T	ACTIV'N ENERGY HIGH T	JUMP	REF.
D F5	MnSiO3	M	6	?	X3.2	4
D F9	BASIC	M	8	4	X3.2	4
D F7	ALUMINA		11	?	X6.2	4
D F2	CaOSiO2	M	6	2	JOINED	4
D F3	MnSiO3	M	7	?	X3.2	4
H/C HI	TITANATE		34	0	JOINED	13
H/C LOW	TITANATE		21	3	X6.2	13
H/C CT	CaOTiO2	1	10	>1	JOINED	13
P 1	PbOSiO2TiO2		260		NONE	14
P 2	PbOSiO2TiO2		230		NONE	14
P 4	PbOSiO2TiO2		200	810	X11	14
P 3	PbOSiO2TiO2		285	470	X20.5	14
P 0	PURE SiO2		2170		NONE	14
P 5	PbOSiO2TiO2		265	1465	X7	14
EUT P	TITANATE	M	25	?	JOINED	22
EUT J	TITANATE	L	6	?	X31	22

LEGEND V RATIO = LIME/SILICA M= MEDIUM
? = TOO FEW DATA
JUMP = VISCOSITY STEP INCREASE
ACTIV'N ENERGY IN kcal

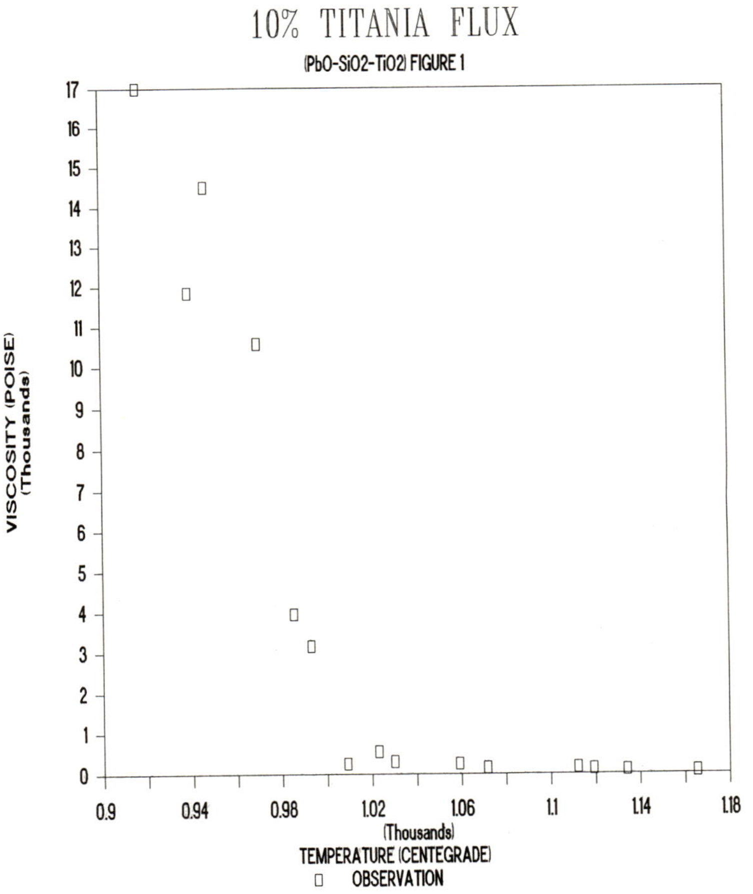

Figure 1. The viscosity/temperature profile of the 60 m/o PbO, 30 m/o SiO_2, 10 m/o TiO_2 flux on continuous cooling.

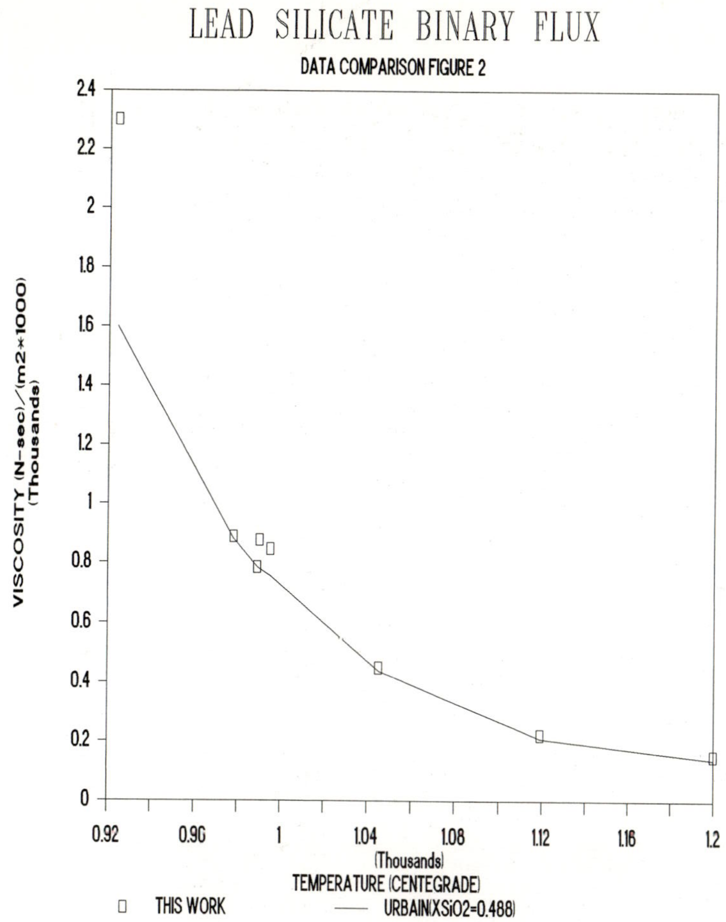

Figure 2. Calibration using a $PbO\text{-}SiO_2$ binary slag.

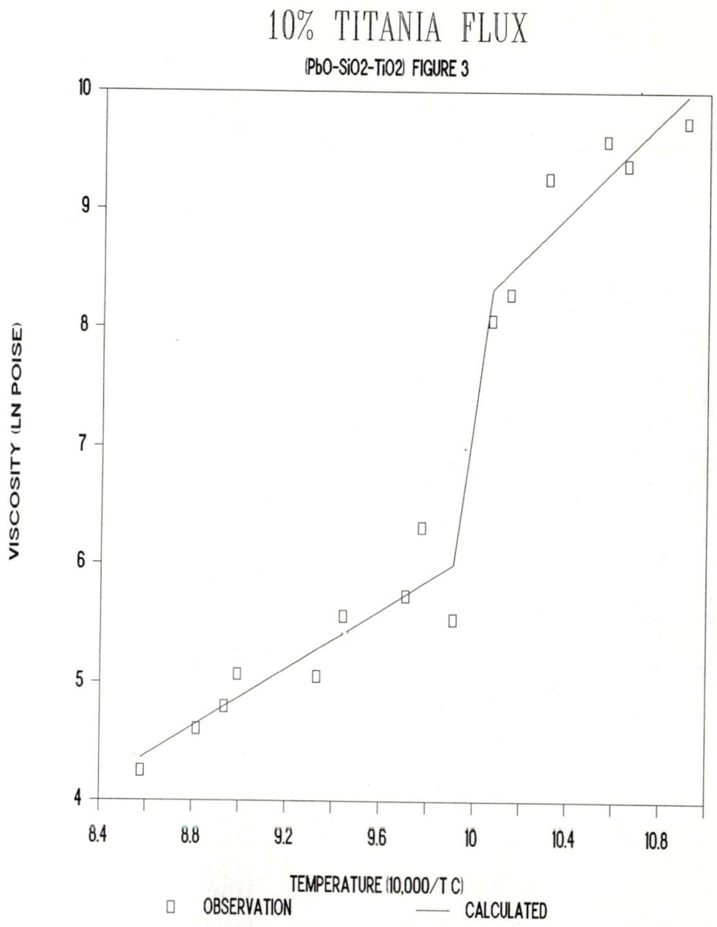

Figure 3. The replot of Figure 1 to determine the activation energy for viscosity. There are two separate activation energies and no common point but over a decade separation.

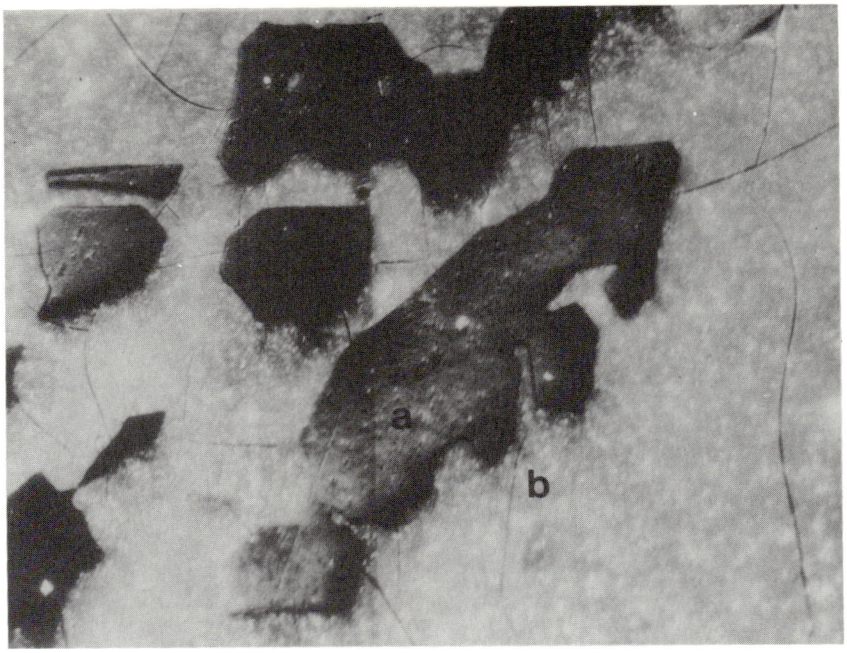

Figure 4a). Micrographs of 2 PbO TiO_2 precipitates in a glassy silicate matrix.

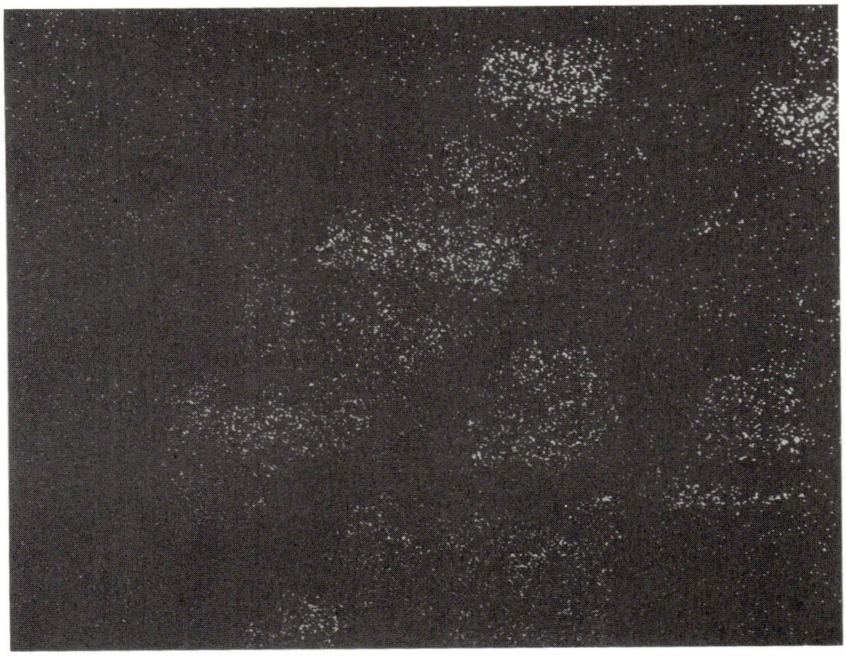

Figure 4b). Ti x-ray map of Figure 4a.

Dissolution of a Transvaal chromite in liquid silicate slags under neutral conditions between 1545 and 1660°C

T R CURR, A WEDEPOHL and R H ERIC

Mr Curr and Mrs Wedepohl are in the Pyrometallurgy and Mineralogy Divisions respectively of the Council for Mineral Technology (Mintek), Randburg, S. Africa; Dr Eric is a Senior Lecturer in the Department of Metallurgy and Materials Engineering, University of the Witwatersrand, Johannesburg, SA.

SYNOPSIS

The compositions of silicate slags, with CaO/SiO_2 mass ratios between 0.03 and 0.55 in equilibrium with a gangue-free Transvaal chromite were determined. The slags were equilibrated with porous chromite crucibles under an argon atmosphere in a vertical-tube resistance-heated furnace. The maximum solubilities of the constituents of chromite spinel (MgO, Al_2O_3, and Cr_2O_3) under these conditions were determined. The solubility of Cr_2O_3 solubility was always less than 1.5 per cent(by mass),and the solubilities of MgO and Al_2O_3 were substantially less than the MgO and Al_2O_3 contents of the slags used industrially for chromite smelting. These two factors indicate that the dissolution of chromite into slags that are tapped from industrial furnaces is likely to be slow. This may in turn be the reason for the observed appearance of undissolved chromite in the discard slags from large submerged-arc furnaces in which chromite is smelted.

Note: all chemical compositions in this paper are expressed in percentages by mass.

INTRODUCTION

Chromite, the ultimate source of all the chromium in stainless steels, is currently smelted in submerged-arc electric furnaces to produce high-carbon ferrochromium for subsequent addition to the steelmaking process. The chromium content of the discard slags from these furnaces effectively determines the chromium recovery, and is therefore an important process variable.

Oosthuyzen and Viljoen[1] recently carried out a mineralogical investigation of the slags from three different plants in South Africa. They found that, at those plants which experienced high losses of chromium (i.e. more than 11 per cent chromium in the slag), more than 60 per cent of this chromium was contained in 'undissolved and partly altered chromite particles'. The presence of these chromite particles in the slags implies that the dissolution of chromite plays an important role in its smelting. Kouroki *et al*[2] have suggested that the dissolution of $MgCr_2O_4$ into molten slags is likely to be the rate-limiting step in the plasma smelting of mixtures of chromite and iron ore. Fujita *et al*[3], using a 550 kg converter, found that slags containing more than 45 per cent MgO plus Al_2O_3 at 1600°C have slower reduction rates, and attribute this to the formation of a layer of MgO- and Al_2O_3-rich spinel around the chromite particles. This layer disrupts the dissolution of the chromite.However, Fukagawa and Shimoda[4] concluded that the dissolution of particles of chromium ore into the molten slag was not the rate-limiting step under static conditions.

The dissolution of chromite is therefore of interest not only for the current submerged-arc process, but also for the new plasma-smelting process[5], and for possible development of a process based on energy from combustion using a combined blown converter[3]. The solubility of chromite in a silicate slag was therefore investigated in order to aquire the basic equilibrium data, a longer-term objective being a study of the kinetics of the dissolution process itself.

EXPERIMENTAL METHOD

A 200 kg sample of chromite from the LG-6 layer of the Bushveld Complex[6] was obtained, and 75 kg of concentrate with a SiO_2 plus CaO content of less than 0.2 per cent was produced by grinding and tabling to remove the gangue (Table 1).

Chromite crucibles (55 mm high and 36 mm in diameter) were made from this material by hand-ramming, drying, and sintering under an argon atmosphere at 1550°C for 1 to 2 hours. An aqueous solution of chromium trioxide was used as a binder for the hand-ramming operation. No significant difference was found between the composition of the chromite in the starting material and in the final sintered crucibles (Table 1). The crucibles had a bulk density of 2.76 g cm^{-3}, a porosity of 38.1 per cent, and a typical mass of 92 g.

A range of synthetic slags was prepared from commercial grades of limestone, quartz, magnesia, and alumina by melting in a small graphite-lined arc-furnace. The molten slags were cooled, ground, and roasted at 550°C for 16 hours to oxidize any graphite or metallic iron present. The compositions of the seven slags are shown in Table 2.

A vertical alumina-tube furnace with a molybdenum-wire heating element was used to maintain the sample at a constant temperature (± 5°C) under an atmosphere of spectrographic-grade argon (total impurities less than 0.002 per cent by volume). The chromite crucibles were filled with approximately 20 g of one of the synthetic slags, and then raised under sealed conditions into the hot zone of the furnace over a period of 45 minutes. The samples were usually held in the hot zone for 1 hour but were occasionally held for up to 12 hours. They were then lowered rapidly (within 30 seconds) into a cold zone (less than 200°C) of the furnace under argon for cooling and final removal. The temperature, which was measured throughout the run, varied by an average of 6°C, while the standard deviation of the change in sample mass after the runs was 0.15 g.

The crucibles and the contained slag (most of which had penetrated the porous crucible walls) were sectioned and examined mineralogically. Phases within the crucible walls were analysed by energy-dispersive spectroscopy (EDS) or by use of an electron microprobe.

RESULTS

The slag that had penetrated the porous walls of the chromite crucibles was observed to have interacted with the chromite grains close to the inner surface of the wall in a variety of ways (Figure 1) depending on the composition of the slag charge. The interaction was usually confined to the edges of the chromite grains, and took the following forms:

(a) entire portions of the original outer surface of the spinel were removed, leaving a chromite grain of drastically altered appearance (slag no.7),

(b) the original surface of the spinel was unchanged, but the centre and edges of the crystals became visibly zoned (slag no.3),

(c) crystals of secondary spinel precipitated on the edges of the chromite grains (slag no.4), or

(d) no visible interaction occurred (slag no.6).

The alteration of the surfaces of the chromite grains decreased with increasing distance from the inner surface of the crucible wall regardless of the form of the alteration.

The compositions of the edges of the chromite grains in contact with a slag phase at inner, middle, and outer positions of the crucible wall showed a convergence towards the original composition in all cases (Figure 2). The centres and edges of chromite grains in the outer portion of the crucible were analysed and compared to detect any concentration gradient within the grains. The differences in the analyses were typically less than one standard deviation. These analyses (centre and edge) were then averaged and compared with the chemical analyses of the original chromite and the average EDS analysis of four chromite grains (centre and edge) in a blank crucible (i.e. one that had undergone the same experimental procedure without a slag charge). The differences between the analyses were found to be within the limits of experimental error for the various methods of determination (Table 1.)

The absence of an experimentally significant concentration gradient in chromite grains in the outer region of the crucible wall, even in the samples that had been held in the hot zone of the furnace for up to 12 hours, led to the conclusion that the slag and chromite at these positions were in chemical equilibrium, and that the slag in contact with the chromite was therefore saturated with respect to constituents in the chrome spinel.

The determination of the equilibrium slag was complicated in many cases by the formation of a secondary crystalline phase upon cooling. The relative proportions of the crystalline and matrix phases present in these cases were estimated visually, and were checked by an EDS image-analysis technique, which indicated that a likely error of ± 10 per cent could be expected. The effect of this error, including the assumption that the density difference was negligible, was frequently reduced because of the similar compositions of the crystalline and matrix phases. A further estimate of the likely error in this procedure was made from a comparison of the CaO/SiO_2 ratio determined in this way with the initial CaO/SiO_2 ratio in the slag (Figure 3).

The estimated equilibrium compositions of the slag are shown in Figures 4 and 5. The data denoted $(FeO)_T$ were calculated from the total iron detected expressed stoichiometrically as FeO, although clearly a significant proportion of Fe_2O_3 should be present. The estimated errors in these compositions were less than 2.0 per cent for MgO, Al_2O_3, and $(FeO)_T$, and less than 0.1 per cent for the Cr_2O_3.

DISCUSSION

The mechanism by which the slag approaches equilibrium with the chromite is presumed to be the repeated transfer of mass between the slag and the edges of the chromite grains as the slag flows outwards through the porous wall of the chromite crucible. Since the phases were analysed by EDS or with the electron microscope, the oxidation state of the iron could not be determined. The usefulness of the technique is therefore limited, although it does provide a useful starting-point for the detailed study of complex systems in which the number of components makes the use of a classical bracketing technique extremely laborious.

The increase in the solubility of MgO, Cr_2O_3, and $(FeO)_T$ with increasing temperature was in accord with expectations. However,the solubility of Al_2O_3 in slags with CaO/SiO_2 ratios of less than 0.3 decreased from 1550°C to 1650°C (Figures 4 and 5). The reason for this behaviour is not known, and further experiments with controlled Fe^{2+}/Fe^{3+} ratios in the slag may be necessary.

The very low solubility of Cr_2O_3 (less than 1.5 per cent) in the slag therefore limits the solubility of the chromite particles. Under conditions of practical smelting, chromite would dissolve only in conditions that were sufficiently reducing to maintain the Cr_2O_3 content of the slag well below this value.The presence of pure solid carbon at 1650°C will satisfy this condition thermodynamically, but the transport of the Cr_2O_3 dissolved in the slag, either by diffusion or by convection, will be constrained by the low solubility of Cr_2O_3, which limits the concentration gradient or driving force. The static slag/chromite conditions in Fukgawa and Shimoda's experiments[4], and the fact that an increase in the MgO and Al_2O_3 contents of the slag-reduced the reduction rate, point towards the rate of diffusion of Cr_2O_3 in the slag as the rate-limiting step in this case.

The solubilities of MgO and Al_2O_3 at 1650°C that were determined in the present study (6.0 and 16.0 per cent respectively) are significantly lower than the MgO and Al_2O_3 contents of the industrial slags (with similar CaO/SiO_2 ratios) investigated by Oosthuyzen and Viljoen[1] (20.0 and 25.0 per cent respectively). It would therefore be expected that chromite entering a similar slag would not dissolve immediately even if the conditions were sufficiently reducing and the slag were sufficiently well stirred to overcome the constraints on the transport of dissolved Cr_2O_3. These constraints were demonstrated experimentally by Fujita *et al*[3] in a well stirred smelting-reduction converter.

CONCLUSIONS

The compositions of silicate slags in equilibrium with gangue-free Transvaal chromite as a function of CaO/SiO_2 at 1550°C and 1650°C have been determined using a porous chromite crucible technique. The low solubility of Cr_2O_3 in these slags (< 1.5 per cent) implies that strongly reducing and well stirred slag conditions would be required to ensure the rapid dissolution of chromite. Another limitation to chromite dissolution would be expected if the MgO and Al_2O_3 contents of the slag were more than 6.0 and 16.0 per cent respectively at 1650°C and at a CaO/SiO_2 ratio of 0.3.

ACKNOWLEDGEMENTS

This paper is published by permission of the Council for Mineral Technology. The contributions of the following people are gratefully acknowledged: Mr C.K. Grassman (laboratory furnace), Mr D.A.Hayman (arc furnace), Mr P. Ellis (EDS analyses), Mrs J.A. Russel (microprobe analysis), Mr R. Guest (chromite tabling),and Dr N.A. Barcza (helpful discussions and suggestions).

REFERENCES

1.Oosthuyzen, E.J. and Viljoen, E.A.Proc. XIV Int. Mineral Processing Congress. Toronto, 1982. Canadian Institute of Metallurgists. pp 6.1-6.17.

2.Kouroki, S., Morita,K. and Sano, N. Proc. Mintek 50 Int. Conf. on Mineral Sci and Technology. Sandton, S. Africa, 1984, Mintek. pp 847-862.

3.Fujita, M. *et al*. Tetsu-to-Hagane, 74 (1988) no 4. pp 680-687.

4.Fukagawa, S. and Shimoda, T. Trans ISIJ. 27 (1987) pp 609-617.

5.Slatter, D *et al*. Infacon 86, 4th Int. Ferro-alloy Congress. Sao Paulo, (1986) Associacao Brasileira dos Productores de Ferro-ligas (ABRAFE), 10 pp.

6.De Waal, S.A. and Hiemstra, S.A. Report 1709 Mintek, Randburg (1975), 86 pp.

Table 1
Comparisons of chromite compositions
(expressed in percentages by mass)

Sample	Head sample	Blank run		Chromite grains in contact with slag	
ANALYTICAL METHOD	XFS	XFS	EDS	EDS and microprobe.	
NUMBER OF SAMPLES	2	2	8	26	26
STATISTIC	Mean	Mean	Mean	Mean	σ
Cr_2O_3	47.5	47.9	48.5	48.3	1.2
FeO	18.5	18.2	ND	ND	ND
$(FeO)_T$	25.3	25.0	26.2	25.1	1.3
MgO	9.8	9.8	11.3	11.0	0.7
Al_2O_3	13.1	12.9	13.0	14.4	1.2
SiO_2	<0.3	<0.3	<0.1	<0.1	ND
CaO	<0.2	<0.2	<0.1	<0.1	ND
TiO_2	0.6	0.5	0.2	0.3	0.1

Size analysis : 27% 150 μm, 16% 75 μm, 0.4% 38 μm

XFS X-ray-fluorescence spectroscopy
EDS Energy dispersive spectroscopy
σ Standard deviation
ND Not determined
FeO Was analyzed by wet chemistry
$(FeO)_T$ Total iron expressed as FeO

Table 2
Compositions of the synthetic slags
(expressed in percentages by mass)

Constituent	Slag no. 1	2	3	4	5	6	7
SiO_2	53.5	52.0	52.7	42.2	44.3	55.1	55.0
CaO	25.6	20.6	2.0	11.4	1.1	4.0	35.9
MgO	13.0	9.8	18.2	26.0	23.1	29.4	4.5
Al_2O_3	4.0	14.4	24.3	18.0	28.8	8.9	1.4
Cr_2O_3	0.02	0.02	0.10	0.22	0.02	0.02	0.09
FeO	0.5	1.0	0.3	0.3	0.8	0.5	0.8
Fe_2O_3	0.1	0.1	0.2	0.1	<0.1	0.1	0.1
Na_2O	0.03	0.02	0.16	0.12	0.09	0.04	0.05
K_2O	0.05	0.04	0.02	0.04	0.05	0.07	0.04
CaO/SiO_2	0.48	0.40	0.04	0.27	0.03	0.07	0.65

Notes
(1) For all slags: MnO < 0.15 %
: TiO_2 < 0.20 %
: ZnO < 0.003 %
(2) Fe_2O_3 = (total iron) - (FeO)
(3) Analyses by X-ray fluorescence spectrometry except
Cr_2O_3 and FeO : wet chemical
Na_2O, K_2O, and ZnO : atomic absorption spectrometry

Slag no. 4

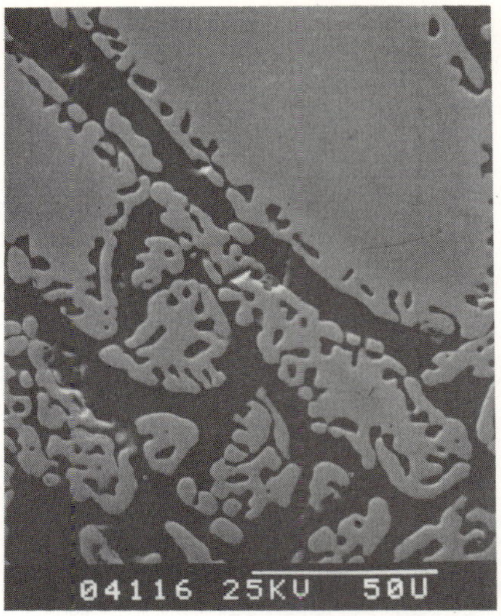

Slag no. 7

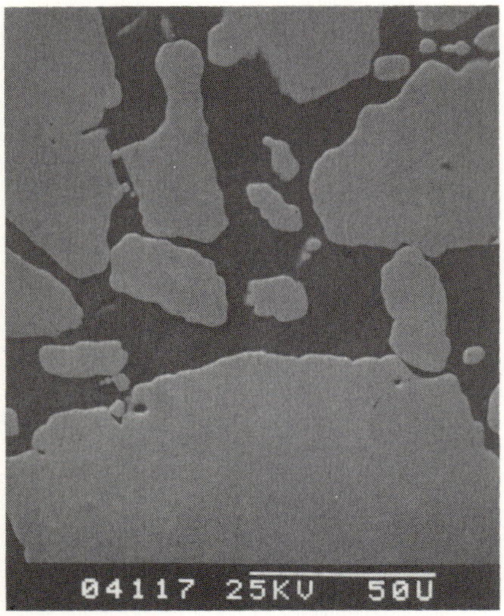

Slag no. 6

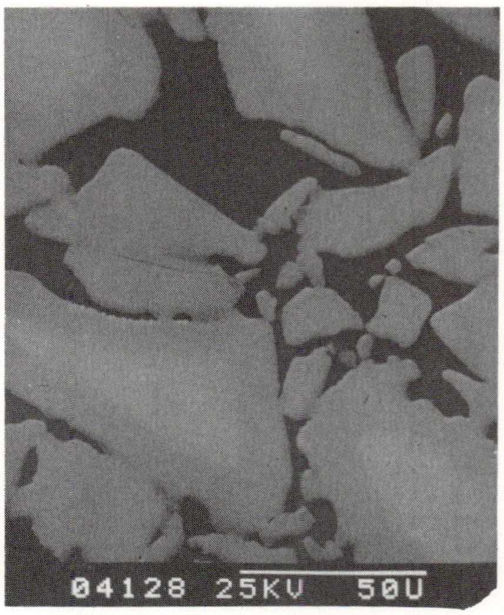

Slag no. 3

FIGURE 1. Interaction of slag with chromite grains at the inner surface of the wall of the chromite crucible.

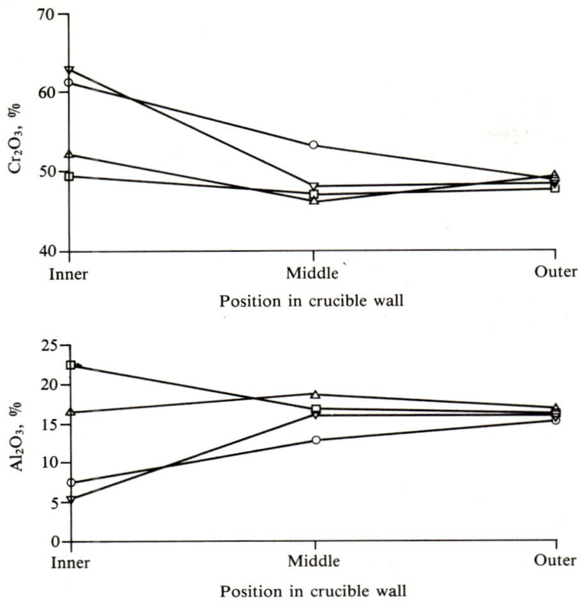

FIGURE 2. Compositional changes in the edges of chromite grains with increasing distance from the inner wall of the crucible.

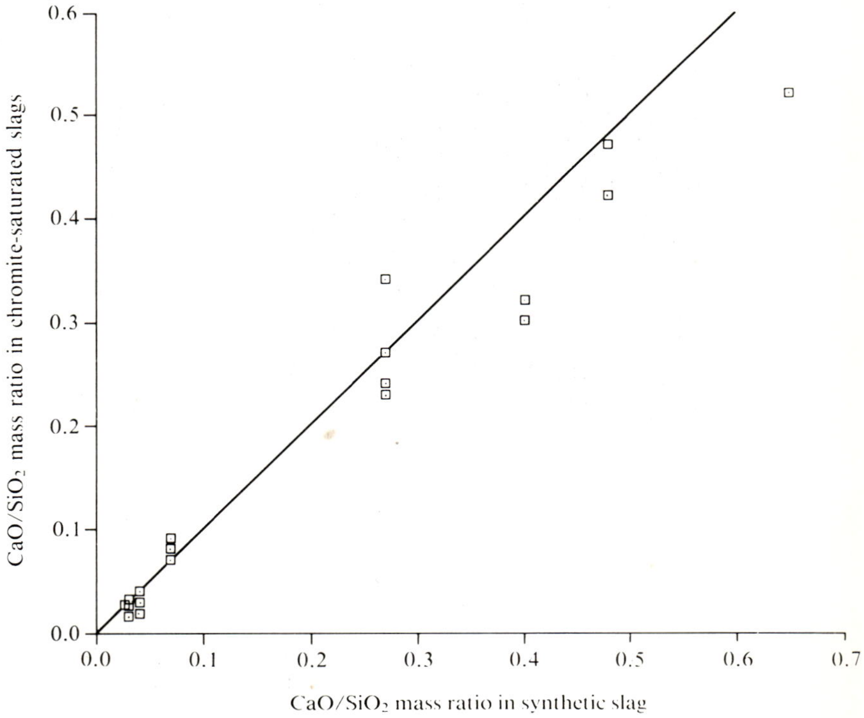

FIGURE 3. Comparison of the CaO/SiO_2 mass ratios in the synthetic slags with those measured in the slag in the outer portions of the crucible wall

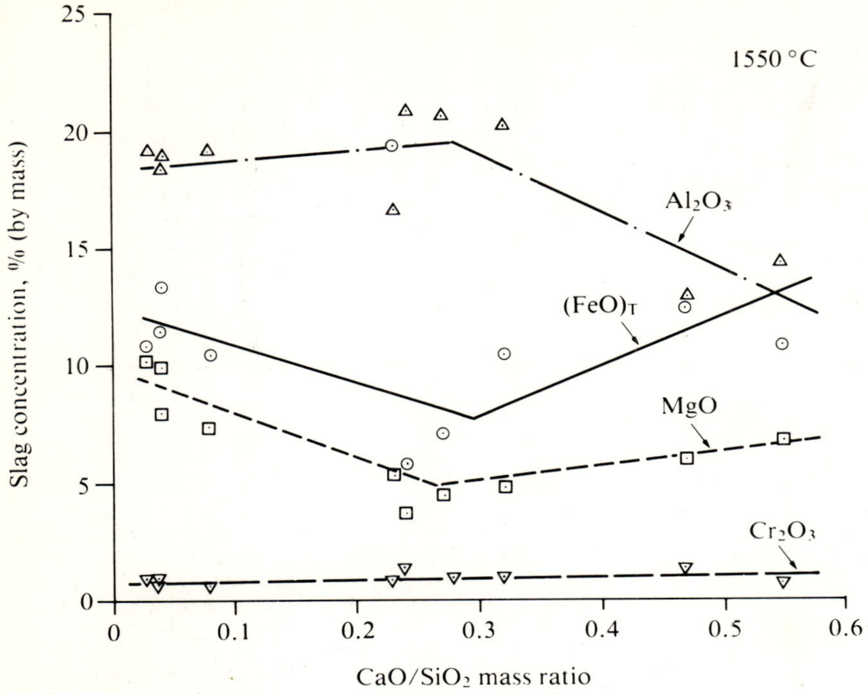

FIGURE 4. Compositions of slags in equilibrium with chromite between 1545 and 1560°C

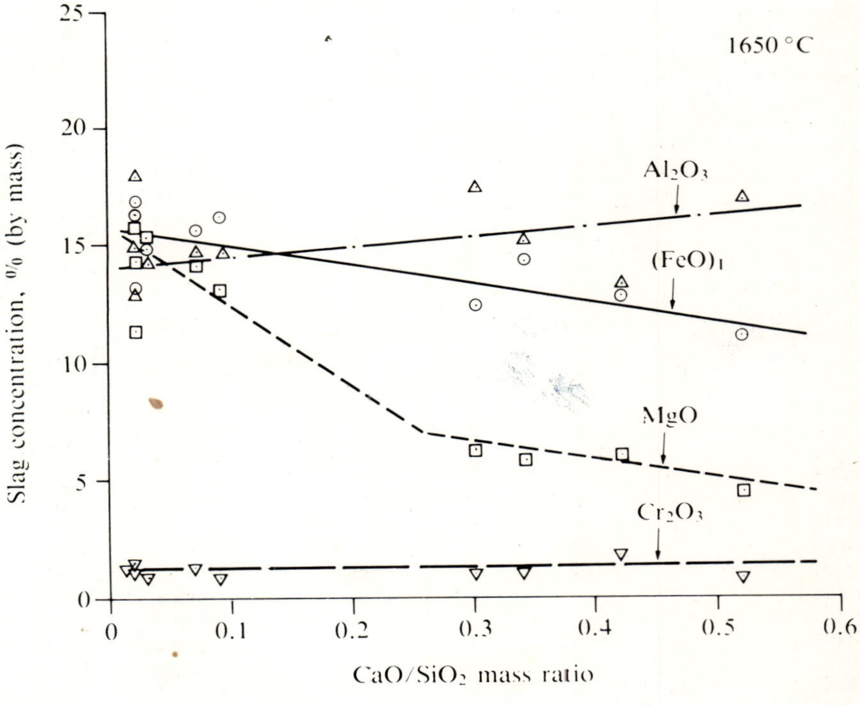

FIGURE 5. Compositions of slags in equilibrium with chromite between 1645 and 1660°C

Simultaneous reaction model for reduction of Cr_2O_3 from slag

C ANYAKWO, S SIMUKANGA and R J POMFRET

Dr Anyakwo, formerly a Research Student at the University of Strathclyde, is now lecturing in Nigeria; Mr Simukanga is a Research Student in the Department of Metallurgy and Materials Engineering at the University of Strathclyde, Glasgow; Dr Pomfret is in the Department of Metallurgy and Materials Engineering at the University of Strathclyde, Glasgow, UK.

SYNOPSIS

The reduction of Cr_2O_3 from a 42% SiO_2, 38% CaO, 20% Al_2O_3 slag by carbon dissolved in iron at 1460°C was studied. Chromium oxide reduction takes place primarily at the slag/metal interface and Cr^{2+} is formed as an intermediary product. The Cr^{2+} reduction is markedly influenced by the furnace atmosphere, where an argon atmosphere gives faster reduction than a CO atmosphere.

Consecutive first order models of increasing complexity are used to analyse the experimental results from which it is found that the reduction reaction follows a consecutive reversible first-order mechanism.

INTRODUCTION

The reduction of Cr_2O_3 from liquid slags takes place as a two stage process. These can be represented as:

$$Cr_2O_3 + \underline{C} \rightleftharpoons 2CrO + CO(g) \quad (1)$$

$$2CrO + 2\underline{C} \rightleftharpoons 2\underline{Cr} + 2CO(g) \quad (2)$$

The reactions suggest that the reduction of Cr_2O_3 from slag follows a consecutive reaction mechanism in which the product of the first reaction becomes itself the reactant of the following reaction. This is evidenced by experimental observations in which there is a colour change in the slag from bright green to deep blue during the initial stages of reaction. This indicates that Cr^{2+} is being formed as an intermediary product.

The present work presents some results of an ongoing study of the reduction of Cr_2O_3 from a 42% SiO_2, 38% CaO, 20% Al_2O_3 slag by carbon dissolved in liquid iron at 1460°C. The experimental results are analysed using consecutive first order reaction models of increasing complexity.

EXPERIMENTAL

The experiments were carried out in a vertical resistance furnace with the slag and metal being contained in a 32 mm diameter recrystallised alumina crucible. The furnace atmosphere was either carbon monoxide or argon. The course of the reaction was followed by slag analysis. The slags were analysed by atomic absorption spectroscopy for total iron and total chromium, and by potentiometric titration for the sum of Fe^{2+} and Cr^{2+}. Cr^{3+} was calculated by difference.

RESULTS

REACTION SITE

There is the possibility of gaseous reduction of Cr_2O_3 taking place at the slag/gas interface, especially in a CO atmosphere. Figure 1 shows the results of two experiments done without any metal present. One was done in a CO atmosphere, the other in argon.

The results indicate that there is very little difference between the results for the different atmospheres, although a small amount of Cr^{3+} is reduced to Cr^{2+} in the first 10 minutes. This indicates that the reduction takes place primarily at the metal/slag interface, and that a metal surface is needed for the reaction to take place at all.

EFFECT OF FURNACE ATMOSPHERE

The furnace atmosphere does, however, have an effect on the reaction between slag and metal. Similar effects have been noted previously in silica reduction studies [1]. As seen in Figures 2 and 3, CO appears to slightly increase the rate of reduction of Cr^{3+}, whilst argon seems to lead to lower Cr^{2+}, by increasing the rate of reduction by a factor of about 1.2. In graphite crucibles silica is reduced faster from calcium silicate slags by carbon saturated iron under an argon atmosphere, than under CO, and it was observed that the slag wetted the crucible much more under argon than under CO. It was proposed that the surface effect changed the reduction reaction which takes place primarily at the slag/crucible/ metal interface [2]. It may be that a similar effect is taking place in these experiments, although the slags appear to wet the crucible under both gases.

CONSECUTIVE REACTION MODEL

Solutions can be obtained for consecutive first order reactions, where the rate equations are solved by using Laplace transforms [3].

The simplest case is of two irreversible reactions.

Case 1: Consecutive irreversible first order reactions:

$$(Cr^{3+}) \xrightarrow{k_1} (Cr^{2+}) \xrightarrow{k_3} [Cr]$$

where the rate equations are:

$$\frac{d(Cr^{3+})}{dt} = -\frac{k_1}{h}(Cr^{3+})$$

$$\frac{d(Cr^{2+})}{dt} = \frac{k_1}{h}(Cr^{3+}) - \frac{k_3}{h}(Cr^{2+})$$

$$\frac{d[Cr]}{dt} = \frac{k_3}{h}(Cr^{2+})$$

where h is the slag height in cm.

Solutions are:

$$(Cr^{3+}) = (Cr^{3+})_o \exp(-K_1 t)$$

$$(Cr^{2+}) = \frac{(Cr^{3+})_o K_1}{K_3 - K_1}[\exp(-K_1 t) - \exp(-K_3 t)]$$

$$[Cr] = (Cr^{3+})_o \left[1 + \frac{1}{K_1 - K_3}[K_3 \exp(-K_1 t) - K_1 \exp(-K_3 t)]\right]$$

where $K_1 = \frac{k_1}{h}$ etc.

For the boundary conditions:

At $t = 0$, $(Cr^{3+}) = (Cr^{3+})_o$; $(Cr^{2+}) = 0$; $[Cr] = 0$.

As can be seen in Figure 4 the fit is not very good, especially for the Cr^{2+} at the longer times. The fit can be improved by considering the first reaction to be reversible.

Case 2: Consecutive first order reactions. 1. reversible, 2. irreversible.

$$(Cr^{3+}) \underset{k_2}{\overset{k_1}{\rightleftharpoons}} (Cr^{2+}) \xrightarrow{k_3} [Cr]$$

Where the rate equations are:

$$\frac{d(Cr^{3+})}{dt} = \frac{k_2}{h}(Cr^{2+}) - \frac{k_1}{h}(Cr^{3+})$$

$$\frac{d(Cr^{2+})}{dt} = \frac{k_1}{h}(Cr^{3+}) - \frac{(k_2 + k_3)}{h}(Cr^{2+})$$

$$\frac{d[Cr]}{dt} = \frac{k_3}{h}(Cr^{2+})$$

Solutions are:

$$(Cr^{3+}) = (Cr^{3+})_o \left[\frac{(K_1 + K_3) - m_1}{m_2 - m_1}\exp(-m_1 t) + \frac{(K_2 + K_3) - m_2}{m_1 - m_2}\exp(-m_2 t)\right]$$

$$(Cr^{2+}) = K_1 (Cr^{3+})_o \left[\frac{1}{m_2 - m_1}\exp(-m_1 t) + \frac{1}{m_1 - m_2}\exp(-m_2 t)\right]$$

$$[Cr] = K_1 K_3 (Cr^{3+})_o \left[\frac{1}{m_1 m_2} - \frac{1}{m_1(m_2 - m_1)}\exp(-m_1 t) - \frac{1}{m_2(m_1 - m_2)}\exp(-m_2 t)\right]$$

where:

$$m_1 = 0.5\left\{(K_1+K_2+K_3) + \sqrt{(K_1+K_2+K_3)^2 - 4K_1K_3}\right\}$$

$$m_2 = 0.5\left\{(K_1+K_2+K_3) - \sqrt{(K_1+K_2+K_3)^2 - 4K_1K_3}\right\}$$

Considering the Cr^{3+} reduction reaction as reversible has given some improvement in the fit, (Fig. 5), especially the Cr^{3+} curve. But as a final stage both reactions can be considered to be reversible.

Case 3: Consecutive reversible first order reactions

$$(Cr^{3+}) \underset{k_2}{\overset{k_1}{\rightleftharpoons}} (Cr^{2+}) \underset{k_4}{\overset{k_3}{\rightleftharpoons}} [Cr]$$

Where the rate equations are:

$$\frac{d(Cr^{3+})}{dt} = \frac{k_2}{h}(Cr^{2+}) - \frac{k_1}{h}(Cr^{3+})$$

$$\frac{d(Cr^{2+})}{dt} = \frac{k_1}{h}(Cr^{3+}) - \frac{(k_2 + k_3)}{h}(Cr^{2+}) + k_4[Cr]$$

$$\frac{d[Cr]}{dt} = \frac{k_3}{h}(Cr^{2+}) - \frac{k_4}{h}[Cr]$$

Solutions are:

$$(Cr^{3+}) = (Cr^{3+})_o \left[\frac{K_2K_4}{m_1 m_2} + \frac{m_1^2 - m_1(K_2 + K_3 + K_4) + K_2K_4}{m_1(m_1 - m_2)}\exp(-m_1 t) + \frac{m_2^2 - m_2(K_2+K_3+K_4) + K_2K_4}{m_2(m_2 - m_1)}\exp(-m_2 t)\right]$$

$$(Cr^{2+}) = K_1(Cr^{3+})_o\left[\frac{K_4}{m_1m_2} + \frac{K_4-m_1}{m_1(m_1-m_2)}\exp(-m_1t) + \frac{K_4-m_2}{m_2(m_2-m_1)}\exp(-m_2t)\right]$$

$$[Cr] = K_1K_3(Cr^{3+})_o\left[\frac{1}{m_1m_2} + \frac{1}{m_1(m_1-m_2)}\exp(-m_1t) + \frac{1}{m_2(m_2-m_1)}\exp(-m_2t)\right]$$

where:

$$m_1 = 0.5\left\{(K_1+K_2+K_3+K_4) + \sqrt{(K_1+K_2+K_3+K_4)^2 - 4(K_1K_3+K_2K_4+K_1K_4)}\right\}$$

$$m_2 = 0.5\left\{(K_1+K_2+K_3+K_4) - \sqrt{(K_1+K_2+K_3+K_4)^2 - 4(K_1K_3+K_2K_4+K_1K_4)}\right\}$$

Using a model which considers both reactions to be reversible gives a good fit, particularly to the Cr^{2+} plateau (Fig. 6).

DISCUSSION

The experimental results appear to fit a consecutive reversible first order model since the sum of squares for the Cr^{3+}, Cr^{2+} and [Cr] points, which is a measure of how close the experimental points are to the theoretical curves, are the lowest compared to the earlier models considered. The model is now being extended to allow for [Cr] to be non-zero at t = 0; this will allow more account to be placed on changing slag volume.

CONCLUSIONS

1) The reduction of Cr_2O_3 from slags takes place primarily at the slag/metal interface.
2) There is an influence of the furnace atmosphere, particularly on the Cr^{2+} reduction reaction, where an argon atmosphere gives faster reduction than a CO atmosphere.
3) The reduction reaction follows a consecutive reversible first order mechanism.

ACKNOWLEDGEMENT

The authors would like to thank the British Council for financial assistance.

REFERENCES

1. R.J. Pomfret and P. Grieveson, "High Temperature Chemical Reaction Engineering" (edited by C.N. Kenny), p.27, Inst. Chem., London (1975).

2. R.J. Pomfret and P. Grieveson, Canadian Metallurgical Quarterly, 22, No. 3, p.287 (1983).

3. C. Capellos and B.H.J. Bielski, "Kinetic Systems - Mathematical Description of Chemical Kinetics in Solutions", New York, Wiley Interscience (1972).

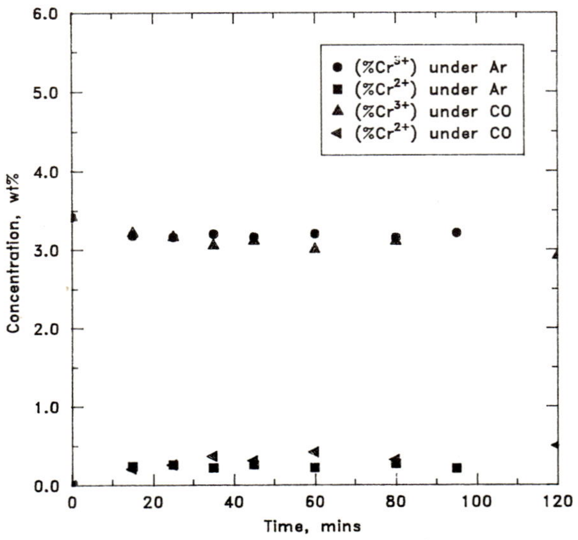

Fig. 1 Reduction of slags in the absence of metal.

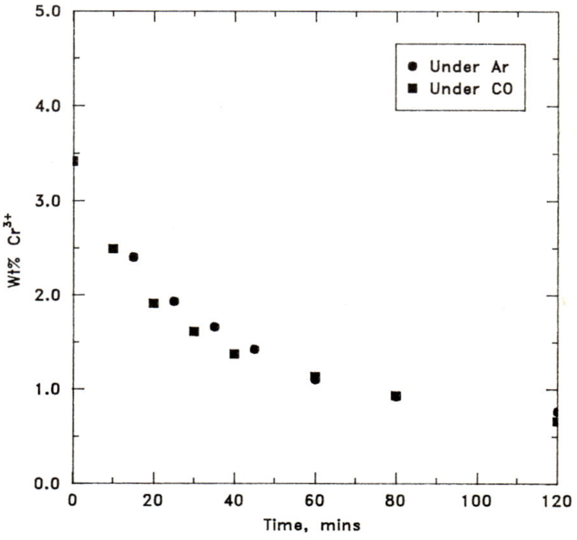

Fig. 2 Effect of furnace atmosphere on Cr^{3+} reduction.

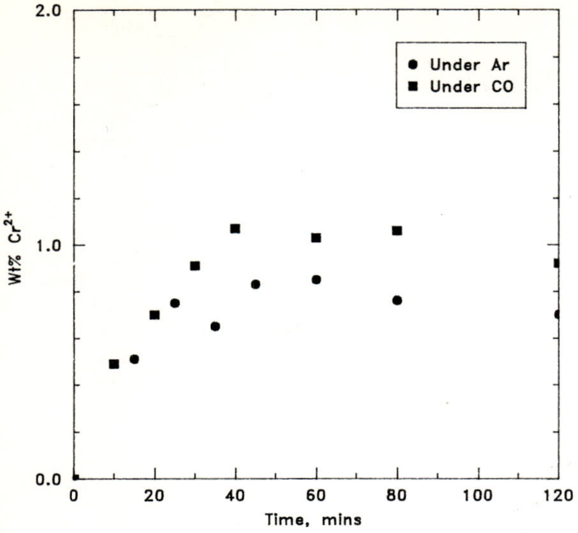

Fig. 3 Effect of furnace atmosphere on Cr^{2+} reduction.

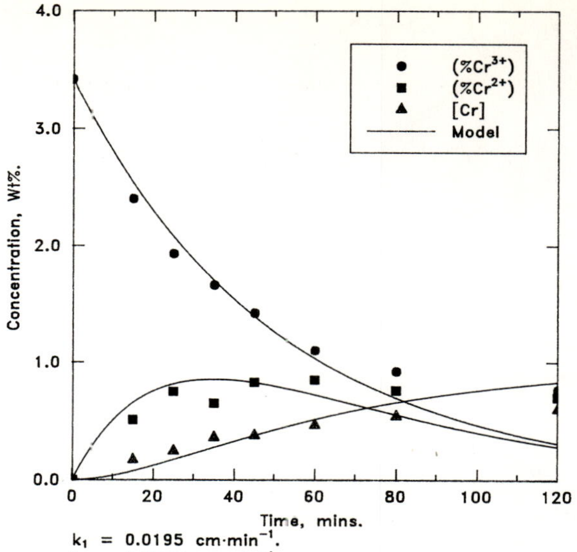

k_1 = 0.0195 cm·min^{-1}.
k_3 = 0.0390 cm·min^{-1}.

Sums of squares of deviations are:
Cr^{3+} = 0.301, Cr^{2+} = 0.303, $Cr_{Met.}$ = 0.091.

Fig. 4 Rate curves for consecutive irreversible first order reactions with experimental results.

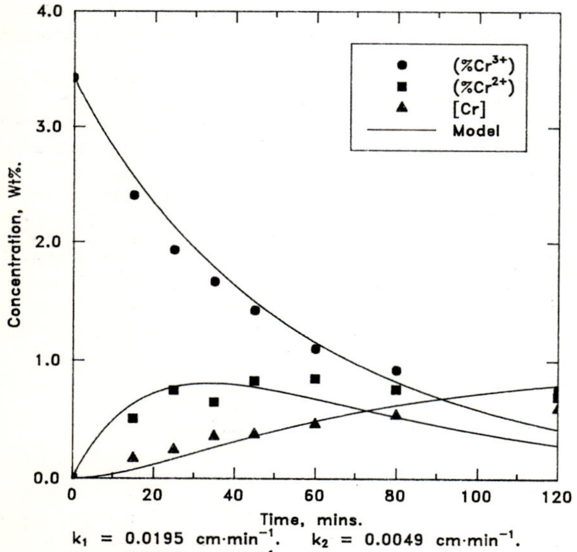

k_1 = 0.0195 cm·min^{-1}. k_2 = 0.0049 cm·min^{-1}.
k_3 = 0.0390 cm·min^{-1}.

Sums of squares of deviations are:
Cr^{3+} = 0.210, Cr^{2+} = 0.298, $Cr_{Met.}$ = 0.071.

Fig. 5 Rate curves for consecutive first order reactions (1 reversible, 2 irreversible) with experimental results.

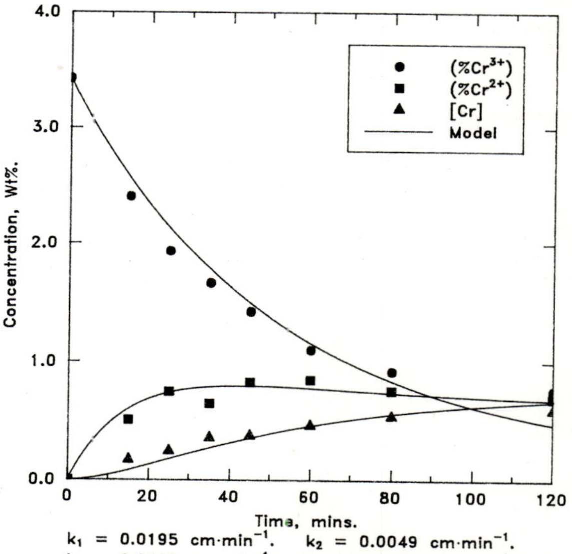

k_1 = 0.0195 cm·min^{-1}. k_2 = 0.0049 cm·min^{-1}.
k_3 = 0.0488 cm·min^{-1}. k_4 = 0.0117 cm·min^{-1}.

Sums of squares of deviations are:
Cr^{3+} = 0.174, Cr^{2+} = 0.037, $Cr_{Met.}$ = 0.025.

Fig. 6 Rate curves for consecutive reversible first order reactions with experimental results.

Lime agglomerates for steelmaking: physical and slag-forming property

R C GUPTA

The author is in the Department of Metallurgical Engineering at the Banaras Hindu University, Varanasi, India.

SYNOPSIS

The lime agglomerates have been developed as an alternative to the lime for use in the LD steelmaking process. These agglomerates possess a regular shape and size, have good strength and provide better storage and slag-forming ability. These are produced by utilising steel plant waste and hence may prove more economical to use than ordinary lime.

The lime sinter is obtained by sintering lime-stone chips coated with iron ore fines utilising coke breeze as fuel. The hollow pellets are high-basicity pellets obtained by two stage pelletisation. First, the iron ore and lime fines are pelletised and then are further coated with iron ore only after firing. The lime briquettes are obtained by compacting lime fines and iron ore fines mixed together, followed by heat hardening.

The paper gives the technique used for their preparation and describes their physical and slag-forming property.

INTRODUCTION

The lime is an important raw material for LD steelmaking. The lime is obtained by calcination of limestone and is added to the melt to produce a specific type of slag. The composition, temperature, fluidity and other parameters of the slag have a decisive effect on the quality of steel produced. The lime is required to the following properties:

(a) A size of 10-50 mm is considered most suitable for the purpose. Since smaller sizes may be carried off by gases whilst bigger sizes require longer dissolution times.

(b) The strength of the lime determines the amount of fines generated during handling. The lime has to be transported from the calcination kilns to the steel shop and storage bins through a system of conveyors. From the bin the materials are discharged through vibrating feeders into weigh hoppers and then fed to the converter through belt, conveyer, hopper and chute. In this process the fines generated are carried away by gases which results in an increase in the pH value of water in the gas cleaning system.

(c) The weather resistance of the lime becomes a significant factor in humid regions. Since lime is hygroscopic in nature, pick-up of moisture from the atmosphere occurs. The moisture in lime results in an increase in the hydrogen content of the steel. Furthermore, as a result of slaking, the strength of lime is reduced so much that it becomes powder after few days of exposure to the atmosphere. Good practice demands that the lime be used as soon as possible after calcination. The scheduling and coordination of lime production, transportation and utilisation is thus critical, if the calcination plant and steel plant are some distance apart.

(d) The composition of lime determines its fluxing ability. Consequently higher CaO contents and lower levels of SiO_2, Al_2O_3, etc. are desirable.

At present most of the lime in steel plants is made by firing limestone in kilns using oil as the fuel. The lime produced is used in the plant. This process has certain limitations:

(1) During calcination nearly 50% lime is generated below 5 mm size and is rejected as waste since it cannot be charged in the LD vessel as such. This wastage is against mineral conservation policy.

(2) Any effort to produce hard lime by increasing the firing temperature reduces its reactivity, whereas, a lower firing temperature reduces the storage life of the lime due to its high sensitivity towards moisture absorption.

(3) With increasing oil costs and its lower availability, the use of solid fuels is becoming a necessity in some parts of the world.

(4) In LD steelmaking the rate of removal of impurities depends upon the early formation of slag. The lime as such takes time to form slag and hence various techniques are being used to accelerate the formation of slag.

Keeping the needs of LD steelmaking in view and considering the limitations of the existing lime preparation technique, the lime agglomerates eg Lime Sinter, Hollow Pellet and Lime Briquettes have been developed at Banaras Hindu University which could be suitable in different working conditions. These lime agglomerates are prepared by different techniques using different types of raw materials.

LIME AGGLOMERATION TECHNIQUES

Lime Sinter

The lime sinter developed by Gupta and Singh[1] consists of agglomerated , calcined, lime particles, bonded together by a coating of calcium ferrite-rich slag. During the sintering process, the limestone gets calcined as well as agglomerated by fusion of the low-melting calcium ferrite present on the surface of lime. This thin coating of fused calcium ferrite also serves as a shield against moisture pick-up by the lime.

The preparation of lime sinter needs limestone chips (5-10 mm size) as basic raw material which may be available as undersized rejects at the mine site. The coke breeze, of 1-2 mm size, is used as fuel in the sinter mixture. The iron ore fines (.003 mm) are utilised to provide a coating on the limestone chips. The chemical composition of typical raw materials used and the product obtained are given in

Table 1. Whilst preparing lime sinter, about 10–15% iron ore fines are mixed with limestone chips using about 6% water. Later, about 15% coke breeze is mixed with these limestone chips coated with ore fines. This mixture is then charged into a pot-type, laboratory sintering pan and the sintering is initiated by igniting the top charge with the application of suction through the grate. The 200 mm thick bed needs 20 minutes to sinter.

Hollow Pellet

This is a high basicity (0.94–1.84) iron ore pellet developed by Gupta and Prakash(2). In conventional pelletisation techniques, when lime is mixed with iron ore it poses a sticking problem during firing due to the fusion of low-melting, calcium ferrite. In order to solve this problem and prepare high basicity pellets, a pellet is made of a mixture of iron ore and lime to give a basicity in the range of 4–5. Then, this highly basic pellet is coated with only iron ore during the second stage of pelletisation. The thickness of the iron ore coating could be adjusted to give the required overall total basicity of 0.94 to 1.84. This double layer pellet when fired does not cause sticking problems due to the presence of iron ore on the surface, which is comparatively refractory in nature. The core pellet of iron ore and lime mixture fuses and melts during firing, developing a cavity in the pellet and hence the name 'hollow pellet'.

Lime Briquette

The lime fines generated at the calcination plant could be used as lumps after briquetting. A study done by Pandey(3) reported that if a high compacting pressure (1500 $kgcm^{-2}$) is used, the briquettes of lime produced have sufficient strength for subsequent handling. The addition of 4–15% iron oxide in lime briquette, followed by firing at 1200°C for one hour, resulted in strong briquettes. In the fired briquette the strength depended on the slag bond and hence a lower briquetting pressure (750 $kgcm^{-2}$) was found to be adequate.

PHYSICAL PROPERTIES OF LIME AND ITS AGGLOMERATES

The use of lime and its agglomerates is affected by its shape, size, strength and weathering nature. These properties are given in Table 2.

Shape and Size

Comparing the handling of the various shapes, the handling is easiest with pellets and most difficult with fines. The hollow pellets can be transported and charged into the converter most efficiently, but the lime sinter and lime briquettes, being lumpy in nature, would need the same handling facility as that of lumpy lime.

The size of hollow pellets and lime briquettes seems to be quite suitable for use in the LD converter. The lime sinter obtained (size 20–100 mm) may need breaking in to particles of 20–50 mm for better dissolution. The shapes and sizes of all the agglomerates are given in Table 2.

Strength and Weathering

Whilst handling, the lime sustains a few knocks and fines are generated. The shatter strength of calcined lime (fresh) was found to be 86% on a 12 mm screen after dropping three times onto a steel plate from a height of 2 meters. The lime sinter, hollow pellet and lime briquette exhibited 54%, 90% and 92% shatter strength, respectively, in the fresh condition. The strength was found to decrease with exposure to humid atmosphere. The lime and lime briquettes lost their strength as a result of weathering, ie 5 days exposure to a humid (70%) atmosphere whereas the lime sinter exhibited a shatter strength of 50% after 5 days exposure which is indicative of very little weathering. The hollow pellets did not indicate any decrease in strength even after a month.

The strength of the agglomerates in fresh condition, depends upon the slag bond. In lime sinter the fused, thin layer of calcium ferrite-rich slag on the lime surface develops bridges between partricles. The shatter index of the lime sinter will depend upon the particle size of limestone chips used, their size distribution and the quantity of iron ore and coke breeze used. During handling, the sinter breaks at the slag bridges and also across some lime particles. When a lime particle is exposed as a result of fracturing, the surface behaves like fresh lime and the strength of the sinter is reduced by weathering. Thus, the strength of lime sinter decreases with atmospheric exposure but does so at a slower rate than lime or lime briquettes, which tend to weather fast. The hollow pellets do not show any decrease in strength since they do not contain any free lime which can weather and affect its strength.

SLAG-FORMING PROPERTY OF LIME AND ITS AGGLOMERATES

During LD steelmaking, molten metal remains in contact with molten slag. The composition, temperature, fluidity and other parameters of the slag have a decisive effect on the course of a heat and the quality of the final metal. The prime object of a steelmaker is to remove harmful impurities (usually phosphorus and sulphur) from the metal. The problem is solved by transferring these impurities into the slag and creating conditions which prevent their passage back into the metal. In the LD converter when lime is added as a flux, a certain length of time is required for slag formation. The process of removing silicon, manganese and phosphorus from the melt occurs at a definite rate. The added flux takes some time to heat and form a slag after which the oxidation of impurities can proceed. Sometimes additives are used to accelerate the formation of slag.

The required composition of slag depends upon the metal composition to be blown. The principal conditions essential for dephosphorisation are:

(a) formation of an oxidising furnace atmosphere and a slag with a high activity of iron oxides,
(b) high basicity of slag and a high activity of CaO,
(c) quick formation of ferruginous lime slag,
(d) a relatively low temperature, and
(e) low activity of phosphorous in slag.

In practice these conditions are provided by:

(1) addition of iron oxide to the melt,
(2) addition of lime,
(3) blowing the bath with oxygen,
(4) removing most of the phosphorous at an early stage of heat when the temperature of the metal is still not very high, and
(5) slag renewal.

The use of lime agglomerates (eg lime sinter, hollow pellet or lime briquette) would be highly effective as they offer compounds of CaO and Fe_2O_3 which are required in the process. The dephosphorisation reaction would commence soon after its melting. The dissolution rate of lime sinter would depend upon the iron oxide content, which increases with increasing iron oxide content(4). Presence of small quantities of silica and alumina in the lime-sinter would help in lowering the fusion temperature. The dissolution behaviour of these agglomerates (size ~ 12 mm) were compared with those of lime by observing the time taken for fusion when added to an unstirred molten steel bath in a laboratory electric arc furnace. The lime sinter and hollow pellets took about 45 seconds for dissolution of lime which took more than 6 minutes and the lime briquettes about 3 minutes. This shows the ability of lime sinter and hollow pellets to form a fusible slag in a short period of time.

Thus, the iron oxide present in the agglomerates plays a triple role. Firstly it assists in the rapid formation of slag, secondly it serves as an oxidant and thirdly it reacts with P_2O_5 to form compounds of the type $(FeO)_m\ P_2O_5$, where $m > 3$, being a base.

On the other hand desulphurisation of the metal proceeds under conditions of:

(a) high activity of CaO in slag,
(b) low activity of FeO in slag, and
(c) low activity of S in slag.

Since these conditions are the opposite to those for phosphorus removal it is necessary to desulphurise the melt before reaching the steelmaking shop. Ladle treatment for desulphurisation is a very common practice. Thus, if lime agglomerates were used in the LD converter then the high iron oxide content in it would not be a disadvantage.

In the case of lime sinter prepared by using coke breeze as the fuel, the fear of adding sulphur to the metal is not justified on the basis of results obtained in industrial trials[5].

USE OF LIME AGGLOMERATES IN INDUSTRIAL PRACTICE

The lime sinter, hollow pellet and lime briquette developed in the laboratory are still awaiting industrial trials. However, in India experiments[5] were conducted using highly-fluxed sinter in LD steelmaking. The maximum CaO content in the fluxed sinter used was 30% cf 57–60% CaO in the lime sinter. The fluxed sinter used during industrial trails at Bokaro was prepared by the conventional techniques except a high percentage of limestone was used in the sinter mixture. Using[5] this fluxed sinter it was concluded that:

(1) When sinter containing about 30% CaO is used there is substantial acceleration of the process of slag formation. Slag with basicity of up to 2.05 can be formed within 9 minutes.
(2) Incidence of slopping, lance jamming, etc. have been reduced considerably and the blowing was smoother.
(3) There was improvement in the yield of metal.
(4) The effect on desulphurisation and dephosphorisation showed that there was no deterioration and one should not have apprehension of any phosphorous and sulphur recovery from sinter.

CHOICE OF LIME AGGLOMERATE FOR LD STEEL

In principle, lime sinter, hollow pellets and lime briquettes can all be used to accelerate the slag formation but the cost, in addition to its physical and chemical properties would be a significant factor in the making a choice. Table 3 gives the process of manufacture and the source of energy used for all the agglomerates and lime. Considering the cost of the fuel and raw materials, the lime sinter may be the cheapest of all the lime products. The high silica in lime sinter tends to lower its basicity. Indian coals contain high mineral matter giving 30% ash in coke breeze. Use of a coke breeze with low ash content would help in decreasing silica levels in the lime sinter.

The hollow pellets would be a possible choice if very good weather resistance is needed for long-duration storage of lime bearing pellets.

Lime briquettes like lime have to be used within a day or two. However it does offer a means of utilising lime fines although its production economics may raise several questions.

CONCLUSIONS

(1) The lime sintering process offers a means of producing a low-cost, weather-resistant flux using cheaper raw materials and energy sources.
(2) Hollow pellets offer a means of producing a highly-basic, weather-resistant pellet from iron ore and lime fines.
(3) Lime briquettes offer a means of using lime fines in lump-form.
(4) Lime agglomerates would offer a means of accelerating slag formation due to their fast dissolution rate.
(5) The high iron oxide content in lime agglomerates may be useful in LD steelmaking due to its triple role as oxidant, base, and a promoter of dissolution rates.

REFERENCES

1. R.C. Gupta and A.P. Singh, Proceedings of 2nd European Electric Steel Congress, Part II, Florence (Italy) 29th Sept-1st Oct 1986, P1.10.

2. R.C. Gupta and B. Prakash, Proceedings of the National Seminar on Raw Materials for Iron Making, Bokaro (India), December 14-16, 1979, IV.32.

3. O.P. Pandey, M. Tech (Material Science) Dissertation, Banaras Hindu University, Veranasi (India) 1984.

4. Masaaki Matsushima, Shgeyuki Yadoomane, Katsumi Mori and Yasuji Kawai, Trans ISIJ, Vol 17, 1977, 442.

5. M.F. Mehta and V.S. Dave, Proceedings of Int. Symposium on Modern Developments in Steel Making, Jamshedpur (India) February 16-18, 1981, 3.1.1.

Table 1 CHEMICAL COMPOSITION OF LIME AND ITS AGGLOMERATES

	Fe_2O_3 %	SiO_2 %	Al_2O_3 %	CaO %	MgO %	P %	S %
Lime							
Raw Material							
LS	1.2	0.5	0.6	40.1	-	-	-
Product							
L	2.8	1.1	1.4	94.5	-	-	-
Lime Sinter							
Raw Materials							
LS	1.2	0.5	0.6	40.1	-	-	-
IO	90.0	2.0	3.6	0.3	-	0.06	-
CBA	14.3	53.6	20.8	8.5	0.86	0.16	0.28
Product							
LSR	25.0	7.0	4.0	57.0	0.15	0.03	0.05
	27.0	11.0	6.0	60.0	0.20	0.05	0.06
Hollow Pellet							
Raw Materials							
IO	90.0	2.0	3.6	0.3	-	0.06	-
L	-	12.5	5.5	81.9	-	-	-
Product							
HP	79.7	3.9	4.1	14.8	-	-	-
Lime Briquette							
Raw Materials							
L	1.5	2.5	0.8	95.1	-	-	-
IO	96.9	1.8	0.8	0.3	-	-	-
Product							
LB	5.3	2.4	0.8	91.2	-	-	-

LS- Lime Stone
L - Lime
IO- Iron Ore
CBA- Coke Breeze(ash)
LSR- Lime Sinter
HP - Hollow Pellet
LB - Lime Briquette

Table 2 PHYSICAL PROPERTIES OF LIME AND ITS AGGLOMERATES

	Lime Sinter	Hollow Pellet	Lime Briquette	Lime
Shape	Lump	Ball	Briquette	Lump
Size mm	20-100	16-20	10-20	20-60
Strength Wt% +12mm Shatter Test				
Fresh	54	90	92	86
After 5 days	50	90	0	0
Storage Life in hours	150	>1000	40-50	40-50

Table 3 PROCESSING SUMMARY

Product	Major Raw Materials	Process	Fuel
Lime	Limestone Lumps	Calcination	Oil
Lime Sinter	Lime stone Chips Iron Ore fines Coke Breeze	Sintering	Coke Breeze
Hollow Pellet	Lime Fines Iron Ore Fines	Pelletisation	Oil
Lime Briquette	Lime Fines Iron Ore Fines	Briquetting	Oil

Some aspects on the basicity of slags

T YOKOKAWA and T MAEKAWA

The authors are in the Department of Chemistry, Faculty of Science at the Hokkaido University, Sapporo, Japan.

SYNOPSIS

The basicities of the slags and glasses are discussed from some different viewpoints. The degree of the electron donating power of the oxide ion, which determines the optical basicity, is reflected also in the X-ray energy shifts and radiation-induced visible absorption. The more rigorous understanding of many experimental findings is achieved by quantum chemical calculations , in which the electronic charge density, as well as the energy of the HOMO levels of the corresponding clusters and the LUMO level of the reagent (Pb^{2+} for example) must be taken into consideration. At present the revised optical basicity defined by an average electron density will provide a new measure for the slags, especially oxide-halide mixtures.

INTRODUCTION

The principle of the acid-base chemistry can be applied to the understanding of the properties of the various slags, as well as the reactions within them. Because of the usefulness of the concept of the basicity of the slags, various scales and definitions have been proposed. The optical basicity based on the electron-donating power of the oxide ions in the slags has been of great advantage.

Sulfide and phosphate capacities are related to the so-called "theoretical" optical basicities. However, due to the lack of the parameters for the anions in the slags, an application of the theoretical optical basicities is limited to the oxide systems and the estimations of the optical basicities of halides or halide+oxide systems have not been successful except in a few cases. The drastic change of the basicity at a particular composition commonly seen in the thermodynamic basicity as well as the experimental optical basicity could not be derived from the original theoretical optical basicities, defined firstly by Duffy and Ingram [1].

In this report some commments, based on our recent experimental findings, will be made on the basicity of the slags.

REVISED OPTICAL BASICITY

Recently a new approach was developed by using the concept of the average electron density (D). By choosing appropriate parameters for the anion species (α), the phosphate capacities of the slags containing fluoride ions could be predicted [2]. The revised optical basicity was defined by the following relations;

$$D = \alpha \ (Z / r^3) \qquad (1)$$

and

$$\gamma = 1.34 \ (D + 0.6), \qquad (2)$$

where Z is the valence of the cation and r is the interionic distance between cation and anion. However, the physical meaning of the parameter, α, remained unknown. In the text below some discussion will be made about α.

The basicity of the oxide slags can be primarily determined by the basicity of the oxide ions. Therefore, the partial negative charge on the oxide ion should become a measure of the basicity of the oxide slags. These were calculated by a molecular orbital method as well as by a rather simple treatment using electronegativities of the individual element in a slag. A linear relation between the optical basicity and the charge on the oxide ion in simple oxyanion was successfully derived [3]. However, in general the basicity must be expressed at least in two terms; one is the electron density and the other is the energy of the levels in which these valence electrons are localized. The HOMO (Highest Occupied Molecular Orbitals) is chosen for this purpose. If the estimation is limited only to the oxide system, the latter can be ignored, because their levels are not so different and can be counted by a number, valency and electronegativity of the counter cations.

However, the experimental optical basicity is in the order,

$$CaO \ (1.0) > CaCl_2 \ (0.72) > CaF_2 \ (0.67). \qquad (3)$$

If only the electron densities on the anion are considered, the above order cannot be predicted. Of course, one cannot predict the basicity of the oxide-halide mixture by using a single parameter such as an original theoretical basicity, in which the parameter concerning the anion is not included.

It is well accepted in aqueous solutions that the strength of the base is defined by the energy gain in the acid-base interaction. This energy can be divided into two: ionic and covalent contributions. The former contribution comes mainly from the electronic charge on the donor and the acceptor. On the other hand, the covalent bond can be strongly made if the energy of the HOMO level of the donor is similar to the energy of the LUMO (Lowest Unoccupied Molecular Orbitals) level of the acceptor. The difference of the anions may be reflected largely on the HOMO level. Therefore the following order can be expected, if the acid (indicator such as Pb^{2+} ion) is kept the same.

$CaCl_2 > CaO > CaF_2$ (covalent contribution) (4)

$CaF_2 > CaO > CaCl_2$ (ionic contribution) (5)

where in reality, the basicity increases with the order cited in equation (3). The parameter, α, must include such a contribution tacitly. One must also note that there are no free oxide or halide ions in the slags. The valence electrons distribute among the constituent elements including cations, so that the use of the cation-anion distance is meaningful.

THE CORRELATION BETWEEN THE BASICITY AND THE CHEMICAL SHIFTS OF THE X-RAY ENERGY

The energy of the X-ray emission spectra changes with the basicity of the slags. For example, the Si Kα energy decreases with an increase of the basicity. This energy shift was induced from the changes of the partial elecronic charge on the corresponding elements; for example, the degree of the covalency of the Si-O bond for the Si Kα line. The correlation between Si Kα and Al Kα energy shifts and the theoretical optical basicity as well as the average electronegativity were observed [4].

The energy shifts of the heavy metal ions such as Tl^+ was also observed. Figure 1 shows the energy shifts of the Tl Lβ_5 (5d - 2p) line in the X Tl_2O (1-X) B_2O_3 glasses [5]. Here the reference material is 30 mol% Tl_2O containing glass. It is seen that the energy decreases with an increase in the Tl_2O content. However, the composition dependence of the energy in the low thallium content is somewhat small.

The NMR studies of the binary thallium borate glasses were carried out [6]. It was noted that the ^{205}Tl chemical shifts remained unchanged up to 18mol% Tl_2O, but decreased rapidly beyond this composition; thus the chemical bond of Tl^+ with the oxygen is predominantly ionic in the glasses of low Tl^+ content and the contributions of the covalent character increases only in the glasses of higher Tl_2O content. The same conclusion were deduced from the spin Hamiltonian parameter of the ESR spectra of the radiation induced Tl^{2+} center [7].

The change of the X-ray energy should be reflected also in the bonding properties. The Tl Lβ_5 energy increases with an increase of the effective positive charge on the Tl^+ ion; thus the Tl Lβ_5 energy decreases with an increase in the basicity of the matrix glasses and slags.

The X-ray energy of thallium ion doped in the various glasses should become a measure of the basicity as seen in the experimental optical basicity. We tried to measure the energy shift of Tl^+ ion (1 mol % doped) in the sodium borate, silicate and phosphate glasses. Although we could not reproduce the characteristic change of the X-ray energy with composition seen in the experimental optical basicity, we could find out the following meaningful energy shifts in each glassy system:

X Na_2O (100-X) B_2O_3 (X = 5 - 30) 0.3 - 0.4 eV
(seems slightly to decrease with X)

3 Na_2O 7 SiO_2 0.76 eV

1 Na_2O 1 P_2O_5 1.24 eV

3 Tl_2O 7 B_2O_3 0.00 eV

The energy shifts are in the same order as that of the Fe K$\beta_{1,3}$ line doped in the corresponding glasses. (phosphate > silicate > borate) [8].

The energy of 6s-6p absorption spectra of Tl^+ ion shifts to longer wavelength with an increase in the basicity. It is understood that the electron donating power of the oxide ion increases with an increase in the basicity. In other words, the electron density of the inner 5d orbital of the Tl^+ ion increases with the basicity, so that the effective positive charge of the Tl^+ ion decreases and produces the energy shifts of the upper 6s and 6p orbitals. The Tl Lβ_5 energy was also affected by this electron flow from the surrounding oxide ion to the Tl^+ ion.

X-RAY INDUCED ABSORPTION SPECTRA

High energy irradiation makes it possible to generate various defect centers in the glasses. As to the alkali silicates, the structure of the centers and their related optical absorption have been studied extensively. The optical absorption spectra are characterized by two peaks in the visible (2 - 3 eV) and one in the ultra violet range (4 eV). The peak energies of the former two , which are assigned to the non-bridging oxygen hole centers, varied systematically with the glass composition (basicity), whereas the latter, assigned to be an electron center, does not show any significant peak energy difference among the glasses.

In figure 2 the absorption spectra of the X Li_2O (30-X) K_2O 70 SiO_2 glasses are shown. It is seen that the absorption intensity around 2.0 eV decreases relative to that around 2.6 eV with an increase of the Li_2O content. The peak energies also shift to the higher values with an increase of the Li_2O content. One of the most interesting results is that the absorption spectrum of the lithium -potassium silicate glass (for example X=15) has almost the same structure as that of the sodium silicate glasses. In figure 3 the relation between the peak position of the visible absorption and the average ionic radius of the alkali metal ions are plotted. The systematic variation of the peak energy with the basicity are deduced.

The absorption energy comes from the transition from the Si-O σ bonding level to the Si-O π non-bonding (almost oxygen p orbital). The electronic field of the alkali cations should cause the variation of the energy of the bonding orbital. Perhaps the more electropositive cation which destabilizes the Si-O bonding, the smaller the absorption energy becomes. This effect is also reflected to the X-ray emission energy (for example, Si Kβ), where Si Kβ energy becomes larger with an increase of the basicity [9]. Although the applicability of this method to determine the basicity is rather limited, it is fruitful to consider the concept of the basicity of the slags from a different viewpoint.

THE BASICITY DEFINED FROM THE ESR SPECTRA OF THE Cu(II) IONS (IMAGAWA'S BASICITY) [10]

This definition is also based on the electron donating power of the oxide ions surrounding Cu(II) ions doped in the glasses. The spin Hamiltonian parameters derived from the hyperfine structure of the ESR signals of the Cu(II) ion can be analyzed by using the LCAO MO method; thus the magnitude of the contribution of the 3d orbitals of the Cu(II) ion determines the degree of the covalency of the Cu – O bond.

Among the several molecular orbitals, the covalency of the B_{2g} π orbital (Cu d_{xy} – O 2p) is important, because it increases with an increase of the basicity of the oxide ion coordinating around the cupric ion. In other words, the covalency of the Cu–O bond increases with an increase in the freedom of the lone pairs on the oxide ion. The correlation of the degree of covalency of the B_{2g} π orbital with the experimental optical basicity is excellent.

QUANTUM CHEMICAL APPROACH TO THE BASICITY

Average electronegativities and their related quantities such as theoretical optical basicities determine the total (average) electron densities of the oxide ions; thus they reflect the Lewis basicity of all kinds of the oxide ions. However, as can be seen in the definition of Imagawa's basicity, the contributions of the electron density to the basicity differ among the various molecular orbitals.

The quantum chemical calculation can be used to predict the basicity of the simple systems. In this approach the perturbation energy of the reagent or the affinity to form the reagent-oxygen chemical bonding of the apporopriate clusters can be used. Such a calculation has been started recently [11]. More precisely, one must include a electron density and the energy of the LUMO level of the cation such as Pb^{2+} ion.

CONCLUDING REMARKS

Aluminium oxide is thought to be an amphotric oxide. However, it reacts preferably with sodium oxide even under the presence of the more acidic oxide such as SiO_2 and B_2O_3 [12]. This is due to the structural requirement, in which the aluminium ion must enter the silicate and borate network joining together with the sodium ion as a charge compensator. In the optical basicity, the Tl^+ and Pb^{2+} ions give different basicity because of the different sites in which they are situated. When the oxide+halide mixture is considered, it may not be clear even whether a unique basicity can be measured by this indicator experiment.

The quantum chemical method is also an approximation. The way of calculation as well as the choice of the clusters should give large effects on the final discussions. It will be required to discuss the basicity by combining the factors which characterize each reaction with the most correlated experimental (optical as well as thermodynamic) observations.

ACKNOWLEDGEMENT

The authors gratefully acknowledge that this work was partially supported by the Nippon Sheet Glass Foundation.

REFERENCES

[1] J. A. Duffy and M. D. Ingram: J. Non-cryst. Solids, 21, 373 (1976).
[2] T. Nakamura, Y. Ueda and J. M. Toguri: J. Japan Inst. Metals, 50, 456 (1986).
[3] J. H. Binks and J. A. Duffy: J. Non-cryst. Solids, 37, 387 (1980).
[4] T. Maekawa and T.Yokokawa: Spectrochim. Acta, part B, 37, 713 (1982).
[5] T. Maekawa, H. Akiyama and T. Yokokawa: to be published.
[6] J. F. Bauger and P. J. Bray: Phys. Chem. Glasses, 10, 77 (1969).
[7] H. Hosono, H. Kawazoe and T. Kanazawa: J. Mat. Science, 16, 57 (1981).
[8] M. Yoshioka, T. Maekawa and T. Yokokawa: Bull. Chem. Soc. Jpn., 57, 2718 (1984).
[9] N. Kikuchi, T. Maekawa and T. Yokokawa: Bull. Chem. Soc. Jpn., 52, 1260 (1979).
[10] H. Imagawa: Phys. Stat. Sol., 30, 469 (1968)
[11] N. Uchida, T. Maekawa and T. Yokokawa: J. Non-cryst. Solids, 85, 290 (1986).
[12] H. Itoh and T. Yokokawa: Trans. JIM, 25,879 (1984).

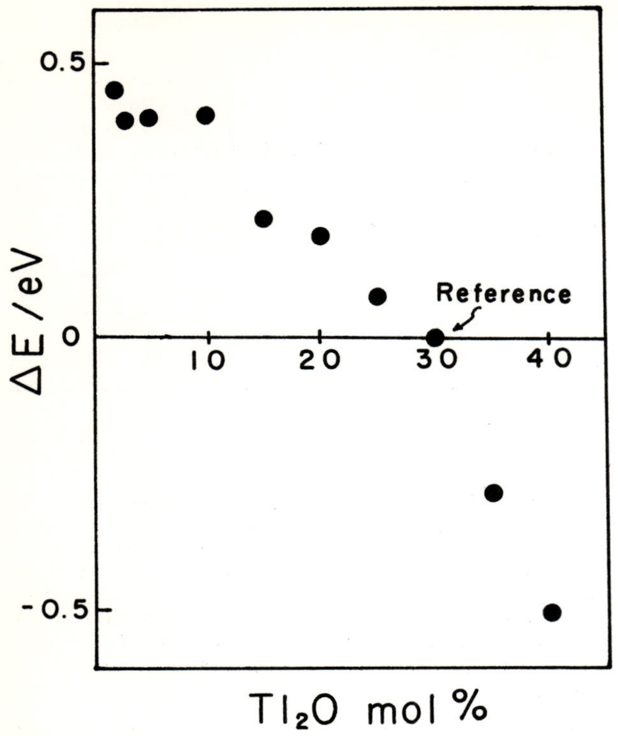

Figure 1 Energy shifts of Tl $L\beta_5$ line of X Tl_2O-(1-X) B_2O_3 glasses.

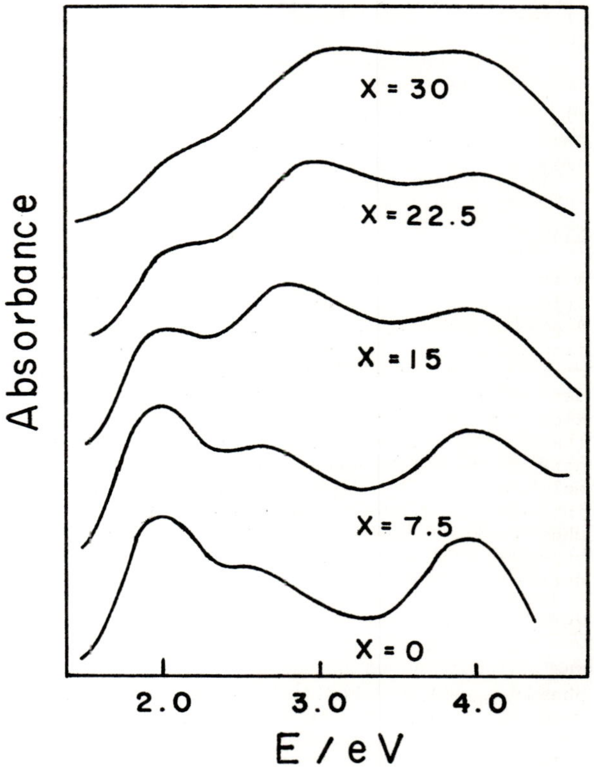

Figure 2 X-ray induced absorption spectra of X Li_2O - (30-X) K_2O - 70 SiO_2 glasses.

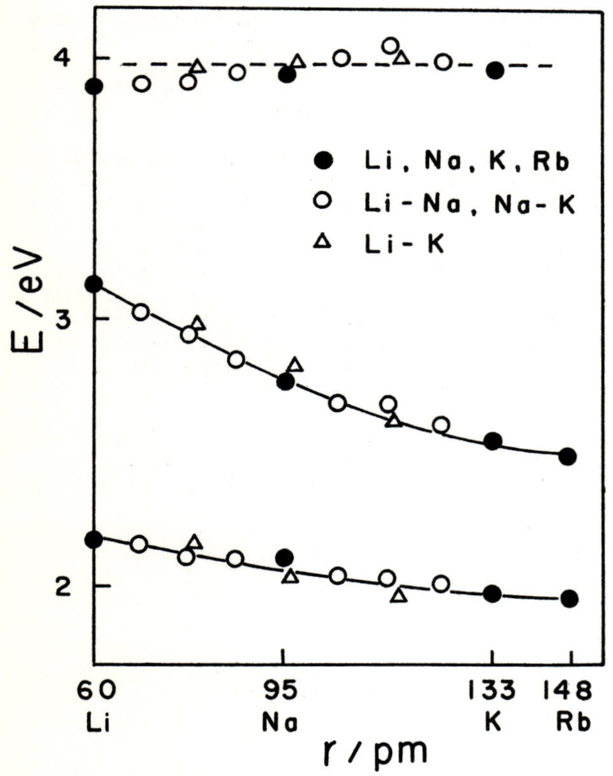

Figure 3 The change of absorption energies with the ionic radii of the alkali metal ions. 30 M_2O-70 SiO_2 glasses with X-ray irradiation.

□

Studies on the MgO solubility in complex steelmaking slags in equilibrium with liquid iron and distribution of phosphorus and vanadium between slag and metal at MgO saturation

R SELIN

The author is at the Department of Production Technology, Mining and Steel Industry, The Royal Institute of Technology, Stockholm, Sweden.

SYNOPSIS

A study was carried out on MgO-saturated slags belonging to the system $CaO-Fe_xO-MgO_{sat}-SiO_2-Al_2O_3-TiO_2-VO_y-P_2O_5$ in equilibrium with liquid iron at 1600°C. MgO solubility, phosphorus distribution and vanadium distribution have been determined for slag compositions relevant to electric steelmaking from Direct-Reduced Iron (DRI). The treated composition range of slag also corresponds to scrap-based electric steelmaking as well as to basic oxygen steelmaking. The influence of the phosphorus content in slag on phosphorus distribution has been studied in a separate series of experiments and is described in detail.

INTRODUCTION

The study of the equilibrium distribution of phosphorus is one of the classical topics of research in steel refining [1-18].

In the recent years the use of basic slag saturated with MgO has been advocated for low refractory consumption. Many investigations have been reported with magnesia crucibles [1-3,8,12-15, 17]. Most of the studies were oriented towards Basic Oxygen Furnace (BOF) steelmaking and consequently the investigated slags consisted mainly of CaO, Fe_xO, MgO and SiO2. In some of the studies, the influence of CaF_2 [1,12] and MnO [1,2,6,14,17] was examined.

In steelmaking processes employing DRI all the gangue oxides present in the iron ore are transferred to the steel furnace. The slag is thus complex and may contain, besides the usual BOF slag components, considerable amounts of oxides as Al_2O_3, TiO_2 and VO_y. The sum of the wt% of Al_2O_3, TiO_2 and VO_y may range from ca. 25 to 120 % of the SiO_2 content. The present work proposes to determine the MgO solubility in complex slags as well as equilibrium distribution of phosphorus and vanadium at MgO saturation.

EXPERIMENTAL METHOD AND RESULTS

The equilibrium experiments were carried out using a $MoSi_2$-heated vertical tube furnace. For each experiment a new dense-sintered MgO crucible (Friedrichsfeld), of inside diameter 36 mm, was used. Synthetic slags were prepared from "pro analysi" grade chemicals. A holding time of $\geq$ 10 hours at equilibrium temperature was found to be adequate for attaining equilibrium. For a detailed description of the method and the experimental results the reader is referred to ref. [19].

THE REFERENCE SYSTEM $CaO-Fe_xO-MgO_{sat}-SiO_2$. 1600°C

General

The system $CaO-Fe_xO-MgO_{sat}-SiO_2$ has been chosen as a reference system. The properties of a more complex slag system, containing in addition Al_2O_3, TiO_2 and VO_y, are determined relative to the corresponding properties of the reference system. At constant pressure and temperature (1600°C) the reference system $CaO-Fe_xO-SiO_2$, when in equilibrium with liquid iron and saturated with MgO, has only 2 degrees of freedom.

In the present system, phosphorus and vanadium distributions are primarily dependent on the oxygen potential and the activity of CaO. Consequently, the independent variables chosen should have a close relationship with these quantities and, for practical convenience, the choice is wt% CaO for the latter one. A suitable variable related to the oxygen potential is $(\%Fe_{tot})$=wt % total iron content in metal free slag, expressed as $(\%FeO^*) = 1.2865 \cdot (\%Fe_{tot})$, i.e. all iron in slag is taken as Fe^{+II}.

It is also found in this work that (%FeO*) and (%CaO) are suitable variables for defining the slag with reference to its solubility of MgO.

Figure 1 shows MgO isosolubility lines based on experimental data of this work and of Suito et al. [12]. Similar diagrams have been drawn also for the distribution of phosphorus and vanadium and for [% O] in metal [19]. Moreover, except for [% O], these diagrams have been converted into a computer model [19]. The influence of temperature has been evaluated from data of Suito et al. [12,20].

Distribution of phosphorus

Assuming all phosphate ions to be monomeric, the phosphorus equilibrium may be written:

$$[P] + 5/2[O] + 3/2(O^{2-}) \quad (PO_4^{3-}) \qquad (1)$$

and the equilibrium constant for this reaction:

$$K_{P1} = (a_{PO_4^{3-}}) / \left([a_P] \cdot [a_O]^{5/2} \cdot (a_{O^{2-}})^{3/2}\right) \quad (2)$$

The hypothetical standard states assumed for phosphorus and oxygen in metal are 1 wt%. From Eq. (2) the phosphorus distribution, L_P. is obtained:

$$L_P = \frac{(\%\ P)}{[\%\ P]} = K_{P1} \cdot \frac{(a_{O^{2-}})^{3/2} \cdot [f_P] \cdot [a_O]^{5/2} \cdot M_P}{(f_{PO_4^{3-}}) \cdot M_{PO_4^{3-}}} \quad (3)$$

For the low [% P] in metal considered here, $[f_P]$ can be assumed to be equal to 1 as a good approximation. For the present system, the remaining terms of the right hand side of Eq. (3) are entirely determined by the slag composition when the system is in equilibrium. If the phosphorus content of the slag is fixed at a constant value, in this case 0.4% P, the slag composition is unambiguously defined at a constant temperature by its content of CaO and FeO*. With %CaO and %FeO* as independent variables, isodistribution curves have been drawn in Fig. 2 for the reference system (1600°C, (% P) 0.40) based on the same experiments as for Fig. 1. The agreement between the experimental values is quite good and the isodistribution curves can be drawn with fairly high precision.

For low phosphorus concentrations in the slag, e.g. (% P)=1.0 and below, it seems reasonable to assume that the phosphorus content itself should not influence the right-hand side of Eq. (3). In order to test this assumption, which is actually the basis for using L_P as a distribution constant, separate experiments were performed. Three starting points in the reference system were chosen and for each point the only parameter to be varied was (% P) in slag, all the other slag component contents being kept constant [19]. To counteract the decrease in basicity that might occur at increased (% P), a constant value of the sum of wt% (SiO_2) and (P_2O_5) was aimed at.

If Lp is a constant, then for each slag composition straight lines with a coefficient of slope equal to 1 should be obtained when log (% P) in slag is plotted against log [% P] in metal.

In literature sometimes one comes across the usage of a phosphorus distribution constant derived from the following reaction:

$$2\ [P] + 5[O] \quad (P_2O_5) \quad (4)$$

Also

$$P + 2 \cdot 5O = PO_{2.5} \quad \text{(See recent work.)}$$

with the equilibrium constant:

$$K_{P2} = (a_{P_2O_5}) / \quad [a_P]^2 \cdot [a_O]^5 \quad (5)$$

assuming a fixed slag composition and inserting fp=1, we obtain the following constant ratio for the distribution of phosphorus:

$$k_4 = (\%\ P)/[\%\ P]^2 \quad (6)$$

From a thermodynamic point of view this way of writing the phosphorus distribution is justified only if phosphorus occurs as dimeric $P_2O_7^{4-}$ ions in the slag or, as was thought earlier, as P_2O_5 molecules. Then the phosphorus distribution will no longer be constant but will instead depend on the total phosphorus content of the system:

$$L_P = \frac{(\%\ P)}{[\%\ P]} = k_4 \quad \cdot \ [\%\ P] = k_5 \quad \cdot \quad (\%\ P) \quad (7)$$

where k_5 is another constant.

Of course there will be considerable differences depending on which of these two expressions for phosphorus distribution is used. Eq. (6) may be rewritten in logarithmic form:

$$\log_{10}\ (\%\ P) = \log_{10}\ k_4 + 2\ \log_{10}\ [\%\ P] \quad (8)$$

i.e. if Eq. (4) is a good description of the actual equilibrium reaction one should obtain straight lines having a slope of about 2 in a log (% P) - log [% P]-diagram.

Figure 3 shows the results plotted in a log-log diagram. Calculating the average slope, weighted against the number of data points for the respective lines, yields a slope equal to 1.20. Thus the slope obtained is somewhat greater than 1 but still much closer to 1 than to 2, and hence Eqs. (1) and (3), i.e. assuming that phosphorus occurs in the form of monomeric phosphate ions, should be preferred to Eqs. (4) and (5). A more detailed evaluation will, however, be done in the following.

A reasonable explanation for the fact that the slope of all the lines in Fig. 3 is greater than 1 may be that the slag contains a certain amount of dimeric phosphate ions in addition to the predominating monomeric ions. For instance, it has been calculated that $P_2O_7^{4-}$ is the predominant ion species in liquid $Ca_2P_2O_7$ at 1800K [21]. At higher basicities, ie.e. higher O^{2-} activity, a shift towards a higher proportion of monomeric ions occurs as demonstrated by the equilibrium:

$$1/2(P_2O_7^{4-}) + 1/2(O^{2-}) \quad (PO_4^{3-}) \quad (9)$$

Assuming the coexistence of monomeric and dimeric phosphate ions as in Eq. (9), an expression will be derived below for the relationship between L_P and total (% P) in slag. The following designations will be used:

$(\%\ P_M)$ = wt % P as monomeric PO_4^{3-} ions.

$(\%\ P_D)$ = wt % P as dimeric $P_2O_7^{4-}$ ions.

(% P) = wt % total P in slag.

For a given slag composition we assume, as above, that the right-hand side of Eq. (3) is constant, which implies that $f_{PO_4^{3-}}$ obeys Henry's law and interaction between the phosphate ions is neglected. The distribution ratio between (% P) in slag as monomeric ions and [% P] in metal will then be constant:

$$k_6 = (\%\ P_M)/[\%\ P] \quad (10)$$

The equilibrium reaction between phosphorus in metal and phosphorus in slag as dimeric phosphate ions may be written in the ionic form as follows:

$$[P] + 5/2[O] + (O^{2-}) \rightleftarrows 1/2(P_2O_7^{4-}) \quad (11)$$

from which the following constant ratio is obtained for phosphorus dimeric phosphate ions in slag using the same assumption, as made above, for

Eq. (10):

$$k_7 = \sqrt{(\%\ P_D)}/[\%\ P] \tag{12}$$

Now, if phosphate ions of higher degrees of polymerization, such as $P_3O_{10}^{5-}$, $P_4O_{13}^{6-}$ etc., are neglected, which seems reasonable as monomeric phosphate ions are expected to be dominant, L_p can be derived from Eqs. (10) and (12):

$$L_p = \frac{(\%P)}{[\%P]} = \frac{(\%P_M) + (\%P_D)}{[\%P]} = k_6 + k_7 \ (\%P_D) \tag{13}$$

The equilibrium constant of reaction (9) for the equilibrium between monomeric and dimeric phosphate ions can be written:

$$k_8 = (\%\ P_M)/\sqrt{(\%\ P_D)} = k_6/k_7 \tag{14}$$

Further, choosing the phosphorus content of the slag phase in the reference system, i.e. (% P = 0.4), as a basis and combining Eqs. (13) and (14) we may express the relative change in L_p as:

$$L_p/L_p(0.4)=[k_8+ \ (\%\ P_D)]/[k_8+ \ (\%\ P_D)(0.4)] \tag{15}$$

This equation enables us to derive a common value of k_8 for the three experimental series. This seems to be a reasonable approach since the difference in the slope of the lines in Fig. 3 should largely fall within the experimental error range. However, in order to introduce into Eq. (15) the analytically determined content, which refers to total (% P) in slag, it is necessary to express (% P_D) as a function of (% P), which can be done in implicit form with the help of Eq. (14):

$$(\%\ P)=(\%\ P_D)+(\%\ P_M)=(\%\ P_D)+k_8 \ \ (\%\ P_D) \tag{16}$$

This expression can be converted into a quadratic equation which can then be solved explicitly with respect to (% P_D):

$$(\%\ P_D) = (\%\ P) + k \ \ [1/2 - \sqrt{1/4 + (\%\ P)/k} \ \ ] \tag{17}$$

Substituting Eq. (17) in to Eq. (15), we can calculate the relative change in phosphorus distribution versus the reference system. $L_p/L_p(0.4)$, for the same experimental points as in Fig. 3. The result is plotted in Fig. 4 against total (% P) in the slag. For the data points, the value on the y-axis thus represents the ratio of observed L_p to the calculated L_p at (% P)=0.4.

The constant k_8 in Eqs. (14)-(17) above can now be evaluated by regression analysis, which yields k_8 = 1.61. For comparison, a number of additional experimental points of similar slag composition from Suito et al. [12.14], with 0.4 and >2 (% P) respectively, have been included in the diagram. Repeating the regression analysis with these points taken into account yields the same value of k_8. Thus despite their higher maximum phosphorus contents, the values of Suito et al. seem to be well described by the same model. The influence of the individual points on the value of k_8 was tested and it was found that k_8 cannot be given to greater accuracy than one decimal, i.e. k_8 = 1.6. This value was used when drawing the curve in Fig. 4. Now, Eq. (15) can be written:

$$L_p/L_p(0.4) = [1.6 + \sqrt{(\%\ P_D)}]/1.820 \tag{18}$$

where

$$(\%\ P_D) = 1.28 + (\%\ P) - 1.6\sqrt{0.64 + (\%\ P)} \tag{19}$$

According to this model the portion of phosphorus as $P_2O_7^{4-}$ ions is 12 % by weight at (% P) = 0.4, increases to 34 % at (% P) = 2.0, and approaches zero when (% P) approaches zero.

The heats in Fig. 4 having about 0.2 % P or more in the slag show a small scatter about the regression line, while unfortunately the scatter for the heats having the lowest phosphorus contents, i.e. less than 0.1 % P, is considerably larger. The larger scatter is likely to be attributable to a slower approach to equilibrium at these low phosphorus contents in both slag and metal as well as to the increased effect of error in the chemical analyses. It is concluded, however, that the suggested model, which considers both monomeric and dimeric phosphate ions, is the one that can account for the experimental results. Further, it is assumed that Eqs. (18) and (19) can be used with reasonable approximation to describe the influence of (% P) on L_p for the entire slag system treated in this work.

ADDITION OF Al_2O_3, TiO_2 AND VO_2 TO THE REFERENCE SYSTEM

It is assumed that the Al_2O_3, TiO_2 and VO_2 substitute SiO_2 on a weight basis. To describe mathematically this substitution, the magnitude % W is introduced such that when adding solely Al_2O_3 to the reference system, W in wt % is defined as:

$$W_{Al} = \frac{(\%\ Al_2O_3)}{(\%Al"22O_3)+(\%\ SiO_2)} \cdot 100 \tag{20}$$

and in the case of a complex slag which also contains TiO_2 and VO2, as:

$$W_{Al}= \frac{(\%\ Al_2O_3)}{(\%Al_2O_3)+(\%SiO_2)+(\%TiO_2)+\%VO_2)*} \cdot 100 \tag{21}$$

The symbol W is used to state the amount of SiO_2 of the reference system that is replaced with Al_2O_3, TiO_2 and VO_2 in slags which contain all oxides, as given by the equation:

$$W = W_{Al} + W_{Ti} + W_V \tag{22}$$

W is not used to define the slags but only to provide a measure of the extent to which a given slag differs from the reference system.

Figure 5 illustrates the results obtained from three experimental series in which, for each series, SiO_2 was gradually replaced with one of Al_2O_3, TiO_2 and VO_2. For each series, a "complex" slag was also made in which equal amounts of these three oxides were added, replacing about half of the SiO_2 in the reference system.

From the experimental results it can be said in summary that the MgO solubility decreases when SiO_2 is replaced by an equal weight of Al_2O_3, TiO_2 and VO_2*. At the same time the phosphorus capacity of the slag increased [19], although this positive effect is counteracted by a pronounced decrease in the oxygen potential of the system apparently resulting from a decrease in the activity coefficient of FeO*. The overall result is a relatively moderate increase in L_p when SiO_2 is replaced by TiO_2 and VO_2 whereas the result varies in the case of Al_2O_3.

It has been found suitable for the purpose of describing the slag system $CaO-Fe_xO-MgO_{sat}-Al_2O_3-TiO_2-VO_y$ in a mathematical form to use the computer model of the reference system as a base. The experimental values of the more complex slag system have been compared with computed values of the reference system with (% CaO) and (% FeO^*) as independent variables. Moreover, relatively simple relationships can be used for converting the output variables of the reference system to the more complex slag system, with W_{Al}, W_{Ti} and W_V as independent variables thus providing a model [19] for the complex system.

Figure 6 shows comparisons made between experimentally derived and computed values of MgO solubility and L_p for all experiments in which one or more of the oxides Al_2O_3, TiO_2 or VO_2 were added. As can be seen, the experimental data are well described by the model. The four heats in which Al_2O_3, TiO_2 and VO_2 were added simultaneously have been marked with filled squares in Fig. 6. Their locations confirm both the solubility of MgO and L_p can be expressed, with good approximation, by functions in which the separately evaluated influences of W_{Al}, W_{Ti} and W_V have been added.

CONCLUSIONS

Based on the experimental data, diagrams have been drawn from which MgO solubility, L_p and L_V for the reference system $CaO-Fe_xO-MgO_{sat}-SiO_2$ are obtained. These quantities can also be calculated by a computer model derived from the diagram.

By using (% CaO) and (% FeO^*) as the main parameters for defining the slag, the corrections which need to be made with respect to MgO solubility, L_p and L_V calculated for the reference system are relatively small when adding Al_2O_3, TiO_2 and VO_y. The influence of these oxides can be described by linear equations thus forming a model for complex slags based on experimental data.

The application of a thermodynamic model which includes the simultaneous occurrence of PO_4^{3-} and $P_2O_7^{4-}$ -ions on the experimental results has led to the conclusion that the monomeric PO_4^{3-} -ions dominate in the investigated MgO-saturated slags at (% P) $<\approx 2.5$. Consequently, (% P)/[% P] is to be preferred to (% P)/[% P]^2 for expressing the phosphorus distribution. Furthermore, an equation has been derived by which correction for the influence of (% P) in slag on L_p can be made.

ACKNOWLEDGEMENTS

This work has been financed by the National Swedish Board for Technical Development (STU) as a part of the long-term research programme "Steelmaking from high phosphorus ores". The author wishes to express his gratitude to Prof. J. O. Edstrom and Drs. S. Seetharaman, S. Afzal and A. Werme for valuable discussions during the course of the work.

SYMBOLS

()	:	solute in slag phase, wt%
[]	:	solute in metal phase, wt%
a	:	activity
f	:	activity coefficient related to wt%
K	:	equilibrium constant
k	:	constant
L_i	:	distribution ratio between slag and metal, (wt% i)/[wt% i]
M	:	atomic, molecular weight

REFERENCES

1. Winkler T. B., Chipman J., Trans. AIME 167(1946), pp. 111-133.
2. Balajiva K., Quarrell A.G., Vajragupta P., J. Iron Steel Inst. 153(1946), pp. 115-150; ibid. 155(1947), pp. 563-567.
3. Vajragupta P., J. Iron Steel Inst. 158(1948), pp. 494-496.
4. Flood H., Grjotheim K., J. Iron Steel Inst. 171(1952), pp. 64-70.
5. Fischer W.A., v Ende H., Stahl u. Eisen 72(1952), No. 23, pp. 1398-1408.
6. Turkdogan E.T., Pearson I., J. Iron Steel Inst. Dec. 1953, pp. 398-401.
7. Knüppel H., Oeters F., Stahl u. Eisen 81(1961), No. 22, pp. 1437-49.
8. Scimar R., Dr. Diss., Sciences Appliquees, University of Liege, 1961.
9. v Ende H., Bardenheuer F., Schürmann E., Stahl u. Eisen 82(1962), No. 15, pp. 1027-35.
10. Bardenheuer F., Oberhäuser P.G., Stahl u. Eisen 89(1969), No. 18, pp. 988-994.
11. Healy G.W., J. Iron Steel Inst. July(1970), pp. 664-668.
12. Suito H., Inoue R., Takada M., Trans. Iron Steel Inst. Japan 21(1981), pp. 250-259.
13. Suito H., Inoue R., ibid. 22(1982), pp. 869-877.
14. Suito H., Inoue R. ibid., 24(1984), pp. 40-46.
15. Suito H., Inoue R. ibid., 24(1984), pp. 47-53.
16. Gaskell D.R., ibid. 22(1982), pp. 997-1000.
17. Guo D., Arch. Eisenhüttenwesen 55(1984), No. 5, pp. 183-188.
18. Turkdogan E.T., Book 298, The Metals Society, London 1983.
19. Selin R., Suppl I, Dr. Diss., royal Inst. of Techn. Stockholm, 1987, TRITA-PT-87-04.
20. Inoue R., Suito H., Trans. Iron Steel Inst. Japan 22(1982), pp. 705-714.
21. Richardson F.D., Physical Chemistry of Metals in Metallurgy, Academic Press, London, 1974.

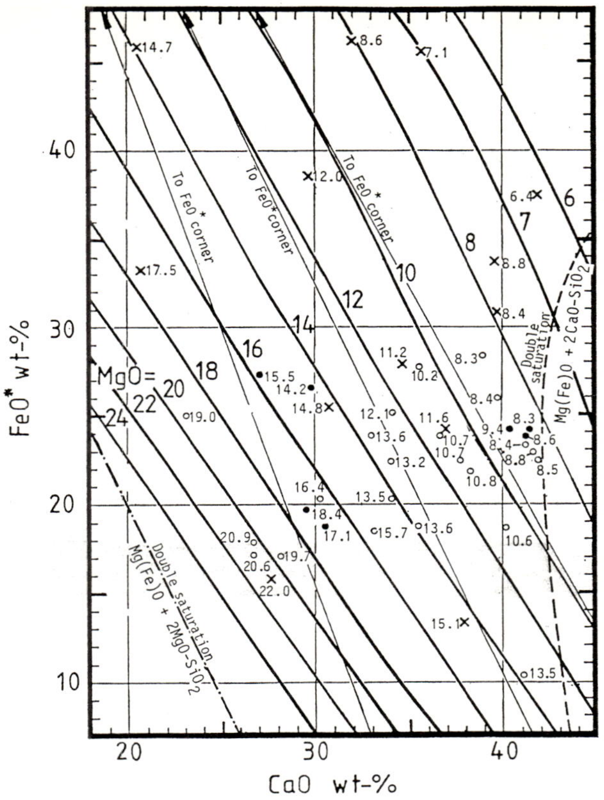

Figure 1: Solubility of MgO in $CaO\text{-}Fe_xO\text{-}MgO_{sat}$-$SiO_2$-slags in equilibrium with liquid iron at 1600°C. (o)(o) This work; (x) Suito etal.[12]. All figures in diagram refer to wt% MgO. FeO^* is $Fe_{tot} \times 1.2865$, i.e. all Fe in slag phase taken as Fe^{+II}.

[P] initial compared to final value			CaO wt %		FeO* wt %	
Below	Close	Above	$\bar{x}$	S_x	$\bar{x}$	S_x
△	◪	▲	33.5	0.2	18.0	0.4
○	◉	●	37.7	0.1	22.3	0.2
□	◪	■	34.5	0.5	21.9	0.9

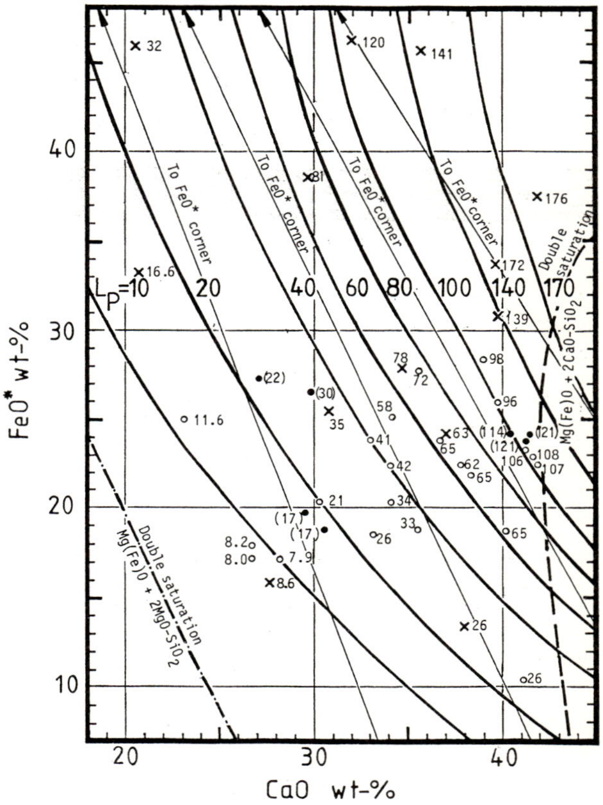

Figure 2: Equilibrium phosphorus distribution between $CaO\text{-}Fe_xO\text{-}MgO_{sat}$-$SiO_2$-0.4wt%P slags and liquid iron at 1600°C. (o)(o) This work; (x) Suito et al. [12]. All figures in the diagram refer to the L_p ratio. FeO^* is $Fe_{tot} \times 1.2865$, i.e. all Fe in slag phase taken as Fe^{+II}.

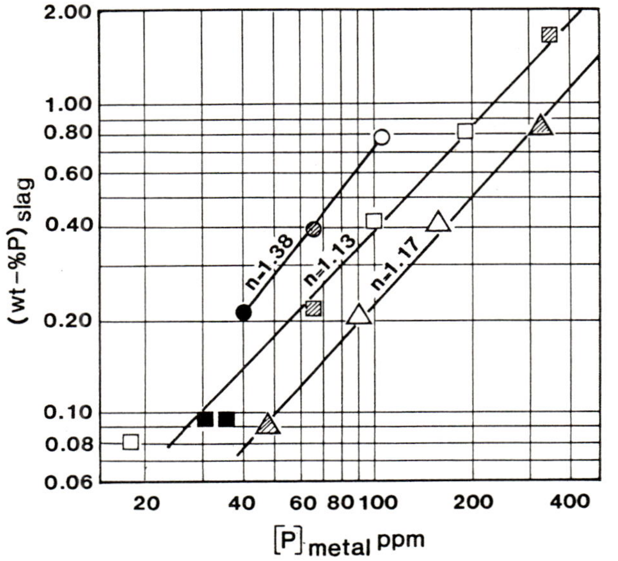

Figure 3: Relation between phosphorus contents in slag and metal for three slag compositions. n denotes the slope of the regression lines.

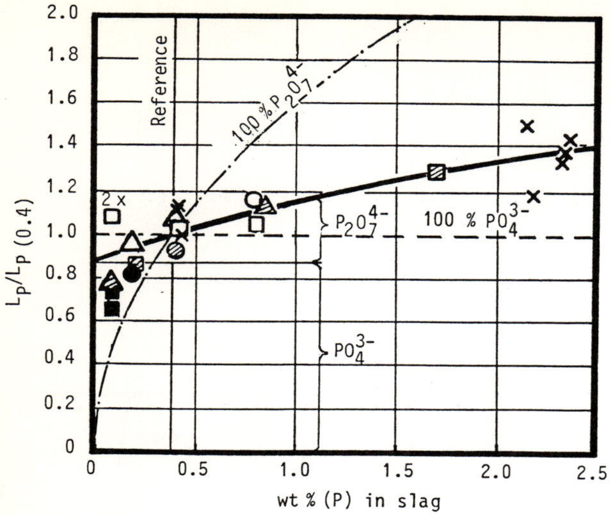

Figure 4: Effect of phosphorus content of slag on the ratio $L_p/L_{p,ref}$. Reference system is CaO-SiO_2-Fe_xO-MgO_{sat}-0.4 wt% P. (, ,) see Fig. 3; (x) Suito et al. [12,14];
(----)   Regression curve according to Eq. (18);
(-.-.-)   Theoretical curve if all phosphorus in slag were present as $P_2O_7^{4-}$;
(- - -)   Ditto for PO_4^{3-}.

			Experimental series	Heat in reference system ($W_i = 0$)	
△	W_{Al}	Addition of Al_2O_3		% CaO	% FeO*
□	W_{Ti}	Addition of TiO_2	△ □ ◐ ⋈ —— C	34.8	19.6
○	W_V	Addition of VO_2	▲ ◪ ◑ ⋈ —·—·— D	41.1	10.3
⋈	W_Σ	Complex slags with Al_2O_3 + TiO_2 + VO_2	▲ ▬ ⬤ ⋈ ---- E	38.1	22.2

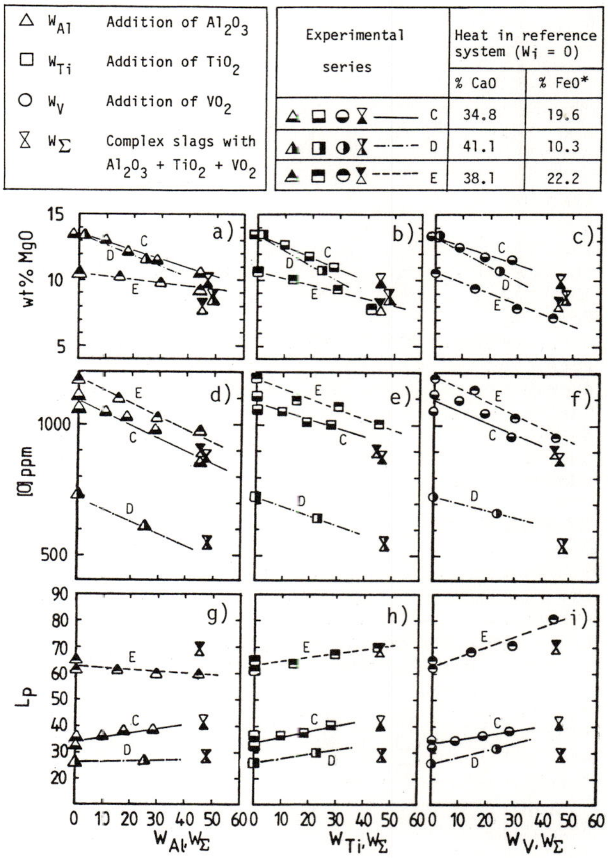

Figure 5: Effect of replacing SiO_2 with Al_2O_3, TiO_2 and VO_2, expressed as W_{Al}, W_{Ti}, W_V and W_Σ, on solubility of MgO in slag, oxygen concentration in metal, and distribution of phosphorus.

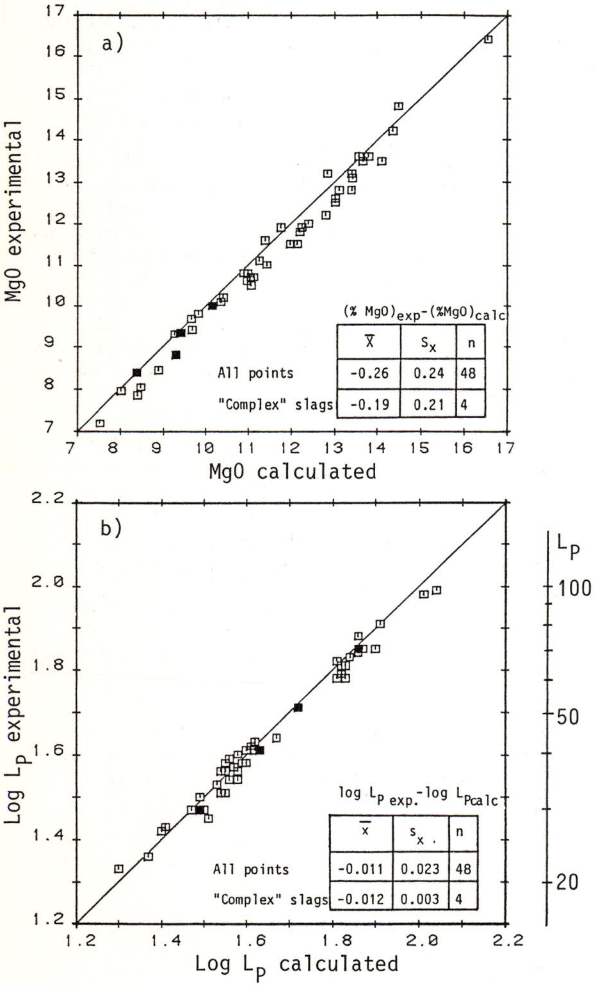

Figure 6: Comparison between experimental and calculated MgO solubility (a) and L_p (b). Filled squares denote complex slags with Al_2O_3 + TiO_2 + VO_2 [19].

Sulphur partition between carbon-saturated iron and CaF_2-CaO-SiO_2 slags

G FERGUSON and R J POMFRET

The authors are with the Division of Metallurgy and Engineering Materials, University of Strathclyde, Glasgow, UK.

SYNOPSIS

The sulphur partition between carbon-saturated iron and CaF_2-CaO-SiO_2 slags was measured at 1500°C and expressed as sulphide capacities. For slags with constant CaO, replacing SiO_2 with CaF_2 increases the sulphide capacity, whereas replacing CaO with CaF_2 for constant SiO_2 content decreases the sulphide capacity. Addition of CaF_2 to slags of higher basicity appears to have no effect on the sulphide capacity; for lower basicity, the sulphide capacity increases for additions of up to 30% CaF_2.

INTRODUCTION

Slags containing CaF_2 have been widely used in the iron and steel industry, particularly for electroslag refining (ESR). CaF_2, on addition to a conventional (e.g. CaO-SiO_2-Al_2O_3) slag will result in the following properties:

*low melting point
*low viscosity
*no reduction of basicity
*high electrical resistivity

Recently, because of these properties, much interest has been shown in the use of CaF_2 as a slag component both for hot metal desulphurisation and dephosphorisation and for secondary steelmaking. In ladle injection processes for example, CaF_2 additions will increase the fluidity of the flux, giving better mixing between slag and metal.

Desulphurisation of steel occurs via the following slag-metal reaction:

$$[S] + (O^{2-}) \rightleftharpoons [O] + (S^{2-}) \qquad (1)$$

It is known that the presence of CaF_2 in a slag can increase the rate of desulphurisation by up to eight times (Figure 1) [1,2,3,4]. It has been suggested, however, that CaF_2 has little effect on the equilibrium sulphur distribution [5,6].

EXPERIMENTAL

The present work is a study of sulphur distribution between CaF_2-CaO-SiO_2 slags and carbon-saturated iron at 1500°C.

The experiments were carried out in a vertical resistance furnace at 1500°C under an atmosphere of high-purity Argon. The corrosive nature of CaF_2-containing slags necessitated the use of graphite crucibles and hence carbon-saturated iron. Initial experiments were carried out to ascertain the optimum equilibration time. Figure 2 shows the sulphur content of the metal after varying reaction times with two CaF_2 and two Al_2O_3-containing slags. As expected, the CaF_2-containing slags reach equilibrium much more quickly. An equilibration time of three hours was found to be sufficient.

After equilibration the metal was analysed for sulphur using the combustion in air method, and the slag analysed for sulphur using the combustion in CO_2 method.

The sulphur partition ratio, L_s, was calculated where

$$L_s = \frac{(\%S)}{[\%S]} \qquad (2)$$

From this the sulphide capacity of the slag can be calculated. The sulphide capacity of the slag, C_s, is defined as

$$C_s = (\%S)\left(\frac{P_{O_2}}{P_{S_2}}\right)^{\frac{1}{2}} \qquad (3)$$

Since this is for gas-slag reactions, a more appropriate way of expressing the sulphide capacity for slag-metal reactions is in terms of activities, i.e.

$$C_s' = (\%S)\frac{h_o}{h_s} \qquad (4)$$

where h_o and h_s are the Henrian activities of oxygen and sulphur respectively. The C_s' values are evaluated as follows:

$$h_s = [\%S].f_s \qquad (5)$$

where f_s, the Henrian activity coefficient, is calculated from:

$$\log f_s = e_s^s[\%S] + e_s^c[\%C] = -0.028[\%S] + 0.11[\%C]$$

h_o is calculated from:

$$C(gr) + \underline{O}(1\%) = CO(g)$$

$$\Delta G° = 660 - 19.81T \text{ cal}$$

$$h_o = \frac{P_{co}}{a_c K} \qquad (6)$$

Assuming $P_{co} = 1$ and $a_c = 1$, since the experiments are carried out under carbon-saturated conditions

$$h_o = \frac{1}{K} = 5.643 \times 10^{-5} \qquad (7)$$

Hence

$$C_s' = L_s \frac{h_o}{f_s} \qquad (8)$$

RESULTS

The variation of C_s' with $\%CaF_2$ for constant %CaO is shown in Figure 3. As CaF_2 replaces SiO_2, the sulphide capacity increases. The effect of CaF_2 replacing CaO (i.e. constant SiO_2) is shown in Figure 4.

As CaF_2 replaces CaO, the sulphide capacity decreases. As would be expected, slags of lower $\%SiO_2$ have higher sulphide capacities. The effect of CaF_2 on the sulphide capacities of slags of constant basicity (i.e. constant $\%CaO/\%SiO_2$) is shown in Figure 5.

Addition of CaF_2 to slags of low basicity increases the sulphide capacity for CaF_2 additions of up to 20-30%. For slags of higher basicity, addition of CaF_2 appears to have little effect on the sulphide capacity.

The effect of the gas atmosphere in the furnace was also investigated. Figure 6 compares the sulphide capacities obtained under Ar with those obtained under CO for the same slags. As can be seen, the gas atmosphere has little effect on the sulphide capacities.

CONCLUSICNS

1. Sulphide capacities of CaF_2-CaO-SiO_2 slags are lower than those of CaF_2-CaO-Al_2O_3 slags.

2. Addition of CaF_2 to slags of low basicity initially increases the sulphide capacity. Addition of CaF_2 to slags of higher basicity appears to have no effect on the sulphide capacity.

3. Replacing CaO with CaF_2 reduces the sulphide capacity.

4. Replacing SiO_2 with CaF_2 increases the sulphide capacity.

5. The gas atmosphere has little effect on the sulphide capacities obtained by this method.

ACKNOWLEDGEMENTS

The authors would like to thank S.E.R.C. for providing funding for this work and the British Steel Corporation's Teesside Laboratories for technical and financial assistance.

REFERENCES

1. Y.H. Kim, PhD thesis, University of Strathclyde (1978)

2. R.G. Ward and K.A. Salmon, J.I.S.I. 196 393 (1960).

3. N. Jones. Unpublished work, University of Strathclyde (1986).

4. T. Takenouchi, K. Suzuki & S. Hara, Trans. I.S.I.J. 19 759 (1979).

5. N.J. Grant & J. Chipman, Trans. A.I.M.E. 185 54 (1946).

6. G.J.W. Kor, Met. Trans. B 8B 107 (1977).

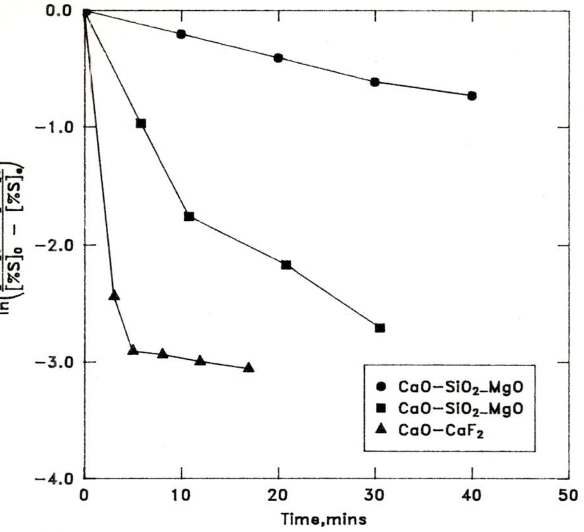

Fig. 1 Effect of CaF_2 on the rate of desulphurisation.

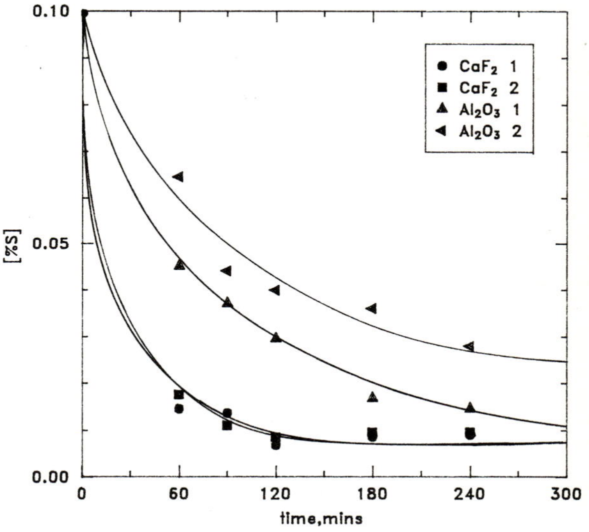

Fig. 2 Effect of equilibration time on final metal sulphur content.

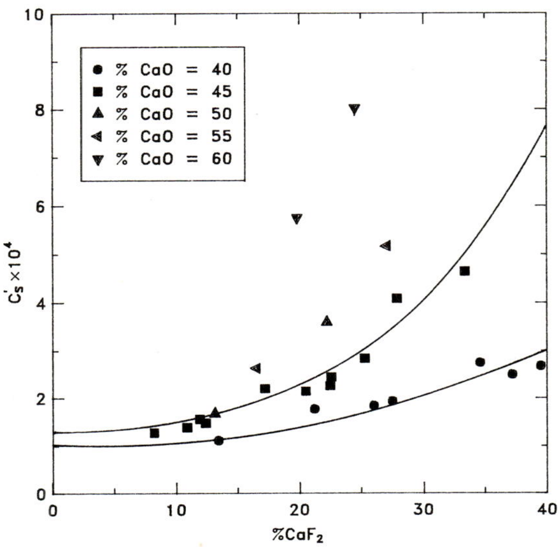

Fig. 3 Effect of CaF_2 on the sulphide capacity for constant CaO content.

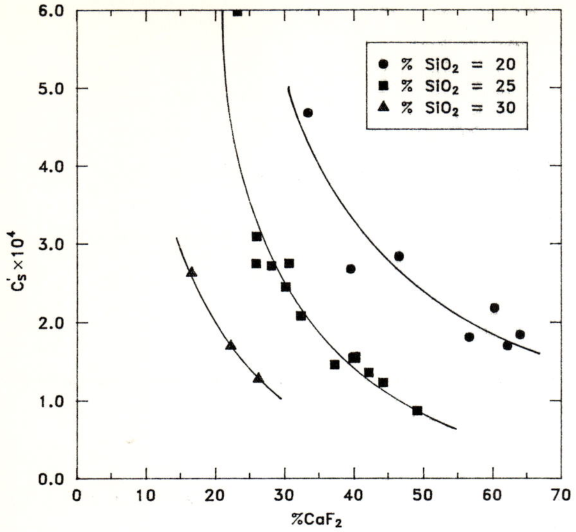

Fig. 4 Effect of CaF_2 on the sulphide capacity for constant SiO_2 content.

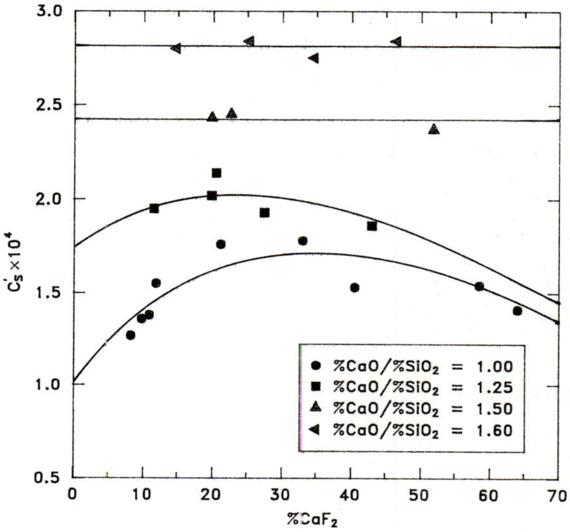

Fig. 5 Effect of CaF_2 on the sulphide capacity for constant ($\%CaO/\%SiO_2$).

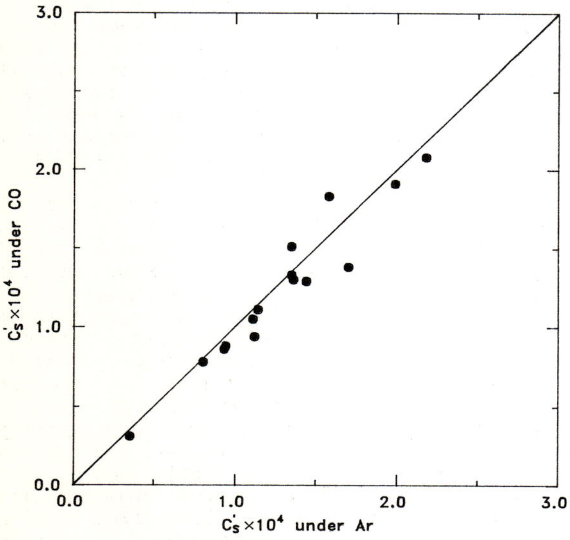

Fig. 6 Effect of furnace atmosphere on the sulphide capacity.

Diverse characteristics of mould powders used during casting of steel

H F MARSTON

The author is at the Swinden Laboratories of British Steel plc, Moorgate, Rotherham.

SYNOPSIS

During casting, mould powders (or casting powders) are routinely used to protect exposed steel from re-oxidation and to provide thermal insulation. The powder forms a slag which assimilates non-metallic inclusions and also infiltrates between the mould wall and the product. In service, the mould powder must not cause excessive erosion of the submerged entry tube in continuous casting, or modify the chemical composition of the steel. This paper briefly compares and contrasts the relative importance of these main characteristics in the selection of a mould powder for the bottom pouring of ingots and the continuous casting of slabs and billets.

INTRODUCTION

Mould powders, alternatively called casting powders, are routinely used in the bottom pouring of ingots and in the continuous casting of slabs, blooms and billets. In service, the mould powder first sinters and then progressively fuses to form a slag. The layers of unreacted and sintered powder and the slag that overlie the steel provide thermal insulation and protection from reoxidation. Where the slag is in contact with the liquid steel, it can assimilate non-metallic inclusions. The slag also infiltrates between the mould wall and the solidifying steel, and in continuous casting this infiltrated slag acts as a lubricant. Amongst the constraints in the design of the mould powder are that it must not cause excessive erosion of the submerged entry tube used in continuous casting, or modify the chemical composition of the steel.

Many ingots are top poured, without a mould powder, and the majority of continuously cast billets are produced with an open stream (optionally gas shrouded) and oil lubrication. However, where a high integrity product is required, ingots are bottom poured and billets are continuously cast using a refractory submerged entry tube. Mould powders are used in these processes and also in the continuous casting of slabs. This paper compares and contrasts the relative importance of the main characteristics of the mould powder for these three casting processes. Bloom casting has not been included because the characteristics required are, in general, intermediate between those of powders for casting billets and slabs.

THE REQUIRED CHARACTERISTICS OF A MOULD POWDER

The bar charts in Figure 1 indicate the relative importance of the identified mould powder characteristics to the three casting processes. It can be seen that the rating may change as the grade of steel or the casting parameters are altered. The relative priorities of these characteristics are compared for each process route in Figure 2. The importance of the different characteristics are discussed in the following sections.

Thermal Insulation (Figure 1(a))

The presence of a partially fused mould powder limits radiative heat loss from the surface of the liquid steel in the mould. This insulating cover allows casting at low superheat without freezing at the meniscus. In continuous casting, good thermal insulation is achieved by maintaining a 'black' practice, by ensuring a full cover of unreacted powder at all times. In ingot casting (Figure 2(a)), the cover must persist throughout casting, without replenishment, requiring a controlled rate of fusion.

The constitution of a mould powder, with regard to the form of the minerals and the nature of the contained carbon, can be as important as the chemical composition in determining its rate of fusion and hence the degree of thermal insulation that it can provide. However, once a fully fluid, homogeneous slag has formed, its properties are defined by its chemical composition; a fluid slag provides minimal thermal insulation.

In continuous casting, adequate thermal insulation is required to prevent the formation of a bridge of solid steel between the submerged entry tube and the mould wall. Such bridging almost invariably leads to a break-out, and termination of the cast. In the submerged casting of billets, this problem of bridging determines the smallest section size that can be produced. The risk of bridging in this process is particularly severe because of the small annulus between the (straight-through) refractory entry tube and the mould wall, and the deep penetration of the incoming superheated liquid metal, away from the surface. Thus, a major consideration when selecting a mould powder for billet casting is that it must provide adequate thermal insulation

(Figure 2(c)), even at the risk of imperfect lubrication or some carbon pick-up, if break-outs are to be avoided.

The fused slag that infiltrates between the cast product and the mould wall controls heat transfer to the mould. Controlling this heat transfer can be beneficial to product quality in each of the casting processes.

Protection from Re-oxidation (Figure 1(b))

Air is in principle excluded from contact with liquid steel by a complete cover of powder or slag, the conditions that also provide good thermal insulation. However, oxygen transfer through the slag can cause reoxidation unless reducing conditions are maintained. Transition metal oxides, such as MnO and Fe_2O_3, are generally detrimental in mould powders as they assist oxygen transfer and hence reoxidation. Edge breaks, which allow lapping as well as reoxidation, can cause surface and sub-surface defects.

Infiltration and Lubrication (Figure 1(c))

Slag infiltration between the product and the mould wall is a feature that is common to all of the casting processes, and is in general beneficial for surface quality. In ingot casting, the infiltrating slag provides a barrier between the rough mould wall and the solidifying shell. In continuous casting, the infiltrating slag provides lubrication between the product and the mould wall. Lubrication is especially important in slab casting because the wide face relaxes onto the mould wall (Figure 2(b)).

Infiltration is a complex process, requiring the mould powder to melt at a sufficient rate to provide a fluid slag layer on the upper surface of the steel, and for this fluid to flow to the meniscus perimeter, where a liquid reservoir must persist to enable the slag to infiltrate between the product and the mould wall. Higher demands are placed on the lubricating properties of the infiltrating slag as the rate of withdrawal of the continuously cast strand increases.

Assimilation of Inclusions (Figure 1(d))

The slag layer must assimilate any inclusions that float out from the body of the steel. In the early days of casting aluminium-killed (AK) steels, copious quantities of alumina inclusions were formed because of inadequate shrouding. The mould powders had to be designed to assimilate these refractory inclusions. Modern clean-steel practices limit the quantity of alumina inclusions, allowing other factors to be given precedence in selecting the mould powder. For example, the lubricative properties can be improved to allow faster casting.

Nevertheless, some grades of steel generate insoluble inclusions as the steel cools to the liquidus temperature, and during solidification. Thus, even if a clean-steel practice is followed, titanium-treated steels may still generate titanium carbo-nitride inclusions that have to be assimilated by the fused layer of the mould powder. Figure 1(d) contrasts the importance of the ability of the mould powder to assimilate inclusions when casting normal clean steels and those that generate copious, insoluble inclusions.

Some mould powders for ingot casting have a very low carbon content in order to fuse completely, with the objective of assimilating non-metallic inclusions. Under oxidising conditions, titanium carbo-nitride inclusions can be oxidised to the more soluble titania. However, thermal insulation and protection from reoxidation are compromised, resulting in inferior surface quality. Lapping may occur, but there is not the risk of a break-out, the penalty for inferior mould powder performance in continuous casting.

Refractory Erosion (Figure 1(e))

Submerged entry tube erosion has to be minimised in continuous casting, especially when sequence casting, and during the casting of small billets with a thin-walled refractory tube. The use of high grade zirconia refractory inserts can relieve but not eliminate the problem; the refractory and the fused mould powder must be compatible. Because erosion predominantly occurs at the slag/steel/refractory interface, the chemical characteristics of the steel can also influence the rate of refractory erosion.

Carbon Pick-up and Compositional Changes (Figure 1(f))

The mould powder must not change the steel composition. Exchange reactions can occur between the steel and the slag produced by the mould powder, but the large volume of steel relative to slag means that compositional changes to the steel are usually inconsequential. However, carbon pick-up can be significant in low carbon steels, such as the ELC (extra low carbon) stainless grades, if an inappropriate powder is selected. Mould powders with a low carbon content are usually employed for these grades, although performance with respect to the other characteristics may be compromised.

Virtually all mould powders contain carbon to inhibit fusion (to enhance thermal insulation) and to provide reducing conditions (to avoid reoxidation). Although some carbon burns out of the powder during the sintering process, much of the carbon from a high carbon mould powder can be transferred to the steel. The balance between carbon pick-up by the steel and the provision of thermal insulation by the powder is particularly severe in billet casting because of the risk of a break-out if thermal insulation is compromised. Alternatives to carbon to improve thermal insulation have been sought, but to date none have become established as viable alternatives, either because of their cost or their own undesirable interaction with steel.

OPTIMISATION OF POWDER CHARACTERISTICS

Mould powder selection is a compromise involving the optimisation of a number of physical and chemical characteristics. Eliminating an undesirable feature in the casting process may allow the balance to be revised, yielding further improvement in performance. As an example, improvements to steel cleanness have reduced the quantity of inclusions that require to be assimilated, and therefore the mould powder can be selected with greater emphasis on its lubricating properties.

Unfortunately, the six characteristics of a mould powder that have been identified here as pertaining to plant performance are not in general directly measurable in the laboratory as physical and chemical properties of the mould powder or the derived slag. In particular, the properties that can be determined may not be representative of the complex multi-phase structure that forms in contact with liquid steel. A characteristic such as lubrication depends on a number of processes, including the rate of fusion of the powder, compositional changes in contact with steel, the viscosity of the molten slag that is produced, the temperature dependence of viscosity, and the solidification characteristics of the slag in contact with the mould wall. Individually, these properties may be more or less clearly defined; for example, the lubricating ability of a pre-fused slag can be measured directly. However, the various stages of fusion, flow and infiltration have to be viewed as a whole to understand the entire 'characteristic' of lubrication.

ACKNOWLEDGEMENTS

The author is grateful to Dr R. Baker (Director, Research and Development) for permission to publish this paper.

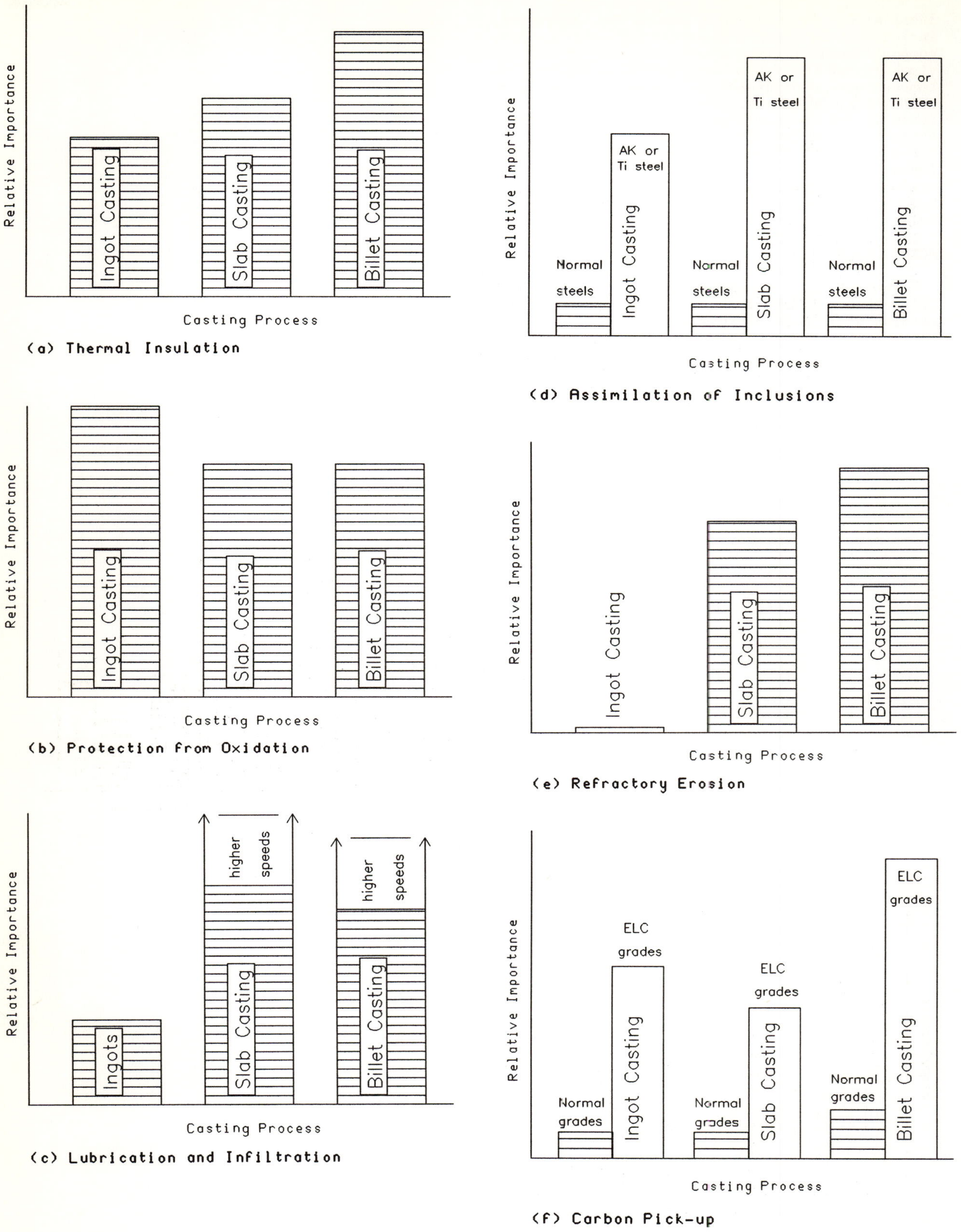

Figure 1. The Relative Importance of the Main Mould Powder Characteristics in Diverse Casting Practices

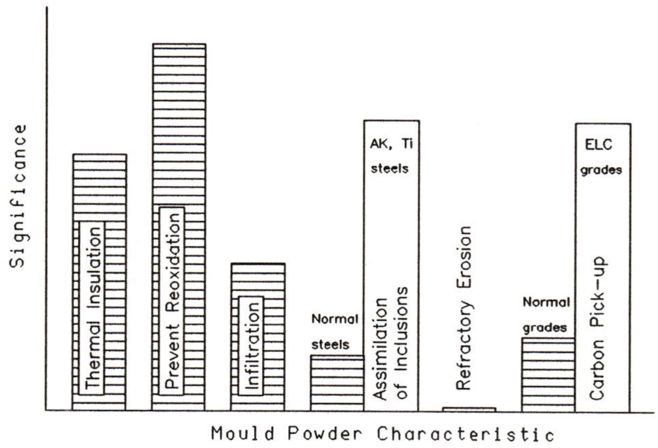

(a) Ingot casting

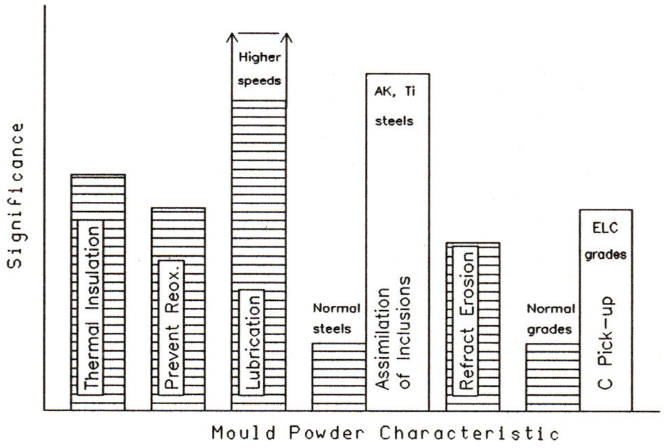

(b) Slab Casting

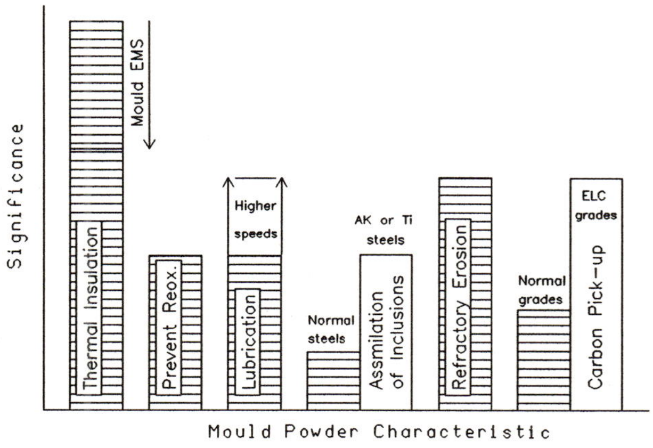

(c) Billet Casting

Figure 2. The Significance of each Mould Powder Characteristic in each Casting Process □

Physical and thermal properties of aluminate slags

F H PONSFORD, K C MILLS, P GRIEVESON, D L CANHAM and C P BROADBENT

Mrs Ponsford and Dr Mills are in the Division of Materials Applications, National Physical Laboratory, Teddington, UK; Professor Grieveson is in the Department of Materials, Imperial College, London; Drs Canham and Broadbent are with Billiton Research BV, Arnhem, The Netherlands.

SYNOPSIS

The heat capacity, enthalpy and enthalpy of fusion of calcium magnesium aluminate slag has been determined using differential scanning calorimetry and drop calorimetry. Estimated values for the solid slag were within 2% of the measured data. The density of the molten slag was determined using the Archimedean method and the estimated value for the density was found to be 2% lower than the measured value. The effective thermal conductivity of two solid aluminate slags was derived from heat flux density measurements recorded in simulation experiments. It was noted that the effective thermal conductivity apparently increased with the thickness of the solidified slag and subsequent analysis of the data yielded values for the heat transfer coefficient of the refractory/slag interface and the actual thermal conductivity of the slag, which represents an average value for the (solid + liquid) slag layers formed on the cold finger.

INTRODUCTION

Data for the physical and thermal properties of calcium magnesium aluminate slags are required to provide an understanding of slag reactions at high temperatures. However, there are few extant data available for aluminates and consequently values of the heat capacity (C_p), enthalpy (H_T-H_{298}), enthalpy of fusion (ΔH^{fus}), density (ρ) and the effective thermal conductivity (λ_{eff}) have been determined in this investigation for relevant slag compositions.

EXPERIMENTAL

Materials

The slags used in this study were prepared by fusion of weighed amounts of pure oxides (Table 1).

TABLE 1. Composition (in mass %) of slags

Slag	% CaO	% MgO	% Al_2O_3
B	7.0	38.7	52.9
C	0.3	44.6	52.3

Heat Capacity Measurements

The heat capacity was measured directly for temperatures between 298 and 1000 K using a Perkin Elmer differential scanning calorimeter (DSC) Model 2. The samples were placed in platinum crucibles and an atmosphere of argon was maintained in the apparatus. Measurements were made for temperature ranges of 200 K using a heating rate of 10 K min^{-1}. Full details of the method have been given elsewhere [1]. Heat capacities obtained with this technique for silver were within 2% of the recommended values.

Enthalpy measurements

Enthalpy (H_T-H_{298}) values were obtained by drop calorimetry for temperatures between 1073 and 1775 K and full details of the method have been reported by Andon *et al* [2]. The samples were sealed in (Pt+10% Rh) crucibles. The performance of the drop calorimeter was checked periodically by measuring the enthalpy of Calorimetry Conference sapphire supplied by the National Bureau of Standards (NBS) and the values obtained were always within 1% of the values recommended by NBS. Enthalpies measured by drop calorimetry were combined with C_p values for the sample obtained by DSC in a computer program and the equations giving the best fit to the experimental data were derived. The temperature dependency of C_p and (H_T-H_{298}) were assumed to have the conventional forms viz as shown in Equations (1) and (2).

$$C_p = a + bT + cT^2 - \frac{d}{T^2} \qquad (1)$$

$$(H_T-H_{298}) = a(T-298) + \frac{b}{2}(T^2-298^2) + \frac{c}{3}(T^3-298^3) + \frac{d}{T} - \frac{d}{298} \qquad (2)$$

Density measurements

Density measurements were performed using the Archimedean method. The slag was melted in a molybdenum crucible in a slightly-reducing atmosphere of argon and carbon monoxide and a molybdenum bob (12.5 mm diam. × 26 mm high) was suspended sequentially in air and then in the molten slag. The buoyancy force (F) was derived from the apparent difference in weight in the two media and the density (ρ) was calculated from Equation (3) where V_0, and v are the volumes of the bob and the immersed wire, respectively, and σ is the force exerted by the surface tension. Calculations of the magnitude of σ showed that it was negligible in comparison with F for the wire dimensions (0.5 mm diam) used in the present experiment

$$\rho = (F + \sigma)(V_0 + v) \qquad (3)$$

Thermal flux measurements

The apparatus used in these measurements was designed to simulate the heat transfer occurring across a water-cooled refractory wall in a high temperature process. It consisted of a water-cooled, copper finger immersed in a bath of slag (or iron covered with molten slag) which was heated inductively to 1550°C using a graphite crucible as a susceptor (Figure 1). A solidified slag layer was formed on the cold finger on immersion and the heat flux was determined by monitoring the temperature rise (ΔT) of the cooling water flowing through the cold finger at a constant known rate. Experiments were also carried out using a cold finger with a sleeve (2.5 mm thick) machined from the furnace bricks. The reproducibility of the heat flux measurements was ± 20%.

After completion of the experiment the cold finger was removed from the crucible and the ease of removal of the solidified slag layer from the cold finger or the sleeve was evaluated qualitatively and the thickness of the solidified slag layer determined.

RESULTS AND DISCUSSION

The values of C_p and $(H_T - H_{298})$ obtained for slag B by both DSC and drop calorimetry are given in Figure 2. The following equations were derived from these results:

$$C_p(\text{Solid}) = 0.991 + 1.92 \times 10^{-4}\,T - 4.26 \times 10^{-8}\,T^2 - 23990\,T^{-2}\ \text{JK}^{-1}\text{g}^{-1} \qquad (4)$$

$$C_p(\text{liquid}) = 2.0\ \text{JK}^{-1}\text{g}^{-1}$$

$$\Delta H^{fus} = 350\ \text{Jg}^{-1}\ ; \quad \Delta S^{fus} = 0.212\ \text{JK}^{-1}\text{g}^{-1} \qquad (5)$$

Estimated values of C_p and $(H_T - H_{298})$ for the solid phase were found to be within ± 2% of the experimental data using a method based on the Kopp-Neumann Rule [3]. The estimated value, C_p=1.5 $JK^{-1}g^{-1}$ for the liquid slag is 25% lower than the experimental value, but the latter could be in error by 20% given the small temperature range of the enthalpy measurements for the liquid phase. Tentative values of 24.7, 24.7 and 50.2 JK^{-1} mol^{-1} have been proposed for the entropy of fusion of CaO, MgO and Al_2O_3, and using these data a value of ΔS^{fus} = 0.51 JK^{-1} g^{-1} was calculated which is appreciably higher than the measured value (0.212 $JK^{-1}g^{-1}$).

The density (ρ) results for slag B are given in Figure 3 and are within 1% of the values reported by investigators [5,6] for slags with similar compositions. Density values for slag B were estimated using a method based on partial molar densities [3] and were found to be about 1% lower than the experimental data. The experimental temperature coefficient ($d\rho/dT$) had a value of 2.085×10^{-7} $kgm^{-3}K^{-1}$.

The heat flux density (Q) data obtained in the simulation experiments were interpreted using the simplified equation (6)* for a plate, where d and λ are the thickness and thermal conductivity, respectively C denotes the refractory material and λ_{eff} represents the effective or average thermal conductivity for the slag layers (solid + liquid) existing between T_{slag} and T_{H_2O}.

$$Q = \frac{(T_{slag} - T_{H_2O})}{[(d/\lambda)_{Cu}] + [(d/\lambda)]_C + [(d/\lambda_{eff})_{slag}]} \qquad (6)$$

The following values were used in the calculations, λ_{Cu} = 380 $Wm^{-1}K^{-1}$, d_{Cu} = 2 mm and d_C = 2.5 mm. A value of λ_C = 66.7 $Wm^{-1}K^{-1}$ was derived experimentally from the heat flux densities measured with sleeved and unsleeved copper fingers immersed in a casting powder [7] of known effective thermal conductivity (λ_{eff}). It was noted that good adhesion between the slag and the finger (or the sleeve) was associated with high values of Q and λ_{eff}. Furthermore, in those experiments where the adhesion was classified as medium, it was observed that λ_{eff} increased with increasing slag thickness, as can be seen from Figure 4.

These observations were consistent with the existence of a significant heat transfer coefficient at the interface between the slag and the wall of the cold finger. A further analysis was carried out assuming that the term (d/λ_{eff}) slag can be divided into two terms, (i) (d/λ_{act}) based on the actual thermal conductivity of the slag (which is an average value for the solid + liquid slag layers in the range T_{slag} to T_{H_2O}) and (ii) a heat transfer coefficient (h) viz:

$$(d/\lambda_{eff}) = (1/h) + (d/\lambda_{act})_{slag} \qquad (7)$$

The results of this analysis of the thermal flux data are given in Figure 5. The slope of the line providing the best fit of the data resulted in a value of λ_{act} = 2 $Wm^{-1}K^{-1}$ for the slag and the intercept yielded a value for h, the heat transfer coefficient, of 1.4×10^3 $Wm^{-2}K^{-1}$ similar to those obtained for casting fluxes which varied between 0.7 and 1.3×10^3 $Wm^{-2}K^{-1}$ [7]. It can also be seen from Figure 5 that for the experiments where good adhesion was obtained, the thermal flux was close to the maximum, which corresponds to an infinite heat transfer coefficient between the slag and the cold finger.

ACKNOWLEDGEMENTS

The careful experimental work carried out by Dr M J Richardson and Dr S Bagha at the National Physical Laboratory and Dr N Machingawuta at Imperial College is gratefully acknowledged.

* Since values obtained using equation (6), which pertains to a plane wall, were within 2% of those calculated using the relationship relevant for cylindrical geometry the simpler equation (6) was used in the subsequent analysis

REFERENCES

1. K.C. MILLS and M.J. RICHARDSON: Thermochemica Acta, 1973, 6, 427.

2. R.J.L. ANDON, J.F. MARTIN and K.C. MILLS: J. Chem. Soc. A, 1971, 17, 1788.

3. K.C. MILLS: Mineral Matter and Ash in Coal, edited K.S. Vorres Am. Chem. Soc. Symp. Ser. 301, 1986, 195.

4. K.C. MILLS and B.J. KEENE: Intl. Materials Rev. 1987, 32, 1.

5. V.D. SMOLYARENKO, A.M. YAKUSEV and F.P. EDNERAL: Izv. VUZ Chern. Met. 1965, 8, (1) 55.

6. P.P. EVSEEV and A.F. FILIPPOV: Izv. VUZ Chern. Met, 1967 10 (3) 55.

7. K.C. MILLS, P. GRIEVESON, A. OLUSANYA and S. BAGHA: Proc. of Continuous Casting '85' held in London, May, 1985 publ. Inst of Metals, Paper 57.

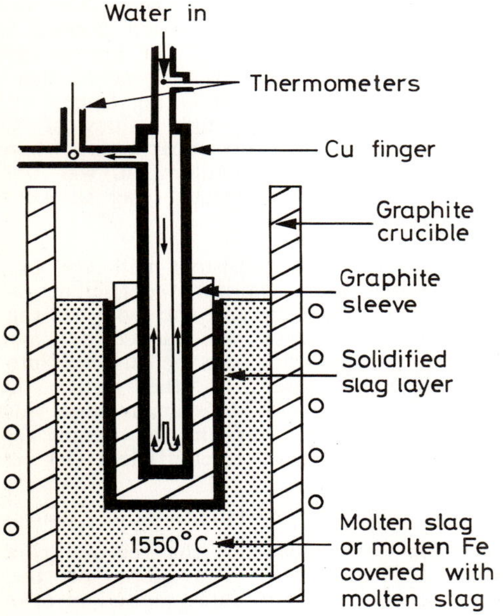

1. Schematic diagram of the apparatus used to measure the heat flux density (Q) with slags B and C.

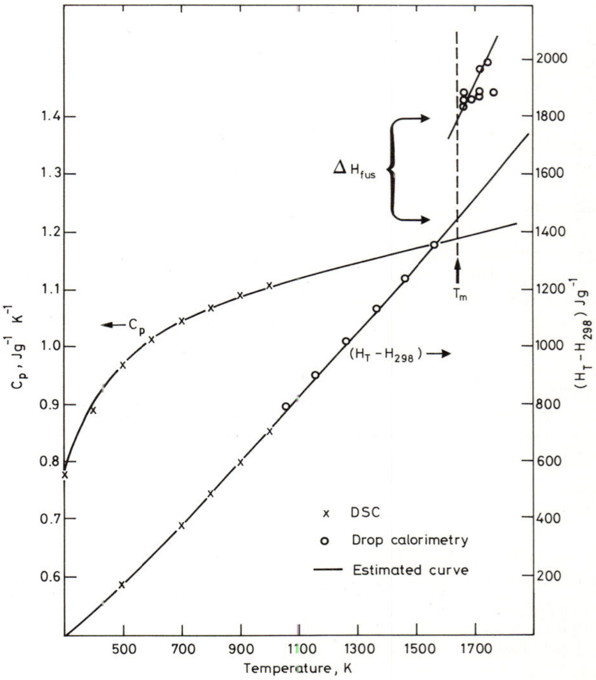

2. The enthalpy and C_p of slag B as a function of temperature.

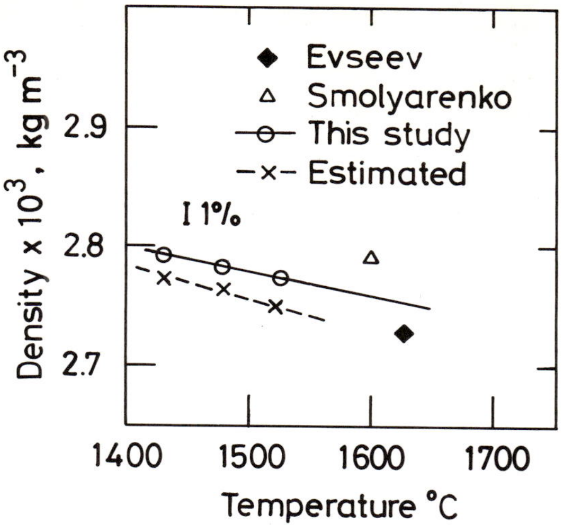

3. The density of slag B as a function of temperature.

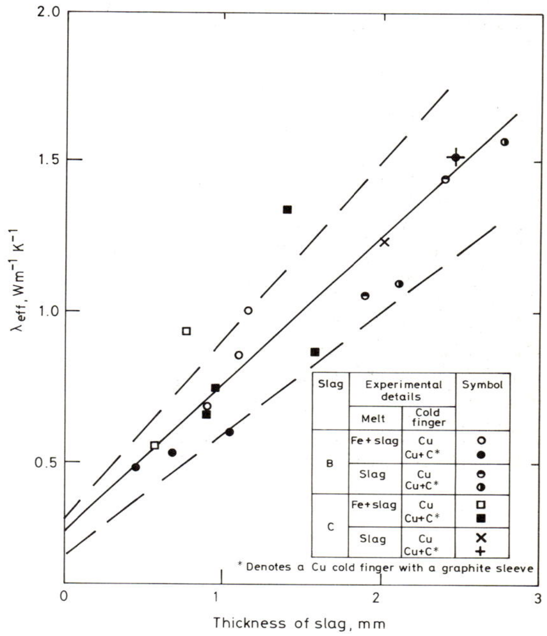

4. The dependence of (λ_{eff}) slag on the thickness of the solidified slag layer.

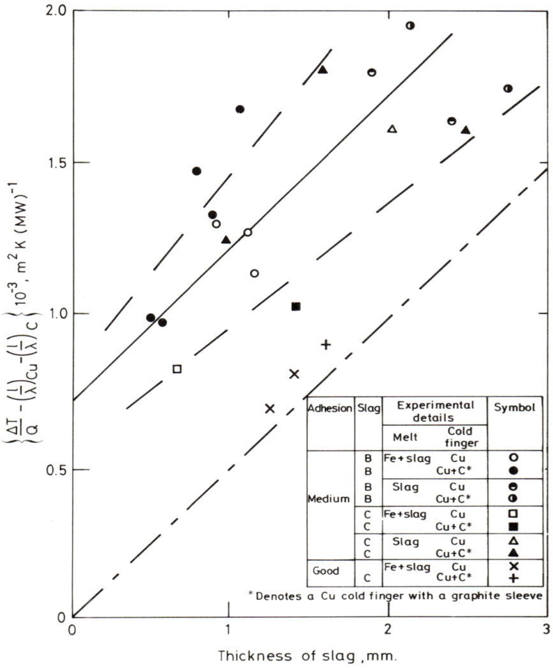

5. The dependence of the parameter (d/λ_{eff}) on the thickness of solidified slag layer.